Topics in Mining, Metallurgy and Materials Engineering

Series editor

Carlos P. Bergmann, Porto Alegre, Brazil

"Topics in Mining, Metallurgy and Materials Engineering" welcomes manuscripts in these three main focus areas: Extractive Metallurgy/Mineral Technology; Manufacturing Processes, and Materials Science and Technology. Manuscripts should present scientific solutions for technological problems. The three focus areas have a vertically lined multidisciplinarity, starting from mineral assets, their extraction and processing, their transformation into materials useful for the society, and their interaction with the environment.

More information about this series at http://www.springer.com/series/11054

Stefan Johann Rupitsch

Piezoelectric Sensors and Actuators

Fundamentals and Applications

 Springer

Stefan Johann Rupitsch
Lehrstuhl für Sensorik
Friedrich-Alexander-Universität
  Erlangen-Nürnberg
Erlangen
Germany

ISSN 2364-3293   ISSN 2364-3307 (electronic)
Topics in Mining, Metallurgy and Materials Engineering
ISBN 978-3-662-58601-3   ISBN 978-3-662-57534-5   (eBook)
https://doi.org/10.1007/978-3-662-57534-5

This Springer imprint is published by the registered company Springer-Verlag GmbH, DE part of Springer Nature
The registered company address is: Heidelberger Platz 3, 14197 Berlin, Germany

Preface

Piezoelectric devices play a major role in our everyday lives. The reason for this lies in the fact that piezoelectric materials enable an efficient conversion from mechanical energy into electrical energy and vice versa. Piezoelectric sensors and actuators represent an important subgroup of piezoelectric devices. Nowadays, the application areas of piezoelectric sensors and actuators range from process measurement technology and nondestructive testing to medicine and consumer electronics.

This book addresses students, researchers as well as industry professionals in the fields of engineering sciences, material sciences, and physics. The author aims at providing information that is important to obtain a deep understanding of piezoelectric sensors and actuators. The book additionally contains selected applications and recent developments (e.g., simulation-based material characterization), which are of great interest to science and industry.

At the beginning, we will study fundamentals of piezoelectric sensors and actuators. The fundamentals include physical basics, the principle of the piezoelectric effect and piezoelectric materials. One focus of the book relates to reliable characterization of sensor and actuator materials by combining numerical simulations with appropriate measurements. Moreover, an efficient phenomenological modeling approach for the large-signal behavior of ferroelectric materials will be presented which facilitates the operation of piezoelectric actuators. A further focus lies on piezoelectric ultrasonic transducers because they are most commonly used in applications like ultrasonic imaging and parking sensors. In this context, a nonreactive measurement approach will be detailed that allows sound field characterization in various media.

The book also deals with piezoelectric sensors and transducers in the large application area of process measurement technology. For example, we will discuss conventional piezoelectric sensors for mechanical quantities (e.g., force) as well as sensor devices for fluid flow measurements. The final part of the book concentrates on piezoelectric positioning systems and motors.

Erlangen, Germany

Stefan Johann Rupitsch

Acknowledgements

This book was written in the course of my habilitation procedure at the Chair of Sensor Technology at the Friedrich-Alexander-University Erlangen-Nuremberg. First of all, I would like to thank Prof. Dr.-Ing. Reinhard Lerch for his valuable support and advice. He has animated me to write the book and has given me the possibility for this time-consuming activity. Moreover, I would like to thank Prof. Dr.-Ing. habil. Paul Steinmann and Prof. Dr.-Ing. habil. Jörg Wallaschek, who served as further members of the mentorship during my habilitation procedure.

I wish to express my gratitude to all the people who have inspired and sustained this work. A special thank goes to the present and former colleagues from the Chair of Sensor Technology. Many topics of the book result from cooperations with these colleagues. The dynamic and stimulating atmosphere at the Chair of Sensor Technology was certainly essential for the preparation of the book. In particular, I would like to thank (in alphabetical order) Dr.-Ing. Lizhuo Chen, Philipp Dorsch, Michael Fink, Dominik Gedeon, Florian Hubert, Dr.-Ing. Jürgen Ilg, Daniel Kiefer, Michael Löffler, Michael Nierla, Dr.-Ing. Peter Ploß, Michael Ponschab, Dr.-Ing. Thomas Scharrer, Prof. Dr.-Ing. Alexander Sutor, Manuel Weiß, Dr.-Ing. Felix Wolf, and Michael Wüst.

I also would like to thank Christine Peter for creating several drawings of the book. I gratefully acknowledge the team of the Springer-Verlag for the excellent cooperation as well as for typesetting and reading the book very carefully.

Finally, I would like to thank my girlfriend Angelina, my parents Rosmarie and Johann and my sister Barbara for their support and understanding all the time.

June 2018   Stefan Johann Rupitsch

Contents

Chapter 1
Introduction

Piezoelectric devices play a major role in our everyday lives. Currently, the global demand for piezoelectric devices is valued at approximately 20 billion euros per year. Piezoelectric sensors and actuators make a substantial contribution in this respect. At the beginning of the opening chapter, we will discuss the fundamentals of sensors and actuators. Section 1.2 addresses the history of piezoelectricity and piezoelectric materials. In Sect. 1.3, application areas as well as application examples of piezoelectricity are listed. The chapter ends with a brief chapter overview of the book.

1.1 Fundamentals of Sensors and Actuators

Sensors and actuators play an important role in various practical applications. Let us start with the fundamentals of sensors. In general, sensors convert measurands into appropriate measurement values. From the system point of view, measurands serve as inputs and measurement values as outputs of sensors. In this book, sensors will be limited to devices that convert mechanical quantities into electrical quantities. The mechanical quantities denote, thus, measurands, while the electrical quantities represent measurement values. Figure 1.1 depicts possible measurands (e.g., mechanical force) and measurement values (e.g., electric voltage).

In contrast to sensors, (electromechanical) actuators convert electrical quantities (e.g., electric voltage) into mechanical quantities (e.g., mechanical force). Hence, actuators operate in opposite direction as sensors (see Fig. 1.1). From the system point of view, electrical quantities serve as inputs, whereas mechanical quantities represent outputs of actuators.

There exist several principles to convert mechanical into electrical quantities and electrical into mechanical quantities. Some conversion principles work in both directions, i.e., it is possible to convert mechanical into electrical quantities and vice versa. Due to this fact, such conversion principles can be exploited for sensors and actu-

© Springer-Verlag GmbH Germany, part of Springer Nature 2019
S. J. Rupitsch, *Piezoelectric Sensors and Actuators*, Topics in Mining, Metallurgy and Materials Engineering, https://doi.org/10.1007/978-3-662-57534-5_1

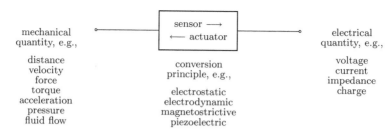

Fig. 1.1 Typical conversion principles as well as input and output quantities of sensors and actuators

ators. The most common bidirectional conversion principles are listed in Fig. 1.1. If a conversion principle allows both working directions, sensors and actuators will often be called *transducers*. As the book title implies, we will focus on the piezo-electric conversion principle. Therefore, the book deals with piezoelectric sensors and piezoelectric actuators, i.e., with piezoelectric transducers.

Apart from the conversion principle, sensors and actuators can be classified according to other aspects. Especially for sensors, one can find several classifications like active/passive sensors. Because piezoelectric sensors do not necessarily need an auxiliary energy, they belong to the group of active sensors.

1.2   History of Piezoelectricity and Piezoelectric Materials

The word *piezoelectricity* originates from the Greek language and means electricity due to pressure. Piezoelectricity was firstly discovered by the Curie brothers in 1880. They recognized that electric charges will arise when mechanical forces are applied to materials like tourmaline, quartz, topaz, and Rochelle salt. This effect is referred to as *direct piezoelectric effect*. In 1881, Lippmann deduced the *inverse piezoelectric effect* from the mathematical point of view. The Curie brothers immediately confirmed the existence of the inverse piezoelectric effect.

The first practical application of piezoelectricity was the sonar, which has been developed during the First World War by Langevin. The main component of the sonar consisted of a thin quartz crystal that was glued between two steel plates. In 1921, Cady invented an electrical oscillator, which was stabilized by a quartz crystal. A few years later, such oscillators were used in all high-frequency radio transmitters. Quartz crystal controlled oscillators are nowadays still the secondary standard for timing and frequency control. The success of sonar and quartz crystal controlled oscillators was responsible that new piezoelectric materials and new applications were explored over the next decades after the First World War. For example, the development of piezoelectric ultrasonic transducers enabled viscosity measurements in fluids and the detection of flaws inside of solids.

During the Second World War, several independent research groups discovered a new class of synthetic materials, which offers piezoelectric constants many times higher than natural materials such as quartz. The synthetically produced polycrystalline ceramic materials were named ferroelectrics and piezoceramic materials. Barium titanate and lead zirconate titanate (PZT) represent two well-known solid solutions that belong to the class of these materials. In 1946, it was demonstrated that barium titanate features pronounced piezoelectric properties after an appropriate poling process. The first commercial use of barium titanate was in phonograph pick-ups. The strong piezoelectric coupling in PZT was discovered in 1954. The intense research in the following decades revealed that the piezoelectric properties of PZT could be controlled by means of doping. In doing so, it is possible to produce ferroelectrically soft and ferroelectrically hard materials. While ferroelectrically soft materials are well suited for piezoelectric actuators and ultrasonic transducers, ferroelectrically hard materials provide an outstanding stability for high power and filter applications. On these grounds, PZT is most commonly used in conventional piezoelectric devices, nowadays.

Even though piezoceramic materials such as PZT feature comparatively high electromechanical coupling factors and can be manufactured in arbitrary shape, quartz crystals still play an important role in practical applications, e.g., for piezoelectric force sensors. There are several reasons for this. For instance, special cuts of quartz crystals lead to material properties that are stable over a wide temperature range as well as almost free of hysteresis. Moreover, quartz crystals can also be synthetically manufactured by the so-called hydrothermal method, which was firstly applied to artificially grow quartz in the 1940s. Aside from quartz, lithium niobate and lithium tantalate are well-known representatives of piezoelectric single crystals. Both materials play a key role in modern telecommunication systems because they often serve as piezoelectric material for surface acoustic wave (SAW) devices.

Over the past decades, the research in the field of piezoelectric materials has concentrated on various topics. Many research groups work on lead-free piezoceramic materials (e.g., sodium potassium niobate) that provide a similar performance as PZT. A further research topic concerns relaxor-based single crystals since the piezoelectric constants of such piezoelectric materials can take values, which greatly exceed those of PZT. Because microelectromechanical systems (MEMS) gain in importance, much research and development are also conducted in the fabrication of thick and thin piezoelectric films. As a last example of research topics, let us mention piezoelectric polymers like polyvinylidene fluoride (PVDF) and cellular polypropylene. If the piezoelectric polymers are produced as thin films, they can be exploited for mechanically flexible sensors and actuators.

1.3 Practical Applications of Piezoelectricity

The application areas of piezoelectricity range from process measurement technology, nondestructive testing and medicine to consumer electronics and sports. Depending on the particular application, one exploits the direct piezoelectric effect, the inverse piezoelectric effect or a combination of both. The following list contains selected applications (e.g., parking sensors) in different application areas.

- process measurement technology and condition monitoring

  – sensors for, e.g., force, torque, acceleration, viscosity
  – measurement of temperature and geometric distance

- automotive industry

  – parking sensors
  – injection systems in diesel engines

- production technology

  – ultrasonic welding
  – ultrasonic cleaning

- nondestructive testing

  – flaw detection
  – material and device characterization

- medicine

  – diagnostics, e.g., pregnancy examinations
  – therapy, e.g., kidney stone fragmentation (lithotripsy)

- consumer electronics

  – loudspeakers
  – inkjet printers
  – lens settings in cameras

- smart materials and structures

  – active noise control
  – structural health monitoring

- sports, e.g., reduction of mechanical vibrations in tennis rackets
- musics, e.g., pickup for guitars
- energy harvesting for local energy supply
- transformers

Even though this list of applications seems to be very long, it could be extended almost indefinitely.

1.4 Chapter Overview

As the title suggests, the book deals with fundamentals and applications of piezo-electric sensors and actuators. According to the list in the previous section, there exists a wide variety of applications of piezoelectricity. In this book, we will concentrate on some selected examples. Many topics refer to research activities, which have been conducted at the Chair of Sensor Technology (Friedrich-Alexander-University Erlangen-Nuremberg) during the last ten years. Apart from the opening chapter, the book is divided into nine chapters.

Chapter 2 addresses the physical basics that are important for piezoelectric sensors and actuators. This includes fundamentals, characteristic quantities as well as basic equations of electromagnetics, continuum mechanics, and acoustics. In Chap. 3, we will study the fundamentals of piezoelectricity. The chapter starts with the principle of the direct and inverse piezoelectric effect. After thermodynamical considerations, the material law of linear piezoelectricity will be derived. Furthermore, the electromechanical coupling inside piezoelectric materials will be classified and quantitatively rated. The chapter ends with a comprehensive overview of piezoelectric materials (e.g., polycrystalline ceramic materials), which are used in practical applications.

Chapter 4 deals with the fundamentals of finite element (FE) simulations since such numerical simulations are nowadays the standard tool for the design and optimization of piezoelectric sensors and actuators. We will start with the basic steps of the FE method. Afterward, the FE method will be applied to the electrostatic field, the mechanical field, and the acoustic field. Due to the fact that piezoelectricity refers to coupling of mechanical and electric quantities, we study the simulation-based coupling of the underlying fields. This also includes the coupling of mechanical and acoustic fields, which is important for piezoelectric ultrasonic transducers.

In Chap. 5, we will discuss the characterization of sensor and actuator materials. The material characterization represents an essential step in the design and optimization because reliable numerical simulations demand precise material parameters. The chapter starts with standard approaches for material characterization. In doing so, a clear distinction between active and passive materials is carried out. Piezoelectric materials are active materials, whereas other materials (e.g., plastics) within piezoelectric sensors and actuators belong to the group of passive materials. The main focus of the chapter lies on the so-called inverse method, which has been developed at the Chair of Sensor Technology. Basically, the inverse method combines FE simulations with measurements. By reducing the deviations between simulation and measurement results, the material parameters get iteratively adjusted in a convenient way. The inverse method is exploited to identify material parameters and properties of selected active and passive materials.

Piezoceramic materials will show a pronounced hysteretic behavior if large electrical excitation signals are used during operation. Chapter 6 details a phenomenological modeling approach, which allows the reliable description of this large-signal behavior. Before the so-called Preisach hysteresis operator is introduced, we will briefly study various modeling approaches on different length scales. Since the Preisach hysteresis operator consists of weighted elementary switching operators, two different weighting procedures are given. The chapter also addressed generalized Preisach hysteresis models, which have been developed at the Chair of Sensor Technology. For instance, the generalization enables the consideration of mechanical stresses that are applied to a piezoceramic material. Finally, we discuss the inversion of the Preisach hysteresis model. The inversion will be of utmost importance when the Preisach hysteresis operator is used for hysteresis compensation.

Chapter 7 treats ultrasonic transducers that exploit piezoelectric materials. The chapter starts with a semi-analytical approach for calculating sound fields and transducer outputs. The approach is based on the so-called spatial impulse response (SIR) of the considered ultrasonic transducer, e.g., a piston-type transducer. Among other things, the SIR is utilized to determine the spatial resolution of spherically focused transducers. Afterward, we will study the general structure and fundamental operation modes of single-element transducers, transducer arrays, and composite transducers. A further section concerns a simple one-dimensional modeling approach that allows analytical description of basic physical relations under consideration of the internal transducer structure. At the end of the chapter, several examples for piezoelectric ultrasonic transducers will be presented.

Practical applications of ultrasonic transducers often demand the characterization of the resulting sound fields. That is the reason why Chap. 8 deals with appropriate measurement principles. The chapter starts with conventional measurement principles such as hydrophones and Schlieren optical methods. The subsequent sections exclusively concentrate on the so-called light refractive tomography (abbr. LRT), which has been realized at the Chair of Sensor Technology. This optical measurement principle enables nonreactive and spatially as well as temporally resolved investigations of sound fields that are generated by piezoelectric ultrasonic transducers, e.g., a cylindrically focused transducer. Exemplary results for sound pressure fields in water and air will be shown. Moreover, LRT is applied to study the wave propagation of mechanical waves in an optically transparent solid.

Piezoelectric sensors are frequently employed for the measurement of physical quantities. In Chap. 9, we will study typical setups of such sensors and their application in the process measurement technology. At the beginning, piezoelectric sensors for the quantities force, torque, pressure, and acceleration are detailed. This includes commonly used readout electronics such as charge amplifiers. Subsequently, a method will be presented which enables the simultaneous determination of thickness and speed of sound for solid plates. The underlying measurement principle is based on ultrasonic waves and has been developed at the Chair of Sensor Technology. The chapter also addresses fluid flow measurements that exploit ultrasonic transducers. We will discuss typical measurement principles as well as a recently suggested modeling approach, which allows efficient estimation of transducer outputs for clamp-on ultrasonic flow meters. At the end, a mechanically flexible cavitation sensor is presented that has been developed at the Chair of Sensor Technology.

The last chapter addresses piezoelectric positioning systems and piezoelectric motors. We will start with piezoelectric stack actuators, which provide much larger strokes than piezoelectric single elements. Because the large strokes call for large electrical excitation signals, Preisach hysteresis modeling from Chap. 6 is applied for a mechanically prestressed stack actuator. The subsequent section details an amplified piezoelectric actuator that was built up at the Chair of Sensor Technology. We will also study model-based hysteresis compensation for a piezoelectric trimorph actuator, which can be used for positioning tasks. The end of the chapter concerns linear and rotary piezoelectric motors.

Chapter 2
Physical Basics

Piezoelectric sensors and actuators connect different physical fields (e.g., electrostatic and mechanical field). With a view to studying the behavior of piezoelectric devices, the fundamentals of those physical fields are indispensable. Therefore, this chapter addresses the physical principles that are important for piezoelectric sensors and actuators. Section 2.1 deals with electromagnetics, especially with the electric field. In Sects. 2.2 and 2.3, the basics of continuum mechanics and acoustics are described, respectively.

2.1 Electromagnetics

In this section, the so-called Maxwell's equations as well as the relevant constitutive equations are introduced allowing the complete description of electromagnetic fields. We will discuss the electrostatic field, which represents a special case of electromagnetic fields. Section 2.1.3 details interface conditions for the electric field between two media exhibiting different material properties. Finally, the lumped circuit element approach is explained that can be exploited to efficiently solve electromagnetic field problems. Further literature concerning the electromagnetic field can be found in [1, 8, 9, 17].

2.1.1 Maxwell's Equations

James Clerk Maxwell published for the first time the full system of partial differential equations, which describe the physical relations in the electromagnetic field [13, 14]. His work relies on previous research performed by Ampère, Gauss, and Faraday. The

© Springer-Verlag GmbH Germany, part of Springer Nature 2019
S. J. Rupitsch, *Piezoelectric Sensors and Actuators*, Topics in Mining, Metallurgy and Materials Engineering, https://doi.org/10.1007/978-3-662-57534-5_2

Table 2.1 Expressions utilized in Maxwell's equations (2.1)–(2.4)

Notation	Description	Unit
H	Magnetic field intensity; vector	$A\,m^{-1}$
E	Electric field intensity; vector	$V\,m^{-1}$
B	Magnetic flux density (magnetic induction); vector	$V\,s\,m^{-2}$; T
D	Electric flux density (electric induction); vector	$A\,s\,m^{-2}$; $C\,m^{-2}$
q_e	Volume charge density; scalar	$A\,s\,m^{-3}$; $C\,m^{-3}$
J	Electric current density; vector	$A\,m^{-2}$
M	Magnetization; vector	$V\,s\,m^{-2}$; T
P	Electric polarization; vector	$A\,s\,m^{-2}$; $C\,m^{-2}$

four Maxwell's equations[1] in differential form are given by (time t; Nabla operator $\nabla = [\partial/\partial x, \partial/\partial y, \partial/\partial z]^t$)

$$\text{Law of Ampère: } \nabla \times \mathbf{H} = \mathbf{J} + \frac{\partial \mathbf{D}}{\partial t} \qquad (2.1)$$

$$\text{Law of Faraday: } \nabla \times \mathbf{E} = -\frac{\partial \mathbf{B}}{\partial t} \qquad (2.2)$$

$$\text{Law of Gauss: } \qquad \nabla \cdot \mathbf{D} = q_e \qquad (2.3)$$

$$\nabla \cdot \mathbf{B} = 0 \qquad (2.4)$$

with the expressions listed in Table 2.1. The general form of Maxwell's equations contains the displacement current($\partial \mathbf{D}/\partial t$) in (2.1) and is, therefore, also applicable in the high-frequency domain of electromagnetic fields. However, for the low-frequency domain (quasi-static case), the wavelength λ of the resulting electromagnetic waves is large compared to the dimensions of conventional electromagnetic devices. That is the reason why (2.1) can be simplified to $\nabla \times \mathbf{H} = \mathbf{J}$.

Several properties of electromagnetic fields can be deduced from Maxwell's equations. The most important findings are listed below.

- Law of Ampère: An electric current density $\mathbf{J}$ generates a magnetic field. The directions of the magnetic field lines relate to the direction of $\mathbf{J}$ according to the so-called right-hand rule (see Fig. 2.1a).
- Law of Faraday: A magnetic flux density $\mathbf{B}$ that is changing with respect to time induces an electric voltage in a conductive loop (see Fig. 2.1b).
- Law of Gauss: Electric charges are the source of the electric field (see Fig. 2.1c).
- Fourth Maxwell's equation: The magnetic field ($\mathbf{B}, \mathbf{H}$) is solenoidal and, therefore, the magnetic field lines are closed (see Fig. 2.1d). Furthermore, magnetic charges do not exist.

---

[1]To achieve a compact form of the subsequent equations, the arguments for both position $\mathbf{r}$ and time t (i.e., $\bullet(\mathbf{r}, t)$) are mostly omitted. Note that this is also done for continuum mechanics as well as for acoustics.

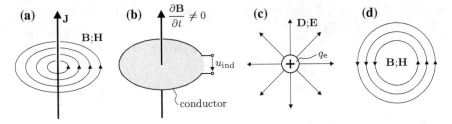

Fig. 2.1 Basic interpretations for **a** Law of Ampère, **b** Law of Faraday (induced electrical voltage u_{ind}), **c** Law of Gauss ($q_e > 0$), **d** fourth Maxwell's equation

Table 2.2 Expressions utilized in constitutive equations (2.5)–(2.7)

Notation	Description	Unit
γ	Electric conductivity; scalar	$\Omega^{-1}\,m^{-1}$
$\mathbf{v}$	Velocity of the volume charges q_e; vector	$m\,s^{-1}$
μ	Magnetic permeability; scalar	$V\,s\,A^{-1}\,m^{-1}$
μ_0	Magnetic permeability of vacuum $(4\pi \cdot 10^{-7})$; scalar	$V\,s\,A^{-1}\,m^{-1}$
μ_r	Relative magnetic permeability; scalar	–
ε	Electric permittivity; scalar	$A\,s\,V^{-1}\,m^{-1}$
ε_0	Electric permittivity of vacuum $(8.854 \cdot 10^{-12})$; scalar	$A\,s\,V^{-1}\,m^{-1}$
ε_r	Relative electric permittivity; scalar	–

In addition to Maxwell's equations, the modeling of media in the electromagnetic field requires constitutive equations, which cover the materials' behavior. For a homogeneous and isotropic material, the constitutive equations read as

$$\mathbf{J} = \gamma(\mathbf{E} + \mathbf{v} \times \mathbf{B}) \tag{2.5}$$
$$\mathbf{B} = \mu\mathbf{H} = \mu_0\mathbf{H} + \mathbf{M} = \mu_0\mu_r\mathbf{H} \tag{2.6}$$
$$\mathbf{D} = \varepsilon\mathbf{E} = \varepsilon_0\mathbf{E} + \mathbf{P} = \varepsilon_0\varepsilon_r\mathbf{E} \tag{2.7}$$

with the expressions listed in Table 2.2. Note that for anisotropic materials, such as piezoceramics, the electric and magnetic material properties (e.g., ε) cannot be completely assigned by single scalar quantities but demand tensors of rank ≥ 2. Table 2.3 contains the electric conductivity γ, the relative magnetic permeability μ_r as well as the relative electric permittivity ε_r of selected media.

2.1.2 Electrostatic Field

In the static case, both electric and magnetic quantities do not depend on time. Electric charges do not move and energy is neither transported nor converted. As a result,

Table 2.3 Electric conductivity γ, relative magnetic permeability μ_r, and relative electric permittivity ε_r of selected media

Media	γ in Ω^{-1} m^{-1}	μ_r	ε_r
Copper	$59 \cdot 10^6$	1	–
Iron	$10 \cdot 10^6$	>300	–
Tungsten	$18 \cdot 10^6$	1	–
PVDF*	10^{-11}	1	6
Polyethylene	10^{-13}	1	2.4
Water	$5 \cdot 10^{-3}$	1	80

* Polyvinylidene fluoride

Maxwell's equations and the constitutive equations can be divided into an electric and a magnetic subsystem. For the so-called electrostatic field (electric subsystem), the relations between the electric quantities can be described with

$$\nabla \times \mathbf{E} = 0 \tag{2.8}$$

$$\nabla \cdot \mathbf{D} = q_e \tag{2.9}$$

$$\mathbf{D} = \varepsilon \mathbf{E} . \tag{2.10}$$

Since the electric field intensity $\mathbf{E}$ is irrotational (see (2.8)), it can be expressed by the so-called electric scalar potential V_e

$$\mathbf{E} = -\nabla V_e . \tag{2.11}$$

Note that (2.8)–(2.11) will also be appropriate for quasi-static electric fields if the resulting magnetic quantities are still negligible. This is the case for materials exhibiting small relative magnetic permeabilities μ_r.

2.1.3 Interface Conditions for Electric Field

At an interface of different media, the quantities of electric and/or magnetic fields may be altered. To study this for the electric field, we assume an interface between two isotropic homogeneous materials showing different electric permittivities (ε_1 and ε_2; Fig. 2.2a). The starting point to derive the interface conditions is Maxwell's equations. The third Maxwell's equation (Law of Gauss; (2.3)) in integral form is given as (volume Ω, surface Γ, surface vector Γ)

$$\int_\Omega (\nabla \cdot \mathbf{D}) \, d\Omega = \oint_\Gamma \mathbf{D} \cdot d\mathbf{\Gamma} = \int_\Omega q_e d\Omega \tag{2.12}$$

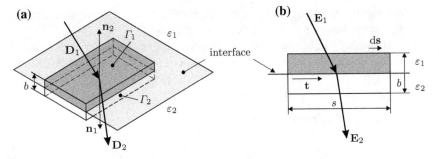

Fig. 2.2 Interface of different media: **a** continuity of electric flux density **D**; **b** continuity of electric field intensity **E**

where the divergence theorem has been applied. Using the relation $d\Omega = b\, d\Gamma$ and the assumption $b \to 0$ results in

$$\lim_{b \to 0} \int_{\Omega} q_e d\Omega = \int_{\Gamma} \sigma_e d\Gamma \ . \tag{2.13}$$

Here, σ_e denotes the surface charge. Furthermore, (2.12) can be rewritten to

$$\lim_{b \to 0} \oint_{\Gamma} \mathbf{D} \cdot d\Gamma \ \longrightarrow \ \mathbf{D}_1 \cdot \mathbf{n}_1 + \mathbf{D}_2 \cdot \mathbf{n}_2 = \mathbf{n}_1(\mathbf{D}_1 - \mathbf{D}_2) = D_{1n} - D_{2n} \tag{2.14}$$

with the normal vectors $\mathbf{n}_1$ and $\mathbf{n}_2$ at the material interface. The expressions D_{1n} and D_{2n} indicate the normal components of $\mathbf{D}_1$ and $\mathbf{D}_2$, respectively. The combination of (2.13) and (2.14) yields the continuity relation for the electric flux density $\mathbf{D} = [D_n, D_t]^t$

$$D_{1n} = D_{2n} + \sigma_e \ . \tag{2.15}$$

Assuming a negligible magnetic field, the second Maxwell's equation (Law of Faraday; (2.2)) in integral form becomes (closed contour $\mathcal{C}$)

$$\int_{\Gamma} (\nabla \times \mathbf{E}) \cdot d\Gamma = \oint_{\mathcal{C}} \mathbf{E} \cdot d\mathbf{s} = 0 \tag{2.16}$$

where Stoke's theorem has been applied. For the material interface shown in Fig. 2.2b and $b \to 0$, we can simplify (2.16) to

$$\lim_{b \to 0} \oint_{\mathcal{C}} \mathbf{E} \cdot d\mathbf{s} \ \longrightarrow \ \mathbf{E}_1 \cdot \mathbf{s} - \mathbf{E}_2 \cdot \mathbf{s} = s\mathbf{t} \cdot (\mathbf{E}_1 - \mathbf{E}_2) = E_{1t} - E_{2t} = 0 \ . \tag{2.17}$$

Consequently, the tangential component E_t of the electric field intensity $\mathbf{E} = [E_n, E_t]^t$ is continuous at the interface of two materials.

By performing similar steps, the continuity relations at a material interface for the magnetic quantities $(\mathbf{B}, \mathbf{H})$ and for the electric current density $\mathbf{J}$ can be deduced.

2.1.4 Lumped Circuit Elements

As previously discussed, we will be able to simplify Maxwell's equations when the device dimensions are much smaller than the wavelength of electromagnetic waves. A further simplification can be performed if either the electric or the magnetic field dominates. The application of the resulting equations is, however, oftentimes still too complicated for various practical situations. Therefore, an alternative approach is commonly utilized yielding reliable approximations of electromagnetic fields. The approach is based on three lumped circuit elements[2]: (i) resistor R, (ii) inductor L, and (iii) capacitor C (graphic symbols in Table 2.4) [1]. While the inductor relates to the magnetic field, the capacitor belongs to the electric field. The inductor and capacitor measure the ability to store magnetic energy and electric charges, respectively. By means of the resistor, we can describe conversions of electromagnetic energy into energy in other physical fields, e.g., kinetic energy in the mechanical field.

The relation between the physical field quantities $(\mathbf{E}; \mathbf{D}; \mathbf{B}; \mathbf{J})$ of electromagnetic fields and the lumped circuit elements $(R; L; C)$ is defined as

$$R = \frac{\int_{\mathbf{r}_1}^{\mathbf{r}_2} \mathbf{E} \cdot d\mathbf{s}}{\int_A \mathbf{J} \cdot d\mathbf{A}} = \frac{U}{I} \qquad \text{unit } \Omega \tag{2.18}$$

$$L = \frac{\int_A \mathbf{B} \cdot d\mathbf{A}}{\int_A \mathbf{J} \cdot d\mathbf{A}} = \frac{\Phi}{I} \qquad \text{unit H} \tag{2.19}$$

$$C = \frac{\oint_S \mathbf{D} \cdot d\mathbf{A}}{\int_{\mathbf{r}_1}^{\mathbf{r}_2} \mathbf{E} \cdot d\mathbf{s}} = \frac{Q}{U} \qquad \text{unit F .} \tag{2.20}$$

Here, U, I, Φ as well as Q stand for scalar quantities that are frequently applied in conjunction with lumped circuit elements in electrical engineering. The electric current $I = \int_A \mathbf{J} \cdot d\mathbf{A}$ (unit A) relates to electric charges flowing through the area A. The expression $U = \int_{\mathbf{r}_1}^{\mathbf{r}_2} \mathbf{E} \cdot d\mathbf{s}$ (unit V) denotes the electric potential difference (electric voltage) from position $\mathbf{r}_2$ and $\mathbf{r}_1$. $\Phi = \int_A \mathbf{B} \cdot d\mathbf{A}$ (unit V s) is the magnetic flux through the area A and $Q = \oint_S \mathbf{D} \cdot d\mathbf{A}$ (unit A s) the electric charge enclosed by the

---

[2] Usually, these lumped circuit elements are time-invariant. Exceptions are configurations changing their geometry with respect to time.

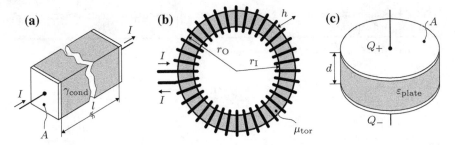

Fig. 2.3 a Electric conductor (length l; area A) with homogeneous conductivity γ_{cond}; **b** toroidal core coil (number of winding N_{wind}; inner radius r_I; outer radius r_O; height h) with relative magnetic permeability μ_{tor}; **c** plate capacitor (plate distance d; area A) containing dielectric medium with relative permittivity ε_{plate}

surface $\mathcal{S}$. Equation (2.18) represents the so-called *Ohm's law*, which is one of the most famous equations in electrical engineering.

For simple configurations such as a conductor, a toroidal core coil (number of windings N_{wind}), and a plate capacitor (see Fig. 2.3), we can approximate the lumped circuit elements with

$$\text{conductor:} \qquad R_{cond} = \frac{l}{\gamma_{cond} A} \qquad (2.21)$$

$$\text{toroidal core coil:} \quad L_{tor} = N_{wind}^2 \frac{\mu_0 \mu_{tor} h}{2\pi} \ln\left(\frac{r_O}{r_I}\right) \qquad (2.22)$$

$$\text{plate capacitor:} \quad C_{plate} = \frac{\varepsilon_{plate} \varepsilon_0 A}{d} \; . \qquad (2.23)$$

The expressions l, A, h, r_O, r_I, and d are geometric dimensions of the components. γ_{cond}, μ_{tor}, and ε_{plate} refer to the decisive material properties. For several configurations of complex geometries, similar approximations can be found.

In contrast to the simplified Maxwell's equations describing purely electric or purely magnetic fields, the lumped elements approach is also applicable in case of spatially separated components of these fields, e.g., a coil in combination with a capacitor. To efficiently investigate the behavior of such combinations with respect to time, an electric circuit containing the lumped elements and sources of electric energy (voltage source and/or current source) is analyzed. For such electric circuits, Kirchhoff's current law and Kirchhoff's voltage law have to be fulfilled at any time

$$\text{Kirchhoff's current law:} \qquad \sum_{k=1}^{n} I_k = 0 \qquad (2.24)$$

$$\text{Kirchhoff's voltage law:} \qquad \sum_{k=1}^{n} U_k = 0 \; . \qquad (2.25)$$

Table 2.4 Fundamental relations and common graphic symbols for lumped elements resistor R, inductor L, and capacitor C

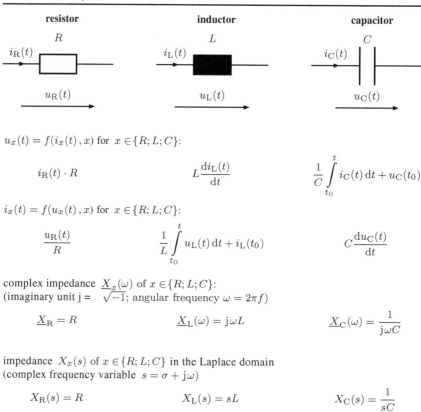

| resistor | inductor | capacitor |

$u_x(t) = f(i_x(t), x)$ for $x \in \{R; L; C\}$:

$$i_R(t) \cdot R \qquad L\frac{di_L(t)}{dt} \qquad \frac{1}{C}\int_{t_0}^{t} i_C(t)\,dt + u_C(t_0)$$

$i_x(t) = f(u_x(t), x)$ for $x \in \{R; L; C\}$:

$$\frac{u_R(t)}{R} \qquad \frac{1}{L}\int_{t_0}^{t} u_L(t)\,dt + i_L(t_0) \qquad C\frac{du_C(t)}{dt}$$

complex impedance $\underline{X}_x(\omega)$ of $x \in \{R; L; C\}$:
(imaginary unit $j = \sqrt{-1}$; angular frequency $\omega = 2\pi f$)

$$\underline{X}_R = R \qquad \underline{X}_L(\omega) = j\omega L \qquad \underline{X}_C(\omega) = \frac{1}{j\omega C}$$

impedance $X_x(s)$ of $x \in \{R; L; C\}$ in the Laplace domain
(complex frequency variable $s = \sigma + j\omega$)

$$X_R(s) = R \qquad X_L(s) = sL \qquad X_C(s) = \frac{1}{sC}$$

Kirchhoff's current law states that at any node of an electric circuit, the sum of the electric currents flowing into the node is equal to the sum of electric currents flowing out of the node. Kirchhoff's voltage law states that the directed sum of the electric voltages U_k around any closed network is zero.

The relation between the time-dependent quantities electric voltage $u(t)$ and electric current $i(t)$ also plays a crucial role in performing circuit analysis. Table 2.4 contains these relations for the different lumped elements. In addition to differential and integral equations in the time domain, the complex impedances $\underline{X}_x(\omega)$ of the lumped circuit elements in the frequency domain as well as those $X_x(s)$ in the Laplace domain are listed. Complex impedances in the frequency domain facilitate the analysis of electric circuits for sinusoidal excitations of frequency $f = \omega/2\pi$, while the approach based on the Laplace domain can also be utilized for certain transient excitation signals [11]. In doing so, the excitation signals have to be transformed into the frequency domain or Laplace domain, respectively. When the (complex) impedances of the components are known, we will then be able to study electric circuits comprising various lumped elements with similar approaches as for resistor networks. However, to obtain electric voltages as well as electric currents with respect to time, appropriate inverse transforms are indispensable, i.e., transform from the Laplace domain to the time domain [4].

At this point, it should be mentioned that a real device cannot be described completely by means of a single lumped element. Reliable modeling requires, strictly speaking, superposition of several lumped elements. Nevertheless, in a certain frequency range, we can approximate the device behavior by a distinct combination of a few lumped elements.

2.2 Continuum Mechanics

Piezoelectric materials are able into convert mechanical into electrical energy and vice versa. Because the mechanical deformations in such solids are mostly rather small during operation, we can describe the mechanical field by linear relations. In this section, the fundamental equations for linear continuum mechanics as well as essential quantities (e.g., mechanical strain) for the mechanical field are detailed. Thereby, a deformable solid body (elastic body) is considered. We start with Navier's equation linking the mechanical stress to both, inner volume forces and time-dependent body displacements. The mechanical strain and the constitutive equations for a deformable solid body will be explained in Sect. 2.2.2 and in Sect. 2.2.3, respectively. At last, we discuss different elastic wave types, which might occur in solid bodies. Further literature concerning continuum mechanics can be found in [2, 3, 18, 20].

2.2.1 Navier's Equation

Navier's equation is a fundamental equation in continuum mechanics. In order to derive this equation, we assume a deformable solid body of arbitrary shape at rest with prescribed volume forces $\mathbf{f}_V$ (given body force per unit of volume; unit $N\,m^{-3}$; e.g., gravity forces) and a support at equilibrium. Hence, the sum of all mechanical forces as well as of all mechanical torques equals zero. If a small part is cut out of the deformable solid body, forces will have to be applied to the cutting planes to still guarantee equilibrium (*Euler–Cauchy stress principle*). These forces correspond to the inner forces of the deformable solid body. Due to the fact that the applied forces are distributed across the cutting planes, it is reasonable to introduce mechanical stresses, which are defined as force per unit area.

In a first step, we cut out a cubical-shaped small part with cutting planes aligned in parallel to the Cartesian coordinate system (see Fig. 2.4a). The expressions $\mathbf{T}_x$, $\mathbf{T}_y$,

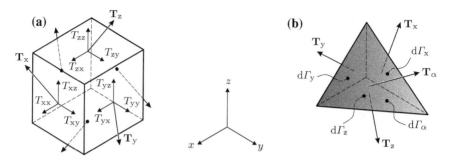

Fig. 2.4 a Mechanical stresses applied to cutting planes of a cubical-shaped part of deformable solid body; **b** mechanical stresses applied to tetrahedral element exhibiting oblique surface

and $\mathbf{T}_z$ denote the mechanical stresses (vector quantity; unit $\mathrm{N\,m^{-2}}$) on the cutting planes, with the index referring to the direction of their normal vector, respectively.

The stress vectors can be split up into scalar components that relate to the Cartesian coordinate system. In doing so, the stress vectors read as

$$\mathbf{T}_x = T_{xx}\mathbf{e}_x + T_{xy}\mathbf{e}_y + T_{xz}\mathbf{e}_z \tag{2.26}$$

$$\mathbf{T}_y = T_{yx}\mathbf{e}_x + T_{yy}\mathbf{e}_y + T_{yz}\mathbf{e}_z \tag{2.27}$$

$$\mathbf{T}_z = T_{zx}\mathbf{e}_x + T_{zy}\mathbf{e}_y + T_{zz}\mathbf{e}_z \tag{2.28}$$

with the unit vector $\mathbf{e}_i$ pointing in direction i. For the scalar components T_{ij}, index i refers to normal vector's direction of the cutting plane and j stands for the direction in which the stress acts. According to this notation, T_{xx}, T_{yy} as well as T_{zz} are normal stresses and T_{xy}, T_{xz}, T_{yx}, T_{yz}, T_{zx} as well as T_{zy} stand for shear stresses.

In a second step, we consider an infinitely small deformable body of tetrahedral shape with three surfaces ($\mathrm{d}\Gamma_x$, $\mathrm{d}\Gamma_y$, and $\mathrm{d}\Gamma_z$) oriented in parallel to the Cartesian coordinate plane (see Fig. 2.4b). If a mechanical force is applied to the oblique surface $\mathrm{d}\Gamma_\alpha$, the equilibrium state will require forces acting on the remaining surfaces

$$\mathrm{d}\Gamma_x\mathbf{T}_x + \mathrm{d}\Gamma_y\mathbf{T}_y + \mathrm{d}\Gamma_z\mathbf{T}_z - \mathrm{d}\Gamma_\alpha\mathbf{T}_\alpha = 0 \,. \tag{2.29}$$

Since the unity normal vector $\mathbf{e}_\alpha$ of the oblique surface can be defined as (Cartesian components n_i)

$$\mathbf{e}_\alpha = n_x\mathbf{e}_x + n_y\mathbf{e}_y + n_z\mathbf{e}_z \,, \tag{2.30}$$

the surface elements $\mathrm{d}\Gamma_x$, $\mathrm{d}\Gamma_y$, and $\mathrm{d}\Gamma_z$ can be written as

$$\mathrm{d}\Gamma_x = \mathrm{d}\Gamma_\alpha n_x \qquad \mathrm{d}\Gamma_y = \mathrm{d}\Gamma_\alpha n_y \qquad \mathrm{d}\Gamma_z = \mathrm{d}\Gamma_\alpha n_z \,. \tag{2.31}$$

Combining this with (2.29) and subsequently with (2.26)–(2.28) yields

$$
\begin{aligned}
\mathbf{T}_\alpha &= \mathbf{T}_x n_x + \mathbf{T}_y n_y + \mathbf{T}_z n_z \\
&= \left(T_{xx} n_x + T_{xy} n_y + T_{xz} n_z \right) \mathbf{e}_x + \left(T_{yx} n_x + T_{yy} n_y + T_{yz} n_z \right) \mathbf{e}_y \\
&\quad + \left(T_{zx} n_x + T_{zy} n_y + T_{zz} n_z \right) \mathbf{e}_z .
\end{aligned}
\tag{2.32}
$$

It is possible to achieve a compact form by exploiting the so-called *Cauchy stress tensor* [**T**] of rank two

$$
[\mathbf{T}] = \begin{bmatrix} T_{xx} & T_{xy} & T_{xz} \\ T_{yx} & T_{yy} & T_{yz} \\ T_{zx} & T_{zy} & T_{zz} \end{bmatrix} .
\tag{2.33}
$$

The mechanical stress acting on the oblique surface $d\Gamma_\alpha$ of the tetrahedral can then be expressed by (transpose t)

$$
\mathbf{T}_\alpha = [\mathbf{T}]^t \, \mathbf{e}_\alpha .
\tag{2.34}
$$

As already mentioned, in case of the equilibrium state, the sums of both mechanical forces and mechanical torques for the deformable solid body at rest are zero. Thus, the equation for translation (surface Γ; surface vector $\boldsymbol{\Gamma}$; volume Ω of the body)

$$
\oint_\Gamma [\mathbf{T}]^t \, d\boldsymbol{\Gamma} + \int_\Omega \mathbf{f}_V d\Omega = 0
\tag{2.35}
$$

and the equation for rotation (position vector $\mathbf{r}$ of a point in the body)

$$
\oint_\Gamma (\mathbf{r} \times [\mathbf{T}]) \, d\boldsymbol{\Gamma} + \int_\Omega (\mathbf{r} \times \mathbf{f}_V) \, d\Omega = 0
\tag{2.36}
$$

have to be fulfilled. From these two equations, it follows that the equation describing the equilibrium state for an infinitely small part of the deformable solid body at rest is given by

$$
\nabla[\mathbf{T}] + \mathbf{f}_V = 0 .
\tag{2.37}
$$

The entries of the Cauchy stress tensor [**T**] feature the properties

$$
T_{xy} = T_{yx} \qquad T_{xz} = T_{zx} \qquad T_{yz} = T_{zy} .
\tag{2.38}
$$

Consequently, [**T**] is symmetric and the nine tensor entries can be reduced to six entries. According to *Voigt notation*, it is convenient to introduce the stress vector **T**

$$\mathbf{T} = \begin{bmatrix} T_{xx} \\ T_{yy} \\ T_{zz} \\ T_{yz} \\ T_{xz} \\ T_{xy} \end{bmatrix} = \begin{bmatrix} T_{11} \\ T_{22} \\ T_{33} \\ T_{23} \\ T_{13} \\ T_{12} \end{bmatrix} = \begin{bmatrix} T_1 \\ T_2 \\ T_3 \\ T_4 \\ T_5 \\ T_6 \end{bmatrix} \qquad (2.39)$$

comprising six components instead of utilizing the tensor notation [**T**]. Note that **T** does not represent a physical stress vector but contains the independent components of the Cauchy stress tensor. If the differential operator $\mathcal{B}$ is introduced in addition

$$\mathcal{B} = \begin{bmatrix} \frac{\partial}{\partial x} & 0 & 0 & 0 & \frac{\partial}{\partial z} & \frac{\partial}{\partial y} \\ 0 & \frac{\partial}{\partial y} & 0 & \frac{\partial}{\partial z} & 0 & \frac{\partial}{\partial x} \\ 0 & 0 & \frac{\partial}{\partial z} & \frac{\partial}{\partial y} & \frac{\partial}{\partial x} & 0 \end{bmatrix}^t, \qquad (2.40)$$

we can rewrite (2.37) in Voigt notation to

$$\mathcal{B}^t \mathbf{T} + \mathbf{f}_V = 0 \qquad (2.41)$$

for the equilibrium of an infinitely small body part. As in the dynamic case, the right-hand side of (2.41) is not zero anymore but has to be the inertia force acting on the body. We finally obtain

$$\mathcal{B}^t \mathbf{T} + \mathbf{f}_V = \varrho_0 \frac{\partial^2 \mathbf{u}}{\partial t^2} \qquad (2.42)$$

where ϱ_0 stands for the material density of the infinitely small body part for equilibrium. The expression $\partial^2 \mathbf{u}/\partial t^2$ denotes the second-order derivate of the body displacement $\mathbf{u} = [u_x, u_y, u_z]^t$ (unit m) with respect to time and, thus, the acceleration of the body. Equation (2.42) is the so-called *Navier's equation* explaining the dynamic behavior of a solid deformable body, strictly speaking of an infinitely small part.

2.2.2 Mechanical Strain

A solid deformable body can be displaced and rotated. Moreover, such a body can be deformed leading to a certain change of its shape. In order to study this deformation for the linear case (i.e., small deformations), let us consider an infinitely small rectangular area (side lengths dx and dy) of a solid body in the two-dimensional space. Due to a mechanical load, the body is displaced and/or rotated and/or deformed. Figure 2.5 depicts the rectangular area for the initial state and resulting from the mechanical load, the parallelogram which is in the following referred to as deformed state. The expressions $u_x(x, y)$ and $u_y(x, y)$ stand for displacements in x- and y-

direction of the edge point $P_0(x, y)$, respectively. α and β denote the angles of the parallelogram in the deformed state. Under the assumption of a small angle α, the side length dx for the initial state changes to

$$\frac{dx + u_x(x + dx, y) - u_x(x, y)}{\cos(\alpha)} \approx dx + u_x(x + dx, y) - u_x(x, y) . \qquad (2.43)$$

To further simplify this expression, we expand $u_x(x + dx, y)$ in a Taylor series (remainder term $\mathcal{O}(n)$)

$$u_x(x + dx, y) = u_x(x, y) + \frac{\partial u_x(x, y)}{\partial x} dx + \mathcal{O}(n) . \qquad (2.44)$$

Inserting (2.44) in (2.43) and neglecting higher order terms finally leads to

$$\frac{dx + u_x(x + dx, y) - u_x(x, y)}{\cos(\alpha)} \approx dx + \frac{\partial u_x(x, y)}{\partial x} dx \qquad (2.45)$$

for the approximated side length of the parallelogram in x-direction. Analogously, the side length in y-direction for the deformed state can be computed. When the differences of the side lengths between deformed and initial state are related to the initial state, we will obtain

$$S_{xx} = \frac{\partial u_x(x, y)}{\partial x} \quad \text{and} \quad S_{yy} = \frac{\partial u_y(x, y)}{\partial y} \qquad (2.46)$$

representing the relative change of dx and dy in x- and y-direction, respectively. S_{xx} and S_{yy} are commonly called normal strains.

In addition to the normal strains yielding elongations, the solid body may be sheared in the deformed state. For the investigated infinitely small rectangle, this shearing is measured by the angles α and β of the parallelogram (see Fig. 2.5). With the displacements and the side length dx, one can deduce the relation

$$\tan(\alpha) = \frac{u_y(x + dx, y) - u_y(x, y)}{dx + u_x(x + dx, y) - u_x(x, y)} . \qquad (2.47)$$

By expanding the displacement expressions $u_y(x + dx, y)$ and $u_x(x + dx, y)$ in Taylor series up to the linear term, (2.47) becomes

$$\tan(\alpha) = \frac{\frac{\partial u_y(x, y)}{\partial x}}{1 + \frac{\partial u_x(x, y)}{\partial y}} . \qquad (2.48)$$

Since small deformations of the rectangle are assumed, $\partial u_y(x, y)/\partial x$ and $\partial u_x(x, y)/\partial y$ as well as α are small compared to 1. Therewith, (2.48) is simplified to

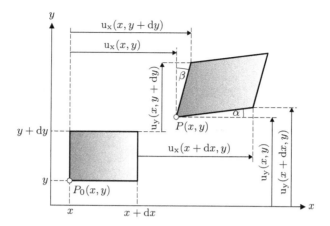

Fig. 2.5 Infinite small rectangle (side lengths dx and dy) in initial state (left) and deformed state (right)

$$\alpha = \frac{\partial u_y(x, y)}{\partial x} \, . \tag{2.49}$$

The same procedure can be utilized to approximate the angle β. To measure the total shearing of the rectangle, we calculate the sum of α and β

$$\alpha + \beta = \frac{\partial u_y(x, y)}{\partial x} + \frac{\partial u_x(x, y)}{\partial y} = 2S_{xy} = 2S_{yx} \, . \tag{2.50}$$

S_{xy} and S_{yx} are so-called shear strains that are equal due to the mentioned definition of total shearing.

For the three-dimensional space, overall nine strains exist in the linear case. S_{xx}, S_{yy} as well as S_{zz} denote normal strains and S_{xy}, S_{yx}, S_{xz}, S_{zx}, S_{yz}, S_{zy} are shear strains. Similar to the mechanical stress in (2.33), it is appropriate to define the strain tensor $[\mathbf{S}]$ of rank two

$$[\mathbf{S}] = \begin{bmatrix} S_{xx} & S_{xy} & S_{xz} \\ S_{yx} & S_{yy} & S_{yz} \\ S_{zx} & S_{zy} & S_{zz} \end{bmatrix} \, . \tag{2.51}$$

The strain tensor comprises overall six independent entries because

$$S_{xy} = S_{yx} \qquad S_{xz} = S_{zx} \qquad S_{yz} = S_{zy} \tag{2.52}$$

holds. Thus, we can reduce the strain tensor to a vector $\mathbf{S}$ containing six components (Voigt notation)

$$\mathbf{S} = \begin{bmatrix} S_{xx} \\ S_{yy} \\ S_{zz} \\ 2S_{yz} \\ 2S_{xz} \\ 2S_{xy} \end{bmatrix} = \begin{bmatrix} S_{11} \\ S_{22} \\ S_{33} \\ 2S_{23} \\ 2S_{13} \\ 2S_{12} \end{bmatrix} = \begin{bmatrix} S_1 \\ S_2 \\ S_3 \\ S_4 \\ S_5 \\ S_6 \end{bmatrix} \qquad (2.53)$$

Again, $\mathbf{S}$ does not represent a physical strain vector. By applying the differential operator $\mathcal{B}$ from (2.40), the relation between the mechanical strain vector and the displacement vector $\mathbf{u} = [u_x, u_y, u_z]^t$ of an arbitrary point within the deformed solid body takes the form

$$\mathbf{S} = \mathcal{B}\mathbf{u} . \qquad (2.54)$$

The mechanical strain is, therefore, uniquely represented through the displacement of the body.

2.2.3 Constitutive Equations and Material Behavior

For the linear case, the relation between mechanical stress and strain for a deformable solid body is known as *Hooke's law*, also called the linear law of elasticity. Hooke's law reads as

$$[\mathbf{T}] = [\mathbf{c}][\mathbf{S}] \qquad (2.55)$$

and alternatively as

$$[\mathbf{S}] = [\mathbf{s}][\mathbf{T}] \qquad (2.56)$$

with the elastic stiffness tensor (elasticity tensor) $[\mathbf{c}]$ and the elastic compliance tensor $[\mathbf{s}] = [\mathbf{c}]^{-1}$ covering the mechanical behavior of the body. Since both $[\mathbf{S}]$ and $[\mathbf{T}]$ are tensors of rank two, $[\mathbf{c}]$ as well as $[\mathbf{s}]$ are tensors of rank four containing 81 entries. Utilizing Einstein summation convention,[3] (2.55) and (2.56) become

$$T_{ij} = c_{ijkl} S_{kl} \qquad \{i, j, k, l\} = \{x, y, z\} \qquad (2.57)$$
$$S_{ij} = s_{ijkl} T_{kl} \qquad (2.58)$$

with the components T_{ij}, S_{kl}, c_{ijkl}, and s_{ijkl} of the tensors. As already mentioned, $[\mathbf{S}]$ and $[\mathbf{T}]$ comprise only six independent entries, respectively. On account of this fact and due to additional symmetry properties, the components of $[\mathbf{c}]$ and $[\mathbf{s}]$ feature the

---

[3]Einstein summation convention: $c_{ijkl} S_{kl} = \sum_{k,l} c_{ijkl} S_{kl}$.

properties

$$c_{ijkl} = c_{jikl} \qquad c_{ijkl} = c_{ijlk} \qquad c_{ijkl} = c_{klij} \tag{2.59}$$

$$s_{ijkl} = s_{jikl} \qquad s_{ijkl} = s_{ijlk} \qquad s_{ijkl} = s_{klij} \tag{2.60}$$

yielding a matrix with 36 entries (6×6), which contains overall 21 independent components.

Now, let us assume a deformable solid body of isotropic and homogeneous material. For such body, the relations between the mechanical stresses and strains are given by (trace of the tensor tr)

$$\text{normal stresses} \quad T_{ii} = 2G\left(S_{ii} + \frac{\nu_P}{1 - 2\nu_P}\text{tr}[\mathbf{S}]\right) \tag{2.61}$$

$$\text{shear stresses} \quad T_{ij} = 2G\,S_{ij} \qquad i \neq j \tag{2.62}$$

where ν_P denotes the so-called Poisson's ratio that measures the ratio of the resulting strain perpendicular to the applied mechanical load to the one in load direction. G stands for the shear modulus (unit $\text{N}\,\text{m}^{-2}$) and relates shear stresses to shear strains. Aside from the scalar quantities Poisson's ratio and shear modulus, an important quantity for an isotropic solid body is the Young's modulus (tensile modulus; unit $\text{N}\,\text{m}^{-2}$) E_M measuring the stiffness of a material. E_M can be calculated from ν_P and G with

$$E_M = 2G(1 + \nu_P) \ . \tag{2.63}$$

Note that apart from the density ϱ_0, two of the three quantities ν_P, G and E_M are sufficient to fully describe the mechanical material properties of the body. These quantities are also used to deduce other essential quantities of continuum mechanics, e.g., the so-called Lamé parameters λ_L and μ_L

$$\lambda_L = \frac{2\nu_P G}{1 - 2\nu_P} = \frac{\nu_P E_M}{(1 + \nu_P)(1 - 2\nu_P)} \tag{2.64}$$

$$\mu_L = G = \frac{E_M}{2(1 + \nu_P)} \ . \tag{2.65}$$

If the symmetry properties of the stiffness tensor $[\mathbf{c}]$ and the material behavior of the isotropic as well as homogeneous solid body are considered, we can rewrite (2.55) to

$$
\begin{bmatrix} T_{xx} \\ T_{yy} \\ T_{zz} \\ T_{yz} \\ T_{xz} \\ T_{xy} \end{bmatrix} =
\begin{bmatrix}
\lambda_L + 2\mu_L & \lambda_L & \lambda_L & 0 & 0 & 0 \\
\lambda_L & \lambda_L + 2\mu_L & \lambda_L & 0 & 0 & 0 \\
\lambda_L & \lambda_L & \lambda_L + 2\mu_L & 0 & 0 & 0 \\
0 & 0 & 0 & \mu_L & 0 & 0 \\
0 & 0 & 0 & 0 & \mu_L & 0 \\
0 & 0 & 0 & 0 & 0 & \mu_L
\end{bmatrix}
\begin{bmatrix} S_{xx} \\ S_{yy} \\ S_{zz} \\ 2S_{yz} \\ 2S_{xz} \\ 2S_{xy} \end{bmatrix} \ . \tag{2.66}
$$

Moreover, in an isotropic and homogeneous solid body, Navier's equation (2.42) incorporating the linear material behavior as well as the previously studied relation between mechanical strain and displacement becomes

$$\mu_L \nabla \cdot \nabla \mathbf{u} + (\lambda_L + \mu_L) \nabla(\nabla \cdot \mathbf{u}) + \mathbf{f}_V = \varrho_0 \frac{\partial^2 \mathbf{u}}{\partial t^2} . \tag{2.67}$$

However, for a general solid body that may be inhomogeneous as well as anisotropic, we are not able to define quantities like the Lamé parameters. By combining (2.42), (2.54) and Hooke's law, Navier's equations can also be expressed as

$$\mathcal{B}^t[\mathbf{c}] \mathcal{B}\mathbf{u} + \mathbf{f}_V = \varrho_0 \frac{\partial^2 \mathbf{u}}{\partial t^2} , \tag{2.68}$$

which is valid for the linear case, i.e., small mechanical deformations. Therein, the expression [$\mathbf{c}$] represents a matrix of dimension 6×6 instead of a tensor quantity comprising 81 entries. This matrix is always referred to as stiffness tensor in the following equations and explanations.

2.2.4 Elastic Waves in Solids

Let us regard a deformable body of infinite extension to discuss different wave types in solids. The wave propagation causes displacements of the infinitely small body fractions depending on both space and time. As it is possible for almost all vector fields, we decompose the displacement vector $\mathbf{u}$ in an irrotational part $\mathbf{u}_{irr}$ and a solenoidal part $\mathbf{u}_{sol}$, for which the following relations are fulfilled

$$\text{irrotational part:} \quad \nabla \times \mathbf{u}_{irr} = 0 \tag{2.69}$$
$$\text{solenoidal part:} \quad \nabla \cdot \mathbf{u}_{sol} = 0 . \tag{2.70}$$

According to the so-called Helmholtz decomposition, we additionally introduce the scalar potential φ and the vector potential $\mathcal{A}$. Therewith, the displacement vector becomes

$$\mathbf{u} = \underbrace{\nabla \varphi}_{\mathbf{u}_{irr}} + \underbrace{\nabla \times \mathcal{A}}_{\mathbf{u}_{sol}} . \tag{2.71}$$

Substitution of $\mathbf{u}$ in the Navier's equation (2.67) by this expression and neglecting the prescribed volume forces $\mathbf{f}_V$ results in

$$\mu_L \nabla \cdot \nabla (\nabla \varphi + \nabla \times \pmb{A}) + (\lambda_L + \mu_L) \nabla \left[\nabla \cdot (\nabla \varphi) + \nabla \cdot (\nabla \times \pmb{A}) \right]$$

$$= \varrho_0 \frac{\partial^2 (\nabla \varphi + \nabla \times \pmb{A})}{\partial t^2} , \qquad (2.72)$$

which can be simplified and rearranged to the form

$$\nabla \left[(\lambda_L + 2\mu_L) \nabla \cdot \nabla \varphi - \varrho_0 \frac{\partial^2 \varphi}{\partial t^2} \right] + \nabla \times \left[\mu_L \nabla \cdot \nabla \pmb{A} - \varrho_0 \frac{\partial^2 \pmb{A}}{\partial t^2} \right] = 0 . \quad (2.73)$$

From (2.73), one equation for φ and $\pmb{A}$ can be deduced, respectively, that has to be satisfied

$$\frac{\partial^2 \varphi}{\partial t^2} = \frac{\lambda_L + 2\mu_L}{\varrho_0} \nabla \cdot \nabla \varphi \qquad (2.74)$$

$$\frac{\partial^2 \pmb{A}}{\partial t^2} = \frac{\mu_L}{\varrho_0} \nabla \cdot \nabla \pmb{A} . \qquad (2.75)$$

To analyze (2.74) and (2.75), appropriate ansatz functions for the scalar and vector potential are required. Here, we choose

$$\varphi = f(\zeta) = f(\mathbf{k} \cdot \mathbf{x} - \omega t) \qquad (2.76)$$

$$\pmb{A} = \mathbf{F}(\zeta) = \mathbf{F}(\mathbf{k} \cdot \mathbf{x} - \omega t) \qquad (2.77)$$

representing elastic waves propagating with the velocity c within the solid body in (positive) direction of the wave vector $\mathbf{k} = [k_x, k_y, k_z]^t$. The expressions $\mathbf{x}$ and ω denote the position of the infinitely small volume fraction within the body and the angular frequency, respectively.

Let us investigate in a first step the equation for the scalar potential φ. From the ansatz function (2.76), the following relations can be derived (k_i component of $\mathbf{k}$; x_i component of $\mathbf{x} = [x, y, z]^t$)

$$\frac{\partial^2 \varphi}{\partial t^2} = \frac{\partial}{\partial t} \left(\frac{\partial \varphi}{\partial \zeta} \frac{\partial \zeta}{\partial t} \right) = \omega^2 \frac{\partial^2 \varphi}{\partial \zeta^2} \quad \text{and} \quad \frac{\partial^2 \varphi}{\partial x_i^2} = \frac{\partial}{\partial x_i} \left(\frac{\partial \varphi}{\partial \zeta} \frac{\partial \zeta}{\partial x_i} \right) = k_i^2 \frac{\partial^2 \varphi}{\partial \zeta^2} .$$

By means of these relations, (2.74) results in

$$\omega^2 \frac{\partial^2 \varphi}{\partial \zeta^2} = \frac{\lambda_L + \mu_L}{\varrho_0} \underbrace{(k_x^2 + k_y^2 + k_z^2)}_{\|\mathbf{k}\|_2^2} \qquad (2.78)$$

where $\|\mathbf{k}\|_2 = k$ stands for the magnitude of the wave vector that is also named wave number k. Because $\omega = c \|\mathbf{k}\|_2$ has to be fulfilled for a propagating wave, we can deduce

$$c_1 = \sqrt{\frac{\lambda_L + 2\mu_L}{\varrho_0}} = \sqrt{\frac{E_M(1 - \nu_P)}{\varrho_0(1 - 2\nu_P)(1 + \nu_P)}} = \sqrt{\frac{2G(1 - \nu_P)}{\varrho_0(1 - 2\nu_P)}} \qquad (2.79)$$

representing the wave propagation velocity of the irrotational displacements $\mathbf{u}_{irr}$. Moreover, (2.71) leads to

$$\begin{aligned} \mathbf{u}_{irr} &= \nabla\varphi = \frac{\partial\varphi}{\partial x}\mathbf{e}_x + \frac{\partial\varphi}{\partial y}\mathbf{e}_y + \frac{\partial\varphi}{\partial z}\mathbf{e}_z \\ &= \underbrace{\frac{\partial\varphi}{\partial\zeta}\frac{\partial\zeta}{\partial x}}_{k_x}\mathbf{e}_x + \underbrace{\frac{\partial\varphi}{\partial\zeta}\frac{\partial\zeta}{\partial y}}_{k_y}\mathbf{e}_y + \underbrace{\frac{\partial\varphi}{\partial\zeta}\frac{\partial\zeta}{\partial z}}_{k_z}\mathbf{e}_z = \mathbf{k}\frac{\partial\varphi}{\partial\zeta}, \end{aligned} \qquad (2.80)$$

which shows that the irrotational part of the displacements is pointing in the direction of wave propagation; i.e., the extension of the volume fractions gets exclusively altered in this direction (see Fig. 2.6a). Against that, the volume fractions remain completely unchanged perpendicular to $\mathbf{k}$. The resulting elastic waves propagating with the velocity c_1 (see (2.79)) are usually referred to as longitudinal or compression waves.

If a similar procedure is applied to the vector potential $\mathcal{A}$ in (2.75) with the ansatz function (2.77), we will obtain for the wave propagation velocity

$$c_t = \sqrt{\frac{\mu_L}{\varrho_0}} = \sqrt{\frac{E_M}{2(1 + \nu_P)\varrho_0}}. \qquad (2.81)$$

Furthermore, inserting the ansatz function (2.77) in (2.71) yields for the solenoidal part $\mathbf{u}_{sol}$ of the displacement

$$\mathbf{u}_{sol} = \nabla \times \mathcal{A} = \mathbf{k} \times \frac{\partial\mathcal{A}}{\partial\zeta}. \qquad (2.82)$$

Hence, $\mathbf{u}_{sol}$ is exclusively perpendicular to the direction $\mathbf{k}$ of wave propagation (see Fig. 2.6b). Such elastic waves propagating with the velocity c_t (see (2.81)) are called transverse or shear waves. The ratio of the propagation velocities for the different waves types, i.e., longitudinal and transverse waves, results in

$$\frac{c_1}{c_t} = \sqrt{\frac{\lambda_L + 2\mu_L}{\mu_L}} = \sqrt{\frac{2(1 - \nu_P)}{1 - 2\nu_P}}, \qquad (2.83)$$

which leads to the inequality $c_1 > \sqrt{2}c_t$. Owing to this fact, it can be stated that the propagation velocity for longitudinal waves in a homogeneous and isotropic solid is always larger than for transverse waves.

Pure longitudinal as well as pure transverse waves can only exist in a solid body of infinite extension. In reality, there always occurs a superposition of these wave

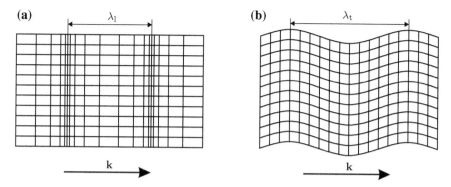

Fig. 2.6 Change of infinitely small fractions of solid body for **a** longitudinal waves and **b** transverse waves; direction of wave propagation **k**; wavelengths of elastic waves λ_l and λ_t

Table 2.5 Typical material parameters for solids in continuum mechanics; equilibrium density ϱ_0; Young's modulus E_M; Poisson's ratio ν_P; Lamé parameters λ_L and μ_L; wave propagation velocity c_l and c_t for longitudinal and transverse elastic waves, respectively

Material	ϱ_0 $\mathrm{kg\,m^{-3}}$	E_M $10^{10}\,\mathrm{N\,m^{-2}}$	ν_P	λ_L $10^{10}\,\mathrm{N\,m^{-2}}$	μ_L $10^{10}\,\mathrm{N\,m^{-2}}$	c_l $\mathrm{m\,s^{-1}}$	c_t $\mathrm{m\,s^{-1}}$
Acrylic glass	1190	0.44	0.39	0.56	0.16	2720	1150
Aluminum	2700	6.76	0.36	6.40	2.49	6490	3030
Cooper	8930	12.62	0.37	13.11	4.61	5000	2270
Iron	7690	20.34	0.29	10.89	7.88	5890	3200
Polyethylene	900	0.08	0.46	0.32	0.03	2030	550
Silver	10600	7.47	0.38	8.57	2.71	3630	1600
Tungsten	19400	41.58	0.27	19.26	16.41	5180	2910

types resulting, for instance, in waves traveling on the surface of a solid, the so-called Rayleigh waves. However, for several practical situations, we are able to approximate the wave propagation in solids by the dominating wave type, i.e., longitudinal or transverse waves.

Table 2.5 contains the most important parameters in continuum mechanics for various solids. Note that the listed values refer to typical parameters, which can be found in literature. Actually, the parameters of a material specimen can strongly deviate from the given values. This mainly stems from differences and uncertainties in the manufacturing process.

2.3 Acoustics

While solid materials work against both, changes of volume and of shape, gases as well as (nonviscous) liquids solely react to changes of volume. This arises from the fact that gases and such liquids are not able to transmit shear forces. Consequently, transverse waves do not propagate. In the following section, the basics of acoustics, i.e., the wave propagation in gases and liquids, will be introduced which are important to understand and to model the behavior of piezoelectric ultrasound transducers. At the beginning, the fundamental quantities (e.g., sound pressure) and the wave theory of sound are discussed. We will subsequently deduce the linear acoustic wave equation, which results from the conservation of mass, the conservation of momentum, and the state equation covering the properties of media. Diffraction and reflection effects at interfaces of different media are studied in Sect. 2.3.4. Finally, we will briefly explain energy absorption mechanisms taking place during the sound propagation in gases and liquids. Further literature concerning acoustics can be found in [5, 10, 12, 16].

2.3.1 Fundamental Quantities

The propagation of sound waves in gases as well as in liquids is accompanied by local and temporal changes of three state variables: (i) density ϱ of the propagation medium; (ii) pressure p (force per area) acting inside the medium, and (iii) velocity $\mathbf{v}$ of the particles, i.e., the volume fractions of the medium. In general, the state variables can be decomposed into (position $\mathbf{r}$ within the medium)

$$\text{density:} \qquad \varrho(\mathbf{r}, t) = \varrho_0 + \varrho_\sim(\mathbf{r}, t) \qquad (2.84)$$

$$\text{pressure:} \qquad p(\mathbf{r}, t) = p_0 + p_\sim(\mathbf{r}, t) \qquad (2.85)$$

$$\text{veloctiy:} \qquad \mathbf{v}(\mathbf{r}, t) = \mathbf{v}_0 + \mathbf{v}_\sim(\mathbf{r}, t) \qquad (2.86)$$

where ϱ_0 (scalar; unit $\mathrm{kg\,m^{-3}}$), p_0 (scalar; unit $\mathrm{N\,m^{-2}}$ and Pa), and $\mathbf{v}_0$ (vector; unit $\mathrm{m\,s^{-1}}$) represent the density, pressure, and particle velocity for the equilibrium state of the medium, respectively. The expressions $\varrho_\sim$, $p_\sim$, and $\mathbf{v}_\sim$ denote fluctuations, which are induced from the propagating sound wave. That is the reason why these quantities are commonly referred to as acoustic density, sound pressure, and acoustic particle velocity. One of them is sufficient to completely specify sound fields and, therefore, $\varrho_\sim$, $p_\sim$ as well as $\mathbf{v}_\sim$ are called sound field quantities. Depending on the frequency f (unit $\mathrm{s^{-1}}$ or Hz) of the fluctuations, we classify acoustics in

- Infrasound: $f \leq 16\,\mathrm{Hz}$.
- Audible sound: $16\,\mathrm{Hz} < f \leq 20\,\mathrm{kHz}$.
- Ultrasound: $20\,\mathrm{kHz} < f \leq 1\,\mathrm{GHz}$.
- Hypersound: $f > 1\,\mathrm{GHz}$.

Apart from $\varrho_\sim$, $p_\sim$, and $\mathbf{v}_\sim$, the acoustic intensity and the acoustic power are important quantities in acoustics. The vectorial acoustic intensity $\mathbf{I}_{\mathrm{ac}}$ is defined by

$$\mathbf{I}_{\mathrm{ac}}(\mathbf{r}, t) = p_\sim(\mathbf{r}, t) \cdot \mathbf{v}_\sim(\mathbf{r}, t) \qquad \text{unit:} \;\; \mathrm{W\,m^{-2}} . \qquad (2.87)$$

Table 2.6 Typical values of averaged sound pressure $\overline{p}_\sim$ and sound pressure level L_p for different sound events; static air pressure $p_0 = 10^5$ Pa

Sound event	$\overline{p}_\sim$ in Pa	L_p in dB
Threshold of hearing	$20 \cdot 10^{-6}$	0
Conversational speech	$< 2 \cdot 10^{-2}$	60
Street noise in a city	$5 \cdot 10^{-2}$	68
Orchestral music	5	108
Noise of a jackhammer	50	128

It measures at position $\mathbf{r}$ the time-dependent sound energy. From the acoustic intensity, we can compute the acoustic power $P_{ac}(t)$ of a sound source

$$P_{ac}(t) = \oint_A \mathbf{I}_{ac}(\mathbf{r}, t) \cdot d\mathbf{A} \quad \text{unit:} \quad W \quad (2.88)$$

with the enveloping surface $\mathbf{A}$, which encloses the sound source.

Due to the large range of the mentioned acoustic quantities (see Table 2.6), it is convenient to introduce normalized logarithmic scaling in Decibel (dB). To obtain meaningful values, $p_\sim$, $\mathbf{I}_{ac}$ as well as P_{ac} have to be averaged over a certain time duration yielding $\overline{p}_\sim$, $\overline{I}_{ac}$, and $\overline{P}_{ac}$. The logarithmic scaling is commonly calculated with ($\|\mathbf{I}_{ac}\|_2 = I_{ac}$)

$$\text{sound pressure level:} \quad L_p = 20 \log_{10}\left(\frac{\overline{p}_\sim}{p_{ref}}\right) ; \quad p_{ref} = 2 \cdot 10^{-5} \text{ Pa} \quad (2.89)$$

$$\text{sound intensity level:} \quad L_I = 10 \log_{10}\left(\frac{\overline{I}_{ac}}{I_{ref}}\right) ; \quad I_{ref} = 10^{-12} \text{ W m}^{-2} \quad (2.90)$$

$$\text{sound power level:} \quad L_P = 10 \log_{10}\left(\frac{\overline{P}_{ac}}{P_{ref}}\right) ; \quad P_{ref} = 10^{-12} \text{ W} . \quad (2.91)$$

p_{ref}, I_{ref}, and P_{ref} represent the values at the threshold of human hearing for sinusoidal sound waves with a frequency of 1 kHz. For example, a sound pressure level of $L_p = 0$ dB in air means that the averaged sound pressure $\overline{p}_\sim$ is equal to 20 μPa. Note that in liquids, other values are chosen for normalization, e.g., $p_{ref} = 1$ μPa in water.

2.3.2 Wave Theory of Sound

The wave theory of sound is directly linked to the conservation of mass, the conservation of momentum, and the state equation.

Conservation of Mass

As already discussed, sound propagation is accompanied by local density variations of the propagation medium. According to the mass conservation, a density variation changes the mass of a spatially fixed volume fraction. The mass change has to be compensated by a certain mass flow through the surface enclosing the volume fraction. In order to give a detailed description, let us consider an arbitrary volume fraction Ω of the medium (see Fig. 2.7a). The mass m of the volume fraction results in

$$m(t) = \int_{\Omega} \varrho(\mathbf{r}, t)\, d\Omega \qquad (2.92)$$

where $\varrho(\mathbf{r}, t)$ is the density depending on both space and time. A positive mass flow through the volume surface Γ during the time interval dt will decrease the mass within the volume fraction. Hence, one can state

$$\underbrace{\oint_{\Gamma} \varrho(\mathbf{r}, t)\, \mathbf{v}(\mathbf{r}, t)\, dt \cdot d\mathbf{\Gamma}}_{\text{mass flow}} = \underbrace{-\int_{\Omega} \Big[\varrho(\mathbf{r}, t + dt) - \varrho(\mathbf{r}, t)\Big]\, d\Omega}_{\text{change of mass}} . \qquad (2.93)$$

The expression $\varrho(\mathbf{r}, t + dt)$ can be approximated by the first-order Taylor series

$$\varrho(\mathbf{r}, t + dt) \approx \varrho(\mathbf{r}, t) + \frac{\partial \varrho(\mathbf{r}, t)}{\partial t} dt . \qquad (2.94)$$

By additionally applying the divergence theorem, (2.93) becomes

$$\int_{\Omega} \nabla \cdot \Big[\varrho(\mathbf{r}, t)\, \mathbf{v}(\mathbf{r}, t)\Big]\, d\Omega = -\int_{\Omega} \frac{\partial \varrho(\mathbf{r}, t)}{\partial t}\, d\Omega . \qquad (2.95)$$

Since this relation has to be fulfilled for each volume fraction within the propagation medium, we are able to rewrite (2.95) in differential form

$$\nabla \cdot \Big[\varrho(\mathbf{r}, t)\, \mathbf{v}(\mathbf{r}, t)\Big] = -\frac{\partial \varrho(\mathbf{r}, t)}{\partial t} , \qquad (2.96)$$

which is usually referred to as *continuity equation*.

Conservation of Momentum

To study the conservation of momentum for sound propagation, we consider at position $\mathbf{r} = [x, y, z]^{t}$ an infinitely small volume $d\Omega = dx\,dy\,dz$ of cubic shape (see Fig. 2.7b) that moves with the medium. In case of a gas or a nonviscous fluid, pressure changes within the propagation medium cause repulsive forces $\mathbf{F}(\mathbf{r}, t) =$

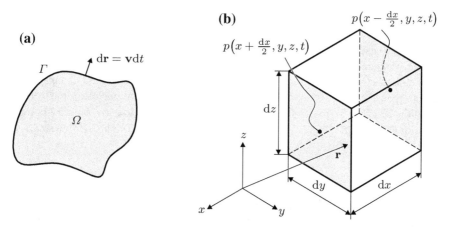

Fig. 2.7 a Considered arbitrary volume fraction Ω (surface Γ) for derivation of continuity equation; **b** infinitely small volume $d\Omega = dxdydz$ of cubic shape, which is considered to deduce Euler's equation

$\left[F_x(\mathbf{r}, t), F_y(\mathbf{r}, t), F_z(\mathbf{r}, t)\right]^t$ acting on this volume. For the considered configuration, the equilibrium of forces in x-direction becomes

$$d F_x(x, y, z, t) + \left[p\left(x + \frac{dx}{2}, y, z, t\right) - p\left(x - \frac{dx}{2}, y, z, t\right)\right] dy \, dz = 0 \ . \tag{2.97}$$

By applying Taylor series expansion up the linear term for $p(x \pm dx/2, y, z, t)$

$$p\left(x \pm \frac{dx}{2}, y, z, t\right) \approx p(x, y, z, t) \pm \frac{\partial p(x, y, z, t)}{\partial x} \frac{dx}{2} \ , \tag{2.98}$$

(2.97) simplifies to

$$d F_x(x, y, z, t) = -\frac{\partial p(x, y, z, t)}{\partial x} d\Omega \ . \tag{2.99}$$

If the same procedure is performed for y- and z-direction, we will obtain

$$d\mathbf{F}(\mathbf{r}, t) = -\nabla p(\mathbf{r}, t) d\Omega \tag{2.100}$$

representing in compact form the equilibrium of forces for the infinitely small volume. Additionally, for the considered volume exhibiting the mass $dm = \varrho(\mathbf{r}, t) d\Omega$, Newton's law has to be fulfilled which reads as

$$d\mathbf{F}(\mathbf{r}, t) = \varrho(\mathbf{r}, t) \, \mathbf{a}(\mathbf{r}, t) \, d\Omega \ . \tag{2.101}$$

The expressions **a** and d**F** stand for the acceleration of the mass and the (repulsive) force acting on the infinitely small volume, respectively. Because the infinitely small volume $d\Omega$ alters its position **r** with respect to time t, its acceleration **a** can be calculated from the particle velocity **v** by (total derivative d/dt)[4]

$$\mathbf{a} = \frac{d\mathbf{v}}{dt} = \frac{\partial \mathbf{v}}{\partial t} + \frac{\partial \mathbf{v}}{\partial x}\frac{\partial x}{\partial t} + \frac{\partial \mathbf{v}}{\partial y}\frac{\partial y}{\partial t} + \frac{\partial \mathbf{v}}{\partial z}\frac{\partial z}{\partial t} = \frac{\partial \mathbf{v}}{\partial t} + (\mathbf{v} \cdot \nabla)\mathbf{v}. \tag{2.102}$$

Utilizing this relation as well as (2.100) and (2.101) finally leads to

$$\varrho \left[\frac{\partial \mathbf{v}}{\partial t} + (\mathbf{v} \cdot \nabla)\mathbf{v} \right] = -\nabla p \,, \tag{2.103}$$

which is the so-called *Euler's equation* in differential form. The expression $(\mathbf{v} \cdot \nabla)\mathbf{v}$ (convective acceleration) accounts for effects of time-independent accelerations in a fluid with respect to space.

State Equation

The state variables (ϱ, p, and **v**) for the description of propagating sound waves are not independent from each other. Actually, the relation between them depends on the properties of the propagation medium. In liquids and gases, pressure is a function of both the density of the medium and its temperature ϑ

$$p = p(\varrho, \vartheta) \,. \tag{2.104}$$

As sound propagation is accompanied by fast local changes within the propagation medium, heat transfer between neighboring volume fractions is in a first step negligible. We can, therefore, assume constant medium temperature during sound propagation as well as adiabatic state changes. Consequently, pressure is exclusively a function of the medium density, i.e., $p = p(\varrho)$. Beyond that, we are able to connect the fluctuations $p_\sim$ and $\varrho_\sim$, which are induced by the propagating sound wave. If the sound pressure is expanded in a Taylor series around the equilibrium state ϱ_0, one will arrive at

$$p_\sim = \frac{A}{1!}\left(\frac{\varrho_\sim}{\varrho_0}\right) + \frac{B}{2!}\left(\frac{\varrho_\sim}{\varrho_0}\right)^2 + \mathcal{O}(n) \tag{2.105}$$

with

$$A = \varrho_0 \left(\frac{\partial p_\sim}{\partial \varrho_\sim}\right)\Bigg|_{\varrho=\varrho_0} \equiv \varrho_0 c_0^2 \quad \text{and} \quad B = \varrho_0^2 \left(\frac{\partial^2 p_\sim}{\partial \varrho_\sim^2}\right)\Bigg|_{\varrho=\varrho_0} \,. \tag{2.106}$$

---

[4]For compactness, arguments position **r** and time t are omitted.

The term c_0 denotes the sound velocity (speed of sound; wave propagation velocity) in the medium. For sound waves exhibiting small fluctuations compared to equilibrium, A is sufficient to describe the relation between sound quantities. This leads to

$$p_\sim = c_0^2 \varrho_\sim \,, \tag{2.107}$$

which is a fundamental equation in acoustics. However, in case of large fluctuations, nonlinear effects (e.g., progressive distortion) occur during sound propagation. Such effects can be modeled by additional consideration of B. The temperature-dependent ratio B/A has become a common expression in nonlinear acoustics and is listed for various liquids and gases in literature [6].

In the following, we assume small values for $p_\sim$ and $\varrho_\sim$ allowing linearization; i.e., only the linear term A of the Taylor series in (2.105) is considered. The linearization is applied to both gases and liquids. For ideal gases, the adiabatic state equation reads as

$$\frac{p_0 + p_\sim}{p_0} = \left(\frac{\varrho_0 + \varrho_\sim}{\varrho_0} \right)^{\kappa} \tag{2.108}$$

where $\kappa = C_p/C_V$ denotes the adiabatic exponent given by the ratio of specific heat C_p at constant pressure and specific heat C_V at constant volume. By means of linear Taylor approximation, (2.108) can be rewritten to

$$p_\sim = \kappa \frac{p_0}{\varrho_0} \varrho_\sim \,. \tag{2.109}$$

Combining this with (2.106) yields for the sound velocity in a gas

$$c_0 = \sqrt{\left(\frac{\partial p_\sim}{\partial \varrho_\sim} \right)\bigg|_{\varrho=\varrho_0}} = \sqrt{\kappa \frac{p_0}{\varrho_0}} \,. \tag{2.110}$$

Since an ideal gas fulfills the relation $p_\sim = \varrho_\sim R_{gas} \vartheta$ (specific gas constant R_{gas}), (2.110) takes the form

$$c_0 = \sqrt{\kappa R_{gas} \vartheta} \,, \tag{2.111}$$

which shows that the sound velocity in gases strongly depends on temperature. In a liquid, (2.109) has to be replaced by

$$p_\sim = K_{liquid} \frac{\varrho_\sim}{p_0} \tag{2.112}$$

where K_{liquid} is the adiabatic bulk modulus

Table 2.7 Typical sound velocities c_0 in $\mathrm{m\,s}^{-1}$ of various liquids and gases for selected temperatures

Acetone	Air	Argon	Benzol	Diesel
1174 at 25 °C	344 at 20 °C	319 at 0 °C	1330 at 25 °C	1250 at 25 °C
	386 at 100 °C			
	553 at 500 °C			
Gallium	Glycerin	Helium	Hydrogen	Water
2870 at 30 °C	1904 at 25 °C	965 at 0 °C	1284 at 0 °C	1480 at 20 °C
				1509 at 30 °C
				1550 at 60 °C

$$K_{\text{liquid}} = \varrho_0 \left(\frac{\partial p_\sim}{\partial \varrho_\sim} \right) \Bigg|_{\varrho = \varrho_0} . \tag{2.113}$$

Therewith, the sound velocity c_0 in liquids computes as

$$c_0 = \sqrt{\frac{K_{\text{liquid}}}{\varrho_0}} . \tag{2.114}$$

Table 2.7 contains sound velocities of various liquids and gases.

2.3.3 Linear Acoustic Wave Equation

With a view to deriving the linear acoustic wave equation, we assume small fluctuations of pressure and density during sound propagation in the propagation medium, i.e.,

$$\varrho_\sim(\mathbf{r}, t) \ll \varrho_0 \quad \text{and} \quad p_\sim(\mathbf{r}, t) \ll p_0 . \tag{2.115}$$

By utilizing this assumption and the fact that the equilibrium quantities ϱ_0, p_0, and $\mathbf{v}_0$ depend neither on position $\mathbf{r}$ nor on time t, the continuity equation (2.96) simplifies to

$$\varrho_0 \nabla \cdot \mathbf{v}_\sim(\mathbf{r}, t) = -\frac{\partial \varrho_\sim(\mathbf{r}, t)}{\partial t} . \tag{2.116}$$

Besides, sound waves of small amplitudes do not cause vortex in gases and nonviscous liquids, which results in an irrotational wave propagation, i.e., pure longitudinal waves. The convective accelerations $(\mathbf{v} \cdot \nabla)\mathbf{v}$ can, thus, be neglected. Together with the assumption of small fluctuations (2.115), Euler's equation (2.103) becomes

$$\varrho_0 \frac{\partial \mathbf{v}_\sim(\mathbf{r}, t)}{\partial t} = -\nabla p_\sim(\mathbf{r}, t) \; . \tag{2.117}$$

If the curl operator is applied to (2.117), one will obtain

$$\nabla \times \mathbf{v}_\sim(\mathbf{r}, t) = 0 \tag{2.118}$$

pointing out again that the particle velocity $\mathbf{v}_\sim$ and, consequently, the wave propagation are irrotational. Now, let us apply an additional time derivative $\partial/\partial t$ to (2.116)

$$\varrho_0 \nabla \cdot \frac{\partial \mathbf{v}_\sim(\mathbf{r}, t)}{\partial t} = -\frac{\partial^2 \varrho_\sim(\mathbf{r}, t)}{\partial t^2} \; . \tag{2.119}$$

Substitution of the expression $\partial \mathbf{v}_\sim/\partial t$ from (2.117) in (2.119) yields then

$$\nabla \cdot \nabla p_\sim(\mathbf{r}, t) = \frac{\partial^2 \varrho_\sim(\mathbf{r}, t)}{\partial t^2} \; . \tag{2.120}$$

By means of the fundamental relation between sound pressure $p_\sim$ and acoustic density $\varrho_\sim$ (see (2.107)), we end up with (Laplace operator $\Delta = \nabla \cdot \nabla$)

$$\Delta p_\sim(\mathbf{r}, t) = \frac{1}{c_0^2} \frac{\partial^2 p_\sim(\mathbf{r}, t)}{\partial t^2} \tag{2.121}$$

representing the linear acoustic wave equation for sound pressure $p_\sim$.

According to the Helmholtz decomposition, we can express the particle velocity $\mathbf{v}_\sim$ by a combination of scalar and vector potential. Since the particle velocity is irrotational (2.118), the decomposition requires only a scalar potential, the so-called acoustic velocity potential Ψ

$$\mathbf{v}_\sim(\mathbf{r}, t) = -\nabla \Psi(\mathbf{r}, t) \; . \tag{2.122}$$

This relation can be inserted in the modified Euler's equation (2.117) leading to

$$p_\sim(\mathbf{r}, t) = \varrho_0 \frac{\partial \Psi(\mathbf{r}, t)}{\partial t} \; . \tag{2.123}$$

The acoustic velocity potential Ψ can, therefore, be easily connected to both quantities: the particle velocity and the sound pressure. Furthermore, if (2.123) is inserted in (2.121), we will obtain

$$\Delta \Psi_\sim(\mathbf{r}, t) = \frac{1}{c_0^2} \frac{\partial^2 \Psi_\sim(\mathbf{r}, t)}{\partial t^2} \; , \tag{2.124}$$

which is the linear acoustic wave equation for the acoustic velocity potential taking the same form as (2.121).

The linear wave equation is restricted to small fluctuations of $p_\sim$ and $\varrho_\sim$ during sound propagation. To consider large fluctuations of those quantities, we need alternative equations that incorporate nonlinear effects, e.g., the KZK (Khokhlov–Zabolotskaya–Kuznetsov) equation [6].

2.3.4 Reflection and Refraction of Sound

So far, we have described propagation of sound waves in homogeneous media. When sound waves impinge on an interface of two media exhibiting different material properties (e.g., density), reflection as well as refraction of the waves may occur. Let us study these effects by regarding plane waves. Such waves feature equal current state quantities (e.g., sound pressure) in each plane aligned perpendicular to the propagation direction (see Fig. 2.8a). For sinusoidal plane waves of frequency f, the geometric distance between identical current states corresponds to the wavelength λ, which is given by fundamental relations in wave propagation (angular frequency ω; wave number k)

$$\lambda = \frac{c_0}{f} = \frac{2\pi c_0}{\omega} = \frac{2\pi}{k} . \tag{2.125}$$

Aside from the acoustic quantities discussed in Sect. 2.3.1, the acoustic impedance Z_{aco} is an essential quantity characterizing the behavior of media with respect to propagating sound waves. In case of a plane sound wave, the acoustic impedance (unit $N s m^{-3}$) of the propagation medium becomes

$$Z_{\text{aco}} = \frac{\hat{p}_\sim}{\hat{v}_\sim} = \varrho_0 c_0 \tag{2.126}$$

where $\hat{p}_\sim$ and $\hat{v}_\sim$ stand for the amplitudes of sound pressure and particle velocity, respectively. Although sound waves represent longitudinal waves and exclusively refer to liquids and gases, the definition of the acoustic impedance $Z_{\text{aco}} = \varrho_0 c_0$ is oftentimes used for solids, which are able to transmit transverse waves in addition. According to the literature in acoustics, we will therefore indicate elastic waves (see Sect. 2.2.4) in solids also as sound waves. Owing to considerably high densities as well as sound velocities of solids, their acoustic impedances (e.g., $Z_{\text{aco}} \approx 50 \cdot 10^6 \, N s m^{-3}$ for iron) are usually many times greater than those of gases and liquids.

Now, let us assume an interface between two different homogeneous materials (see Fig. 2.8b), medium 1 and medium 2, exhibiting the equilibrium densities ϱ_1 and ϱ_2 as well as the sound velocities c_1 and c_2, respectively. A sound wave shall impinge in medium 1 on this interface. Because the interface is located in the yz-plane and the plane sound waves are supposed to propagate in the xy-plane, it is sufficient to discuss the configuration in the two-dimensional space, i.e., in the xy-plane. The sinusoidal incident wave $p_{I\sim}$ depending on both position and time can be written

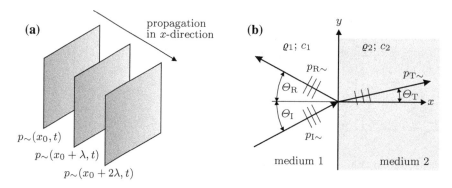

Fig. 2.8 a Plane pressure wave propagating in x-direction; gray planes depict areas of equal current state; **b** reflection and transmission of plane pressure wave impinging on interface of different media

as (real part $\Re\{\cdot\}$; imaginary unit $j = \sqrt{-1}$)

$$p_{I\sim}(x, y, t) = \Re\left\{\hat{\underline{p}}_{I\sim} e^{j k_1 [x \cos \Theta_1 + y \sin \Theta_1]} e^{j \omega t}\right\} . \qquad (2.127)$$

Θ_I stands for the incident angle with respect to the normal direction of the interface and k_1 is the wave number of the incident pressure wave in medium 1. The complex-valued amplitude $\hat{\underline{p}}_{I\sim} = \hat{p}_{I\sim} e^{j \phi_I}$ comprises the amplitude $\hat{p}_{I\sim}$ of the wave as well as its phase angle ϕ_I for $t = 0$. To achieve compact formulations for plane waves, they are further given as complex-valued representation. In doing so, (2.127) becomes

$$\underline{p}_{I\sim}(x, y) = \hat{\underline{p}}_{I\sim} e^{j k_1 [x \cos \Theta_1 + y \sin \Theta_1]} . \qquad (2.128)$$

For the reflected wave $p_{R\sim}$ propagating in medium 1 and the transmitted wave $p_{T\sim}$ propagating in medium 2, the complex-valued representation leads to (complex-valued amplitudes $\hat{\underline{p}}_{R\sim}$ and $\hat{\underline{p}}_{T\sim}$)

$$\underline{p}_{R\sim}(x, y) = \hat{\underline{p}}_{R\sim} e^{j k_1 [-x \cos \Theta_R + y \sin \Theta_R]} \qquad (2.129)$$

$$\underline{p}_{T\sim}(x, y) = \hat{\underline{p}}_{T\sim} e^{j k_2 [x \cos \Theta_T + y \sin \Theta_T]} \qquad (2.130)$$

where k_2 is the wave number in medium 2. Θ_R and Θ_T are the angles of the reflected and transmitted waves (see Fig. 2.8b), respectively. At the material interface, the summed current pressure values of incident and reflected wave coincide with the transmitted wave, i.e.,

$$\underline{p}_{I\sim}(x = 0, y) + \underline{p}_{R\sim}(x = 0, y) = \underline{p}_{T\sim}(x = 0, y) \qquad (2.131)$$

has to be fulfilled. Since (2.131) holds for each y-position on the interface, we can deduce with (2.128)–(2.130) the relations

$$k_1 \sin \Theta_I = -k_1 \sin \Theta_R \qquad (2.132)$$
$$k_1 \sin \Theta_I = k_2 \sin \Theta_T \,. \qquad (2.133)$$

The angles of incident waves Θ_I and reflected waves Θ_R are, thus, equal which is commonly referred to as law of reflection for acoustics. Due to the fact that the frequency f and, consequently, the angular frequency ω of the propagating pressure waves coincide in both media, we may rewrite (2.133) by utilizing (2.125) to

$$\frac{\sin \Theta_I}{c_1} = \frac{\sin \Theta_T}{c_2} \,. \qquad (2.134)$$

This fundamental equation is called the law of refraction for acoustics, which is especially familiar in optics as Snell's law [7]. In case of material combinations featuring the properties $c_2 > c_1$, an incident angle

$$\Theta_I > \arcsin\left(\frac{c_1}{c_2}\right) \qquad (2.135)$$

causes total reflection at the interface. Instead of the propagating pressure wave p_T, the pressure distribution decreases exponentially in medium 2 for such material combination.

Neither the law of reflection nor the law of refraction provides information about the pressure amplitudes of the reflected and transmitted wave with respect to the incident one. To compute the ratio of the amplitudes, we take a closer look at the particle velocities on the material interface. From the physical point of view, the velocities of the sound waves have to be equal at the interface, e.g., the x-components of the particle velocities $\mathbf{v}_\sim = [v_{x\sim}, v_{y\sim}, v_{z\sim}]$. Therefore, we obtain by additionally using the acoustic impedances Z_{aco1} and Z_{aco2} of medium 1 and medium 2 (see (2.126))

$$\underbrace{\frac{p_{I\sim}}{Z_{aco1}} \cos \Theta_I}_{v_{I,x\sim}} - \underbrace{\frac{p_{R\sim}}{Z_{aco1}} \cos \Theta_R}_{v_{R,x\sim}} = \underbrace{\frac{p_{T\sim}}{Z_{aco2}} \cos \Theta_T}_{v_{T,x\sim}} \,. \qquad (2.136)$$

Inserting of (2.131) in (2.136) yields

$$r_p = \frac{p_{R\sim}}{p_{I\sim}} = \frac{Z_{aco2} \cos \Theta_I - Z_{aco1} \cos \Theta_T}{Z_{aco1} \cos \Theta_T + Z_{aco2} \cos \Theta_I} \qquad (2.137)$$

$$t_p = \frac{p_{T\sim}}{p_{I\sim}} = \frac{2 Z_{aco2} \cos \Theta_I}{Z_{aco1} \cos \Theta_T + Z_{aco2} \cos \Theta_I} \,. \qquad (2.138)$$

The expressions r_p and t_p denote the reflection coefficient and transmission coefficient for the incident pressure waves, respectively. In case of plane pressure waves impinging perpendicular to the material interface, i.e., $\Theta_I = \Theta_R = \Theta_T = 0°$, (2.137) and (2.138) simplify to

$$r_p = \frac{Z_{aco2} - Z_{aco1}}{Z_{aco1} + Z_{aco2}} \quad \text{and} \quad t_p = \frac{2Z_{aco2}}{Z_{aco1} + Z_{aco2}} \, . \tag{2.139}$$

Equation (2.139) states that a great difference in the acoustic impedances results in a high reflection coefficient $r_p \approx 1$ and, thus, in a small transmission coefficient $t_p \approx 0$. Such great differences occur for the combination solid/gas. If medium 1 is a solid and medium 2 a gas ($Z_{aco1} \gg Z_{aco2}$), almost the complete impinging pressure wave will be reflected at the interface. For the alternative configuration, i.e., medium 1 is a gas and medium 2 a solid ($Z_{aco1} \ll Z_{aco2}$), the impinging wave is also almost completely reflected. The first material configuration poses a so-called acoustically soft interface, while the second configuration refers to an acoustically hard interface.

At the end of this subsection, a special case of reflection and refraction at material interfaces should be pointed out in addition. As discussed in Sect. 2.2.4, both longitudinal and transverse waves can propagate in solids. Hence, at interfaces of different solids as well as liquids and solids, the wave types may change. For instance, a longitudinal plane wave impinging on a liquid/solid interface can generate both wave types (i.e., longitudinal and transverse) for the transmitted waves propagating in the solid. The angles for reflection and transmission of longitudinal and transverse waves result from (2.134) by applying the sound velocities of the wave types.

2.3.5 Sound Absorption

Sound propagation is always accompanied by absorption mechanisms, whereby sound energy is predominately converted into heat energy. This conversion leads to attenuation (damping) of acoustic quantities for increasing propagation paths. Especially in gases and liquids, which are, strictly speaking, viscous, attenuation strongly alters amplitudes and even waveforms of propagating sound waves. For example, a sinusoidal plane wave of frequency $f = \omega/2\pi$ propagating in x-direction is then given as

$$p_\sim(x, t) = \Re\left\{ \hat{p}_\sim \cdot e^{-\alpha_{at}x} \cdot e^{j[kx - \omega t]} \right\} = \hat{p}_\sim \cdot e^{-\alpha_{at}x} \cos(kx - \omega t) \, . \tag{2.140}$$

$p_\sim(x, t)$ is the current sound pressure at position x and $\hat{p}_\sim$ stands for the pressure amplitude at position $x = 0$. By means of the expression $e^{-\alpha_{at}x}$, the attenuation of the pressure wave is modeled. Since the attenuation coefficient α_{at} is positive, the sound pressure amplitude decreases exponentially for increasing distances.

Sound absorption mechanisms in liquids as well as in gases are commonly subdivided into the so-called classical absorption and molecular absorption. Moreover, the classical absorption is subdivided into effects arising from inner friction and thermal conductance. These three sound absorption mechanisms are briefly discussed below.

- **Absorption due to inner friction**: Contrary to the assumption in Sect. 2.3.2, liquids as well as gases are capable to act against deformation to a small extent. This can be ascribed to inner friction. A quantity that measures inner friction within liquids and gases is the dynamic viscosity η_L (unit $N\,s\,m^{-2}$). The attenuation coefficient α_η attributed to the inner friction computes as

$$\alpha_\eta = \frac{2\eta_L}{3\varrho_0} \frac{\omega^2}{c_0^3} .$$

(2.141)

Thus, α_η highly depends on both the frequency of the pressure wave and the sound velocity.

- **Absorption due to thermal conductance**: Up to now, we neglected the thermal conductance in liquids and gases for deriving the fundamental relations in acoustics. Actually, the differing compression of neighboring volume fractions during sound propagation causes local variations in temperature. Since liquids and gases exhibit nonzero thermal conductance, a certain thermal flow occurs. This thermal flow reduces the energy of the sound wave. The attenuation coefficient α_{th} originating from thermal conductance can be calculated with

$$\alpha_{th} = \frac{\kappa - 1}{\kappa} \frac{\nu_{th}}{\varrho_0 C_V} \frac{\omega^2}{c_0^3} .$$

(2.142)

Here, ν_{th} denotes the thermal conductivity (unit $W\,K^{-1}\,m^{-1}$) of the medium. Again, the attenuation coefficient strongly depends on both the frequency of the pressure wave and the sound velocity.

- **Molecular absorption**: Sound absorption due to inner friction and thermal conductance (i.e., classical absorption mechanism) is sufficient to account for the attenuation in monoatomic liquids and gases. However, most liquids and gases comprise polyatomic molecules featuring complicate structures. A propagating sound wave generates oscillations (e.g., rotation) of molecules' atoms leading to heating of propagation medium as well as attenuation of sound energy. This effect is referred to as thermal relaxation. Besides thermal relaxation, other molecular absorption mechanisms such as structure and chemical relaxation processes take place during sound propagation in polyatomic liquids and gases.

The attenuation coefficient α_{at}, which measures sound absorption during propagation, results from the sum

$$\alpha_{at} = \underbrace{\alpha_\eta + \alpha_{th}}_{\alpha_{cl}} + \alpha_{mol} .$$

(2.143)

Table 2.8 Attenuation factors D_{at} (see (2.144)) in dB m^{-1} for sound waves propagating in water and air in case of varying frequencies f; temperature of water and air 20 °C; relative air humidity 40%

	1 kHz	10 kHz	100 kHz	1 MHz	10 MHz
water	2×10^{-7}	2×10^{-5}	0.002	0.22	22
air	0.004	0.18	3.3	160	–

Thereby, α_{cl} and α_{mol} are the attenuation coefficients arising from classical and molecular absorption, respectively. From (2.141) and (2.142), we can deduce that α_η and α_{th} increase quadratically with the frequency of the sound waves. Below the relaxation frequencies, this is also approximately valid for the molecular attenuation coefficient α_{mol} of various media.

Sound absorption causes remarkable distortions of broadband ultrasound pulses, which are propagating over a long distance. Table 2.8 contains attenuation factors D_{at} (unit dB m^{-1}) defined as

$$D_{at} = 20 \cdot \log_{10}\left(\frac{\hat{p}_\sim|_{x=0\,m}}{\hat{p}_\sim|_{x=1\,m}}\right) = 20 \cdot \alpha_{at} \cdot \log_{10}(e) \approx 8.69 \cdot \alpha_{at} \qquad (2.144)$$

for sound waves propagating in water as well as in air. It can be clearly seen that the attenuation in both media heavily increases with frequency. Moreover, the sound attenuation in air is much higher than in water.

Similar pulse distortions like those resulting from sound absorption may occur if sound propagation is accompanied by velocity dispersion. In this context, velocity dispersion means that the sound velocity c_0 depends on the frequency f of the sound waves, i.e., $c_0 = c_0(f)$. Note that velocity dispersion is here assumed to be linear and does not belong to effects in nonlinear acoustics where high-pressure amplitudes lead to different sound velocities. However, according to the Kramers–Kronig relations,[5] velocity dispersion will not occur if α_{at} increases quadratically with frequency [15, 19]. Because α_d of water mostly features this behavior (i.e., $\alpha_{at} \propto f^2$) over a wide frequency range, dispersion is not present yielding a constant sound velocity c_0. In the sound propagation medium air, the relation between attenuation coefficient α_{at} and sound frequency f strongly depends on temperature, ambient pressure as well as humidity. That is the reason why velocity dispersion in air is much more important for technical applications than in water.

References

1. Albach, M.: Elektrotechnik. Pearson Studium, München (2011)
2. Auld, B.A.: Acoustic Fields and Waves in Solids, 2nd edn. Krieger Publishing Company, USA (1973)

---

[5]The Kramers–Kronig relations are a special case of the Hilbert transform (cf. Sect. 5.4.1).

3. Brekhovskikh, L., Gonacharov, V.: Mechanics of Continua and Wave Dynamics. Springer, Berlin (1985)
4. Bronstein, I.N., Semendjajew, K.A., Musiol, G., Mühlig, H.: Handbook of Mathematics, 6th edn. Springer, Berlin (2015)
5. Crocker, M.J.: Handbook of Acoustics. Wiley, New York (1998)
6. Hamilton, M., Blackstock, D.T.: Nonlinear Acoustics: Theory and Applications. Academic Press Inc, Cambridge (2009)
7. Hecht, E.: Optics, 5th edn. Pearson, London (2016)
8. Ida, N., Bastos, J.P.A.: Electromagnetics and Calculation of Fields, 2nd edn. Springer, Berlin (1997)
9. Jackson, J.D.: Classical Electrodynamics, 3rd edn. Wiley, New York (1998)
10. Kuttruff, H.: Acoustics: An Introduction. Routledge Chapman and Hall, London (2006)
11. Lerch, R.: Elektrische Messtechnik, 7th edn. Springer, Berlin (2016)
12. Lerch, R., Sessler, G.M., Wolf, D.: Technische Akustik: Grundlagen und Anwendungen. Springer, Berlin (2009)
13. Maxwell, J.C.: A Treatise on Electricity and Magnetism, vol. 1. Cambridge University Press, Cambridge (2010)
14. Maxwell, J.C.: A Treatise on Electricity and Magnetism, vol. 2. Cambridge University Press, Cambridge (2010)
15. O'Donnell, M., Jaynes, E.T., Miller, J.G.: Kramers-Kronig relationship between ultrasonic attenuation and wave velocity. J. Acoust. Soc. Am. **69**(3), 696–701 (1981)
16. Rossing, T.D.: Springer Handbook of Acoustics. Springer, Berlin (2007)
17. Smythe, W.B.: Static and Dynamic Electricity, 3rd edn. CRC Press, Boca Raton (1989)
18. Timoshenko, S.P., Goodier, J.N.: Theory of Elasticity, 3rd edn. Mcgraw-Hill Higher Education, USA (1970)
19. Waters, K.R., Hughes, M.S., Mobley, J., Brandenburger, G.H., Miller, J.G.: On the applicability of Kramer-Krnig relations for ultrasonic attenuation obeying a frequency power law. J. Acoust. Soc. Am. **108**(2), 556–563 (2000)
20. Ziegler, F.: Mechanics of Solids and Fluids, 2nd edn. Springer, Berlin (1995)

Chapter 3
Piezoelectricity

In this chapter, we will discuss the physical effect of piezoelectricity, which describes the interconnection of mechanical and electrical quantities within materials. Sect. 3.1 details the principle of the piezoelectric effect. Thereby, a clear distinction is made between the direct and the inverse piezoelectric effect. Since different coupling mechanisms take place within piezoelectric materials, we will conduct in Sect. 3.2 thermodynamical considerations allowing a distinct separation of the coupling mechanisms. Subsequently, the material law for linear piezoelectricity will be derived that is given by the constitutive equations for piezoelectricity. By means of these equations, one is able to connect mechanical and electrical quantities. In Sect. 3.4, the electromechanical coupling within piezoelectric materials is classified. This includes intrinsic and extrinsic effects as well as different modes of piezoelectricity. Afterward, we introduce electromechanical coupling factors, which rate the efficiency of energy conversion within piezoelectric materials, i.e., from mechanical to electrical energy and vice versa. Section 3.6 finally concentrates on the internal structure of various piezoelectric materials (e.g., piezoceramic materials), the underlying manufacturing process as well as typical material parameters. Further literature concerning piezoelectricity can be found in [14, 15, 17, 28, 40, 42].

3.1 Principle of Piezoelectric Effect

Basically, the piezoelectric effect is understood as the linear interaction between mechanical and electrical quantities. Materials offering a pronounced interaction are usually referred to as piezoelectric materials. A mechanical deformation of such material due to an applied mechanical load results in a macroscopic change of the electric polarization. In case of appropriate electrodes covering the material, we can measure electric voltages or electric charges that are directly related to the mechanical deformation. On the other hand, an electric voltage applied to the electrodes yields

© Springer-Verlag GmbH Germany, part of Springer Nature 2019
S. J. Rupitsch, *Piezoelectric Sensors and Actuators*, Topics in Mining, Metallurgy and Materials Engineering, https://doi.org/10.1007/978-3-662-57534-5_3

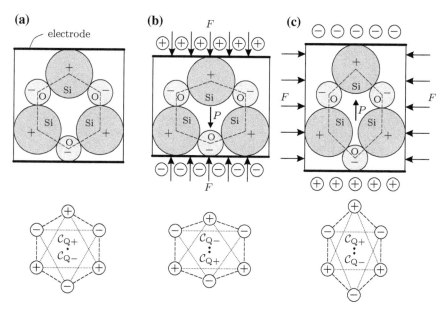

Fig. 3.1 Simplified inner structure of quartz crystal SiO$_2$ covered with electrodes at top and bottom surface; **a** original state of crystal without any mechanical loads; **b** longitudinal and **c** transverse mode of direct piezoelectric effect due to mechanical forces F; electric polarization $P = \|\mathbf{P}\|_2$ pointing from center $\mathcal{C}_{Q-}$ of negative charges to center $\mathcal{C}_{Q+}$ of positive ones; bottom panels illustrate locations of $\mathcal{C}_{Q+}$ and $\mathcal{C}_{Q-}$ within structure for the three states, respectively

a mechanical deformation of the piezoelectric material. Both conversion directions are, therefore, possible, i.e., from mechanical input to electrical output and from electrical input to mechanical output, respectively. Strictly speaking, the conversion from mechanical to electrical quantities is given by the *direct piezoelectric effect*, while the *inverse piezoelectric effect*[1] describes the conversion from electrical to mechanical quantities. Because the direct and inverse piezoelectric effect require changes of electric polarization, piezoelectric materials do not contain any free electric charges and, thus, these materials are electrical insulators.

With a view to studying the principles of the (direct) piezoelectric effect in more detail, let us consider as piezoelectric material the naturally occurring quartz crystal SiO$_2$ consisting of the chemical elements silicon Si and oxygen O. Figure 3.1 depicts a simplified setup of the quartz crystal in different states. The quartz crystal is covered with electrodes at the top and bottom surface. There is no force acting on the material in the original state (see Fig. 3.1a), while the quartz crystal is mechanically loaded with the force F in the deformed state (see Fig. 3.1b and c) in different directions. These forces lead to certain mechanical deformations of the material.

In the original state, the center $\mathcal{C}_{Q+}$ of positive charges (silicon ions) geometrically coincides with the center $\mathcal{C}_{Q-}$ of negative charges (oxygen ions). As a result,

---

[1] Sometimes, the inverse piezoelectric effect is named reverse or converse piezoelectric effect.

the material is electrically neutral to the outside. In contrast, the mechanical deformations in Fig. 3.1b and c imply that the centers of charges do not coincide anymore. Consequently, electric dipole moments arise pointing from $\mathcal{C}_{Q-}$ to $\mathcal{C}_{Q+}$. The dipole moment is characterized by the electric polarization $\mathbf{P}$. The greater the geometric distance between $\mathcal{C}_{Q-}$ and $\mathcal{C}_{Q+}$, the higher the magnitude $\|\mathbf{P}\|_2$ of the electric polarization will be. To compensate the electric polarization within the material, which represents an electric imbalance, charges are electrostatically induced on the electrodes. According to the origin, this effect is also referred to as displacement polarization. If the electrodes are electrically short-circuited, there will occur a charge flow, i.e., an electric current. Alternatively, one can measure an electric voltage between the electrically unloaded electrodes.

For the inverse piezoelectric effect, the same processes take place within the piezoelectric material but in reverse direction. If an electric voltage is applied to the electrodes, charges will be electrostatically induced on them. These charges constitute an electric imbalance that is compensated by a dipole moment within the material. Hence, the centers of positive and negative charges ($\mathcal{C}_{Q+}$ and $\mathcal{C}_{Q-}$) have to differ geometrically, which implies a mechanical deformation of the piezoelectric material.

Depending on the directions of the applied mechanical force and the resulting electric polarization, we can distinguish between different modes of piezoelectricity (see Sect. 3.4.3). For instance, Fig. 3.1b shows the longitudinal mode and Fig. 3.1c the transverse mode. The same will hold for the inverse piezoelectric effect when the applied electric voltage is related to the direction of the resulting mechanical deformation.

3.2 Thermodynamical Considerations

According to the first law of thermodynamics, the change $d\mathcal{U}$ (per unit volume[2]) of the internal energy in a closed system results from the work dW (per unit volume) done on the system and the heat energy dQ (per unit volume) added to the system. In case of piezoelectric systems, the work W can be split up into mechanical energy W_{mech} and electrical energy W_{elec}. Therewith, the first law of thermodynamics reads as

$$d\mathcal{U} = dW + dQ = dW_{\text{mech}} + dW_{\text{elec}} + dQ . \tag{3.1}$$

Now, let us introduce state variables describing the energy of the physical fields, respectively. These quantities are listed below.

- *Mechanical energy*: Mechanical stress S_{ij} and mechanical strain T_{ij}; both components of tensors (rank 2).
- *Electrical energy*: Electric field intensity E_m and electric flux density D_m; both components of vectors.

---

[2]Energy per unit volume is equivalent to energy density.

- *Heat energy*: The second law of thermodynamics states that the change dQ of heat energy is given by the temperature ϑ and the change ds of entropy per unit volume; both scalar quantities.

Under the assumption of small changes, the superposition of the state variables in (3.1) yields[3]

$$d\mathcal{U} = E_m dD_m + T_{ij} dS_{ij} + \vartheta ds .\tag{3.2}$$

Thus, $d\mathcal{U}$ results from the changes of D_m, S_{ij} and s, which represent extensive state variables. However, in practical applications of piezoelectric materials, the intensive state variables E_m, T_{ij}, and ϑ are prescribed. That is the reason why we use a special thermodynamical potential, the so-called Gibbs free energy $\mathcal{G}$

$$\mathcal{G} = \mathcal{U} - E_m D_m - T_{ij} S_{ij} - \vartheta s \tag{3.3}$$

instead. When the independent quantities E_m, T_{ij}, and ϑ are specified, the closed system will arrive at the thermodynamic equilibrium in such a way that $\mathcal{G}$ is minimized. Therefore, the total derivative of $\mathcal{G}$ has to be zero, i.e.,

$$d\mathcal{G} \equiv 0 = -D_m dE_m - S_{ij} dT_{ij} - s d\vartheta .\tag{3.4}$$

From this relation, we can compute the resulting extensive state variables by fixing selected intensive state variables, which leads to

$$D_m = -\left.\frac{\partial \mathcal{G}}{\partial E_m}\right|_{T,\vartheta} , \quad S_{ij} = -\left.\frac{\partial \mathcal{G}}{\partial T_{ij}}\right|_{E,\vartheta} , \quad s = -\left.\frac{\partial \mathcal{G}}{\partial \vartheta}\right|_{T,E} .\tag{3.5}$$

For instance, T_{ij} as well as ϑ are fixed to calculate D_m. But strictly speaking, each extensive state variable depends on all intensive ones. Nevertheless, small changes are assumed and, consequently, we are able to terminate the Taylor series expansion after the linear part. In doing so, one ends up with linearized state equations for the extensive state variables D_m, S_{ij}, and s

$$dD_m = \underbrace{\left.\frac{\partial D_m}{\partial E_n}\right|_{T,\vartheta}}_{\varepsilon_{mn}^{T,\vartheta}} \overbrace{dE_n}^{\text{dielectric material law}} + \underbrace{\left.\frac{\partial D_m}{\partial T_{kl}}\right|_{E,\vartheta}}_{d_{mkl}^{E,\vartheta}} \overbrace{dT_{kl}}^{\text{direct piezoelectric effect}} + \underbrace{\left.\frac{\partial D_m}{\partial \vartheta}\right|_{E,T}}_{p_m^{E,T}} \overbrace{d\vartheta}^{\text{pyroelectric effect}} \tag{3.6}$$

---

[3]In accordance with the relevant literature, the space directions are denoted by $\{1, 2, 3\}$ instead of $\{x, y, z\}$; $\{i, j, k, l, m, n\} = \{1, 2, 3\}$; Einstein summation convention, i.e., $T_{ij} S_{ij} = \sum_{i,j} T_{ij} S_{ij}$.

$$
\mathrm{d}S_{ij} = \underbrace{\left.\frac{\partial S_{ij}}{\partial E_n}\right|_{T,\vartheta}}_{d_{ijn}^{T,\vartheta}} \overbrace{\mathrm{d}E_n}^{\text{inverse piezoelectric effect}} + \underbrace{\left.\frac{\partial S_{ij}}{\partial T_{kl}}\right|_{E,\vartheta}}_{s_{ijkl}^{E,\vartheta}} \overbrace{\mathrm{d}T_{kl}}^{\text{Hooke's law}} + \underbrace{\left.\frac{\partial S_{ij}}{\partial \vartheta}\right|_{E,T}}_{\alpha_{ij}^{E,T}} \overbrace{\mathrm{d}\vartheta}^{\text{thermal expansion}} \quad (3.7)
$$

$$
\mathrm{d}\mathfrak{s} = \underbrace{\left.\frac{\partial \mathfrak{s}}{\partial E_n}\right|_{T,\vartheta}}_{\rho_n^{T,\vartheta}} \overbrace{\mathrm{d}E_n}^{\text{electrocaloric effect}} + \underbrace{\left.\frac{\partial \mathfrak{s}}{\partial T_{kl}}\right|_{E,\vartheta}}_{\alpha_{kl}^{E,\vartheta}} \overbrace{\mathrm{d}T_{kl}}^{\text{piezocaloric effect}} + \underbrace{\left.\frac{\partial \mathfrak{s}}{\partial \vartheta}\right|_{E,T}}_{C^{E,T}} \overbrace{\mathrm{d}\vartheta}^{\text{specific heat}} . \quad (3.8)
$$

These three equations contain the connections between electrical, mechanical, and thermal quantities within piezoelectric materials. Each partial derivative represents a material parameter characterizing a specific linearized coupling mechanism. The coupling mechanisms are named in (3.6)–(3.8), and the utilized notation is summarized in Table 3.1.

Table 3.1 Expressions used in (3.6)–(3.8) and for Heckmann diagram Fig. 3.2

Notation	Description	Unit
Intensive state variables		
E_n	electric field intensity; vector	$V\,m^{-1}$
T_{kl}	mechanical stress; tensor rank 2	$N\,m^{-2}$
ϑ	temperatur; scalar	$K; {}^\circ C$
Extensive state variables		
D_m	electric flux density; vector	$C\,m^{-2}$
S_{kl}	mechanical strain; tensor rank 2	–
$\mathfrak{s}$	entropy per unit volume; scalar	$J\,m^{-3}\,K^{-1}$
Material parameters		
$\varepsilon_{mn}^{T,\vartheta}$	electric permittivities; tensor rank 2	$A\,s\,V^{-1}\,m^{-1}; F\,m^{-1}$
$s_{ijkl}^{E,\vartheta}$	elastic compliance constants; tensor rank 4	$m^2\,N^{-1}$
$C^{E,T}$	heat per unit volume; scalar	$J\,m^{-3}\,N^{-1}$
$d_{ijn}^{T,\vartheta}$; $d_{mkl}^{E,\vartheta}$	piezoelectric strain constants; tensor rank 3	$m\,V^{-1}; C\,N^{-1}$
e_{nkl}^{ϑ} ; e_{mij}^{ϑ}	piezoelectric stress constants; tensor rank 3	$C\,m^{-2}; N\,V^{-1}\,m^{-1}$
$\rho_n^{T,\vartheta}$; $\rho_m^{E,T}$	pyroelectric coefficients; vector	$C\,m^{-2}\,K^{-1}$
π_n^{T} ; π_m^{T}	pyroelectric coefficients; vector	$V\,m^{-1}\,K^{-1}$
$\alpha_{kl}^{E,\vartheta}$; $\alpha_{ij}^{E,T}$	thermal expansion coefficients; tensor rank 2	K^{-1}
τ_{ij}^{E} ; τ_{kl}^{E}	thermal stress coefficients; tensor rank 2	$N\,m^{-2}\,K^{-1}$

The superscripts of the material parameters point out which physical quantities are presumed to stay constant in the framework of parameter identification. This is particularly crucial as only in that way, different coupling mechanisms can be separated. For each material parameter, there are two state variables listed, e.g., T and ϑ for $\varepsilon_{mn}^{T,\vartheta}$. However, to some extent, the amount of constant quantities can be reduced from two to one. Let us conduct the reduction for the direct piezoelectric effect (material parameter $d_{mkl}^{E,\vartheta}$) and the inverse piezoelectric effect (material parameter $d_{ijn}^{T,\vartheta}$). If both, the electric flux density D_m in (3.6) and the mechanical strain S_{ij} in (3.7) are replaced by the Gibbs free energy $\mathcal{G}$ according to (3.5), we will obtain

$$\underbrace{\left.\frac{\partial D_n}{\partial T_{kl}}\right|_{E,\vartheta}}_{d_{nkl}^{E,\vartheta}} = -\left.\frac{\partial^2 \mathcal{G}}{\partial E_n \partial T_{kl}}\right|_{E,\vartheta} = -\left.\frac{\partial^2 \mathcal{G}}{\partial T_{kl}\partial E_n}\right|_{T,\vartheta} = \underbrace{\left.\frac{\partial S_{kl}}{\partial E_n}\right|_{T,\vartheta}}_{d_{kln}^{T,\vartheta}} \mathrel{\widehat{=}} d_{nkl}^{\vartheta} . \qquad (3.9)$$

Hence, the piezoelectric strain constants d_{nkl}^{ϑ} depend neither on the electric field intensity E_n nor on the mechanical stress T_{kl}. The same procedure can be applied for the pyroelectric and electrocaloric effect as well as for thermal expansion and the

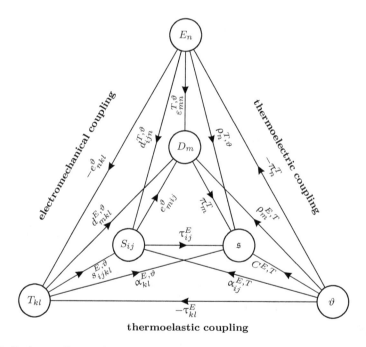

Fig. 3.2 Heckmann diagram demonstrating coupling mechanisms within piezoelectric materials; intensive and extensive state variables at conners of outer and inner triangle, respectively; utilized notation is given in Table 3.1

piezocaloric effect. Moreover, (3.9) demonstrates that there exist several symmetries, which we can utilize to reduce tensors to matrices by means of Voigt notation.

The so-called *Heckmann diagram* (Fig. 3.2) is a descriptive representation of coupling mechanisms taking place within piezoelectric materials (notation in Table 3.1). The diagram is composed of an outer and an inner triangle. While the corners of the outer triangle contain intensive state variables, the extensive state variables are placed at the corners of the inner triangle. As Eqs. (3.6)–(3.8), the Heckmann diagram shows the different coupling mechanisms, which can be categorized into electromechanical, thermoelastic, and thermoelectric interconnections.

3.3 Material Law for Linear Piezoelectricity

To derive the material law for linear piezoelectricity that is given by the constitutive equations for piezoelectricity, we start with the linearized state Eqs. (3.6) and (3.7) for the electric flux density D_m and the mechanical strain S_{ij}

$$\mathrm{d}D_m = \varepsilon_{mn}^{T,\vartheta}\mathrm{d}E_n + d_{mkl}^{\vartheta}\mathrm{d}T_{kl} + \rho_m^T\mathrm{d}\vartheta \tag{3.10}$$

$$\mathrm{d}S_{ij} = d_{ijn}^{\vartheta}\mathrm{d}E_n + s_{ijkl}^{E,\vartheta}\mathrm{d}T_{kl} + \alpha_{ij}^E\mathrm{d}\vartheta \ . \tag{3.11}$$

If temperature changes $\mathrm{d}\vartheta$ are neglected (i.e., isothermal change in state), the linearized state equations will become

$$\mathrm{d}D_m = \varepsilon_{mn}^T\mathrm{d}E_n + d_{mkl}\mathrm{d}T_{kl} \tag{3.12}$$

$$\mathrm{d}S_{ij} = d_{ijn}\mathrm{d}E_n + s_{ijkl}^E\mathrm{d}T_{kl} \tag{3.13}$$

with the electric permittivities ε_{mn}^T for constant mechanical stress, the elastic compliance constants s_{ijkl}^E for constant electric field intensity and the piezoelectric strain constants d_{mkl}. Under the assumption that D_m, E_n, S_{ij} as well as T_{kl} are zero in the initial state, (3.12) and (3.13) can be written as

$$D_m = \varepsilon_{mn}^T E_n + d_{mkl}T_{kl} \tag{3.14}$$

$$S_{ij} = d_{ijn}E_n + s_{ijkl}^E T_{kl} \ , \tag{3.15}$$

which represents the *d*-form (strain-charge form) of the material law for linear piezoelectricity. In contrast, the so-called *e*-form (stress-charge form) reads as

$$D_m = \varepsilon_{mn}^S E_n + e_{mkl}S_{kl} \tag{3.16}$$

$$T_{ij} = -e_{ijn}E_n + c_{ijkl}^E S_{kl} \tag{3.17}$$

with the electric permittivities ε_{mn}^S for constant mechanical strain, the elastic stiffness constants c_{ijkl}^E for constant electric field intensity and the piezoelectric stress constants e_{mkl}. Alternatively to the constitutive equations in *d*-form and *e*-form, the g-form

$$E_m = \beta_{mn}^T D_n - g_{mkl} T_{kl} \tag{3.18}$$

$$S_{ij} = g_{ijn} D_n + s_{ijkl}^D T_{kl} \tag{3.19}$$

and h-form

$$E_m = \beta_{mn}^S D_n - h_{mkl} S_{kl} \tag{3.20}$$

$$T_{ij} = -h_{ijn} D_n + c_{ijkl}^D S_{kl} . \tag{3.21}$$

can be sometimes found in the relevant literature. Here, g_{mkl} denotes piezoelectric voltage constants (unit $\mathrm{V\,m\,N^{-1}}$; $\mathrm{m^2\,C^{-1}}$) and h_{mkl} the piezoelectric h constants (unit $\mathrm{V\,m^{-1}}$; $\mathrm{N\,C^{-1}}$). The expression $\beta_{ij}^{T,S}$ indicates electric impermittivities (unit $\mathrm{V\,m\,A^{-1}\,s^{-1}}$; $\mathrm{m\,F^{-1}}$) for constant stress and constant strain, respectively. The Eqs. (3.14), (3.16), (3.18), and (3.20) relate to the direct piezoelectric effect, whereas the inverse piezoelectric effect is explained by means of Eqs. (3.15), (3.17), (3.19), and (3.21).

Due to symmetries within the tensors of rank four for the mechanical field (s_{ijkl}, c_{ijkl}) as well as within the tensors of rank three for piezoelectric coupling (d_{mkl}, e_{mkl}, g_{mkl}, and h_{mkl}), the number of independent components in (3.14)–(3.21) is reduced significantly. We are, therefore, able to transform the tensor equations to matrix equations. In Voigt notation, the constitutive equations for piezoelectricity become (transpose t)

$$\begin{aligned}
d\text{-form} \quad \mathbf{D} &= \left[\varepsilon^T\right]\mathbf{E} + [\mathbf{d}]\,\mathbf{T} & (3.22)\\
\mathbf{S} &= [\mathbf{d}]^t\,\mathbf{E} + \left[\mathbf{s}^E\right]\mathbf{T} & (3.23)\\
e\text{-form} \quad \mathbf{D} &= \left[\varepsilon^S\right]\mathbf{E} + [\mathbf{e}]\,\mathbf{S} & (3.24)\\
\mathbf{T} &= -[\mathbf{e}]^t\,\mathbf{E} + \left[\mathbf{c}^E\right]\mathbf{S} & (3.25)\\
g\text{-form} \quad \mathbf{E} &= \left[\beta^T\right]\mathbf{D} - [\mathbf{g}]\,\mathbf{T} & (3.26)\\
\mathbf{S} &= [\mathbf{g}]^t\,\mathbf{D} + \left[\mathbf{s}^D\right]\mathbf{T} & (3.27)\\
h\text{-form} \quad \mathbf{E} &= \left[\beta^S\right]\mathbf{D} - [\mathbf{h}]\,\mathbf{S} & (3.28)\\
\mathbf{T} &= -[\mathbf{h}]^t\,\mathbf{D} + \left[\mathbf{c}^D\right]\mathbf{S} . & (3.29)
\end{aligned}$$

While the vectors for the electrical field ($\mathbf{D}$, $\mathbf{E}$) contain three components, those for the mechanical field ($\mathbf{T}$, $\mathbf{S}$) consist of six independent components (cf. Sects. 2.1 and 2.2). The reduced tensors describing mechanical properties ($[\mathbf{s}]$, $[\mathbf{c}]$) and piezoelectric coupling ($[\mathbf{d}]$, $[\mathbf{e}]$, $[\mathbf{g}]$ and $[\mathbf{h}]$) exhibit the dimensions 6×6 and 3×6, respectively. On the contrary, the tensors for electrical properties ($[\varepsilon]$, $[\beta]$) are of dimension 3×3. In d-form, the reduced set of constitutive equations for piezoelectricity is in component notation given by

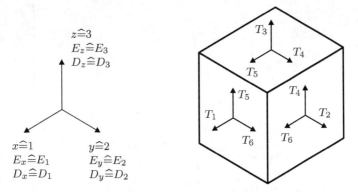

Fig. 3.3 Common notation in constitutive equations (reduced set) for piezoelectricity with respect to Cartesian coordinate system xyz in three-dimensional space (cf. Fig. 2.4 in Sect. 2.2)

$$
\begin{bmatrix} D_1 \\ D_2 \\ D_3 \end{bmatrix} = \begin{bmatrix} \varepsilon_{11}^T & \varepsilon_{12}^T & \varepsilon_{13}^T \\ \varepsilon_{21}^T & \varepsilon_{22}^T & \varepsilon_{23}^T \\ \varepsilon_{31}^T & \varepsilon_{32}^T & \varepsilon_{33}^T \end{bmatrix} \begin{bmatrix} E_1 \\ E_2 \\ E_3 \end{bmatrix} + \begin{bmatrix} d_{11} & d_{12} & d_{13} & d_{14} & d_{15} & d_{16} \\ d_{21} & d_{22} & d_{23} & d_{24} & d_{25} & d_{26} \\ d_{31} & d_{32} & d_{33} & d_{34} & d_{35} & d_{36} \end{bmatrix} \begin{bmatrix} T_1 \\ T_2 \\ T_3 \\ T_4 \\ T_5 \\ T_6 \end{bmatrix}
$$

$$(3.30)$$

$$
\begin{bmatrix} S_1 \\ S_2 \\ S_3 \\ S_4 \\ S_5 \\ S_6 \end{bmatrix} = \begin{bmatrix} d_{11} & d_{21} & d_{31} \\ d_{12} & d_{22} & d_{32} \\ d_{13} & d_{23} & d_{33} \\ d_{14} & d_{24} & d_{34} \\ d_{15} & d_{25} & d_{35} \\ d_{16} & d_{26} & d_{36} \end{bmatrix} \begin{bmatrix} E_1 \\ E_2 \\ E_3 \end{bmatrix} + \begin{bmatrix} s_{11}^E & s_{12}^E & s_{13}^E & s_{14}^E & s_{15}^E & s_{16}^E \\ s_{21}^E & s_{22}^E & s_{23}^E & s_{24}^E & s_{25}^E & s_{26}^E \\ s_{31}^E & s_{32}^E & s_{33}^E & s_{34}^E & s_{35}^E & s_{36}^E \\ s_{41}^E & s_{42}^E & s_{43}^E & s_{44}^E & s_{45}^E & s_{46}^E \\ s_{51}^E & s_{52}^E & s_{53}^E & s_{54}^E & s_{55}^E & s_{56}^E \\ s_{61}^E & s_{62}^E & s_{63}^E & s_{64}^E & s_{65}^E & s_{66}^E \end{bmatrix} \begin{bmatrix} T_1 \\ T_2 \\ T_3 \\ T_4 \\ T_5 \\ T_6 \end{bmatrix} .
$$

$$(3.31)$$

Figure 3.3 depicts the utilized notations for the electric field intensity E_i, the electric flux density D_i, and the mechanical stress T_p. The notation for the mechanical strain S_p is given in Sect 2.2.2. Note that in the remaining part of this book, all constitutive equations for piezoelectricity refer to this reduced notation of the tensor relations.

There exist further symmetries within piezoelectric materials, which are accompanied by a remarkable reduction of independent components in addition. Besides, several entries of the reduced tensors are zero. For instance, the component notation in d-form for a piezoelectric material of crystal class 6mm results in

$$
\begin{bmatrix} D_1 \\ D_2 \\ D_3 \end{bmatrix} = \begin{bmatrix} \varepsilon_{11}^T & 0 & 0 \\ 0 & \varepsilon_{11}^T & 0 \\ 0 & 0 & \varepsilon_{33}^T \end{bmatrix} \begin{bmatrix} E_1 \\ E_2 \\ E_3 \end{bmatrix} + \begin{bmatrix} 0 & 0 & 0 & 0 & d_{15} & 0 \\ 0 & 0 & 0 & d_{15} & 0 & 0 \\ d_{31} & d_{31} & d_{33} & 0 & 0 & 0 \end{bmatrix} \begin{bmatrix} T_1 \\ T_2 \\ T_3 \\ T_4 \\ T_5 \\ T_6 \end{bmatrix}
$$

$$(3.32)$$

$$
\begin{bmatrix} S_1 \\ S_2 \\ S_3 \\ S_4 \\ S_5 \\ S_6 \end{bmatrix} = \begin{bmatrix} 0 & 0 & d_{31} \\ 0 & 0 & d_{31} \\ 0 & 0 & d_{33} \\ 0 & d_{15} & 0 \\ d_{15} & 0 & 0 \\ 0 & 0 & 0 \end{bmatrix} \begin{bmatrix} E_1 \\ E_2 \\ E_3 \end{bmatrix} + \begin{bmatrix} s_{11}^E & s_{12}^E & s_{13}^E & 0 & 0 & 0 \\ s_{12}^E & s_{11}^E & s_{13}^E & 0 & 0 & 0 \\ s_{13}^E & s_{13}^E & s_{33}^E & 0 & 0 & 0 \\ 0 & 0 & 0 & s_{44}^E & 0 & 0 \\ 0 & 0 & 0 & 0 & s_{44}^E & 0 \\ 0 & 0 & 0 & 0 & 0 & 2\left(s_{11}^E - s_{12}^E\right) \end{bmatrix} \begin{bmatrix} T_1 \\ T_2 \\ T_3 \\ T_4 \\ T_5 \\ T_6 \end{bmatrix}
$$

and, thus, contains only 10 independent quantities. Nevertheless, we are confronted with an anisotropic material behavior that is for crystal class 6mm prevalently referred to as transversely isotropic behavior. In case of such transversely isotropic behavior, a plane within the material can be found in which the material parameters are identical for all directions. As a consequence, five parameters are required to describe the mechanical properties instead of two parameters for an isotropic material (cf. Sect. 2.2.3).

If the material parameters of a single set of constitutive equations (e.g., in d-form) are known, one will be able to determine the parameters for all other forms. The underlying parameter conversion between the forms reads as

$$
\begin{aligned}
c_{pr}^E s_{qr}^E &= \delta_{pq} & c_{pr}^D s_{qr}^D &= \delta_{pq} \\
\beta_{ik}^S \varepsilon_{jk}^S &= \delta_{ij} & \beta_{ik}^T \varepsilon_{jk}^T &= \delta_{ij} \\
c_{pq}^D &= c_{pq}^E + e_{kp} h_{kq} & s_{pq}^D &= s_{pq}^E - d_{kp} g_{kq} \\
\varepsilon_{ij}^T &= \varepsilon_{ij}^S + d_{iq} e_{jq} & \beta_{ij}^T &= \beta_{ij}^S - g_{iq} h_{jq} \\
e_{ip} &= d_{iq} c_{qp}^E & d_{ip} &= \varepsilon_{ik}^T g_{kp} \\
g_{ip} &= \beta_{ik}^T d_{kp} & h_{ip} &= g_{iq} c_{qp}^D
\end{aligned} \qquad (3.33)
$$

with $\{i, j, k\} = \{1, 2, 3\}$ and $\{p, q, r\} = \{1, 2, 3, 4, 5, 6\}$. Figure 3.4 shows the interconnections of the state variables within the constitutive equations in the different forms. However, since both the g-form and the h-form are extremely rare, we restrict the following explanations to the d-form and e-form.

At the end of this section, let us take a closer look at the superscripts (T, S, E and D) in the constitutive equations. Especially in the context of parameter identification, these superscripts are of utmost importance. The superscript T states that the material parameters (e.g., ε_{ij}^T) have to be identified for the case of free mechanical vibrations, i.e., the piezoelectric sample must not be clamped. In contrast, the

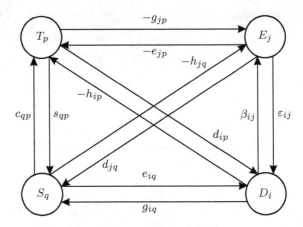

Fig. 3.4 Interconnections of state variables within constitutive equations for piezoelectricity (Voigt notation); $\{i, j\} = \{1, 2, 3\}$; $\{p, q\} = \{1, 2, 3, 4, 5, 6\}$

superscript S refers to clamped arrangements and, consequently, mechanical vibrations have to be prohibited during parameter identification. It seems only natural that the latter condition can hardly be fulfilled. For the mechanical parameters (e.g., s_{pq}^E), the superscripts E and D occur in the constitutive equations. Parameters with superscript E result from electrically short-circuited piezoelectric samples and those with D for samples, which are electrically unloaded.

3.4 Classification of Electromechanical Coupling

The electromechanical coupling within piezoelectric materials can be attributed to different effects. In particular, we distinguish between intrinsic and extrinsic effects, which will be discussed in Sects. 3.4.1 and 3.4.2, respectively. Thereby, the piezoelectric material is assumed to consist of many identical areas, the so-called unit cells, featuring defined polarization states. The possible modes of piezoelectricity are shown in Sect. 3.4.3.

3.4.1 Intrinsic Effects

Basically, intrinsic effects of electromechanical coupling take place on the atomistic level. If sufficiently small mechanical or electrical loads are applied, the structure of the piezoelectric material and, therefore, the geometric arrangement of the unit cells within the material will remain unchanged. However, the positions of atoms within the unit cells are altered, which also yields changes of the centers of positive and negative charges, i.e., electric polarization (see Sect. 3.1). The direct and inverse piezoelectric effect capture this material behavior. For that reason, both effects are intrinsic effects. We are able to describe them with the aid of the material law for linear piezoelectricity. By utilizing (2.7, p. 9) and (3.22), the electric polarization $\mathbf{P}^{\text{rev}}$ becomes

$$
\begin{aligned}
\mathbf{P}^{\text{rev}} = \mathbf{D} - \varepsilon_0 \mathbf{E} &= [\mathbf{d}]\,\mathbf{T} + \left[\varepsilon^T\right]\mathbf{E} - \varepsilon_0 \mathbf{E} \\
&= [\mathbf{d}]\,\mathbf{T} + \varepsilon_0 \underbrace{\left(\left[\varepsilon_r^T\right] - 1\right)}_{\left[\chi_e^T\right]} \mathbf{E}
\end{aligned}
\tag{3.34}
$$

with the tensor $\left[\varepsilon_r^T\right]$ (matrix of dimension 3×3) of the relative permittivities for constant mechanical stress. Since the displacement polarization will return to zero if there is neither a mechanical load nor an electrical load, the variable $\mathbf{P}$ is equipped with the superscript rev standing for reversible. The expression $\left[\chi_e^T\right]$ in (3.34) is the electric susceptibility (matrix of dimension 3×3), which rates the polarizability of a dielectric medium in response to an applied electric field intensity $\mathbf{E}$.

In case of the direct piezoelectric effect, we can deduce from (3.23) and (3.34) the relations $\mathbf{S} \propto \mathbf{T}$ as well as $\mathbf{P}^{\text{rev}} \propto \mathbf{T}$. Consequently, the mechanical strain $\mathbf{S}$ of the piezoelectric material and the electric polarization $\mathbf{P}^{\text{rev}}$ within linearly depend on the applied mechanical stress $\mathbf{T}$. For the inverse piezoelectric effect, one obtains $\mathbf{S} \propto \mathbf{E}$ and $\mathbf{P}^{\text{rev}} \propto \mathbf{E}$, which means that both quantities linearly depend on the applied electric field intensity.

Apart from the direct and inverse piezoelectric effect, there exists a further intrinsic effect of electromechanical coupling, the so-called *electrostriction*. This effect arises in every material featuring dielectric properties. When we apply an electric field to such a material, the opposite sides of the unit cells will be differently charged causing attraction forces. As a result, the material thickness is reduced in the direction of the applied electric field. Due to the fact that an electric field pointing in opposite direction reduces material thickness in the same manner, electrostriction is a quadratic effect, i.e., $\mathbf{S} \propto \mathbf{E}^2$. With the exception of relaxor ferroelectrics (e.g., lead magnesium niobate), electrostriction is always weakly pronounced in dielectric materials. Note that the piezoelectric coupling normally will dominate electrostriction if a material offers piezoelectric properties.

Because intrinsic effects commonly occur for small inputs, the underlying linear material behavior is referred to as *small-signal behavior*. These effects will become particularly important when a piezoelectric system is excited resonantly, which is the case in various sensor and actuator applications.

3.4.2 Extrinsic Effects

When we apply large mechanical loads or large electrical loads to a piezoelectric material, extrinsic effects may additionally arise within the material. Contrarily to intrinsic effects, the geometric arrangement of the unit cells within the material is modified which macroscopically leads to the so-called remanent electric polarization.[4] Extrinsic effects are irreversible since these modifications will remain approximately the same if the material is mechanically and electrically unloaded. However, by applying sufficiently large mechanical or electrical loads, we are able to alter the geometric arrangement of the unit cells again and, therefore, the remanent electric polarization. In accordance with ferromagnetism as a magnetic phenomenon, piezoelectric materials showing extrinsic effects are frequently named ferroelectric materials (e.g., piezoceramics). The behavior of such materials is known as ferroelectric behavior.

To describe the impact of extrinsic effects mathematically in ferroelectric materials, let us take a look at the constitutive equations for piezoelectricity (see Sect. 3.3). Strictly speaking, the constitutive equations in their original form exclusively cover the direct and inverse piezoelectric effect, i.e., intrinsic effects. Nevertheless, we can consider extrinsic effects in those equations by appropriate extensions. In particular, the aimed state variables (e.g., $\mathbf{D}$ and $\mathbf{S}$ in d-form) are divided into reversible and irreversible parts. The reversible parts (superscript rev) characterize intrinsic effects of electromechanical coupling, whereas the irreversible parts (superscript irr) account for extrinsic effects. Therewith, the d-form of the constitutive equations reads as [19, 20]

$$\mathbf{D} = \mathbf{D}^{\text{rev}} + \mathbf{D}^{\text{irr}} = \left[\varepsilon^T\left(\mathbf{P}^{\text{irr}}\right)\right]\mathbf{E} + \left[\mathbf{d}\left(\mathbf{P}^{\text{irr}}\right)\right]\mathbf{T} + \mathbf{P}^{\text{irr}} \qquad (3.35)$$

$$\mathbf{S} = \mathbf{S}^{\text{rev}} + \mathbf{S}^{\text{irr}} = \left[\mathbf{d}\left(\mathbf{P}^{\text{irr}}\right)\right]^t\mathbf{E} + \left[\mathbf{s}^E\left(\mathbf{P}^{\text{irr}}\right)\right]\mathbf{T} + \mathbf{S}^{\text{irr}}. \qquad (3.36)$$

Here, $\mathbf{P}^{\text{irr}}$ stands for the remanent electric polarization and $\mathbf{S}^{\text{irr}}$ for the mechanical strain that is resulting from the modified geometric arrangement of the unit cells. Depending on the remanent electric polarization, the properties of the ferroelectric material change. This fact is considered in (3.35) and (3.36) by the argument $\mathbf{P}^{\text{irr}}$ of the material tensors $\left[\mathbf{s}^E\right], \left[\varepsilon^T\right]$, and $[\mathbf{d}]$, respectively.

On the basis that extrinsic effects are mainly arising in case of large inputs, the resulting material behavior is known as *large-signal behavior*. The large-signal behavior usually goes hand in hand with nonlinear responses of ferroelectric materials such as hysteresis curves, which is decisive for several actuator applications (see Chap. 10). At this point, it should be mentioned again that ferroelectric materials belong to the group of piezoelectric materials. Therefore, ferroelectric materials also show intrinsic effects of electromechanical coupling, i.e., the direct and inverse piezoelectric effect.

3.4.3 Modes of Piezoelectric Effect

According to the constitutive equations for piezoelectricity (3.22)–(3.29), there exist 18 possibilities in total (e.g., $d_{11}, \ldots, d_{36}$) to couple the components of electrical and

---

[4]The remanent electric polarization is also known as orientation polarization.

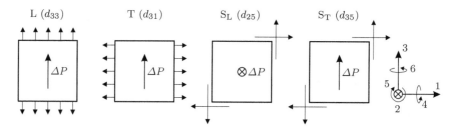

Fig. 3.5 Examples for longitudinal (L), transverse (T), longitudinal shear (S_L), and transverse shear modes (S_T) of piezoelectricity within cubical-shaped piezoelectric material; macroscopic change ΔP of electric polarization

Table 3.2 Assignment of piezoelectric strain constants d_{ip} to four modes of piezoelectricity; L, T, S_L, and S_T refer to longitudinal, transverse, longitudinal shear, and transverse shear modes, respectively

	T_1	T_2	T_3	T_4	T_5	T_6
D_1	d_{11}	d_{12}	d_{13}	d_{14}	d_{15}	d_{16}
	L	T	T	S_L	S_T	S_T
D_2	d_{21}	d_{22}	d_{23}	d_{24}	d_{25}	d_{26}
	T	L	T	S_T	S_L	S_T
D_3	d_{31}	d_{32}	d_{33}	d_{34}	d_{35}	d_{36}
	T	T	L	S_T	S_T	S_L

mechanical fields within piezoelectric materials. Each possibility belongs to one out of four specific modes of piezoelectric coupling. These modes are named longitudinal, transverse, longitudinal shear, and transverse shear mode. In the following, the different modes are discussed for the direct piezoelectric effect, i.e., a mechanical input is converted into an electrical output. In particular, we consider the piezoelectric strain constants d_{ip} that link components of the mechanical stress T_p to those of the electric flux density D_i. The mechanical stress yields an electric flux density causing a certain macroscopic change ΔP of the electric polarization in an individual direction. Figure 3.5 illustrates on basis of a cubical-shaped piezoelectric material the four modes of the piezoelectric effect, which are described below in detail (summary in Table 3.2).

- *Longitudinal mode* L: d_{11}, d_{22}, and d_{33}
  A normal stress (e.g., T_3 in Fig. 3.5) is accompanied by a change of electric polarization in the same direction.
- *Transverse mode* T: d_{12}, d_{13}, d_{21}, d_{23}, d_{31}, and d_{32}
  In contrast to the longitudinal mode, the change of electric polarization occurs perpendicular to the applied mechanical load.
- *Longitudinal shear mode* S_L: d_{14}, d_{25}, and d_{36}
  If a shear stress (e.g., T_5 in Fig. 3.5) is applied, the polarization will change

perpendicular to the plane (e.g., 13-plane in Fig. 3.5) in which the piezoelectric material is sheared.

- *Transverse shear mode* S_T: d_{15}, d_{16}, d_{24}, d_{26}, d_{34}, and d_{35}
  Contrary to the longitudinal shear mode, the electric polarization changes in the plane in which the piezoelectric material is sheared.

This classification of modes also holds for the inverse piezoelectric effect as well as for all other forms of the constitutive equations for piezoelectricity.

3.5 Electromechanical Coupling Factors

As mentioned in Sect. 3.3, we can fully describe the linearized behavior of piezoelectric materials by means of appropriate constitutive equations. Thereby, the material parameters have to be known, e.g., s_{pq}^E, ε_{ij}^T, and d_{ip} in d-form. Nevertheless, so-called electromechanical coupling factors k are oftentimes introduced in addition rating the efficiency of energy conversion within piezoelectric materials. This concerns the conversion of mechanical into electrical energy as well as of electrical into mechanical energy, which are here given by ($1 \leq i \leq 3$; $1 \leq p \leq 6$)

$$\text{mechanical energy per unit volume:} \quad W^{\text{mech}} = \frac{S_p T_p}{2} \tag{3.37}$$

$$\text{electrical energy per unit volume:} \quad W^{\text{elec}} = \frac{D_i T_i}{2}. \tag{3.38}$$

To obtain a dimensionless measure for k, the converted energy is related to the input energy. If mechanical energy acts as input on the piezoelectric material, the electromechanical coupling factor k (strictly speaking k^2) will result from

$$k^2 = \frac{\text{mechanical energy converted into electrical energy}}{\text{mechanical input energy}} \tag{3.39}$$

and in the case of electrical input energy from

$$k^2 = \frac{\text{electrical energy converted into mechanical energy}}{\text{electrical input energy}}. \tag{3.40}$$

Actually, the conversion of mechanical into electrical energy and vice versa is always incomplete. Therefore, k^2 as well as k are smaller than 1. In the following two subsections, we will study both coupling directions separately by means of a lossless piezoelectric cylinder (base area A_S; thickness l_S) covered with electrodes at the bottom and top surface. The cylinder features piezoelectric properties in thickness direction, which coincide with the 3-direction.

3.5.1 Conversion from Mechanical into Electrical Energy

In order to quantify the conversion from mechanical into electrical energy within the piezoelectric cylinder, let us consider a single conversion cycle consisting of three subsequent states A, B, and C. Figure 3.6 shows a sketch of the arrangement for the three states as well as diagrams for the relevant mechanical and electrical state variables, i.e., S_3, T_3, D_3, and E_3. In state A, the piezoelectric material is neither mechanically nor electrically loaded. As a result, the state variables in 3-direction are zero, i.e., $S_3 = T_3 = D_3 = E_3 = 0$.

- A $\Rightarrow$ B: From state A to state B, the piezoelectric cylinder is electrically short-circuited (i.e., $E_3 = 0$) and mechanically loaded in negative 3-direction with the force F. Due to this force, mechanical stresses T_3 occur in the cylinder, which are accompanied by a negative deformation of the cylinder in thickness direction. At state B, the force reaches its highest value F_{max}. By utilizing the d-form of the

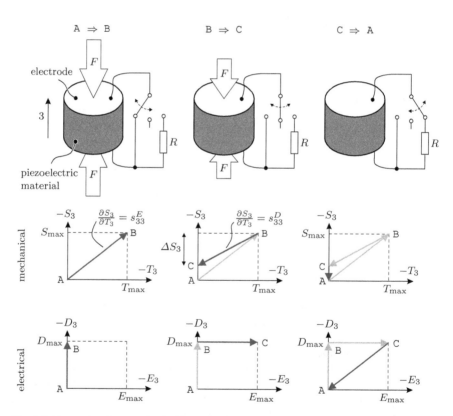

Fig. 3.6 Conversion from mechanical into electrical energy within piezoelectric materials demonstrated by cylinder [14]; A, B, and C represent three defined states during conversion cycle; S_3, T_3, D_3 as well as E_3 denote decisive state variables

constitutive equations, the state variables become

$$S_3 = S_{max} = s_{33}^E T_{max} \qquad T_3 = T_{max}$$
$$D_3 = D_{max} = d_{33} T_{max} \qquad E_3 = 0 .$$

The expression $T_{max} = -F_{max}/A_S$ denotes the peak value of the mechanical stress. Overall, the mechanical energy (per unit volume) done on the piezoelectric cylinder computes as

$$W_{AB}^{mech} = \frac{S_3 T_3}{2} = \frac{s_{33}^E T_{max}^2}{2} . \qquad (3.41)$$

- B $\Rightarrow$ C: During this state change, the piezoelectric cylinder gets electrically unloaded. The applied mechanical force is, moreover, reduced to zero. Therefore, the electric flux density D_3 remains constant. At state C, the state variables take the form

$$S_3 = \frac{d_{33}^2 T_{max}}{\varepsilon_{33}^T} \qquad T_3 = 0$$
$$D_3 = D_{max} = d_{33} T_{max} \qquad E_3 = E_{max} = \frac{d_{33} T_{max}}{\varepsilon_{33}^T} .$$

Although there do not occur mechanical stresses in 3-direction, the cylinder thickness changes (i.e., $S_3 \neq 0$) with respect to state A, which is a consequence of piezoelectric coupling. The mechanical energy per unit volume released from the cylinder is given by

$$W_{BC}^{mech} = \frac{S_3 T_3}{2} = \frac{\Delta S_3 T_{max}}{2} = \frac{T_{max}}{2} \underbrace{T_{max} \left[s_{33}^E - \frac{d_{33}^2}{\varepsilon_{33}^T} \right]}_{\Delta S_3} \qquad (3.42)$$

with the deformation change ΔS_3 from state B to state C. We can simplify this relation by using the parameter conversion $s_{33}^D = s_{33}^E - d_{33}^2/\varepsilon_{33}^T$ (see (3.33)) to

$$W_{BC}^{mech} = \frac{s_{33}^D T_{max}^2}{2} . \qquad (3.43)$$

- C $\Rightarrow$ A: Finally, we return to the initial state A of the conversion cycle. In doing so, the piezoelectric cylinder is electrically loaded with a resistor R and mechanically unloaded (i.e., $T_3 = 0$). At state A, the state variables become

$$S_3 = 0 \qquad T_3 = 0$$
$$D_3 = 0 \qquad E_3 = 0 .$$

From state C to state A, the cylinder releases electrical energy per unit volume to the resistor, which computes as

$$W_{CA}^{elec} = \frac{E_3 D_3}{2} = \frac{d_{33}^2 T_{max}^2}{2\varepsilon_{33}^T} . \tag{3.44}$$

Let us now take a look at the energy balance for the entire conversion cycle. From state A to state B, the piezoelectric cylinder is mechanically loaded with W_{AB}^{mech} and from state B to state C, the cylinder releases W_{BC}^{mech}. Thus, the stored energy in the cylinder is given by

$$W_{AB}^{mech} - W_{BC}^{mech} = \frac{T_{max}^2}{2}\left(s_{33}^E - s_{33}^D\right) . \tag{3.45}$$

In accordance with the conservation of energy, the stored energy has to correspond to the released electrical energy W_{CA}^{elec} from state C to the initial state A. This is also reflected in the parameter conversion $s_{33}^E - s_{33}^D = d_{33}^2/\varepsilon_{33}^T$. The electromechanical coupling factor k_{33} for the conversion from mechanical into electrical energy results in

$$k_{33}^2 = \frac{W_{CA}^{elec}}{W_{AB}^{mech}} = \frac{W_{AB}^{mech} - W_{BC}^{mech}}{W_{AB}^{mech}} = \frac{s_{33}^E - s_{33}^D}{s_{33}^E} = \frac{d_{33}^2}{\varepsilon_{33}^T s_{33}^E} . \tag{3.46}$$

The indices of k_{pq} refer to the direction of the applied mechanical loads and that of the electrical quantities, respectively.

3.5.2 Conversion from Electrical into Mechanical Energy

As for the previous conversion direction, we consider a single conversion cycle consisting of three subsequent states A, B, and C (see Fig. 3.7). Again, the piezoelectric material is neither mechanically nor electrically loaded in state A, i.e., $S_3 = T_3 = D_3 = E_3 = 0$.

- A $\Rightarrow$ B: From state A to state B, the piezoelectric cylinder is mechanically unloaded (i.e., $T_3 = 0$) and electrically loaded in negative 3-direction with the electrical voltage U. This voltage causes an electric field intensity E_3, which is due to piezoelectric coupling responsible for a certain deformation of the cylinder. At state B, the voltage reaches its highest value U_{max}. The d-form of the constitutive equations for piezoelectricity yields the state variables

$$\begin{aligned} S_3 &= S_{max} = d_{33}E_{max} & T_3 &= 0 \\ D_3 &= D_{max} = \varepsilon_{33}E_{max} & E_3 &= E_{max} \end{aligned}$$

with the peak value $E_{max} = -U_{max}/l_S$ of electric field intensity. The electrical energy per unit volume stored in the cylinder at state B computes as

$$W_{AB}^{elec} = \frac{E_3 D_3}{2} = \frac{\varepsilon_{33}^T E_{max}^2}{2} . \tag{3.47}$$

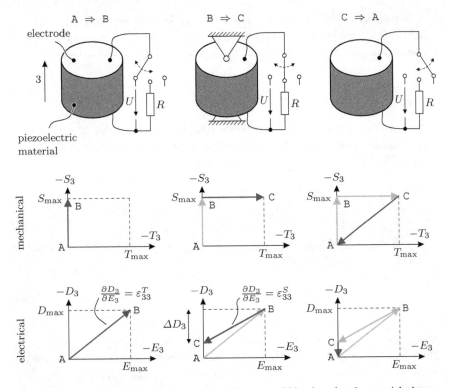

Fig. 3.7 Conversion from electrical to mechanical energy within piezoelectric materials demonstrated by cylinder; A, B, and C represent three defined states during conversion cycle; S_3, T_3, D_3 as well as E_3 denote decisive state variables

- B $\Rightarrow$ C: During this state change, the piezoelectric cylinder is electrically loaded with a resistor R. Besides, we prohibit mechanical movements in thickness directions, which can be arranged by an appropriate clamping of the cylinder. Consequently, S_3 stays constant and the mechanical stress T_3 changes. At state C, the state variables are given by

$$S_3 = S_{max} = d_{33} E_{max} \qquad T_3 = T_{max} = \frac{d_{33} E_{max}}{s^E_{33}}$$
$$D_3 = D_{max} = d_{33} T_{max} \qquad E_3 = 0\,.$$

Since stresses occur in the cylinder, there is a remaining flux density (i.e., $D_3 \neq 0$) on the electrodes at state C. The electrical energy (per unit volume) released by the resistor results in

$$W^{elec}_{BC} = \frac{E_3 D_3}{2} = \frac{E_{max} \Delta D_3}{2} = \frac{E_{max}}{2} \underbrace{E_{max} \left[\varepsilon^T_{33} - \frac{d^2_{33}}{s^E_{33}} \right]}_{\Delta D_3} \qquad (3.48)$$

with flux density change ΔD_3 from state B to state C. Similar to the previous conversion direction, we are able to simplify this relation by using the parameter conversion $\varepsilon_{33}^S = \varepsilon_{33}^T - d_{33}^2/s_{33}^E$ (see (3.33)) to

$$W_{BC}^{\text{elec}} = \frac{\varepsilon_{33}^S E_{\max}^2}{2} . \tag{3.49}$$

- C $\Rightarrow$ A: At the end of the conversion cycle, we return to state A by removing the mechanical clamping. Furthermore, the piezoelectric cylinder is electrically unloaded. Finally, the state variables become

$$S_3 = 0 \qquad T_3 = 0$$
$$D_3 = 0 \qquad E_3 = 0 .$$

During this state change, the mechanical energy per unit volume that is released from the cylinder computes as

$$W_{CA}^{\text{mech}} = \frac{S_3 T_3}{2} = \frac{d_{33}^2 E_{\max}^2}{2 s_{33}^E} . \tag{3.50}$$

Again, we take into account the energy balance for the entire conversion cycle. The piezoelectric cylinder is electrically loaded from state A to state B with W_{AB}^{elec}. Subsequently, the cylinder releases W_{BC}^{elec} from state B to state C leading to the stored energy

$$W_{AB}^{\text{elec}} - W_{BC}^{\text{elec}} = \frac{E_{\max}^2}{2}\left(\varepsilon_{33}^T - \varepsilon_{33}^S\right) , \tag{3.51}$$

which has to correspond to the released mechanical energy W_{CA}^{mech} from state C into state A. This can also be seen from the parameter equation $\varepsilon_{33}^T - \varepsilon_{33}^S = d_{33}^2/s_{33}^E$. The electromechanical coupling factor k_{33} characterizing the conversion from electrical into mechanical energy reads as

$$k_{33}^2 = \frac{W_{CA}^{\text{mech}}}{W_{AB}^{\text{elec}}} = \frac{W_{AB}^{\text{elec}} - W_{BC}^{\text{elec}}}{W_{AB}^{\text{elec}}} = \frac{\varepsilon_{33}^T - \varepsilon_{33}^S}{\varepsilon_{33}^T} = \frac{d_{33}^2}{\varepsilon_{33}^T s_{33}^E} . \tag{3.52}$$

As the comparison of (3.46) and (3.52) reveals, k_{33} is identical for both directions of energy conversion. In other words, it does not matter for conversion efficiency if mechanical energy is converted into electrical energy or vice versa. Note that this is an essential property of the piezoelectric coupling mechanism.

Apart from k_{33}, we can deduce various other electromechanical coupling factors for piezoelectric materials. For the crystal class 6mm, k_{31} and k_{15} are such factors that relate to the transverse and transverse shear mode of piezoelectricity, respectively. These electromechanical coupling factors are given by ($s_{44}^E = s_{55}^E$)

$$k_{31}^2 = \frac{d_{31}^2}{\varepsilon_{33}^T s_{11}^E} \quad \text{and} \quad k_{15}^2 = \frac{d_{15}^2}{\varepsilon_{11}^T s_{55}^E} \ . \tag{3.53}$$

Depending on the piezoelectric material, the electromechanical coupling factors differ significantly. For instance, several piezoceramic materials offer remarkable values for k_{33}, k_{31}, and k_{15}. On the other hand, certain materials (e.g., cellular polymers) provide large values for k_{33} but only a small coupling factor k_{31}.

3.6   Piezoelectric Materials

Piezoelectric materials exhibit either a crystal structure or, at least, areas with a crystal-like structure. In general, a crystal is characterized by a periodic repetition of the atomic lattice structure in all directions of space. The smallest repetitive part of the crystal is termed unit cell. Depending on the symmetry properties of the unit cell, we can distinguish between 32 crystal classes that are also termed crystallographic point groups (see Fig. 3.8) [28, 42]. Piezoelectric properties will only arise when the structure of the unit cell is asymmetric. 21 out of the 32 crystal classes fulfill this property because they are noncentrosymmetric which means that they do not have a center of symmetry. 20 out of these 21 crystal classes show piezoelectricity. While 10 crystal classes are pyroelectric, the remaining 10 crystal classes are nonpyroelectric. The 20 piezoelectric crystal classes can be grouped into the seven crystal systems: triclinic, monoclinic, orthorhombic, tetragonal, rhombohedral, hexagonal, and cubic crystal system. Table 3.3 contains the abbreviations of the 20 piezoelectric crystal classes according to the Hermann–Mauguin notation. As a matter of course, the material tensors (e.g., $[\mathbf{d}]$, $[\varepsilon^T]$ and $[\mathbf{s}^E]$ in d-form) of piezoelectric materials differ for the crystal classes regarding the number of independent material parameters as well as the entries being nonzero.

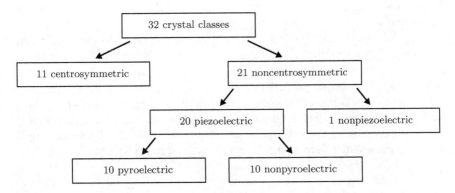

Fig. 3.8 Classification of the 32 crystal classes (crystallographic point groups)

Table 3.3 Abbreviations of piezoelectric crystal classes according to Hermann–Mauguin notation; crystal classes grouped into seven crystal systems

Crystal system	Abbreviations of crystal classes
Triclinic	1
Monoclinic	2 and m
Orthorhombic	222 and mm2
Tetragonal	$4, \bar{4}, 422, 4mm$ and $\bar{4}2m$
Rhombohedral (trigonal)	3, 32 and 3m
Hexagonal	$6, \bar{6}, 622, 6mm$ and $\bar{6}m2$
Cubic	23 and $\bar{4}3m$

The choice of the used piezoelectric material always depends on the applications. Several piezoelectric sensors and actuators call for materials that provide high piezoelectric strain constants d_{ip} and high electromechanical coupling factors. A wide variety of applications demand piezoelectric materials that are free from hysteretic behavior and offer high mechanical stiffness, i.e., small values for the elastic compliance constants s_{pq}^E. On the other hand, there also exist applications requiring mechanically flexible materials with piezoelectric properties. It is, therefore, impossible to find a piezoelectric material that is most appropriate for all kinds of piezoelectric sensors and actuators. However, for each application, one can define a specific figure of merit, which comprises selected material parameters. In the following, we will study different piezoelectric materials, their main properties, and the manufacturing process. This includes single crystals such as quartz (see Sect. 3.6.1), polycrystalline ceramic materials such as lead zirconate titanate (see Sect. 3.6.2) and polymers such as PVDF (see Sect. 3.6.3).

3.6.1 Single Crystals

There exists a large number of piezoelectric single crystals. In general, they can be divided into naturally occurring (e.g., quartz) and synthetically produced materials (e.g., lithium niobate). Table 3.4 lists well-known representatives for both groups including chemical formulas as well as crystal classes. Below, we will concentrate on quartz because this piezoelectric single crystal plays still an importance role in practical applications like piezoelectric sensors. The reason for its importance is not least due to the possibility that quartz can be manufactured synthetically by artificial growth. Furthermore, lithium niobate will be briefly discussed since such piezoelectric single crystals are often utilized in surface acoustic wave (SAW) devices. At the end, a short introduction to relaxor-based single crystals is given.

Quartz

Quartz crystals at room temperature are commonly named α-quartz. At the temperature of 573 °C, a structural phase transition takes place inside the quartz crystal, which is stable in the temperature range from 573 °C to 870 °C. The resulting crystal

Table 3.4 Chemical formula and crystal class of selected naturally occurring and synthetically produced piezoelectric single crystals

Material	Group	Chemical formula	Crystal class
α-quartz	Natural	SiO_2	32
β-quartz	Natural	SiO_2	622
Tourmaline	Natural	$(Na,Ca)(Mg,Fe)_3B_3Al_6$ $Si_6(O,OH,F)_{31}$	3m
CGG	Synthetic	$Ca_3Ga_2Ge_4O_{14}$	32
Lithium niobate	Synthetic	$LiNbO_3$	3m
Lithium tantalate	Synthetic	$LiTaO_3$	3m

is named β-quartz and differs considerably from the α-quartz [14, 42]. This does not only refer to the material parameters but also to the crystal class. α- and β-quartzes belong to crystal class 32 and 622, respectively. Since the temperatures in most of the practical applications are below $500\,^\circ$C, let us study α-quartzes in more detail. In the d-form, the material tensors of the crystal class 32 feature the structure

$$\left[\mathbf{s}^E\right] = \begin{bmatrix} s_{11}^E & s_{12}^E & s_{13}^E & s_{14}^E & 0 & 0 \\ s_{12}^E & s_{11}^E & s_{13}^E & -s_{14}^E & 0 & 0 \\ s_{13}^E & s_{13}^E & s_{33}^E & 0 & 0 & 0 \\ s_{14}^E & -s_{14}^E & 0 & s_{44}^E & 0 & 0 \\ 0 & 0 & 0 & 0 & s_{44}^E & 2s_{14}^E \\ 0 & 0 & 0 & 0 & 2s_{14}^E & 2\left(s_{11}^E - s_{12}^E\right) \end{bmatrix}, \quad (3.54)$$

$$\left[\mathbf{\varepsilon}^S\right] = \begin{bmatrix} \varepsilon_{11}^T & 0 & 0 \\ 0 & \varepsilon_{11}^T & 0 \\ 0 & 0 & \varepsilon_{33}^T \end{bmatrix}, \quad [\mathbf{d}] = \begin{bmatrix} d_{11} & -d_{11} & 0 & d_{14} & 0 & 0 \\ 0 & 0 & 0 & 0 & -d_{14} & -d_{11} \\ 0 & 0 & 0 & 0 & 0 & 0 \end{bmatrix} \quad (3.55)$$

and, thus, contain altogether 10 independent entries.

α-quartz exists in two crystal shapes, namely the left-handed and right-handed α-quartz. The naming originates from the rotation of linearly polarized light that propagates through the quartz crystal along its optical axis. While right-handed quartzes rotate the polarization plane clockwise, left-handed quartzes perform a counterclockwise rotation. The difference of both shapes also appears in the sign of the material parameters, e.g., d_{14} is positive for a left-handed α-quartz and negative for a left-handed α-quartz. Figure 3.9 illustrates the fundamental structure of a left-handed α-quartz, whereby the z-axis coincides with the optical axis of the crystal.

It seems only natural that α-quartzes in the original structure as shown in Fig. 3.9 cannot be used in piezoelectric devices. Owing to this fact, one has to cut out parts such as thin slices. The most well-known cuts are listed hereinafter [14, 28].

Fig. 3.9 Fundamental
structure of left-handed
α-quartz

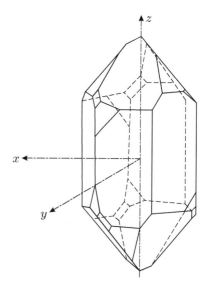

- X-cut: plate perpendicular to x-axis.
- Y-cut: plate perpendicular to y-axis.
- Z-cut: plate perpendicular to z-axis.
- Rotated Y-cuts: plate perpendicular to yz-plane.

By means of these cuts, it is possible to create piezoelectric elements, which offer distinct modes of piezoelectricity. For example, the X-cut will lead to the longitudinal mode if the cutted plate is covered with electrodes at the top and bottom surface.

Even though quartz is one of the most frequent minerals on earth, the great demand for quartz crystals in desired size and quality makes synthetic production indispensable. Synthetic quartz crystal can be artificially grown by the so-called hydrothermal method [21, 24]. Thereby, the crystal growth is conducted in a thick-walled autoclave at very high pressure up to 200 MPa and temperatures of $\approx$400 °C. Water with a small amount of sodium carbonate (chemical formula Na_2CO_3) or sodium hydroxide (chemical formula NaOH) serves as a solvent. The synthetic production of quartz crystals weighting more than 1 kg takes several weeks.

Quartz crystals are often used as piezoelectric elements in practical applications (e.g., for force and torque sensors; see Sect. 9.1) because such single crystals offer a high mechanical rigidity as well as a high electric insulation resistance. They are almost free of hysteresis and allow an outstanding linearity in practical use. Moreover, there exist temperature-compensated cuts that exhibit constant material parameters over a wide temperature range. Compared to other piezoelectric material like piezoceramic materials, quartz crystals provide, however, only small piezoelectric strain constants d_{ip}. This is a big disadvantage regarding efficient piezoelectric actuators. Besides, piezoelectric elements cannot be cut out from quartz crystals in any shape. The main material parameters of a left-handed α-quartz are listed in Table 3.5.

Table 3.5 Decisive material parameters of selected piezoelectric materials; material density ϱ_0 in 10^3 kg m^{-3}; elastic compliance constants s_{pq}^E in 10^{-12} m N^{-2}; relative permittivity $\varepsilon_{ii}^T/\varepsilon_0$; piezoelectric strain constants d_{ip} in 10^{-12} m V^{-1}; table entries '−' irrelevant; given data represent averaged values from literature and manufacturer [5, 10, 14, 30, 35, 42]

Material	ϱ_0	s_{11}^E	s_{12}^E	s_{13}^E	s_{14}^E	s_{33}^E	s_{44}^E	$\varepsilon_{11}^T/\varepsilon_0$	$\varepsilon_{33}^T/\varepsilon_0$
Quartz	2.65	12.8	−1.8	−1.2	4.5	9.7	20.0	4.5	4.6
Lithium niobate	4.63	5.8	−1.0	−1.5	−1.0	5.0	17.0	84	30
PZT-5A (soft)	7.75	16.4	−5.7	−7.2	−	18.8	47.5	1730	1700
PZT-5H (soft)	7.50	16.5	−4.8	−8.5	−	20.7	43.5	3130	3400
PIC155 (soft)	7.76	16.2	−4.8	−7.1	−	17.8	52.4	1500	1350
PIC255 (soft)	7.80	15.9	−5.7	−7.4	−	21.0	44.9	1650	1750
Pz29 (soft)	7.45	17.0	−5.8	−8.8	−	22.9	54.1	2440	2870
PIC181 (hard)	7.85	11.8	−4.1	−5.0	−	14.1	35.3	1220	1140
PIC300 (hard)	7.78	11.1	−4.8	−3.7	−	11.8	28.2	960	1030
Pz24 (hard)	7.70	10.4	−3.0	−7.6	−	23.4	23.0	810	407

Material	d_{11}	d_{14}	d_{15}	d_{22}	d_{31}	d_{33}	k_{15}	k_{31}	k_{33}
Quartz	2.3	0.7	−	−	−	−	*	**	***
Lithium niobate	−	−	26	8.5	−3.0	9.2	0.23	0.08	0.25
PZT-5A (soft)	−	−	584	−	−171	374	0.68	0.34	0.75
PZT-5H (soft)	−	−	741	−	−274	593	0.68	0.39	0.75
PIC155 (soft)	−	−	539	−	−154	307	0.65	0.35	0.66
PIC255 (soft)	−	−	534	−	−174	393	0.66	0.35	0.69
Pz29 (soft)	−	−	724	−	−243	574	0.67	0.37	0.75
PIC181 (hard)	−	−	389	−	−108	253	0.63	0.32	0.66
PIC300 (hard)	−	−	155	−	−82	154	0.32	0.26	0.46
Pz24 (hard)	−	−	151	−	−58	194	0.37	0.30	0.67

* $k_{11} = 0.09$ for X-cut; ** $k_{66} = 0.14$ for Y-cut; *** $k_{44} = 0.03$ for Z-cut

Lithium Niobate

Lithium niobate is a well-known synthetically produced piezoelectric single crystal, which exhibits a high Curie temperature ϑ_C of 1210 °C. That is the reason why this piezoelectric single crystal is mainly used for high-temperature sensors and ultrasonic transducers [2, 42]. Lithium niobate belongs to the crystal class 3m. In the d-form, the material tensors of this crystal class are given by

$$
[\mathbf{s}^E] =
\begin{bmatrix}
s_{11}^E & s_{12}^E & s_{13}^E & s_{14}^E & 0 & 0 \\
s_{12}^E & s_{11}^E & s_{13}^E & -s_{14}^E & 0 & 0 \\
s_{13}^E & s_{13}^E & s_{33}^E & 0 & 0 & 0 \\
s_{14}^E & -s_{14}^E & 0 & s_{44}^E & 0 & 0 \\
0 & 0 & 0 & 0 & s_{44}^E & 2s_{14}^E \\
0 & 0 & 0 & 0 & 2s_{14}^E & 2\left(s_{11}^E - s_{12}^E\right)
\end{bmatrix}, \tag{3.56}
$$

$$
\left[\varepsilon^{S}\right] = \begin{bmatrix} \varepsilon_{11}^{T} & 0 & 0 \\ 0 & \varepsilon_{11}^{T} & 0 \\ 0 & 0 & \varepsilon_{33}^{T} \end{bmatrix}, \quad [\mathbf{d}] = \begin{bmatrix} 0 & 0 & 0 & 0 & d_{14} & -2d_{22} \\ -d_{22} & d_{22} & 0 & d_{15} & 0 & 0 \\ d_{31} & d_{31} & d_{33} & 0 & 0 & 0 \end{bmatrix} \quad (3.57)
$$

and, thus, contain altogether 13 independent entries.

Single crystals of lithium niobate can be artificially grown with the aid of the so-called Czochralski process [8]. This manufacturing process starts with the molten state of the desired material (i.e., here lithium niobate), which is placed in a melting pot. In a next step, a slowly rotating metal rod with a seed crystal at its lower end gets immersed from above into the melt. If the seed crystal is immersed in the right way, a homogeneous boundary layer will develop between the melt and the crystal's solid part. Subsequently, the rotating combination of metal rod and seed crystal has to be slowly pulled upwards. During this step, the melt solidifies at the boundary layer, which leads to crystal growth. By varying the rate of pulling and rotation, one can extract long single crystals that are commonly named ingots. The resulting single crystals of lithium niobate are finally brought into the desired shape through suitable crystal cuts, e.g., X-cut.

In comparison with quartz crystals, lithium niobate crystals provide significantly higher piezoelectric strain constants d_{ip}. Typical material parameters are listed in Table 3.5.

Relaxor-Based Single Crystals

As mentioned in Sect. 3.4.1, relaxor ferroelectrics offer pronounced electrostriction. Relaxor-based single crystals are based on such materials. Nowadays, solid solutions of lead magnesium niobate and lead titanate (PMN-PT) as well as solid solutions of lead zinc niobate and lead titanate (PZN-PT) represent the most widely studied material compositions for relaxor-based single crystals [33, 48]. The chemical formulas of both material compositions read as

- PMN-PT: $(1\text{-}x)\text{Pb}(\text{Mg}_{1/3}\text{Nb}_{2/3})\text{O}_3 - x\text{PbTiO}_3$ and
- PZN-PT: $(1\text{-}x)\text{Pb}(\text{Zn}_{1/3}\text{Nb}_{2/3})\text{O}_3 - x\text{PbTiO}_3$,

whereby x ranges from 0 to 1. In principle, these material compositions can be used for polycrystalline ceramic materials (see Sect. 3.6.2). The resulting piezoceramic materials show excellent piezoelectric properties like high piezoelectric strain constants d_{ip}. However, by growing single crystals, the available properties are improved significantly. Crystal growth of PMN-PT and PZN-PT is often conducted by the high-temperature flux technique and the Bridgman growth technique [23, 25, 26]. After growing, the single crystals have to be polarized in an appropriate direction.

The electromechanical coupling factor k_{33} of PMN-PT and PZN-PT single crystals usually exceeds 0.9. In case of special material compositions and additional doping, one can reach extremely high electric permittivities as well as d_{33}-values $> 2000\,\text{pm V}^{-1}$. These outstanding properties are responsible for the great interest in PMN-PT and PZN-PT single crystals for practical applications [6, 18], e.g., they should be ideal candidates for efficient piezoelectric components in ultrasonic transducers.

3.6.2 Polycrystalline Ceramic Materials

Polycrystalline ceramic materials are the most important piezoelectric materials for practical applications because they offer outstanding piezoelectric properties. Moreover, several of these so-called *piezoceramic materials* can be manufactured in a cost-effective manner. Barium titanate and lead zirconate titanate (PZT) represent two well-known solid solutions that belong to the group of piezoceramic materials. Hereafter, we will discuss the manufacturing process, the basic molecular structure, the poling process, and the hysteretic behavior of piezoceramic materials. The focus lies on PZT since it is frequently used in practical applications. At the end, a brief introduction to lead-free piezoceramic materials will be given.

Manufacturing Process

The manufacturing process of piezoceramic materials mostly comprises six successive main steps, namely (i) mixing, (ii) calcination, (iii) forming, (iv) sintering, (v) applying of electrodes, and (vi) poling [12, 17]. At the beginning, powders of the raw materials (e.g., zirconium) are mixed. The powder mixture gets then heated up to temperatures between 800 and 900 °C during calcination. Thereby, the raw materials react chemically with each other. The resulting polycrystalline substance is ground and mixed with binder. Depending on the desired geometry of the piezoelectric element, there exist various processes for forming. The most widely used process is cold pressing. During the subsequent sintering, the so-called green body gets bound as well as compressed at temperatures of $\approx 1200\,°C$. To achieve a higher material density, sintering is commonly carried out in an oxygen atmosphere. The sintered blanks are, sometimes, cut and polished. In the fifth step, the blanks get equipped with electrodes. This can be done by either screen printing or sputtering. The obtained materials are usually referred to as *unpolarized ceramics*. The poling as final process step yields the piezoceramic material, i.e., a ceramic material featuring piezoelectric properties. Poling is usually carried out by applying strong electric fields in the range of $2\text{--}8\,\mathrm{k\,V\,mm^{-1}}$ in a heated oil bath. Alternatively, one can use corona discharge for poling.

It is possible to produce various shapes of piezoceramic elements by means of the common manufacturing process. Piezoceramic disks, rings, plates, bars as well as cylinders are standard shapes of piezoceramic elements. However, slightly modified manufacturing processes additionally allow the production of other shapes like thin piezoceramic fibers that can be used for piezoelectric composite transducers (see Sect. 7.4.3). With the aid of screen printing, physical vapor deposition (e.g., sputtering), chemical vapor deposition (e.g., metal-organic CVD) as well as chemical solution deposition (e.g., solution-gelation), one can also generate films of piezoceramic materials [4, 16, 31, 43].

Molecular Structure

Piezoceramic materials such as barium titanate and PZT consist of countless crystalline unit cells, which exhibit the so-called *perovskite structure*, being named after

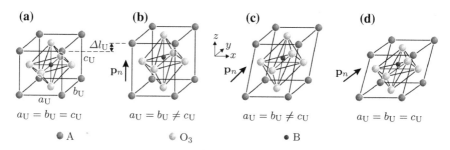

Fig. 3.10 Unit cell of perovskite structure ABO$_3$ in **a** cubic phase, **b** tetragonal phase, **c** orthorhombic phase, and **d** rhombohedral phase; spontaneous electric polarization $\mathbf{p}_n$; edge lengths a_U, b_U and c_U of unit cell

the mineral perovskite [17]. In general, piezoceramic materials featuring this structure can be described by the chemical formula ABO$_3$. While A and B are two cations (i.e., positive ions) of different size, O$_3$ is an anion (i.e., negative ion) that bonds to both. The large cations A are located on the corners of the unit cell and the oxygen anions O$_3$ in the center of the faces, respectively. Depending on the state of the unit cell, the small cation B is located in or close to the center. The edge lengths of a unit cell amounts a few angstroms Å (1 Å $\hat{=}$ 0.1 nm).

The unit cells of many piezoceramic materials take predominantly one of four phases, namely (i) the cubic, (ii) the tetragonal, (iii) the orthorhombic, or (iv) the rhombohedral phase (see Fig. 3.10). In the cubic phase, the center $\mathcal{C}_{Q+}$ of positive charges (cations) geometrically coincides with the center $\mathcal{C}_{Q-}$ of negative charges (anions). A single unit cell behaves, thus, electrically neutral. Note that this paraelectric phase of the unit cells only exists above the Curie temperature ϑ_C. When the temperature falls below ϑ_C, the cubic unit cell will undergo a phase transition from cubic to another phase, i.e., to the tetragonal, the orthorhombic, or the rhombohedral phase. Thereby, the unit cell gets deformed. For example, the deformation from the cubic to the tetragonal phase is Δl_U in a single direction. In any case, the small cation B leaves the center of the unit cell. Consequently, $\mathcal{C}_{Q+}$ does not geometrically coincide with $\mathcal{C}_{Q-}$ anymore. That is the reason why there arises a dipole moment, which is termed spontaneous electric polarization $\mathbf{p}_n$. If the positive charges amount overall q_n and the distance from $\mathcal{C}_{Q-}$ to $\mathcal{C}_{Q+}$ is defined by the vector $\mathbf{r}_n$, the spontaneous polarization of a single unit cell will become $\mathbf{p}_n = q_n \mathbf{r}_n$.

The direction of the spontaneous electric polarization $\mathbf{p}_n$ inside a unit cell differs for the individual phases. With regard to the local coordinate system of a unit cell, $\mathbf{p}_n$ exhibits different proportions in x-, y- and z-direction. The so-called *Miller index* $[hkl]$ describes the direction of $\mathbf{p}_n$ in the local coordinate system [28, 42]. For the tetragonal, the orthorhombic and the rhombohedral phase, the Miller index takes the form [001], [011], and [111], respectively. During the phase transition from the cubic to the tetragonal phase, the central cation B can move into six directions. This is accompanied by 6 possible directions of $\mathbf{p}_n$ in a global coordinate system. In contrast, there exist 8 possible directions for the rhombohedral phase and 12 possible directions for the orthorhombic phase. The total electric polarization $\mathbf{P}$ of a piezoceramic element containing N unit cells results from the vectorial sum of all spontaneous electric polarizations related to the element's volume V, i.e.,

$$\mathbf{P} = \frac{1}{V} \sum_{n=1}^{N} \mathbf{p}_n = \frac{1}{V} \sum_{n=1}^{N} q_n \mathbf{r}_n \ . \tag{3.58}$$

Here, the individual polarizations $\mathbf{p}_n$ of the unit cells have to be regarded in a global coordinate system.

Phase diagrams represent an appropriate way to illustrate the phases of piezoceramic materials with respect to both temperature ϑ and material composition. Figure 3.11 illustrates the phase diagram of PZT, which features the chemical formula $Pb(Ti_xZr_{1-x})O_3$ [17, 40]. The abscissa starts with pure $PbZrO_3$ and ends with pure $PbTiO_3$. In between, the molar amount x of $PbTiO_3$ increases linearly. Since the cubic phase is dominating above the Curie temperature ϑ_C, PZT does not behave like a piezoelectric material for temperatures $\vartheta > \vartheta_C$. It can also be seen that ϑ_C increases with increasing x. According to Fig. 3.11, there exist three stable phases of the unit cells below ϑ_C. The tetragonal phase and rhombohedral phase of PZT are also named ferroelectric phases, while the orthorhombic phase that dominates for high contents of zirconium is referred to as antiferroelectric phase. The phase boundary between the rhombohedral and tetragonal phase is of great practical importance. In the vicinity of this phase boundary, which is termed morphotropic phase boundary, a PZT element contains roughly the same number of unit cells in rhombohedral and tetragonal phase. The almost equal distribution of both phases is frequently assumed to be the origin for the outstanding properties of PZT like high electromechanical coupling factors. At room temperature, the morphotropic phase boundary is located at x = 48, i.e., the chemical formula reads as $Pb(Ti_{48}Zr_{52})O_3$. A further advantage of PZT lies in the relatively low temperature dependence, which stems from the vertical progression of the morphotropic phase boundary.

Poling Process and Hysteresis

As already mentioned, piezoceramic materials consist of countless unit cells. Now, let us take a closer look at the internal structure of such materials, which is also illustrated in Fig. 3.12. Neighboring unit cells that exhibit the same directions of spontaneous electric polarization $\mathbf{p}_n$ form a so-called domain [14, 17]. Several of those domains form a single grain, which represents the smallest interrelated part of a piezoceramic material. The boundaries between neighboring grains are named grain boundaries, whereas the boundaries between neighboring domains inside a grain are termed domain walls. In the tetragonal phase, the directions of $\mathbf{p}_n$ between two neighboring domains can be differ by either 90 or 180°. By contrast, this angle can take the values 71, 109, or 180° in case of the rhombohedral phase of PZT.

It is not surprising that apart from the material composition, the current domain configuration determines the properties of a piezoceramic material. When such a material is cooled down below the Curie temperature (e.g., after sintering), the unit

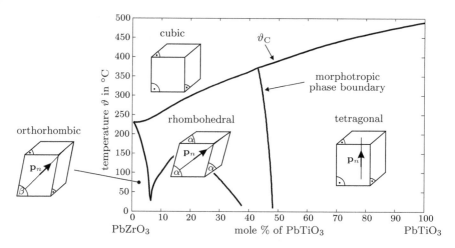

Fig. 3.11 Phase diagram showing dominating phases of PZT (chemical formula $Pb(Ti_xZr_{1-x})O_3$) with respect to temperature ϑ and material composition; Curie temperature ϑ_C; spontaneous electric polarization $\mathbf{p}_n$ of unit cell

Fig. 3.12 Internal structure of piezoceramic materials; arrows indicate direction of spontaneous electric polarization $\mathbf{p}_n$ of unit cells inside single domain

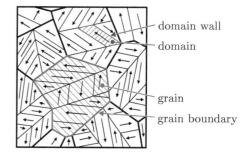

cells will undergo a phase transition from the cubic phase to another one. This goes hand in hand with a spontaneous electric polarization and mechanical deformations of the unit cells yielding mechanical stresses. Owing to the fact that each closed system wants to minimize its free energy (see Sect. 3.2), the unit cells align themselves in a way that both the electrical and the mechanical energy take a minimum. The resulting characteristic domain configuration (cf. Fig. 3.12) leads, thus, to a vanishing total electric polarization $\mathbf{P}$. That is the reason why the material, the unpolarized ceramics, does not offer piezoelectric properties in this state from the macroscopic point of view.

To activate the piezoelectric coupling which means $\|\mathbf{P}\|_2 \neq 0$, we have to appropriately align the unit cells and, therefore, the domains within the unpolarized ceramics. The underlying activation process is known as poling [12, 14, 17]. The alignment is mostly conducted by applying strong electric fields to the unpolarized ceramics. In doing so, the spontaneous electric polarizations $\mathbf{p}_n$ of the unit cells get aligned along the crystal axis that is nearest to the direction of the electrical field lines. The number of neighboring unit cells exhibiting the same direction of $\mathbf{p}_n$ increases which reduces the number of domains inside a single grain. This implies a growth of the still existing domains as well as a movement of the domain walls. Apart from that, the alignment of the unit cells causes a deformation of the grains, which becomes visible as mechanical strain of the piezoceramic element.

In the following, we will regard a mechanically unloaded thin piezoceramic disk, which is completely covered with electrodes at the top and bottom surface. Therefore, it is possible to reduce the vectors of electric polarization $\mathbf{P}$ and mechanical strain $\mathbf{S}$ to scalars, i.e., to P and S. Figure 3.13a and b depict $P(E)$ and $S(E)$ of the disk as a function of the applied electric field intensity E, respectively. Exemplary domain configurations for the states A to G are shown in Fig. 3.13c. These states are also marked in $P(E)$ as well as $S(E)$. In the initial state A (e.g., after sintering), the disk shall be unpolarized which means $P = 0$. Without limiting the generality, the mechanical strain is assumed to be zero, i.e., $S = 0$. The domains will get aligned when a strong positive electric field is applied. While doing so, the so-called virgin curve is passed through until the positive saturation of the electric polarization P_{sat}^+ and the mechanical strain S_{sat}^+ is reached at state B. In this state, the spontaneous electric polarizations of the unit cells are almost perfectly aligned, which can also be seen in high values of P_{sat}^+ and S_{sat}^+. A further increase of E does not considerably increase both values because the intrinsic effects of electromechanical coupling (see Sect. 3.4.1) are dominating then. In contrast, steep gradients in $P(E)$ and $S(E)$ indicate extrinsic effects (see Sect. 3.4.2).

If E is reduced to zero (state C), the majority of unit cells will stay aligned, i.e., the domain configuration barely changes. Only a few unit cells belonging to unstable domains switch back to the original state, whereby P as well as S get slightly reduced. The resulting positive values at $E = 0$ are termed remanent electric polarization P_r^+ and remanent mechanical strain S_r^+, respectively.[5] When E is reduced to negative values, state D will be passed through. At this state, the electric polarization of the disks equals zero, which is arranged by the negative coercive field intensity E_c^-. A further reduction of E leads to the negative saturation of the electric polarization P_{sat}^- and the mechanical strain S_{sat}^- at state E. Thereby, the domains get aligned in the opposite direction to that in case of positive saturation (cf. Fig. 3.13c). The electric polarization takes, thus, the negative value of P_{sat}^+. In contrast, S_{sat}^- coincides with S_{sat}^+ because it does not make a difference for the mechanical deformation of a unit cell whether $\mathbf{p}_n$ points in positive or negative direction. The remaining parts of the curves $P(E)$ and $S(E)$ describe the same material behavior as the previous ones. State F refers to the negative remanent electric polarization $P_r^- = -P_r^+$ and remanent mechanical strain $S_r^+ = S_r^-$, while state G with the positive coercive field E_c^+ corresponds to state D. Note that the initial domain configuration of state A can be retrieved by heating up the piezoceramic material above the Curie temperature.

As can be seen in Fig. 3.13, piezoceramic materials exhibit a strongly pronounced hysteretic behavior, which is commonly termed *ferroelectric behavior*. For high electrical excitation signals, the electric polarization and mechanical strain differ,

---

[5] P_r and S_r coincide with $\|\mathbf{P}^{\text{irr}}\|_2$ and $\|\mathbf{S}^{\text{irr}}\|_2$ from (3.35) and (3.36).

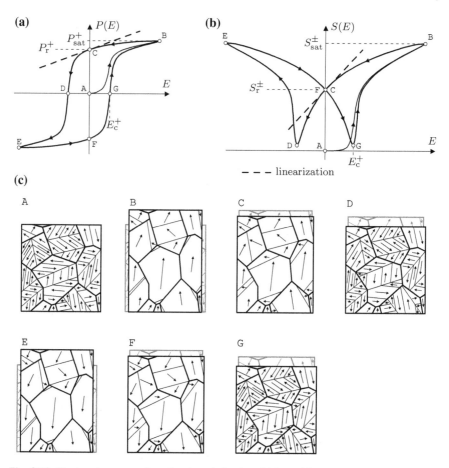

Fig. 3.13 Hysteresis curves of **a** electric polarization $P(E)$ and **b** mechanical strain $S(E)$ of piezoceramic disk; linearization refers to small-signal behavior; **c** exemplary domain configurations for states A to G [46]

therefore, considerably for increasing and decreasing inputs, respectively. The phenomenological modeling of this large-signal behavior will be addressed in Chap. 6. Besides the ferroelectric behavior, piezoceramic materials also show *ferroelastic behavior*. This means for a piezoceramic disk that both the mechanical strain S and the electric polarization P are altered by the mechanical stress T. Figure 3.14a depicts exemplary curves $P(T)$ and $S(T)$ of the resulting hysteretic behavior.

It seems only natural that the material composition remarkably affects the ferroelectric behavior of piezoceramic materials. By means of doping, one can change properties of piezoceramic materials such as electromechanical coupling factors in a specific manner [14, 17, 41]. A general distinction is made between acceptor doping and donor doping. In case of acceptor doping, the remanent electric polarization $\left|P_r^\pm\right|$, the electromechanical coupling factors and the elastic compliance constants s_{pq}^E get

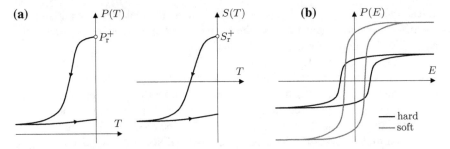

Fig. 3.14 **a** Ferroelastic behavior of electric polarization $P(T)$ and mechanical strain $S(T)$; mechanical stress T; **b** hysteresis curve of $P(E)$ for ferroelectrically hard and soft piezoceramic materials, respectively

reduced, while the coercive field intensity $|E_c^\pm|$ increases. The opposite behavior results from donor doping of the piezoceramic material, i.e., $|P_r^\pm|$, the electromechanical coupling factors and s_{pq} are reduced, while $|E_c^\pm|$ decreases. In accordance with the behavior of ferromagnetic materials, acceptor and donor doping of piezoceramic materials leads to *ferroelectrically hard* and *ferroelectrically soft materials*, respectively. Figure 3.14b demonstrates the basic difference of both material types in the hysteresis curve $P(E)$. The larger width of the hysteresis curve of ferroelectrically hard materials stems from the lower domain wall mobility compared to ferroelectrically soft materials. Because the area within the hysteresis curve rates thermal losses, ferroelectrically hard materials usually take in piezoelectric actuators lower temperatures during operation than ferroelectrically soft materials.

Piezoelectric sensors and actuators that contain piezoceramic materials are mainly operated in the polarized state, i.e., state C in Fig. 3.13a and b. The small-signal behavior refers to the linearization. We can describe the linearized behavior of almost any piezoceramic material by the constitutive equations for piezoelectric materials of crystal class 6mm (see (3.32)). Consequently, the three material tensors contain altogether 10 independent entries. The identification of these entries is detailed in Chap. 5. Typical material parameters of piezoceramic materials are listed in Table 3.5. It can be clearly seen that the piezoelectric strain constants are much higher than those of α-quartz.

Lead-Free Materials

Most of the practically used piezoceramic materials like PZT contain a considerable amount of lead, which belongs to hazardous substances. However, the European Union has adopted the directives *Waste Electrical and Electronic Equipment* (WEEE) and *Restriction of the use of certain Hazardous Substances in electrical and electronic equipment* (RoHS) in 2003 [13]. These directives aim at protecting human health and environment by substitution of hazardous substances through safe or safer materials. Similar directives were also adopted in other countries. Some defined applications containing hazardous substances are exempt from the directives because the elimination is technically impracticable. The list of exempted applications, which is reviewed at least every four years, includes the use of lead in electronic ceramic parts such as piezoelectric devices. Nevertheless, the use will be prohibited as soon as there exists a suitable substitution for lead-based materials.

During the last two decades, there has been a lot of research to find lead-free piezoceramic materials, which provide similar performance as lead-based piezoceramic materials in practical applications. The most important lead-free piezoceramic materials are nowadays based on sodium potassium niobate, bismuth sodium titanate and bismuth potassium titanate [11, 38, 39, 47]. The chemical formula of sodium potassium niobate takes the form $K_{1-x}Na_xNbO_3$ (KNN or NKN). With the aid of appropriate doping and modifications of material composition, the sintering of KNN-based piezoceramic materials is facilitated. To some extent, KNN-based materials offer a piezoelectric strain constant $d_{33} > 200\,pm\,V^{-1}$, which is comparable to PZT. This also refers to the electromechanical coupling factors.

The chemical formula of bismuth sodium titanate and bismuth potassium titanate reads as $Bi_{0.5}Na_{0.5}TiO_3$ (BNT) and $Bi_{0.5}K_{0.5}TiO_3$ (BKT), respectively. Pure BNT-based and BKT-based piezoceramic materials are difficult to sinter and exhibit high coercive field intensities $\left|E_c^\pm\right|$ of more than $5\,kV\,mm^{-1}$, which implies a complicated poling process [38]. Furthermore, they show comparatively low piezoelectric properties. Just as in case of KNN-based materials, one can modify BNT-based and BKT-based piezoceramic materials in a way that sintering is facilitated and piezoelectric properties get improved. The binary Bi-based material compositions BNT-BT, BKT-BT, and BNT-BKT as well as the ternary material compositions BNT-BT-BKT and BNT-BT-KNN represent well-known modifications. Here, BT stands for barium titanate, i.e., $BaTiO_3$.

Although the progress in competitive KNN-based, BNT-based, and BKT-based piezoceramic materials is promising, the great breakthrough has not been made yet. This mainly stems from the costly manufacturing process and the worse piezoelectric properties compared with lead-based materials. PZT is, therefore, still dominating the market. However, lead-free piezoceramic materials are already commercially available, e.g., the BNT-based material PIC700 from the company PI Ceramic GmbH [35].

Apart from piezoelectric single crystals (e.g., quartz) and the just mentioned materials (e.g., KNN), there exists a variety of other lead-free materials, which offer piezoelectric properties. Aluminum nitride (AlN) and zinc oxide (ZnO) are prominent representatives for such lead-free materials. Both show a hexagonal crystal structure and belong to the group of wurtzite-structured materials [40]. The polarization direction of AlN and ZnO is set by the crystal orientation. Regardless of whether these materials are used as single crystals or oriented polycrystalline ceramic, the polarization direction cannot be altered after the manufacturing process. That is the reason why both materials do not belong to ferroelectric materials. Compared to other lead-based piezoceramic materials, the achieved electromechanical coupling is much less pronounced. The piezoelectric strain constant d_{33} of AlN and ZnO amounts $\approx 5\,pm\,V^{-1}$ and $\approx 10\,pm\,V^{-1}$, respectively [34]. Nevertheless, by modified material compositions like scandium aluminum nitride (chemical formula $Sc_xAl_{1-x}N$), d_{33} reaches values $>20\,pm\,V^{-1}$ [1, 29]. AlN and its modifications can be sputter-deposited on silicon substrates at moderate temperatures [27]. Because this procedures allows the

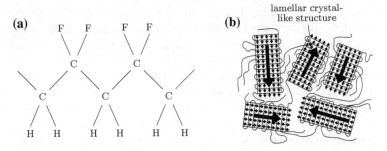

Fig. 3.15 **a** Chain molecules CH_2 and CF_2 of PVDF; **b** arrangement of chain molecules within PVDF material [22]; dipole moments inside lamellar crystal-like structures point in same direction

fabrication of thin piezoelectric films with thicknesses less than 100 nm, AlN-based materials are often utilized in micro electromechanical systems (MEMS).

3.6.3 Polymers

There exist some polymers, which show piezoelectric properties after activation. If such polymers are produced in thin foils, they can be used as mechanically flexible piezoelectric sensors and actuators. Below, we will concentrate on two piezoelectric polymers, namely polyvinylidene fluoride (PVDF) and cellular polypropylene.

PVDF

PVDF materials consist of long-chain molecules (see Fig. 3.15a) that are alternately composed of methylene (chemical formula CH_2) and fluorocarbon (chemical formula CF_2) [36, 44]. Partly, the chain molecules are arranged in thin lamellar crystal-like structures (see Fig. 3.15b). The remaining chain molecules are irregularly arranged which leads to amorphous regions. PVDF belongs, therefore, to the group of semi-crystalline piezoelectric polymers. Inside a single lamellar crystal-like structure, the dipole moments of the molecules point in the same direction. However, since the lamellar crystal-like structures are orientated randomly, PVDF does not offer piezoelectric properties in the initial state.

PVDF can be synthesized from the gaseous vinylidene fluoride via a free-radical polymerization [9]. The polymerization process takes place in suspensions and emulsions at temperatures from 10–150 °C and pressures up to 30 MPa. Melt casting, solution casting, and spin coating represent typical subsequent process steps. The result of the manufacturing process is usually a PVDF film with a thickness in the range of a few microns up to more than 100 μm.

Just as in case of polycrystalline ceramic materials, one has to activate the electromechanical coupling of PVDF films with the aid of an appropriate poling process [14, 36]. The basic idea of poling lies in aligning the lamellar crystal-like structures and, thus, in aligning the dipole moments. When the PVDF film is equipped with electrodes, this can be conducted by applying high electric field intensities up to $100\,\text{kV}\,\text{mm}^{-1}$. The process should be carried out in a heated vacuum chamber or inside an electrically insulating fluid to avoid electrical breakdowns. Poling of PVDF materials is also possible by exploiting corona discharge. In the course of this, the film does not have to be equipped with electrodes. Regardless of the used process, mechanical stretching of PVDF during poling enhances the alignment of the lamellar crystal-like structures and, therefore, the piezoelectric properties get improved. Similar to polycrystalline ceramic materials, PVDF exhibits a certain hysteretic large-signal behavior of the total electric polarization and mechanical strain. That is the reason why PVDF is a ferroelectric polymer.

The piezoelectric properties of PVDF materials are less pronounced than of piezoceramic materials like PZT. Typically, uniaxially stretched PVDF films provide the piezoelectric strain constants $d_{33} \approx -20\,\text{pm}\,\text{V}^{-1}$ and $d_{31} \approx 15\,\text{pm}\,\text{V}^{-1}$ [14]. Note that the signs of d_{33} and d_{31} differ from those of other piezoelectric materials (see Table 3.5). The relative electric permittivity and the electromechanical coupling factors of PVDF films amount $\varepsilon_r = 12$, $k_{33} \approx 0.3$ and $k_{31} \approx 0.1$ [36]. Copolymers of PVDF with trifluoroethylene (P(VDF-TrFE)) offer slightly higher $|d_{ip}|$-values and better electromechanical coupling factors. The comparatively large piezoelectric voltage constant $g_{33} \approx 0.2\,\text{V}\,\text{m}\,\text{N}^{-1}$ and low acoustic impedance[6] of $Z_{\text{aco}} \approx 4 \cdot 10^6\,\text{N}\,\text{s}\,\text{m}^{-3}$ make PVDF films interesting as broadband piezoelectric components for hydrophones and ultrasonic transducers (see, e.g., Sect. 8.1.1).

Cellular Polypropylene

Cellular polypropylene (PP) is currently the best-known representative of the so-called *cellular ferroelectrets*, which are also termed *cellular piezoelectrets* and *voided charged polymers* [3, 36]. It features piezoelectric properties after an appropriate activation procedure. The conventional manufacturing process of cellular PP films starts with inclusions such as air or small grains of sand, which are blown into PP in the molten state. During the following cooling, spherical cavities with approximately $10\,\mu\text{m}$ in diameter emerge around the inclusions. The modified PP material is, subsequently, heated up and extruded in films of $70–100\,\mu\text{m}$ thickness by means of two-dimensional shaking motions below atmospheric pressure. In doing so, the spherical cavities get deformed to lenticular voids with diameters ranging from $10\,\mu\text{m}$ up to $100\,\mu\text{m}$ and heights in the range of $2–10\,\mu\text{m}$ (see Fig. 3.16a). The surface roughness of the PP film is minimized by sealing both the top and bottom surface with homogeneous PP. The geometric dimensions of the lenticular voids (especially their height) are often additionally increased by exposing the PP film to a remarkable overpressure (e.g., 2 MPa) for several hours [45, 49]. When the overpressure is rapidly reduced, the size of the voids will increase permanently.

To activate electromechanical coupling in the resulting PP film, corona discharge is usually applied for poling. This procedure leads to permanent electric charges arising at the interfaces of lenticular voids and surrounding PP (see Fig. 3.16b).

---

[6]The acoustic impedance of piezoceramic materials exceeds $20 \cdot 10^6\,\text{N}\,\text{s}\,\text{m}^{-3}$.

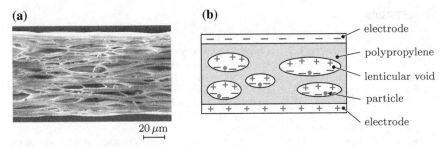

Fig. 3.16 a Scanning electron micrograph of PP film; **b** charge distribution inside PP film at interface of lenticular voids and surrounding PP [37]

Since the electric charges exhibit opposite signs at the top and bottom surface of the interfaces, an electric field is generated inside the voids. The electric field yields an electrostatic force, which reduces the void heights. The film thickness gets, thus, slightly reduced during poling. In the final step of the manufacturing process, the top and bottom surface of the PP film is metallized with thin layers (e.g., aluminum) that serve as electrodes. The reason for the piezoelectric properties of polarized PP films lies in the fact that each charged lenticular void represents a dipole with a certain dipole moment. Just as several other piezoelectric materials, PP films also show a hysteretic large-signal behavior.

A well-established cellular PP film for piezoelectric transducers is the electromechanical film (EMFi) material, which is produced by the company Emfit Ltd [7, 32]. The piezoelectric strain constant d_{33} of PP films can reach values of more than 600 pm V^{-1}, which is comparable to piezoceramic materials (see Table 3.5). In contrast, d_{31} takes extremely low values. Both the relative electric permittivity and the electromechanical coupling factor $k_{33} \approx 0.1$ of PP films are much smaller than that of piezoceramic materials and PVDF. Nevertheless, PP films are well suited for airborne ultrasonic transducers (see Sect. 7.6.1). Besides the high d_{33}-value, this can be ascribed to the outstanding piezoelectric voltage constant $g_{33} \approx 30$ V m N^{-1} and the extremely small acoustic impedance $Z_{\mathrm{aco}} \approx 3 \cdot 10^4$ N s m^{-3} of cellular PP. The main problem of conventional PP films in practical applications lies in the strongly pronounced temperature sensitivity because temperatures higher than 60 °C cause an irreversible change of the piezoelectric properties. However, it is possible to relocate this change to higher temperatures by using cellular ferroelectrets that are based on layered structures of polytetrafluoroethylene (PTFE) and fluoroethylene-propylene (FEP) [50]. The resulting cellular fluorocarbon film shows d_{33}-values of almost 600 pm V^{-1} at a temperature of 120 °C.

References

1. Akiyama, M., Kamohara, T., Kano, K., Teshigahara, A., Takeuchi, Y., Kawahara, N.: Enhancement of piezoelectric response in scandium aluminum nitride alloy thin films prepared by dual reactive cosputtering. Adv. Mater. **21**(5), 593–596 (2009)
2. Baba, A., Searfass, C.T., Tittmann, B.R.: High temperature ultrasonic transducer up to 1000 °C using lithium niobate single crystal. Appl. Phys. Lett. **97**(23) (2010)
3. Bauer, S., Gerhard-Multhaupt, R., Sessler, G.M.: Ferroelectrets: soft electroactive foams for transducers. Phys. Today **57**(2), 37–43 (2004)
4. Budd, K.D., Dey, S.K., Payne, D.A.: Sol-gel processing of PbTiO3, PbZrO3, PZT, and PLZT thin films. Br. Ceram. Proc. **36**, 107–121 (1985)
5. CeramTec GmbH: Product portfolio (2018). https://www.ceramtec.com
6. Chen, Y., Lam, K.H., Zhou, D., Yue, Q., Yu, Y., Wu, J., Qiu, W., Sun, L., Zhang, C., Luo, H., Chan, H.L.W., Dai, J.: High performance relaxor-based ferroelectric single crystals for ultrasonic transducer applications. Sensors (Switzerland) **14**(8), 13730–13758 (2014)
7. Emfit Ltd: Manufacturer of electro-mechanical films (2018). https://www.emfit.com
8. Evers, J., Klüfers, P., Staudigl, R., Stallhofer, P.: Czochralski's creative mistake: a milestone on the way to the gigabit era. Angewandte Chemie - International Edition **42**(46), 5684–5698 (2003)
9. Gallantree, H.R.: Review of transducer applications of polyvinylidene fluoride. IEE Proc. I: Solid State Electron Devices **130**(5), 219–224 (1983)
10. Gautschi, G.: Piezoelectric Sensorics. Springer, Heidelberg (2002)
11. Guo, Y., Kakimoto, K.I., Ohsato, H.: Phase transitional behavior and piezoelectric properties of (Na0.5K0.5)NbO3-LiNbO3 ceramics. Appl. Phys. Lett. **85**(18), 4121–4123 (2004)
12. Haertling, G.H.: Ferroelectric ceramics: History and technology. J. Am. Ceram. Soc. **82**(4), 797–818 (1999)
13. Hedemann-Robinson, M.: The EU directives on waste electrical and electronic equipment and on the restriction of use of certain hazardous substances in electrical and electronic equipment: adoption achieved. Eur. Environ. Law Rev. **12**(2), 52–60 (2003)
14. Heywang, W., Lubitz, K., Wersing, W.: Piezoelectricity: Evolution and Future of a Technology. Springer, Heidelberg (2008)
15. Ikeda, T.: Fundamentals of Piezoelectricity. Oxford University Press (1996)
16. Izyumskaya, N., Alivov, Y.I., Cho, S.J., Morkoç, H., Lee, H., Kang, Y.S.: Processing, structure, properties, and applications of PZT thin films. Crit. Rev. Solid State Mater. Sci. **32**(3–4), 111–202 (2007)
17. Jaffe, B., Cook, W.R., Jaffe, H.: Piezoelectric Ceramics. Academic Press Limited (1971)
18. Jiang, X., Kim, J., Kim, K.: Relaxor-PT single crystal piezoelectric sensors. Crystals **4**(3), 351–376 (2014)
19. Kaltenbacher, M.: Numerical Simulation of Mechatronic Sensors and Actuators - Finite Elements for Computational Multiphysics, 3rd edn. Springer, Heidelberg (2015)
20. Kamlah, M., Böhle, U.: Finite element analysis of piezoceramic components taking into account ferroelectric hysteresis behavior. Int. J. Solids Struct. **38**(4), 605–633 (2001)
21. Laudise, R.A., Nielsen, J.W.: Hydrothermal crystal growth. Solid State Phys. - Adv. Res. Appl. **12**(C), 149–222 (1961)
22. Lerch, R., Sessler, G.M., Wolf, D.: Technische Akustik: Grundlagen und Anwendungen. Springer, Heidelberg (2009)
23. Li, X., Luo, H.: The growth and properties of relaxor-based ferroelectric single crystals. J. Am. Ceram. Soc. **93**(10), 2915–2928 (2010)
24. Liebertz, J.: Synthetic precious stones. Angewandte Chemie International Edition in English **12**(4), 291–298 (1973)
25. Lim, L.C., Shanthi, M., Rajan, K.K., Lim, C.Y.H.: Flux growth of high-homogeneity PMN-PT single crystals and their property characterization. J. Cryst. Growth **282**(3–4), 330–342 (2005)
26. Luo, J., Zhang, S.: Advances in the growth and characterization of relaxor-pt-based ferroelectric single crystals. Crystals **4**(3), 306–330 (2014)

27. Martin, F., Muralt, P., Dubois, M.A., Pezous, A.: Thickness dependence of the properties of highly c-axis textured ain thin films. J. Vac. Sci. Technol. A: Vac. Surf. Films **22**(2), 361–365 (2004)

28. Mason, W.P.: Piezoelectric Crystals and Their Application to Ultrasonics. D. van Nostrand, New York (1949)

29. Mayrhofer, P.M., Euchner, H., Bittner, A., Schmid, U.: Circular test structure for the determination of piezoelectric constants of ScxAl1-xN thin films applying Laser Doppler Vibrometry and FEM simulations. Sens. Actuators A: Phys. **222**, 301–308 (2015)

30. Meggitt Sensing Systems: Product portfolio (2018). https://www.meggittsensingsystems.com

31. Muralt, P., Kohli, M., Maeder, T., Kholkin, A., Brooks, K., Setter, N., Luthier, R.: Fabrication and characterization of PZT thin-film vibrators for micromotors. Sens. Actuators A: Phys. **48**(2), 157–165 (1995)

32. Paajanen, M., Lekkala, J., Kirjavainen, K.: Electromechanical Film (EMFi) - a new multipurpose electret material. Sens. Actuators A: Phys. **84**(1), 95–102 (2000)

33. Park, S.E., Shrout, T.R.: Ultrahigh strain and piezoelectric behavior in relaxor based ferroelectric single crystals. J. Appl. Phys. **82**(4), 1804–1811 (1997)

34. Patel, N.D., Nicholson, P.S.: High frequency, high temperature ultrasonic transducers. NDT Int. **23**(5), 262–266 (1990)

35. PI Ceramic GmbH: Product portfolio (2018). https://www.piceramic.com

36. Ramadan, K.S., Sameoto, D., Evoy, S.: A review of piezoelectric polymers as functional materials for electromechanical transducers. Smart Mater. Struct. **23**(3) (2014)

37. Rupitsch, S.J., Lerch, R., Strobel, J., Streicher, A.: Ultrasound transducers based on ferroelectret materials. IEEE Trans. Dielectr. Electr. Insul. **18**(1), 69–80 (2011)

38. Rödel, J., Jo, W., Seifert, K.T.P., Anton, E.M., Granzow, T., Damjanovic, D.: Perspective on the development of lead-free piezoceramics. J. Am. Ceram. Soc. **92**(6), 1153–1177 (2009)

39. Rödel, J., Webber, K.G., Dittmer, R., Jo, W., Kimura, M., Damjanovic, D.: Transferring lead-free piezoelectric ceramics into application. J. Eur. Ceram. Soc. **35**(6), 1659–1681 (2015)

40. Safari, A., Akdogan, E.K.: Piezoelectric and Acoustic Materials for Transducer Applications. Springer, Heidelberg (2010)

41. Safari, A., Allahverdi, M., Akdogan, E.K.: Solid freeform fabrication of piezoelectric sensors and actuators. J. Mater. Sci. **41**(1), 177–198 (2006)

42. Tichy, J., Erhart, J., Kittinger, E., Privratska, J.: Fundamentals of Piezoelectric Sensorics. Springer, Heidelberg (2010)

43. Torah, R.N., Beeby, S.P., Tudor, M.J., White, N.M.: Thick-film piezoceramics and devices. J. Electroceram. **19**(1), 95–110 (2007)

44. Ueberschlag, P.: PVDF piezoelectric polymer. Sens. Rev. **21**(2), 118–125 (2001)

45. Wegener, M., Wirges, W., Gerhard-Multhaupt, R., Dansachmüller, M., Schwödiauer, R., Bauer-Gogonea, S., Bauer, S., Paajanen, M., Minkkinen, H., Raukola, J.: Controlled inflation of voids in cellular polymer ferroelectrets: Optimizing electromechanical transducer properties. Appl. Phys, Lett. **84**(3), 392–394 (2004)

46. Wolf, F.: Generalisiertes Preisach-Modell für die Simulation und Kompensation der Hysterese piezokeramischer Aktoren. Ph.D. thesis, Friedrich-Alexander-University Erlangen-Nuremberg (2014)

47. Wu, J., Xiao, D., Zhu, J.: Potassium-sodium niobate lead-free piezoelectric materials: past, present, and future of phase boundaries. Chem. Rev. **115**(7), 2559–2595 (2015)

48. Zhang, S., Shrout, T.R.: Relaxor-PT single crystals: observations and developments. IEEE Trans. Ultrason. Ferroelectr. Freq. Control **57**(10), 2138–2146 (2010)

49. Zhang, X., Hillenbrand, J., Sessler, G.M.: Piezoelectric d33 coefficient of cellular polypropylene subjected to expansion by pressure treatment. Appl. Phys. Lett. **85**(7), 1226–1228 (2004)

50. Zhang, X., Hillenbrand, J., Sessler, G.M.: Ferroelectrets with improved thermal stability made from fused fluorocarbon layers. J. Appl. Phys. **101**(5) (2007)

Chapter 4
Simulation of Piezoelectric Sensor and Actuator Devices

Nowadays, computer simulations play a key role in the design, optimization, and characterization of piezoelectric sensor and actuator devices. The primary reason for this lies in the fact that simulations as an important step in computer-aided engineering (CAE) allow to predict the device behavior without fabricating expensive prototypes. Consequently, we can accelerate the device design, which goes hand in hand with reduced development costs and a reduced time to market. Simulations allow, furthermore, to determine quantities (e.g., inside a material), which cannot be measured at reasonable expense.

There exist various approaches for simulations of technical devices. The most important are finite difference methods [17], finite element methods [27], boundary element methods [4] as well as approaches that are based on lumped circuit elements (see Sect. 7.5) [15]. Hereafter, we will exclusively concentrate on the finite element (FE) method because this method is very well suited for numerical simulation of piezoelectric sensor and actuator devices. The main advantages of the FE method are listed below [14].

- *Numerical efficiency:* The FE method yields sparsely populated and symmetric matrices for the resulting algebraic system of equations. Hence, the storage and solution of the algebraic system of equations can be conducted in an efficient way.
- *Complex geometry:* We are able to discretize complex two-dimensional and three-dimensional computational domains with the aid of appropriate finite elements, e.g., triangles and tetrahedron elements.
- *Analysis possibilities:* The FE method can be used for static, transient, harmonic, and eigenfrequency analysis of the investigated problem.

However, the FE method also exhibits certain disadvantages. For instance, the FE method may lead to a considerable discretization effort, especially for large computational domains. Another inherent disadvantage is that each computational domain has to be spatially bounded. If an open domain is required for the numerical simulation

© Springer-Verlag GmbH Germany, part of Springer Nature 2019
S. J. Rupitsch, *Piezoelectric Sensors and Actuators*, Topics in Mining, Metallurgy
and Materials Engineering, https://doi.org/10.1007/978-3-662-57534-5_4

(e.g., free-field radiation of an ultrasonic transducer), we will need special techniques such as absorbing boundary conditions.

One can choose from a large number of FE software packages (FE solvers), which are commercially available. The software packages ANSYS [1], COMSOL Multiphysics [6], NACS [22], and PZFlex [19] are some well-known representatives. They differ in the supported physical fields as well as in the coupling of these fields.

In this chapter, we will study the fundamentals of the FE method, which are important for simulating the behavior of piezoelectric sensors and actuators. The focus lies on linear FE simulations. Section 4.1 deals with the basic steps of the FE method, e.g., Galerkin's method. Subsequently, the FE method will be applied to electrostatics (see Sect. 4.2), the mechanical field (see Sect. 4.3), and the acoustic field (see Sect. 4.4). At the end, we will discuss the coupling of different physical fields because this represents a decisive step for reliable FE simulations of piezoelectric sensors and actuators. For a better understanding, the chapter also contains several simulation examples. Further literature concerning the FE method can be found in [3, 12, 14, 20, 23, 27].

4.1 Basic Steps of Finite Element Method

Figure 4.1 illustrates the basic steps of the FE method. The starting point are always partial differential equations (PDEs) for the physical fields within the investigated technical problem, e.g., the physical fields that are involved in the piezoelectric sensor and actuator. In a next step, this so-called *strong formulation* of the PDE gets multiplied by an appropriate test function, which yields a variational form. After partial integration (integration by parts) of the resulting product over the whole computational domain, we obtain the so-called *weak formulation* of the PDE. In the final step, we apply *Galerkin's method* by approximating both the aimed quantity and the test function with finite elements. This leads to an algebraic system of equations.

In the following subsection, we will detail the basic steps of the FE method for a one-dimensional PDE. Section 4.1.2 treats spatial discretization of computational domains and efficient computation. In Sect. 4.1.3, the difference between Lagrange and Legendre ansatz functions will be discussed. Finally, we introduce an appropriate scheme for time discretization, which is important for transient FE simulations.

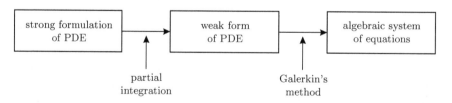

Fig. 4.1 Basic step of FE method

4.1.1 Finite Element Method for a One-Dimensional Problem

In order to demonstrate the idea of the FE method, let us consider a one-dimensional (1-D) hyperbolic partial differential equation. Such partial different equation commonly occurs for mechanical problems. It comprises derivations with respect to time as well as to space. The starting point of the FE method is the strong formulation of the PDE

$$-\frac{\partial^2 u(x,t)}{\partial x^2} + c\frac{\partial^2 u(x,t)}{\partial t^2} = f(x,t) \; . \tag{4.1}$$

Here, the expression $u(x,t)$ represents the aimed quantity (e.g., mechanical displacement) depending on both space x and time t. The term c stands for a constant, and $f(x,t)$ is a known excitation (source) term that varies with space and time. In addition to (4.1), boundary and initial conditions are required to uniquely solve the hyperbolic PDE. For the spatial computational domain $x \in [a,b]$ and the investigated time interval $t \in [0,T]$, appropriate conditions are

boundary conditions: $u(a,t) = u_a$ and $u(b,t) = u_b$

initial conditions: $u(x,0) = u_0(x)$ and $\left.\dfrac{\partial u(x,t)}{\partial t}\right|_{t=0} = \dot{u}_0(x)$

and $\left.\dfrac{\partial^2 u(x,t)}{\partial t^2}\right|_{t=0} = \ddot{u}_0(x) \; .$

u_a and u_b refer to constant boundary conditions for the aimed quantity $u(x,t)$, which have to be fulfilled at any time. Oftentimes, these conditions are called Dirichlet boundary conditions. In case of $u_a = 0$, we have a homogeneous and, otherwise, i.e., $u_a \neq 0$, an inhomogeneous Dirichlet boundary conditions. Apart from such boundary conditions, homogeneous or inhomogeneous Neumann boundary conditions can be specified defining the first-order derivative $\partial u/\partial x$ with respect to space at the boundary of the spatial computational domain. The initial conditions $u_0(x)$, $\dot{u}_0(x)$, and $\ddot{u}_0(x)$ indicate $u(x,t)$, its first-order and second-order derivate with respect to time at $t=0$ in the spatial computational domain $x \in [a,b]$.

To achieve a clearly arranged form of partial differential equations in strong formulation including boundary conditions (BC) as well as initial conditions (IC), we introduce a compact scheme, which is also used later on. For the 1-D hyperbolic PDE, this scheme reads as

PDE $-\dfrac{\partial^2 u(x,t)}{\partial x^2} + c\dfrac{\partial^2 u(x,t)}{\partial t^2} = f(x,t)$

$x \in [a,b] \; ; \quad [a,b] \subset \mathbb{R}$

$t \in [0,T] \; ; \quad [0,T] \subset \mathbb{R}$

Given	$f : [a, b] \times [0, T] \to \mathbb{R}$
	$c = \text{const.}$
BC	$u(a, t) = u_a \quad \text{on} \quad a \times [0, T]$
	$u(b, t) = u_b \quad \text{on} \quad b \times [0, T]$
IC	$u(x, 0) = u_0(x) \quad \forall x \in [a, b]$
	$\dot{u}(x, 0) = \dot{u}_0(x) \quad \forall x \in [a, b]$
	$\ddot{u}(x, 0) = \ddot{u}_0(x) \quad \forall x \in [a, b]$
Find	$u(x, t) : (a, b) \times (0, T] \to \mathbb{R} \ .$

Weak Formulation of the PDE

A fundamental step of the FE method is to transform the PDE from its strong formulation into the weak formulation. Thereby, we multiply the original PDE with an arbitrary test function $w(x)$ and integrate the resulting product over the whole spatial computational domain. The test function has to fulfill only two criteria: (i) $w(x)$ vanishes at Dirichlet boundaries and (ii) the first-order derivative of $w(x)$ with respect to space exists in the weak sense. For the hyperbolic PDE in (4.1), the multiplication by the test function $w(x)$ and integration over the spatial computational domain $x \in [a, b]$ results in

$$\int_a^b w(x) \left[-\frac{\partial^2 u(x, t)}{\partial x^2} + c \frac{\partial^2 u(x, t)}{\partial t^2} - f(x, t) \right] dx = 0 \ . \tag{4.2}$$

The first term can be simplified by means of partial integration, which has to be replaced for higher dimensional PDEs with Green's first integration theorem

$$\int_a^b w(x) \frac{\partial^2 u(x, t)}{\partial x^2} dx = \left[w(x) \frac{\partial u(x, t)}{\partial x} \right]_a^b - \int_a^b \frac{\partial w(x)}{\partial x} \frac{\partial u(x, t)}{\partial x} dx \ . \tag{4.3}$$

Since the test function $w(x)$ vanishes at Dirichlet boundaries (i.e., $w(a) = w(b) = 0$), the weak formulation of (4.1) finally becomes

$$\int_a^b \left[\frac{\partial w(x)}{\partial x} \frac{\partial u(x, t)}{\partial x} + c w(x) \frac{\partial^2 u(x, t)}{\partial t^2} \right] dx = \int_a^b w(x) f(x) dx \ . \tag{4.4}$$

In contrast to the strong formulation, the dimension of the spatial derivative of the aimed quantity $u(x, t)$ has been reduced by one. As the weak formulation incorporates Neumann boundary conditions $\partial u(x, t) / \partial x$, they are called natural conditions.

Dirichlet boundary conditions demand additional consideration in further steps of the FE method and are, therefore, frequently referred to as essential conditions.

Galerkin's Method

Within Galerkin's Method, the spatial computational domain is subdivided into cells, the so-called *finite elements*. In case of the studied 1-D hyperbolic PDE, we divide the domain $[a, b]$ into M sufficiently small intervals $[x_{i-1}, x_i] \ \forall i = 1, \ldots, M$ where each interval border x_i is a node. The chosen intervals have to satisfy the properties:

- Ascending order of node positions, i.e., $x_{i-1} < x_i \ \forall i = 1, \ldots, M$.
- Computational domain is completely covered, i.e., $[a, b] = \bigcup_{i=1}^{M} [x_{i-1}, x_i]$.
- No intersection of intervals, i.e., $[x_{i-1}, x_i] \cap [x_{j-1}, x_j] = 0 \ \forall i \neq j$.

Without limiting the generality, the 1-D computational domain can be equidistantly discretized yielding the node positions (interval width h)

$$x_i = a + ih \ \forall i = 0, \ldots, M \quad \text{with } h = \frac{b-a}{M}. \tag{4.5}$$

Based on the spatial discretization of the computational domain, we subsequently approximate both the aimed quantity $u(x, t)$ and the test function $w(x)$. Since, here, only the spatial properties of $u(x, t)$ are investigated, the dependency on time t is omitted.

Spatial approximation in the FE method is conducted with a linear combination of ansatz functions[1] featuring local support. For the 1-D problem, each of these ansatz functions is just different from zero in the interval $[x_{i-1}, x_{i+1}]$. Due to the fact that the weak formulation contains solely spatial derivatives up to one, several types of ansatz functions are suitable. For the sake of simplicity, let us choose piecewise linear hat functions $N_i(x)$ defined as (see Fig. 4.2)

$$N_i(x) = \begin{cases} 0 & \text{for} \quad x_0 \leq x \leq x_{i-1} \\ \dfrac{x - x_{i-1}}{h} & \text{for} \quad x_{i-1} < x \leq x_i \\ \dfrac{x_{i+1} - x}{h} & \text{for} \quad x_i < x \leq x_{i+1} \\ 0 & \text{for} \quad x_{i+1} < x \leq x_M \end{cases} \tag{4.6}$$

for $i = 1, \ldots, M - 1$. The functions $N_0(x)$ and $N_M(x)$ at the boundary of the computational domain are defined in a similar manner.

The linear hat functions fulfill the required properties $N_i(x_j) = 1 \ \forall \ i = j$ and $N_i(x_j) = 0 \ \forall \ i \neq j$. By means of these functions, the approximations $\mathsf{u}(x)$ and $\mathsf{w}(x)$ of $u(x)$ and $w(x)$, respectively, are given by

---

[1] Ansatz functions are also called shape, basis, interpolation, or finite functions.

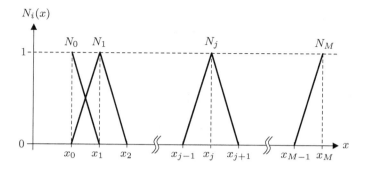

Fig. 4.2 Piecewise linear hat functions $N_i(x)$ utilized to approximate aimed quantity of considered 1-D hyperbolic PDE ($x_0 = a$ and $x_M = b$)

$$u(x) \approx \mathsf{u}(x) = \sum_{i=1}^{M-1} N_i(x)\,\mathsf{u}_i + N_0(x)\,u_a + N_M(x)\,u_b \qquad (4.7)$$

$$w(x) \approx \mathsf{w}(x) = \sum_{i=1}^{M-1} N_i(x)\,\mathsf{w}_i \qquad (4.8)$$

where $\mathsf{u}_i = \mathsf{u}(x_i)$ and $\mathsf{w}_i = \mathsf{w}(x_i)$ represent approximated values at node x_i. Note that between two neighboring nodes, the approximation of $u(x)$ and $w(x)$ depends on the chosen ansatz functions. In the particular case, the interim values are evaluated according to linear equations. However, because we are able to apply ansatz functions of higher order (see Sect. 4.1.3), the approximation can be performed more precisely.

Inserting (4.7) and (4.8) in the weak formulation (4.4) of the PDE leads to the terms (argument x omitted)

$$\int_a^b \frac{\partial w}{\partial x}\frac{\partial u}{\partial x}\mathrm{d}x = \int_a^b \frac{\partial}{\partial x}\left[\sum_{i=1}^{M-1} N_i\mathsf{w}_i\right]\frac{\partial}{\partial x}\left[\sum_{j=1}^{M-1} N_j\mathsf{u}_j + N_0 u_a + N_M u_b\right]\mathrm{d}x$$

$$= \sum_{i=1}^{M-1}\mathsf{w}_i\left\{\sum_{j=1}^{M-1}\mathsf{u}_j\int_a^b \frac{\partial N_i}{\partial x}\frac{\partial N_j}{\partial x}\mathrm{d}x\right.$$

$$\left. + \int_a^b \frac{\partial N_i}{\partial x}\left[\frac{\partial N_0}{\partial x}u_a + \frac{\partial N_M}{\partial x}u_b\right]\mathrm{d}x\right\} \qquad (4.9)$$

$$\int_a^b cw\frac{\partial^2 u}{\partial t^2}\mathrm{d}x = \int_a^b c\left[\sum_{i=1}^{M-1} N_i\mathsf{w}_i\right]\left[\sum_{j=1}^{M-1} N_j\frac{\partial^2\mathsf{u}_j}{\partial t^2}\right]\mathrm{d}x$$

$$= \sum_{i=1}^{M-1} \mathsf{w}_i \left\{ \sum_{j=1}^{M-1} \frac{\partial^2 \mathsf{u}_j}{\partial t^2} \int_a^b c N_i N_j \mathrm{d}x \right\} \tag{4.10}$$

$$\int_a^b w f \mathrm{d}x = \int_a^b \left[\sum_{i=1}^{M-1} N_i \mathsf{w}_i \right] f \mathrm{d}x = \sum_{i=1}^{M-1} \mathsf{w}_i \left\{ \int_a^b N_i f \mathrm{d}x \right\} . \tag{4.11}$$

Thereby, integrals and sums were interchanged which is possible since u_j and w_i are constants and, therefore, do not depend on space. We may also omit sums over the approximated test function (i.e., $\sum_{i=1}^{M-1} \mathsf{w}_i$) due to the fact that $w(x)$ can be chosen almost arbitrarily and these sums appear identical in all terms. As a result, the expressions in the bracket $\{\bullet\}$ exclusively remain from (4.9)–(4.11). By additionally introducing the matrix and vector components

$$\mathsf{M}_{ij} = \int_a^b c N_i(x) N_j(x) \, \mathrm{d}x \tag{4.12}$$

$$\mathsf{K}_{ij} = \int_a^b \frac{\partial N_i(x)}{\partial x} \frac{\partial N_j(x)}{\partial x} \mathrm{d}x \tag{4.13}$$

$$\mathsf{f}_i = \int_a^b N_i(x) f \mathrm{d}x - \int_a^b \frac{\partial N_i(x)}{\partial x} \left[\frac{\partial N_0(x)}{\partial x} u_a + \frac{\partial N_M(x)}{\partial x} u_b \right] \mathrm{d}x , \tag{4.14}$$

one is able to rewrite the resulting *algebraic system of equations* in matrix form (second-order time derivative $\ddot{\mathbf{u}} = \partial^2 \mathbf{u}/\partial t^2$)

$$\mathbf{M} \ddot{\mathbf{u}} + \mathbf{K} \mathbf{u} = \mathbf{f} , \tag{4.15}$$

which is still continuous in time. Herein, $\mathbf{M}$ and $\mathbf{K}$, both of dimension $(M-1) \times (M-1)$, stand for the mass matrix and stiffness matrix, respectively. Since the ansatz functions feature local support, the matrices are sparsely populated. The vector $\mathbf{f}$ with length $M-1$ is the right-hand side of the algebraic system of equations. The solution of (4.15) provides the vector $\mathbf{u} = [\mathsf{u}_1, \mathsf{u}_2, \ldots, \mathsf{u}_{M-1}]^t$ with length $M-1$ containing approximated results of the aimed quantity $u(x_i)$ for every node x_i.

4.1.2 Spatial Discretization and Efficient Computation

Usually, one has to deal with two-dimensional (2-D) and three-dimensional (3-D) problems in practice, e.g., numerical simulations for sensors and actuators. Thus, the spatial computational domain Ω cannot be subdivided into line intervals but demands

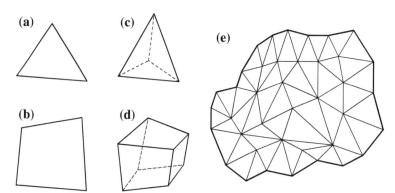

Fig. 4.3 Finite elements to spatially discretize 2-D and 3-D computational domains Ω: **a** triangular element for $\mathbb{R}^2$; **b** quadrilateral element for $\mathbb{R}^2$; **c** tetrahedron element for $\mathbb{R}^3$; **d** hexahedron element for $\mathbb{R}^3$; **e** spatial discretization (also denoted as mesh or computational grid) of a 2-D computational domain by means of triangles

alternative finite elements in $\mathbb{R}^2$ and $\mathbb{R}^3$. In the same manner as the line intervals for 1-D problems, the finite elements have to satisfy the properties (i) complete covering of the computational domain and (ii) no intersection of elements. Figure 4.3 shows appropriate finite elements for 2-D (triangular and quadrilateral elements) and 3-D (tetrahedron and hexahedron elements) spaces. Due to local support of the ansatz functions on the nodes within the elements, they are frequently referred to as nodal (Lagrangian) finite elements.[2]

With a view to assembling **f**, **M**, and **K** of the algebraic system of equations (4.15), we have to calculate spatial derivatives of the ansatz functions and integrals over the subdomain of a finite element (see, e.g., (4.12)). Especially for fine spatial discretizations in 3-D problems, this procedure provokes remarkable computational effort. To optimize the assembly, the so-called parent elements are introduced which exhibit uniform geometric dimensions. The parent elements are defined in a local coordinate system (see Fig. 4.4). For such elements, we are able to efficiently evaluate both, spatial derivatives of ansatz functions and their numerical integrations. By means of a unique transform, the parent element defined in local coordinates gets subsequently transformed into global coordinates of the spatial computational domain Ω. After assembling the equations for all finite elements to global system matrices, the resulting algebraic system of equations in (4.15) can finally be efficiently solved with problem-specific algebraic methods, e.g., multigrid methods [5, 14].

---

[2]For 3-D electromagnetic problems, edge (Nédélec) finite elements are often applied instead of nodal (Lagrangian) elements.

4.1.3 Ansatz Functions

As a matter of principle, FE simulations require ansatz functions featuring local support. This means that each ansatz function has to be just different from zero in the considered finite element and in the immediately neighboring ones. Lagrange and Legendre ansatz functions provide local support, and therefore, they are applicable for the FE method [2, 11, 14]. In the following, let us take a closer look at these most important categories of ansatz functions.

Lagrange Ansatz Functions

Lagrange ansatz functions are widely utilized for FE simulations in all sectors of engineering. The underlying procedure is commonly referred to as h-version of the FE method or in abbreviated form h-FEM. For the 1-D case (see Sect. 4.1.1), the Lagrange ansatz function $N_i^{p_d}(\xi)$ for node i is defined as

$$N_i^{p_d}(\xi) = \prod_{\substack{j=1 \\ j \neq i}}^{p_d+1} \frac{\xi - \xi_j}{\xi_i - \xi_j} \tag{4.16}$$

leading to $N_i^{p_d}(\xi_i) = 1$. Here, p_d stands for the order (i.e., polynomial degree) of the Lagrange polynomial and ξ_i is the position of the ith node within the finite element. As (4.16) shows, each node has its own ansatz function. If we assume 1-D parent elements (see Fig. 4.4a) ranging from -1 to 1 and nodes that are equally distributed, the node positions will result in

$$\xi_i = -1 + \frac{2(i-1)}{p_d} \quad i = 1, \ldots, p_d + 1 . \tag{4.17}$$

Thus, the number of nodes $n_{\text{nodes}} = p_d + 1$ within a single finite element increases for increasing order of the Lagrange polynomial. Moreover, the relation

$$\sum_{i=1}^{p_d+1} N_i^{p_d}(\xi) = 1 \quad \forall \, \xi \in [-1, 1] . \tag{4.18}$$

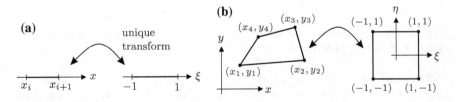

Fig. 4.4 a Original finite element and parent element for 1-D; global coordinate x; local coordinate ξ; **b** original quadrilateral finite element and parent element for 2-D; global coordinate system (x, y); local coordinate system (ξ, η)

Table 4.1 Lagrange ansatz functions $N_i^{p_d}(\xi)$ and node positions ξ_i for polynomial degree 1 and 2 in case of 1-D parents elements, i.e., $\xi \in [-1, 1]$

	$N_1^{p_d}(\xi)$	$N_2^{p_d}(\xi)$	$N_3^{p_d}(\xi)$	Node positions ξ_i
$p_d = 1$	$\dfrac{1-\xi}{2}$	$\dfrac{1+\xi}{2}$		$[-1; 1]$
$p_d = 2$	$\dfrac{\xi(\xi-1)}{2}$	$(1-\xi)(1+\xi)$	$\dfrac{\xi(\xi+1)}{2}$	$[-1; 0; 1]$

Lagrange ansatz functions
h-FEM

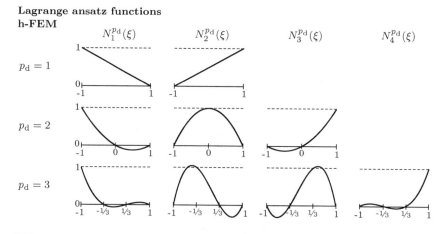

Legendre ansatz functions
p-FEM

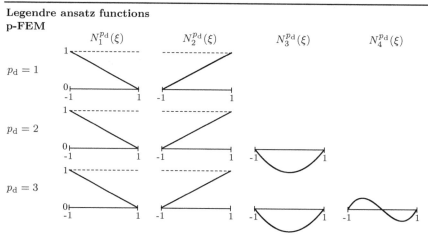

Fig. 4.5 Lagrange and Legendre ansatz functions $N_i^{p_d}(\xi)$ for 1-D up to polynomial degree $p_d = 3$ and node positions ξ_i within parent element $\xi \in [-1, 1]$

holds which means, in other words, that the complete set of ansatz functions is needed to compute the aimed quantity between the nodes. Table 4.1 contains the resulting Lagrange ansatz functions $N_i^{p_d}(\xi)$ and the node positions ξ_i for $p_d = 1$ as well as $p_d = 2$. Figure 4.5 displays the ansatz function up to $p_d = 3$.

By means of increasing order p_d in h-FEM, we can choose coarser spatial discretization for the computational domain without losing precision in simulations. In doing so, the total number of finite elements n_{elem} is reduced and one would expect remarkably reduced computational effort. However, due to additional nodes within the finite elements for $p_d > 1$, the number of unknown quantities for each element becomes larger which has to be considered in the view of computation time.

Legendre Ansatz Functions

FE simulations utilizing Legendre ansatz functions are oftentimes called p-version of the FE method (p-FEM). The Legendre ansatz functions $N_i^{p_d}(\xi)$ for a parent element ($\xi \in [-1, 1]$; see Fig. 4.4) are defined as

$$N_1^{p_d}(\xi) = \frac{1-\xi}{2} \quad N_2^{p_d}(\xi) = \frac{1+\xi}{2} \tag{4.19}$$

$$N_i^{p_d}(\xi) = \phi_{i-1}(\xi) \quad \forall \, i = 3, \ldots, p_d + 1 . \tag{4.20}$$

The expression $\phi_i(\xi)$ represents the integrated Legendre polynomial L_i and results from

$$\phi_i(\xi) = \int\limits_{-1}^{\xi} L_{i-1}(x) \, dx \tag{4.21}$$

$$L_i(x) = \frac{1}{2^i i!} \frac{\partial^i}{\partial x^i} (x^2 - 1)^i . \tag{4.22}$$

Similar to Lagrange ansatz functions, the amount of Legendre ansatz functions rises for increasing polynomial degree p_d. The ansatz functions of lower order remain, however, unchanged which means that the set of ansatz functions for $p_d + 1$ includes all ansatz functions of order p_d. On account of this fact, they are also named hierarchical ansatz functions.

There are various benefits of p-FEM over h-FEM. For instance, we can use different orders of ansatz functions in p-FEM for neighboring finite elements. Besides, the polynomial degree p_d may be altered in different spatial directions for 2-D and 3-D simulations, which is especially useful to avoid the so-called locking effects in FE simulations for thin mechanical structures (e.g., cantilevers). In case of equal polynomial degrees in different spatial directions, the simulation procedure is referred to as isotropic p-FEM and otherwise anisotropic p-FEM.

The accuracy of numerical simulations in p-FEM is basically achieved by increasing p_d instead of choosing a finer computational grid. As a result, p-FEM requires a less amount of nodes to spatially discretize the computational domain than h-FEM. Nevertheless, we have to handle an increasing number of unknowns for the nodes since each ansatz function is weighted individually. Furthermore, p-FEM will only make sense if the investigated geometry allows a coarse computational grid. Despite these shortcomings, the targeted use of p-FEM leads to strongly reduced calculation times for smooth and high-frequency problems, e.g., numerical simulation of high-frequency fields in simply shaped mechanical structures.

4.1.4 Time Discretization

Up to now, we have treated solely the dependence on space for the FE method. However, one is mainly concerned with physical processes depending also on time. In order to incorporate time in the FE procedure, an appropriate time discretization is required. Let us discuss the fundamentals of time discretization for the 1-D hyperbolic PDE (4.1). Similar to 1-D space discretization, the investigated time interval $[0, T]$ is subdivided into N sufficiently small subintervals (time step t_i)

$$[0, T] = \bigcup_{i=1}^{N} [t_{i-1}, t_i] \qquad \text{with } 0 < t_1 < t_2 < \ldots < t_{N-1} < t_N = T . \qquad (4.23)$$

Without limiting the generality, we may assume equidistant time sampling, i.e., a constant time step size Δt given by

$$\Delta t = t_i - t_{i-1} = \frac{T}{N} . \qquad (4.24)$$

The evaluation of the time discretization within the FE method is prevalently performed according to the so-called *Newmark scheme* [12, 14]. For the spatially discretized hyperbolic PDE in matrix form (4.15), the Newmark scheme contains three substeps, which are briefly explained below. To achieve compact expressions, we use the nomenclature

$$\mathbf{u}(t_i) = \left[u_1(t_i), \ldots, u_{M-1}(t_i) \right]^t = \mathbf{u}^{(i)} = \left[u_1^{(i)}, \ldots, u_{M-1}^{(i)} \right]^t . \qquad (4.25)$$

1. **Compute predictor step**: Starting from the known quantities $\mathbf{u}^{(i)}$, $\dot{\mathbf{u}}^{(i)}$, and $\ddot{\mathbf{u}}^{(i)}$ for time step t_i, the predicted values result from

$$\tilde{\mathbf{u}} = \mathbf{u}^{(i)} + \Delta t \, \dot{\mathbf{u}}^{(i)} + \frac{(\Delta t)^2}{2}(1 - 2\beta_N) \, \ddot{\mathbf{u}}^{(i)} \qquad (4.26)$$

$$\tilde{\dot{\mathbf{u}}} = \dot{\mathbf{u}}^{(i)} + (1 - \gamma_N) \, \Delta t \, \ddot{\mathbf{u}}^{(i)} . \qquad (4.27)$$

2. **Solve algebraic system of equations**: The predicted value $\tilde{\mathbf{u}}$ is then utilized to pose an algebraic system of equations

$$\mathbf{M}^{\star} \ddot{\mathbf{u}}^{(i+1)} = \mathbf{f}^{(i+1)} - \mathbf{K} \tilde{\mathbf{u}} \qquad (4.28)$$

$$\mathbf{M}^{\star} = \mathbf{M} + \beta_N (\Delta t)^2 \, \mathbf{K} . \qquad (4.29)$$

Here, $\mathbf{M}^\star$ represents the effective mass matrix.[3] The solution of (4.28) yields $\ddot{\mathbf{u}}^{(i+1)}$ for the subsequent time step t_{i+1}.

3. **Perform corrector step**: By means of $\ddot{\mathbf{u}}^{(i+1)}$, we are able to correct the predicted values $\tilde{\mathbf{u}}$ and $\dot{\tilde{\mathbf{u}}}$

$$\mathbf{u}^{(i+1)} = \tilde{\mathbf{u}} + \beta_{\mathrm{N}}(\Delta t)^2\,\ddot{\mathbf{u}}^{(i+1)} \tag{4.30}$$

$$\dot{\mathbf{u}}^{(i+1)} = \dot{\tilde{\mathbf{u}}} + \gamma_{\mathrm{N}}\Delta t\,\ddot{\mathbf{u}}^{(i+1)}\,. \tag{4.31}$$

As a result, $\mathbf{u}^{(i+1)}$, $\dot{\mathbf{u}}^{(i+1)}$ as well as $\ddot{\mathbf{u}}^{(i+1)}$ are now known and the predicted values can be calculated for time step t_{i+2}.

The parameters β_{N} and γ_{N} determine the type of integration with respect to time and, moreover, the stability of the integration procedure. For example, $\beta_{\mathrm{N}} = 0$ and $\gamma_{\mathrm{N}} = 0.5$ yield an explicit time integration. $\beta_{\mathrm{N}} = 0.25$ and $\gamma_{\mathrm{N}} = 0.5$ result in an implicit time integration, which is unconditionally stable (A-stable) for all choices of time step sizes Δt. Note that stability of the integration procedure is necessary but not sufficient for precise simulation results. In case of a rough time discretization, one is concerned with numerical dispersion yielding, e.g., distorted pulses for transient simulations. To avoid such numerical dispersions, the time step size has to be chosen sufficiently small. In summary, FE simulations incorporating time discretization provide an approximation of the aimed quantity with respect to both space and time.

4.2 Electrostatics

In order to apply the FE method to quasi-static electric or electrostatic fields (see Sect. 2.1.2), an appropriate PDE is required which results from the combination of (2.9, p. 10)–(2.11). The PDE reads as

$$-\nabla \cdot \varepsilon \nabla V_{\mathrm{e}} = q_{\mathrm{e}} \tag{4.32}$$

where ε is the electric permittivity, V_{e} the electric scalar potential, and q_{e} the volume charge density, respectively. In compact form, the strong formulation of the PDE in 3-D becomes (computational domain Ω)[4]

PDE	$-\nabla \cdot \varepsilon \nabla V_{\mathrm{e}} = q_{\mathrm{e}}$
	$\Omega \subset \mathbb{R}^3$
Given	$q_{\mathrm{e}} : \Omega \to \mathbb{R}$
	$\varepsilon : \Omega \to \mathbb{R}$

---

[3] Alternatively to the effective mass matrix $\mathbf{M}^\star$, the Newmark scheme can be defined for the effective stiffness matrix $\mathbf{K}^\star$.

[4] The argument for the position $\mathbf{r}$ is mostly omitted in the following.

$$\text{BC} \qquad \frac{\partial V_e}{\partial \mathbf{n}} = 0 \;\; \text{on} \;\; \partial \Omega$$

$$\text{Find} \qquad V_e : \Omega \to \mathbb{R} \,.$$

The term $\mathbf{n}$ indicates the normal vector with respect to the boundary $\partial\Omega$ of Ω. Due to the fact that electrostatic fields do not depend on time, initial conditions are useless here. This also holds for quasi-static electric fields.

According to the first basic step of the FE method, the strong form (4.32) is transformed into weak form with a scalar test function $w(\mathbf{r})$. After applying Green's first integration theorem, the weak form results in

$$\int_{\Omega} \varepsilon \nabla w \cdot \nabla V_e d\Omega - \int_{\Omega} w q_e d\Omega = 0 \,. \tag{4.33}$$

Spatial discretization (Galerkin's method) of $w(\mathbf{r})$ and $V_e(\mathbf{r})$ yields then the algebraic system of equations

$$\mathbf{K}_{V_e} \mathbf{v}_e = \mathbf{f}_{V_e} \,. \tag{4.34}$$

If we utilize Lagrange ansatz functions, the vector $\mathbf{v}_e$ will contain the approximated values of V_e at the nodes of the spatially discretized computational domain Ω. The stiffness matrix $\mathbf{K}_{V_e}$ and the right-hand side $\mathbf{f}_{V_e}$ of (4.34) are given by (ansatz functions N_i)

$$\mathbf{K}_{V_e} = \bigwedge_{l=1}^{n_{elem}} \mathbf{K}^l \;; \qquad \mathbf{K}^l = \left[k_{ij}^l \right] \;; \qquad k_{ij}^l = \int_{\Omega^l} \varepsilon (\nabla N_i)^t \, \nabla N_j d\Omega \tag{4.35}$$

$$\mathbf{f}_{V_e} = \bigwedge_{l=1}^{n_{elem}} \mathbf{f}^l \;; \qquad \mathbf{f}^l = \left[f_i^l \right] \;; \qquad f_i^l = \int_{\Omega^l} \nabla N_i q_e d\Omega \,. \tag{4.36}$$

n_{elem} stands for the number of finite elements (e.g., hexahedra in $\mathbb{R}^3$) used to spatially discretize Ω. For each finite element (index l), the element matrix $\mathbf{K}^l$ is composed of the components k_{ij}^l, which result from the integral over the element domain Ω^l. The same procedure is carried out to calculate the right-hand side $\mathbf{f}^l$. At this point, it should be emphasized that the dependence of ε as well as of q_e on space has to be considered. Finally, $\mathbf{K}^l$ and $\mathbf{f}^l$ are fully assembled (assembling operator $\bigwedge$) for all elements leading to $\mathbf{K}_{V_e}$ and $\mathbf{f}_{V_e}$, respectively.

Example

As a practical example for electrostatics, let us study a plate capacitor with two identical electrodes of rectangular shape, which are assumed to be infinitely thin. The dielectric medium between the electrodes exhibits the relative permittivity $\varepsilon_{plate} = 20$. Figure 4.6 displays the geometric arrangement in two views. By means of the plate length $l_{plate} = 10\,\text{mm}$, the plate width $w_{plate} = 5\,\text{mm}$, and the plate distance

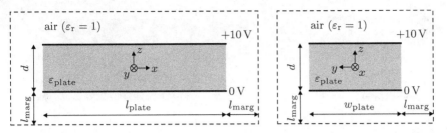

Fig. 4.6 Two views showing geometric arrangement of rectangular plate capacitor with dielectric material between electrodes (relative permittivity ε_r); drawings are not in scale

$d = 1\,\mathrm{mm}$, we can analytically approximate the capacity value C_{plate} with (2.23, p. 13), which leads to

$$C_{\mathrm{plate}} = \frac{\varepsilon_{\mathrm{plate}}\,\varepsilon_0\,l_{\mathrm{plate}}\,w_{\mathrm{plate}}}{d} = 8.854\,\mathrm{pF}\ . \tag{4.37}$$

If one applies this simple analytical approximation, stray fields outside the electrodes will be neglected. Consequently, the approximated C_{plate} is too small. The impact of these stray fields on C_{plate} can be determined with the aid of the FE method for the 3-D case. In doing so, the computational domain Ω has to contain a certain border area surrounding the plate capacitor (see Fig. 4.6) because the boundary condition $\partial V_e/\partial \mathbf{n}$ implies field lines that are in parallel to the boundary $\partial\Omega$. The border area is assumed to be air. Stray fields will not be formed in the simulation results when we do not use such border area. Without limiting the generality, the computational domain of the plate capacitor is extended by the margin l_{marg} on each side and in each spatial direction. The overall computational domain comprising plate capacitor as well as border area features, thus, the geometric dimensions $(l_{\mathrm{plate}} + 2l_{\mathrm{marg}}) \times (w_{\mathrm{plate}} + 2l_{\mathrm{marg}}) \times (d + 2l_{\mathrm{marg}})$. The following FE simulations were performed with quadratic Lagrange ansatz functions, i.e., h-FEM with $p_d = 2$.

Figure 4.7 shows the simulation result for the electric scalar potential $V_e(x, z)$ in the xz-plane. Thereby, the margin was set to $l_{\mathrm{marg}} = 5\,\mathrm{mm}$. The bottom electrode was set to ground (i.e., $0\,\mathrm{V}$), whereas the electric potential of the top electrode amounts $+10\,\mathrm{V}$. It can be clearly seen that there arise considerable electric potentials in the border area of the plate capacitor. Owing to this fact, the stray fields in the border area are not negligible.

The obtained simulation result also enables the calculation of C_{plate}. The calculation is based on the energy density w_{elec} of the electric field, which is with the electric field intensity $\mathbf{E} = -\nabla V_e$ and the electric flux density $\mathbf{D}$ given by

$$w_{\mathrm{elec}} = \frac{1}{2}\mathbf{E} \cdot \mathbf{D} = \frac{1}{2}\varepsilon_r\varepsilon_0\mathbf{E} \cdot \mathbf{E} = \frac{1}{2}\varepsilon_r\varepsilon_0\,\|\mathbf{E}\|_2 = \frac{1}{2}\varepsilon_r\varepsilon_0\,\|\nabla V_e\|_2\ . \tag{4.38}$$

Fig. 4.7 Simulated electric potential $V_e(x, z)$ in xz-plane (i.e., $y = 0$ mm) of plate capacitor; margin $l_{marg} = 5$ mm; 3-D computational gird comprises 161693 tetrahedron elements

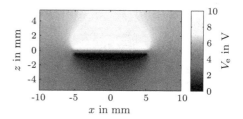

Table 4.2 Resulting capacity value C_{plate} from FE simulation with respect to considered margin l_{marg} of plate capacitor (see Fig. 4.6)

l_{marg} in mm	0.1	0.5	1	5	10
C_{plate} in pF	8.880	8.950	8.989	9.055	9.070

Since the total electric energy W_{elec} depends on C_{plate} and the potential difference U between top and bottom electrodes, we can exploit w_{elec} to determine C_{plate}. The underlying relations read as

$$W_{elec} = \frac{1}{2} C_{plate} U^2 \longrightarrow C_{plate} = \frac{2 W_{elec}}{U^2} . \tag{4.39}$$

W_{elec} results from summing the energy W_{elec}^i of all finite elements within Ω, i.e.,

$$W_{elec} = \sum_{i=1}^{n_{elem}} W_{elec}^i \quad \text{with} \quad W_{elec}^i = \frac{1}{2} \int_{\Omega^i} \varepsilon_r \varepsilon_0 \| \nabla V_e \|_2 \, d\Omega . \tag{4.40}$$

Table 4.2 lists calculated capacity values for different margins. If the margin is small (e.g., $l_{marg} = 0.1$ mm), C_{plate} will be close to the approximation in (4.37). This follows from the neglected stray fields. As expected, C_{plate} increases for increasing l_{marg}. For the considered configuration, margins greater than 10 mm cause only a slight increase in C_{plate}. In general, the computation time of the FE simulation drastically increases for increasing l_{marg} because the number of finite elements n_{elem} grows rapidly.

4.3   Mechanical Field

The PDE in linear continuum mechanics describing the mechanical field of a solid deformable body in an infinitely small fraction at position[5] $\mathbf{r}$ is defined as (see 2.68, p. 23 in Sect. 2.2.3)

---

[5]For the sake of clarity, the arguments for both space and time are mostly omitted in the following equations of this chapter.

$$\mathcal{B}^t[\mathbf{c}]\,\mathcal{B}\mathbf{u} + \mathbf{f}_V = \varrho_0\ddot{\mathbf{u}} \qquad (4.41)$$

with the displacement vector $\mathbf{u}(\mathbf{r}, t)$, the differential operator $\mathcal{B}$ (2.40, p. 18), the stiffness tensor $[\mathbf{c}]$, the prescribed volume force $\mathbf{f}_V$, and the material density ϱ_0, respectively. In the 3-D case, the compact form of (4.41) for the computational domain Ω reads as (boundary $\partial\Omega$ of Ω)

PDE	$\mathcal{B}^t[\mathbf{c}]\,\mathcal{B}\mathbf{u} + \mathbf{f}_V = \varrho_0\ddot{\mathbf{u}}$
	$\Omega \subset \mathbb{R}^3$
	$t \in [0, T]\,; \quad [0, T] \subset \mathbb{R}$
Given	$c_{ij} : \Omega \to \mathbb{R}$
	$\varrho_0 : \Omega \to \mathbb{R}$
	$\mathbf{f}_V : \Omega \times [0, T] \to \mathbb{R}^3$
BC	$\mathbf{u} = \mathbf{u}_e$ on $\Gamma_e \times [0, T]$
	$[\mathbf{T}]^t\,\mathbf{n} = \mathbf{T}_n$ on $\Gamma_n \times [0, T]$
	$\Gamma_e \cup \Gamma_n = \partial\Omega$
IC	$\mathbf{u}(\mathbf{r}, 0) = \mathbf{u}_0(\mathbf{r}) \quad \forall \mathbf{r} \in \Omega$
	$\dot{\mathbf{u}}(\mathbf{r}, 0) = \dot{\mathbf{u}}_0(\mathbf{r}) \quad \forall \mathbf{r} \in \Omega$
	$\ddot{\mathbf{u}}(\mathbf{r}, 0) = \ddot{\mathbf{u}}_0(\mathbf{r}) \quad \forall \mathbf{r} \in \Omega$
Find	$\mathbf{u}(\mathbf{r}, t) : \overline{\Omega} \times (0, T] \to \mathbb{R}^3\,.$

$\overline{\Omega}$ represents the computational domain without its Dirichlet boundary. The expression $\mathbf{n}$ denotes the normal vector with respect to the boundary Γ_n of Ω and $\mathbf{T}_n$ the resulting mechanical stress perpendicular to this boundary. Either $\mathbf{T}_n$ or $\mathbf{u}_e$ have to be assigned as boundary condition to obtain a unique solution of the PDE. Without limiting the generality, we set both boundary conditions to zero, i.e., $\mathbf{T}_n = 0$ and $\mathbf{u}_e = 0$. Hence, the weak form of (4.41) after utilizing Green's first integration theorem is given by

$$\int_{\Omega} \varrho_0\mathbf{w} \cdot \ddot{\mathbf{u}}\ \mathrm{d}\Omega + \int_{\Omega} (\mathcal{B}\mathbf{w})^t[\mathbf{c}]\,\mathcal{B}\mathbf{u}\ \mathrm{d}\Omega = \int_{\Omega} \mathbf{w} \cdot \mathbf{f}_V\ \mathrm{d}\Omega \qquad (4.42)$$

where $\mathbf{w}(\mathbf{r})$ is an appropriate test function. Since the displacement $\mathbf{u}$ is a vector quantity, $\mathbf{w}$ has also to be a vector quantity. Moreover, in contrast to the electrostatic field, the ansatz functions N_i for spatial discretization of Ω need to be applied for each component within Galerkin's method. For Lagrange ansatz functions, the approximation $\mathbf{u}$ of the displacement vector in 3-D (space dimensions $n_d = 3; \{1, 2, 3\} \cong \{x, y, z\}$) computes as

$$\mathbf{u}(\mathbf{r}) \approx \mathbf{u}(\mathbf{r}) = \sum_{j=1}^{n_d} \sum_{i=1}^{n_{\text{nodes}}} N_i(\mathbf{r})\, u_{i,j}\, \mathbf{e}_j \qquad \text{with } u_{i,j} = \mathbf{u}(\mathbf{r}_i) \cdot \mathbf{e}_j \qquad (4.43)$$

or alternatively by introducing the approximated vector $\mathbf{u}_i = \left[u_{i,x}, u_{i,y}, u_{i,z} \right]^t = \mathbf{u}(\mathbf{r}_i)$ at node i as

$$\mathbf{u}(\mathbf{r}) = \sum_{i=1}^{n_{\text{nodes}}} \mathbf{N}_i(\mathbf{r})\, \mathbf{u}_i \; ; \quad \mathbf{N}_i(\mathbf{r}) = \begin{bmatrix} N_i(\mathbf{r}) & 0 & 0 \\ 0 & N_i(\mathbf{r}) & 0 \\ 0 & 0 & N_i(\mathbf{r}) \end{bmatrix} . \qquad (4.44)$$

$\mathbf{e}_j$ stands for the unit vector pointing in direction j and n_{nodes} is the total amount of all nodes within the n_{elem} finite elements, which are used to spatially discretize the computational domain $\overline{\Omega}$. Finally, the algebraic system of equations in matrix form becomes

$$\mathbf{M_u}\ddot{\mathbf{u}} + \mathbf{K_u}\mathbf{u} = \mathbf{f_u} . \qquad (4.45)$$

The mass matrix $\mathbf{M_u}$, stiffness matrix $\mathbf{K_u}$ and right-hand side $\mathbf{f_u}$ are assembled according to

$$\mathbf{M_u} = \bigwedge_{l=1}^{n_{\text{elem}}} \mathbf{M}^l \; ; \qquad \mathbf{M}^l = \left[\mathbf{m}_{ij}^l \right] \; ; \qquad \mathbf{m}_{ij}^l = \int_{\Omega^l} \varrho_0 \mathbf{N}_i^t \mathbf{N}_j \mathrm{d}\Omega \qquad (4.46)$$

$$\mathbf{K_u} = \bigwedge_{l=1}^{n_{\text{elem}}} \mathbf{K}^l \; ; \qquad \mathbf{K}^l = \left[\mathbf{k}_{ij}^l \right] \; ; \qquad \mathbf{k}_{ij}^l = \int_{\Omega^l} \mathcal{B}_i^t [\mathbf{c}]\, \mathcal{B}_j \mathrm{d}\Omega \qquad (4.47)$$

$$\mathbf{f_u} = \bigwedge_{l=1}^{n_{\text{elem}}} \mathbf{f}^l \; ; \qquad \mathbf{f}^l = \left[\mathbf{f}_i^l \right] \; ; \qquad \mathbf{f}_i^l = \int_{\Omega^l} \mathbf{N}_i^t \mathbf{f}_V \mathrm{d}\Omega \qquad (4.48)$$

with

$$\mathcal{B}_i = \begin{bmatrix} \frac{\partial N_i}{\partial x} & 0 & 0 & 0 & \frac{\partial N_i}{\partial z} & \frac{\partial N_i}{\partial y} \\ 0 & \frac{\partial N_i}{\partial y} & 0 & \frac{\partial N_i}{\partial z} & 0 & \frac{\partial N_i}{\partial x} \\ 0 & 0 & \frac{\partial N_i}{\partial z} & \frac{\partial N_i}{\partial y} & \frac{\partial N_i}{\partial x} & 0 \end{bmatrix}^t . \qquad (4.49)$$

The assembling procedure is similar to the previously discussed FE method for electrostatics. Note that here, the vector $\mathbf{u}$ of unknowns takes the form

$$\mathbf{u} = \left[u_{1,x}, u_{1,y}, u_{1,z}, u_{2,x}, \ldots, u_{n_{\text{nodes}},x}, u_{n_{\text{nodes}},y}, u_{n_{\text{nodes}},z} \right]^t \qquad (4.50)$$

and, thus, contains three times as much components as in case of scalar quantities.

In many practical situations, the FE method for mechanical problems can be considerably simplified. The (i) plane strain state, the (ii) plane stress state, and the

(iii) axisymmetric stress–strain relations are three fundamental simplifications in continuum mechanics.

- *Plane strain state:* Let us assume an elastic body, which is large in one direction (e.g., z-direction) and features equal cross sections (e.g., in xy-plane) perpendicular to this dimension. If the boundary conditions and forces acting on the body are identical for each cross section, the dependence of the displacements (e.g., u_z) and strains (e.g., S_{yz}) on the dominating body dimension can be neglected.
- *Plane stress state:* We will be able to utilize this state if, for instance, the considered elastic body represents a thin plate (e.g., in xy-plane) made of homogeneous isotropic material, which is loaded by forces acting within the plate plane. For such configurations, several components of the stress tensor (e.g., T_{zz}) and the strain tensor (e.g., S_{yz}) can be set to zero.
- *Axisymmetric stress–strain relation:* This simplification can be applied when the investigated geometry and the material arrangement are axisymmetric. In that case, a cylindrical-coordinate system (radius r, height z, angle Θ) can be introduced, where both, displacements (e.g., u_Θ) and strains (e.g., $S_{r\Theta}$), do not depend on Θ.

As a result of the three fundamental simplifications, the original mechanical problem in 3-D changes to a 2-D problem. The required mesh to spatially discretize the computational domain is substantially reduced yielding a smaller number of nodes n_{nodes} and, thus, an algebraic system of equations with less unknown quantities.

4.3.1 Types of Analysis

Several different types of analysis are commonly utilized in numerical simulations based on the FE method. To discuss the basic types of analysis, we start with an extended version of (4.45) for the algebraic system of equations in mechanics

$$\mathbf{M_u}\ddot{\mathbf{u}} + \mathbf{D_u}\dot{\mathbf{u}} + \mathbf{K_u}\mathbf{u} = \mathbf{f_u} \ . \tag{4.51}$$

Here, the (damping) matrix $\mathbf{D_u}$ accounts for attenuation within the investigated elastic body.

Static Analysis

In case of static analysis, we presume that the aimed quantity (i.e., $\mathbf{u}$), the boundary conditions as well as the right-hand side $\mathbf{f_u}$ do not depend on time t. Hence, one can state $\ddot{\mathbf{u}} = \dot{\mathbf{u}} = 0$ and, therewith (4.51), takes the form

$$\mathbf{K_u}\mathbf{u} = \mathbf{f_u} \ . \tag{4.52}$$

So, the mass matrix $\mathbf{M_u}$ and damping matrix $\mathbf{D_u}$ have no influence on the result. As for the electrostatic field in Sect. 4.2, initial conditions are useless for the static analysis.

Transient Analysis

Both the external loads and the aimed quantity $\mathbf{u}$ may vary with respect to time in case of a transient analysis. Therefore, we are not able to simplify (4.51). In addition to the spatial computational domain, the investigated time interval $[0, T]$ is discretized. For each time step t_i,

$$\mathbf{M_u}\ddot{\mathbf{u}}^{(i)} + \mathbf{D_u}\dot{\mathbf{u}}^{(i)} + \mathbf{K_u}\mathbf{u}^{(i)} = \mathbf{f_u}^{(i)} \tag{4.53}$$

has to be fulfilled. According to the Newmark scheme (see Sect. 4.1.4), the solution $\mathbf{u}^{(i+1)}$ for the subsequent time step t_{i+1} results from three computation substeps.

1. Compute predictor step:

$$\tilde{\mathbf{u}} = \mathbf{u}^{(i)} + \Delta t\, \dot{\mathbf{u}}^{(i)} + \frac{(\Delta t)^2}{2}(1 - 2\beta_\mathrm{N})\, \ddot{\mathbf{u}}^{(i)} \tag{4.54}$$

$$\tilde{\dot{\mathbf{u}}} = \dot{\mathbf{u}}^{(i)} + (1 - \gamma_\mathrm{N})\, \Delta t\, \ddot{\mathbf{u}}^{(i)} . \tag{4.55}$$

2. Solve algebraic system of equations:

$$\mathbf{M_u^\star}\ddot{\mathbf{u}}^{(i+1)} = \mathbf{f_u}^{(i+1)} - \mathbf{D_u}\tilde{\dot{\mathbf{u}}} - \mathbf{K_u}\tilde{\mathbf{u}} \tag{4.56}$$

$$\mathbf{M_u^\star} = \mathbf{M_u} + \gamma_\mathrm{N}\Delta t\, \mathbf{D_u} + \beta_\mathrm{N}(\Delta t)^2\, \mathbf{K_u} . \tag{4.57}$$

3. Perform corrector step:

$$\mathbf{u}^{(i+1)} = \tilde{\mathbf{u}} + \beta_\mathrm{N}(\Delta t)^2\, \ddot{\mathbf{u}}^{(i+1)} \tag{4.58}$$

$$\dot{\mathbf{u}}^{(i+1)} = \tilde{\dot{\mathbf{u}}} + \gamma_\mathrm{N}\Delta t\, \ddot{\mathbf{u}}^{(i+1)} . \tag{4.59}$$

The parameters β_N and γ_N determine the type of integration, i.e., explicit or implicit integration.

It seems only natural that the transient analysis needs much more computational effort than the static analysis. Especially for large computational grids and long periods of time, this may lead to numerical simulations, which cannot be solved in a reasonable amount of time anymore.

Harmonic Analysis

If the behavior of a system in case of harmonic excitation with frequency f has to be figured out, we can perform a transient analysis with an appropriate excitation signal. However, to achieve the steady state of the system, a sufficiently long period of time $[0, T]$ is required for the simulation, which is usually accompanied by an unacceptable computational effort. On account of this fact, a harmonic analysis should be carried out instead. In doing so, the algebraic system of equations (4.53) in time domain is transformed into the complex frequency domain. The time-dependent expressions $\mathbf{f_u}$ and $\mathbf{u}$ as well as the time derivatives are replaced by (angular frequency $\omega = 2\pi f$)

$$\mathbf{f_u} \longrightarrow \hat{\mathbf{f}}_\mathbf{u} \cdot e^{j\omega t} \ ; \qquad\qquad \mathbf{u} \longrightarrow \hat{\mathbf{u}} \cdot e^{j\varphi_\mathbf{u}} \cdot e^{j\omega t} \qquad (4.60)$$

$$\frac{\partial}{\partial t} \longrightarrow j\omega \ ; \qquad\qquad \frac{\partial^2}{\partial t^2} \longrightarrow -\omega^2 \qquad (4.61)$$

$\hat{\mathbf{f}}_\mathbf{u}$ and $\hat{\mathbf{u}}$ stand for the amplitudes of both the right-hand side $\mathbf{f_u}$ and the mechanical displacements $\mathbf{u}$ at the nodes of the spatially discretized computational domain, respectively. $\varphi_\mathbf{u}$ is a vector containing the phase for the displacement components at each node.[6] So, the vectors $\hat{\mathbf{u}}$ and $\varphi_\mathbf{u}$ feature the same length. The combination of (4.51), (4.60), and (4.61) leads to

$$\left(-\omega^2 \mathbf{M_u} + j\omega \mathbf{D_u} + \mathbf{K_u}\right) \hat{\mathbf{u}} \cdot e^{j\varphi_\mathbf{u}} = \hat{\mathbf{f}}_\mathbf{u} \ , \qquad (4.62)$$

which is in contrast to (4.51) a complex-valued algebraic system of equations. The solution of this system of equations provides for each node both the amplitude and the phase of the displacement components at frequency f.

Eigenfrequency Analysis

To predict the resonance behavior of a system by means of harmonic analysis, one has to investigate a certain frequency range in which resonance is expected. This procedure may take very long computation time. A time-efficient alternative to the harmonic analysis is the eigenfrequency analysis. Thereby, the system behavior is studied without considering the damping matrix $\mathbf{D_u}$ and the right-hand side $\mathbf{f_u}$ of (4.51). As for the harmonic analysis, we perform a transform into the complex frequency domain yielding the eigenvalue equation

$$\left[-(2\pi f)^2 \mathbf{M_u} + \mathbf{K_u}\right] \hat{\mathbf{u}} \cdot e^{j\varphi_\mathbf{u}} = 0 \ . \qquad (4.63)$$

The solution of this equation is a pairwise combination of the so-called eigenfrequencies $f_{(i)}$ representing the eigenvalues and eigenvectors $\hat{\mathbf{u}}_{(i)}e^{j\varphi_{\mathbf{u},(i)}}$. In other words, for each eigenfrequency, one obtains amplitude and phase for the aimed quantity at the nodes of the computational domain. Each eigenvector indicates an eigenmode of the mechanical system.

From the physical point of view, excitations at the eigenfrequency result in a system behavior according to the eigenvectors. In case of an undamped mechanical systems (i.e., $\mathbf{D_u} = 0$), the displacements and, consequently, vibrations at this frequency might be rather high. Note that for such a system, the eigenfrequencies coincide with the resonance frequencies of the system.

---

[6]Alternatively to the approach with amplitude and phase, the complex frequency domain can be represented by real and imaginary parts.

4.3.2 Attenuation within Mechanical Systems

Each mechanical system is subject to a certain attenuation, which arises due to inner friction. For example, when a one-sided clamped beam is excited by a pulse, the resulting mechanical vibrations will decay. To incorporate attenuation within the FE method, we often add an expression that is proportional to the mechanical velocity $\dot{\mathbf{u}}$ of the mechanical system. The so-called *damping matrix* $\mathbf{D_u}$ represents the proportionality factor. The resulting algebraic system of equation was already shown in (4.51).

It seems only natural that attenuation can only be considered in a realistic manner if an appropriate damping matrix $\mathbf{D_u}$ is used. Many FE formulations exploit the *Rayleigh damping model* to determine $\mathbf{D_u}$. The idea of this damping model lies in linearly combining mass matrix $\mathbf{M_u}$ and stiffness matrix $\mathbf{K_u}$ of the system. Therefore, $\mathbf{D_u}$ is given by

$$\mathbf{D_u} = \alpha_M \mathbf{M_u} + \alpha_K \mathbf{K_u} \tag{4.64}$$

with the mass proportional damping coefficient α_M and the stiffness proportional damping coefficient α_K. According to [3], a mode superposition analysis including attenuation yields (*i*th eigenfrequency $f_{(i)}$)

$$\alpha_M + \alpha_K \left[2\pi f_{(i)}\right]^2 = 4\pi f_{(i)} \xi_{d,i} \ . \tag{4.65}$$

The expression $\xi_{d,i}$ denotes the modal damping ratio for the *i*th eigenmode (i.e., at the *i*th eigenfrequency) and computes as

$$\xi_{d,i} = \frac{\alpha_M + \alpha_K \left[2\pi f_{(i)}\right]^2}{4\pi f_{(i)}} \ . \tag{4.66}$$

The frequency-dependent damping ratio $\xi_d(f)$ results from the same formula by replacing $f_{(i)}$ with f. $\xi_d(f)$ takes a minimum at $f_{(i)}$ and increases exponentially for $f < f_{(i)}$ and linearly for $f > f_{(i)}$.

The Rayleigh damping model is applicable for transient as well as harmonic analysis based on FE method. However, strictly speaking, the frequency-dependent damping ratio $\xi_d(f)$ will only lead to a good approximation of attenuation if the considered frequency is close to a eigenfrequency. That is the reason why other damping models are applied instead. A common damping model assumes constant attenuation; i.e., ξ_d does not depend on frequency. We can achieve a constant value of ξ_d by means of $\alpha_M = 0$ and $\alpha_K = \alpha_d/(2\pi f)$ (see (4.65)) with the damping coefficient α_d. By inserting this in (4.64), the damping matrix results in $\mathbf{D_u} = \alpha_d \mathbf{K_u}/(2\pi f)$. For harmonic FE simulations (4.62), the influence of attenuation remains, thus, constant because the frequency f cancels out in the expression $j\omega \mathbf{D_u}$. Therefore, attenuation exclusively depends on the product $\alpha_d \mathbf{K_u}$.

Instead of directly introducing a damping matrix $\mathbf{D_u}$ in the FE method, it is possible to consider attenuation for harmonic simulations by using complex-valued material parameters. In case of a mechanical system, this means that the stiffness tensor changes to the complex-valued version

$$\left[\underline{\mathbf{c}}^E\right] = \left[\mathbf{c}^E\right]_{\Re} + j\left[\mathbf{c}^E\right]_{\Im} = \left[\mathbf{c}^E\right]_{\Re}\left[1 + j\alpha_d\right] . \tag{4.67}$$

While the real part $\left[\mathbf{c}^E\right]_{\Re}$ coincides with the original material parameters, the imaginary part $\left[\mathbf{c}^E\right]_{\Im}$ rates attenuation. The damping coefficient α_d in (4.67) corresponds to the previous definition with $\mathbf{D_u}$ since the imaginary part yields again the expression $\alpha_d \mathbf{K_u}$ in the algebraic system of equations.

4.3.3 Example

Let us study the mechanical behavior of a one-sided clamped copper beam by means of the FE method. The decisive material parameters density, Young's modulus, and Poisson's ratio were set to $\varrho_0 = 8930 \, \text{kg m}^{-3}$, $E_M = 126.2 \, \text{GPa}$, and $\nu_P = 0.37$, respectively (see Table 2.5 on p. 26). Figure 4.8a shows the geometric setup for this cantilever beam (length $l_{\text{beam}} = 10 \, \text{mm}$; height $h_{\text{beam}} = 0.5 \, \text{mm}$) in the xy-plane. For the sake of simplicity, we assume that the geometric dimension in z-direction is very large compared to h_{beam} as well as l_{beam}. Moreover, the boundary conditions and external forces acting on the structure are supposed to be equal for each cross section in parallel to the xy-plane. The mechanical problem can, therefore, be treated as plane strain state; i.e., it is sufficient to apply the FE method for the 2-D case (see Fig. 4.8b). The clamping on the left hand side implies that there both the displacements u_x in x-direction and the displacements u_y in y-direction equal zero. At the right upper end, the beam is loaded with an external static force of $F = 10 \, \text{N}$ in negative y-direction. The following FE simulations were performed with quadratic Lagrange ansatz functions, i.e., h-FEM with $p_d = 2$.

Since the quantities do not depend on time, we can conduct a static analysis according to (4.52). Figure 4.9 displays the computed bending line $u_y(x)$, which refers

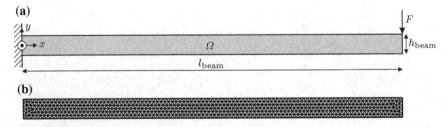

Fig. 4.8 **a** Geometric setup of cantilever beam with length $l_{\text{beam}} = 10 \, \text{mm}$ and height $h_{\text{beam}} = 0.5 \, \text{mm}$ in xy-plane; **b** 2-D computational grid comprising 1208 triangles

Fig. 4.9 Simulation result
for bending line $u_y(x)$ of
beam's centerline

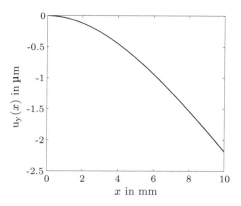

to the centerline of the cantilever beam, i.e., at $y = 0$. As expected, the beam deflects
in negative y-direction. The deflection at the right beam end amounts 2.18 μm.

For such a simple arrangement, the beam deflection can also be approximated in
an analytical manner. The simplest approximation for the bending line results from
the Euler–Bernoulli beam theory, which supposes small mechanical deformations,
negligible shear deformations as well as plane cross sections of the beam that are
always perpendicular to the beam's centerline during deformation [24, 26]. According to the Euler–Bernoulli beam theory, the deflection $u_y(x = l_{beam})$ of the right beam
end computes as

$$u_y(x = l_{beam}) = \frac{4Fl_{beam}^3}{E_M h_{beam}^3} \qquad (4.68)$$

for the plane strain state. This leads to the approximated deflection of 2.54 μm. The
deviation from the FE simulation stems from the simplifications in the course of the
Euler–Bernoulli beam theory.

In Fig. 4.10a and b, one can see the simulated displacement $u_x(x, y)$ in x-direction
and the simulated displacement $u_y(x, y)$ in y-direction along the cantilever beam
for $F = 10$ N. Not surprisingly, the maximum of $|u_x(x, y)|$ is much smaller than
of $|u_y(x, y)|$. Figure 4.10c depicts the simulated von Mises stress $T_{mis}(x, y)$, which
is here defined as (arguments x and y omitted)

$$T_{mis} = \sqrt{\left(T_{xx} + T_{yy}\right)^2 \left(\nu_P^2 - \nu_P + 1\right) + T_{xx} T_{yy} \left(2\nu_P^2 - 2\nu_P - 1\right) + 3T_{xy}^2} \qquad (4.69)$$

with the normal stresses T_{xx} and T_{yy} and the shear stress T_{xy}. Especially close to the
clamping (i.e., the left beam end), $T_{mis}(x, y)$ takes high values at the top and bottom
sides of the beam.

Finally, we regard an eigenfrequency analysis (see (4.63)). Figure 4.11 shows the
simulated first five eigenmodes and eigenfrequencies $f_{(i)}$ for transverse vibrations of
the cantilever beam. As expected, higher eigenfrequencies are accompanied by an

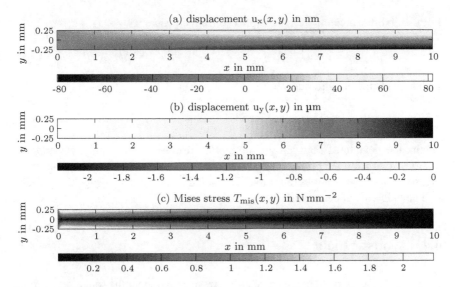

Fig. 4.10 Simulation result for **a** displacement $u_x(x, y)$ in x-direction, **b** displacement $u_y(x, y)$ in y-direction, and **c** von Mises stress $T_{mis}(x, y)$ along cantilever beam; color bars relate directly to figure above

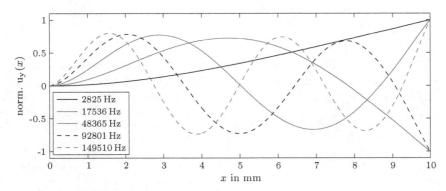

Fig. 4.11 Simulated normalized eigenmodes $u_y(x)$ along cantilever beam and eigenfrequencies $f_{(i)}$ for first five eigenmodes

increasing amount of local minima and maxima along the beam. The eigenfrequencies for transverse vibrations can also be analytically approximated by [24]

$$f_{(i)} = \frac{\lambda_{(i)}^2}{2\pi l_{beam}^2} \sqrt{\frac{E_M h_{beam}^2}{12 \varrho_0}} \,. \tag{4.70}$$

For the first five eigenmodes, the expression $\lambda_{(i)}$ equals

$$\lambda_{(i)} = \{1.875; 4.694; 7.855; 10.996; 14.137\} , \tag{4.71}$$

which yields the eigenfrequencies

$$f_{(i)} = \{3036; 19028; 53284; 104417; 172591\} \, \text{Hz} . \tag{4.72}$$

Again, the deviations between the simulation results and the analytical approximations stem from simplifications that are performed for the approximation. If the Poisson's ratio ν_P is set to zero for the FE simulations, the obtained eigenfrequencies will coincide much better with the approximations in (4.72).

4.4 Acoustic Field

In Sect. 2.3.3, we derived the linear acoustic wave equation for the sound pressure $p_\sim$ and the acoustic velocity potential Ψ. For the sound pressure, the wave equation at position $\mathbf{r}$ reads as

$$\frac{1}{c_0^2} \ddot{p}_\sim - \Delta p_\sim = f_p \tag{4.73}$$

with the sound velocity c_0 and the excitation function f_p generating acoustic waves in the medium. The compact form of (4.73) for the 3-D computational domain Ω including boundary as well as initial conditions becomes

PDE	$\dfrac{1}{c_0^2} \ddot{p}_\sim - \Delta p_\sim = f_p$
	$\Omega \subset \mathbb{R}^3$
	$t \in [0, T] ; \quad [0, T] \subset \mathbb{R}$
Given	$c_0 : \Omega \to \mathbb{R}$
	$f_p : \Omega \times [0, T] \to \mathbb{R}$
BC	$p_\sim = p_e$ on $\Gamma_e \times [0, T]$
	$\dfrac{\partial p_\sim}{\partial \mathbf{n}} = p_n$ on $\Gamma_n \times [0, T]$
	$\Gamma_e \cup \Gamma_n = \partial \Omega$
IC	$p_\sim(\mathbf{r}, 0) = p_0(\mathbf{r}) \quad \forall \mathbf{r} \in \Omega$
	$\dot{p}_\sim(\mathbf{r}, 0) = \dot{p}_0(\mathbf{r}) \quad \forall \mathbf{r} \in \Omega$
	$\ddot{p}_\sim(\mathbf{r}, 0) = \ddot{p}_0(\mathbf{r}) \quad \forall \mathbf{r} \in \Omega$
Find	$p_\sim(\mathbf{r}, t) : \overline{\Omega} \times (0, T] \to \mathbb{R} .$

Similar to the FE method for the mechanical field (see Sect. 4.3), either p_e or p_n have to be prescribed as boundary condition of the PDE. To simplify the following expressions, we set these boundary conditions to zero. Therewith, the weak form after applying Green's first integration theorem results in (scalar test function $w(\mathbf{r})$)

$$\int\limits_{\Omega} \frac{1}{c_0^2} w \ddot{p}_{\sim} \mathrm{d}\Omega + \int\limits_{\Omega} \nabla w \cdot \nabla p_{\sim} \mathrm{d}\Omega - \int\limits_{\Omega} w f \mathrm{d}\Omega = 0 . \tag{4.74}$$

The subsequent spatial discretization of $w(\mathbf{r})$ and $p_{\sim}(\mathbf{r})$ according to Galerkin's method leads to the algebraic system of equations in matrix form

$$\mathbf{M}_p \ddot{\mathbf{p}} + \mathbf{K}_p \mathbf{p} = \mathbf{f}_p . \tag{4.75}$$

For Lagrange ansatz functions, the vector $\mathbf{p}$ contains approximated values of the sound pressure $p_{\sim}$ at the nodes of the spatially discretized computational domain Ω. The mass matrix $\mathbf{M}_p$, stiffness matrix $\mathbf{K}_p$, and right-hand side $\mathbf{f}_p$ are given by (number n_{elem} of finite elements; ansatz function N_i)

$$\mathbf{M}_p = \bigwedge_{l=1}^{n_{\mathrm{elem}}} \mathbf{M}^l ; \qquad \mathbf{M}^l = \left[m_{ij}^l \right] ; \qquad m_{ij}^l = \int\limits_{\Omega^l} \frac{1}{c_0^2} N_i N_j \mathrm{d}\Omega \tag{4.76}$$

$$\mathbf{K}_p = \bigwedge_{l=1}^{n_{\mathrm{elem}}} \mathbf{K}^l ; \qquad \mathbf{K}^l = \left[k_{ij}^l \right] ; \qquad k_{ij}^l = \int\limits_{\Omega^l} (\nabla N_i)^{\mathrm{t}} \nabla N_j \mathrm{d}\Omega \tag{4.77}$$

$$\mathbf{f}_p = \bigwedge_{l=1}^{n_{\mathrm{elem}}} \mathbf{f}^l ; \qquad \mathbf{f}^l = \left[f_i^l \right] ; \qquad f_i^l = \int\limits_{\Omega^l} N_i f_p \mathrm{d}\Omega . \tag{4.78}$$

To account for attenuation during the wave propagation, an appropriate damping matrix $\mathbf{D}_p$ has to be introduced in addition.

Again, different types of analysis can be performed. In contrast to the mechanical field, the static analysis makes no sense because sound pressure is an alternating quantity. For the transient analysis, we subdivide the investigated time interval $[0, T]$ into sufficiently small subintervals and apply the Newmark scheme according to Sect. 4.1.4. In case of harmonic and eigenfrequency analysis, the algebraic system of equations (4.75) in matrix form has to be transformed into the complex frequency domain.

Alternatively to the sound pressure $p_{\sim}$, one is able to conduct the FE method for the acoustic field by means of the acoustic velocity potential Ψ. The decision if $p_{\sim}$ or Ψ is utilized primarily depends on the prescribed boundary conditions of the investigated acoustic problem. Principally, we distinguish between three different cases.

• Pressure as boundary condition, i.e., $p_{\sim} = p_e$ on $\partial\Omega$: The acoustic problem should be studied with the PDE for $p_{\sim}$.

- Normal component of the particle velocity as boundary condition, i.e., $\mathbf{n} \cdot \mathbf{v}_\sim = v_n$ on $\partial\Omega$: Since there is a unique and simple relation between particle velocity and acoustic velocity potential (2.122, p. 34), the acoustic field should be calculated with the PDE for Ψ. The sound pressure results from (2.123).
- Mixed boundary conditions, i.e., $p_\sim = p_e$ on Γ_e and $\mathbf{n} \cdot \mathbf{v}_\sim = v_n$ on Γ_n: Both quantities are appropriate to solve the acoustic problem. However, we have to convert the boundary conditions into the used quantity.

4.4.1 Open Domain Problems

When the acoustic field of a sound source is studied, we will be mostly interested in the free-field radiation. Actually, the computational domain for numerical simulations based on FE is always limited. On account of this fact, boundary conditions or methods mimicking the so-called open domain are indispensable. The previously discussed boundary conditions cause, however, reflections of the impinging acoustic waves. Consequently, they are not appropriate for free-field simulations. To obtain an open computational domain, several techniques have been developed. In the following, we briefly discuss two famous approaches, namely (i) the absorbing boundary condition (ABC) and (ii) the perfectly matched layer (PML).

Absorbing Boundary Conditions

Let us consider a sinusoidal 1-D pressure wave propagating in x-direction with the sound velocity c_0. The acoustic wave may travel in both positive x-direction and negative x-direction. In the complex domain, the solution of the linear wave equation for these waves is given by

$$\underline{p}_\sim^+(x, t) = \hat{p}_\sim e^{\mathrm{j}(\omega t - kx)} \tag{4.79}$$

$$\underline{p}_\sim^-(x, t) = \hat{p}_\sim e^{\mathrm{j}(\omega t + kx)} \tag{4.80}$$

with the sound pressure amplitude $\hat{p}_\sim$, the angular frequency ω, and the wave number $k = \omega/c_0$, respectively. Furthermore, we assume at x_{bound} a virtual boundary Γ_{bound} where the relation

$$\left(\frac{\partial}{\partial t} + c_0 \frac{\partial}{\partial x} \right) = 0 \tag{4.81}$$

has to be fulfilled. Inserting (4.79) and (4.80) in (4.81) reveals that the relation is only satisfied for waves traveling in positive x-direction, i.e., $\underline{p}^+_\sim$. Hence, these waves can pass the boundary (see Fig. 4.12). In contrast, waves traveling in negative x-direction are totally reflected at the boundary.

If (4.81) is applied to the FE method at the boundary $\partial\Omega$ of the computational domain Ω, one will achieve an open computational domain. We are, thus, able to

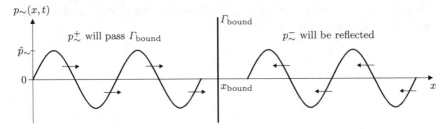

Fig. 4.12 According to ABC, acoustic waves $p_\sim^+$ traveling in positive x-direction will pass virtual boundary Γ_{bound} at x_{bound} and waves $p_\sim^-$ traveling in negative x-direction will be totally reflected there

simulate free-field radiation of an acoustic source. Since this procedure solely affects entries in the damping matrix $\mathbf{D}_p$, which refer to $\partial\Omega$, the relation (4.81) is usually named absorbing boundary condition.

It is important to mention that an ABC will only work perfectly when the pressure waves impinge perpendicularly onto the boundary. However, in many practical situations, it is impossible to choose a boundary geometry ensuring perpendicular sound incidence for each excitation signal. That is why alternative methods are oftentimes demanded to mimic an open computational domain, e.g., the perfectly matched layer technique.

Perfectly Matched Layer

With a view to explaining the basic idea of the PML technique, we assume a plane acoustic wave propagating in positive x-direction. The wave impinges perpendicularly onto an interface at $x = 0$ of two media (medium 1 and medium 2; see Fig. 4.13a), which feature different acoustic properties. For this configuration, the reflection coefficient r_{pres} for the incident pressure wave becomes (see (2.139, p. 38))

$$r_{\text{pres}} = \frac{Z_{\text{aco2}} - Z_{\text{aco1}}}{Z_{\text{aco1}} + Z_{\text{aco2}}} \tag{4.82}$$

with the acoustic impedances Z_{aco1} and Z_{aco2} of medium 1 and medium 2, respectively. The acoustic impedances for plane waves are defined as

$$Z_{\text{aco1}} = \varrho_1 c_1 \quad \text{and} \quad Z_{\text{aco2}} = \varrho_2 c_2 . \tag{4.83}$$

From (4.82), it can be easily deduced that there will not occur any reflection at the interface if $Z_{\text{aco1}} = Z_{\text{aco2}}$ holds. We can fulfill this condition by choosing appropriate combinations of sound velocities and densities for the two media. Without limiting the generality, let us consider the following combination

$$\varrho_2 = \varrho_1 \left(1 - j\alpha_{\xi x}\right) \tag{4.84}$$

$$c_2 = \frac{c_1}{1 - j\alpha_{\xi x}} \tag{4.85}$$

where $\alpha_{\xi x}$ is an arbitrary positive number. Therewith, the (complex-valued) wave number $\underline{k}_2$ in medium 2 results in ($k_1 = \omega/c_1$)

$$\underline{k}_2 = \frac{\omega}{c_2} = k_1\left(1 - \mathrm{j}\alpha_{\xi x}\right) . \tag{4.86}$$

Replacing k_2 in the solution of the linear wave equation for sound pressure waves in medium 2, which travel in positive x-direction (cf. (4.79)), with (4.86) yields

$$\underline{p}^+_{\sim}(x, t) = \hat{p}_{\sim}\mathrm{e}^{\mathrm{j}(\omega t - \underline{k}_2 x)} = \hat{p}_{\sim}\mathrm{e}^{\mathrm{j}(\omega t - k_1 x)} \underbrace{\mathrm{e}^{-\alpha_{\xi x} x}}_{\text{damping}} . \tag{4.87}$$

The expression $\mathrm{e}^{-\alpha_{\xi x} x}$ causes attenuation due to the fact that $\alpha_{\xi x}$ (attenuation coefficient) was assumed to be a positive number. As a result, the amplitude of the propagating pressure wave gets exponentially reduced in medium 2 (see Fig. 4.13a).

To utilize this principle in the FE method for 2-D as well as 3-D acoustic problems, the original computational domain Ω_{orig} has to be surrounded by an additional computational region Ω_{PML}, the so-called perfectly matched layer (see Fig. 4.13c). In this layer, propagating acoustic waves are attenuated until they are reflected at the outer boundary $\partial\Omega_{\mathrm{PML}}$ of Ω_{PML}. During the propagation of the reflected waves back to the interface $\partial\Omega_{\mathrm{orig}}$ of Ω_{PML} and Ω_{orig}, further attenuation is present. The intensity of the reflected acoustic wave will, therefore, be negligible if a sufficient thickness of the PML as well as a proper attenuation coefficient is chosen.

Note that the PML technique does not only require acoustic impedances matching of Ω_{orig} and Ω_{PML} for sound pressure waves impinging perpendicular onto their

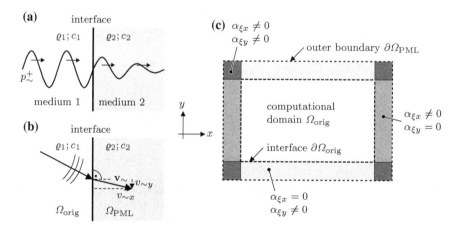

Fig. 4.13 a Sound pressure wave $p_{\sim}(x, t)$ propagating in x-direction and impinging perpendicularly onto interface of medium 1 and medium 2; **b** refraction at interface of Ω_{orig} and Ω_{PML}; particle velocity $\mathbf{v}_{\sim}$; **c** setting of attenuation coefficients $\alpha_{\xi x}$ and $\alpha_{\xi y}$ in 2-D within PML layer that surrounds Ω_{orig}

interface $\partial\Omega_{\text{orig}}$ but also for oblique incident sound. In order to arrange impedance matching, we locally split the incident wave $p_\sim$ into plane waves $p_{\sim x}$, $p_{\sim y}$ and $p_{\sim z}$ propagating in x-, y-, and z-direction, respectively, i.e.,

$$p_\sim = p_{\sim x} + p_{\sim y} + p_{\sim z} \,. \tag{4.88}$$

The splitting is conducted according to the components of the particle velocity $\mathbf{v}_\sim = \begin{bmatrix} v_{\sim x}, v_{\sim y}, v_{\sim z} \end{bmatrix}^t$ at the interface in the PML layer (see Fig. 4.13b). Moreover, individual attenuation coefficients $\alpha_{\xi x}$, $\alpha_{\xi y}$, and $\alpha_{\xi z}$ are applied in the different spatial directions (see Fig. 4.13c). By means of this technique, reflections of oblique incident pressure waves are avoided at the interface $\partial\Omega_{\text{orig}}$, which leads to an open computational domain. It is possible to incorporate the PML technique in the FE method for analysis in the complex frequency domain (i.e., harmonic and eigenfrequency) and in modified form also for transient analysis [13]. However, the additional region Ω_{PML} implies increasing computational effort.

4.4.2   Example

As a practical example for acoustics, let us study the sound field, which is generated by a piston-type ultrasonic transducer featuring a circular active surface (radius $R_T = 10\,\text{mm}$) and a uniform surface normal velocity. The ultrasonic transducer operates in water with a sound velocity of $c_0 = 1500\,\text{ms}^{-1}$. Owing to the symmetry of the transducer, we can restrict the computational domain Ω to a rotationally symmetric configuration. Figure 4.14 illustrates the considered 2-D geometric arrangement, whereby the rotation axis coincides with the z-axis. Ω is a quarter circle with radius $R_\Omega = 100\,\text{mm}$.

Fig. 4.14 Rotationally symmetric configuration of computational domain Ω with radius $R_\Omega = 100\,\text{mm}$ for piston-type ultrasonic transducer; circular active surface with radius $R_T = 10\,\text{mm}$; amplitude $\hat{v}_n = 1\,\text{mm s}^{-1}$ of surface normal velocity; absorbing boundary conditions at Γ_{ABC}

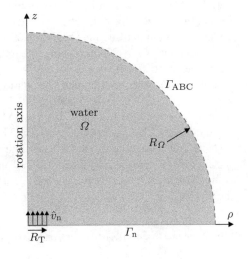

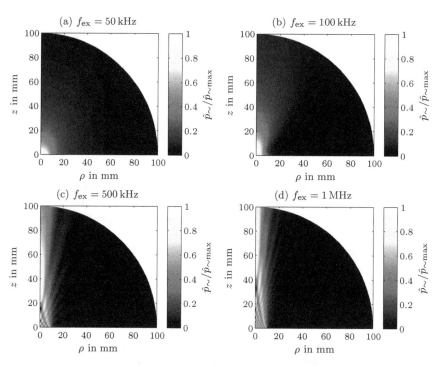

Fig. 4.15 Simulated normalized sound pressure distribution $\hat{p}_\sim(\rho, z)$ for excitation frequency f_{ex} **a** 50 kHz, **b** 100 kHz, **c** 500 kHz, and **d** 1 MHz; normalization with respect to maximum amplitude $\hat{p}_{\sim max}$

The active surface of the piston-type transducer gets emulated by a line of radial dimension R_T that oscillates sinusoidally with the velocity amplitude $\hat{v}_n = 1 \text{ mm s}^{-1}$ at the frequency f_{ex}. Therefore, we have to prescribe $\hat{v}_n$ along R_T. The remaining part Γ_n of the lower limit of Ω was assumed to be acoustically hard, i.e., $\hat{v}_n = 0$. Due to the rotationally symmetric configuration, this also refers to the z-axis. With a view to simulating free-field radiation, absorbing boundary conditions were used at the boundary Γ_{ABC}. The following FE simulations were performed with quadratic Lagrange ansatz functions, i.e., h-FEM with $p_d = 2$.

Figure 4.15 contains simulation results for the normalized sound pressure distribution[7] $\hat{p}_\sim(\rho, z)$, which were obtained by a harmonic FE analysis. The excitation frequency f_{ex} was varied from 50 kHz up to 1 MHz. Not surprisingly, $\hat{p}_\sim(\rho, z)$ strongly depends on f_{ex}. The underlying causes will be thoroughly investigated in Sect. 7.2.

Note that the required computation time of the harmonic FE simulation varies widely for the considered values of f_{ex}. This originates from the size of the used computational grid because reliable FE simulations call for a sufficiently fine grid. In the present case, one wavelength $\lambda_{aco} = c_0/f_{ex}$ of the sound wave was discretized by

---

[7]The sound pressure distribution corresponds to the spatially resolved sound pressure magnitudes.

triangular elements with side lengths smaller than $\lambda_{aco}/10$. As a result, Ω comprises 2469 triangles for $f_{ex} = 50\,kHz$ and 977043 triangles for $f = 1\,MHz$, respectively. It seems only natural that such an increasing amount of finite elements has a significant impact on the required computation time. Especially when a transient FE analysis is desired, this fact may lead to an unacceptable computational effort.

4.5 Coupled Fields

If numerical simulations are carried out for piezoelectric sensors and actuators, one will always be concerned with the coupling of different physical fields. The (quasi-static) electric field is coupled to the mechanical field inside the piezoelectric material. For ultrasonic transducers, we have to additionally regard the coupling of mechanical and acoustic fields. In the following, the relevant coupling conditions as well as their incorporation into the FE method will be discussed.

4.5.1 Piezoelectricity

As for single physical fields (e.g., mechanical field), the numerical simulation of piezoelectricity demands appropriate PDEs. These equations are obtained by combining the material law of piezoelectricity with fundamental relations of mechanical and (quasi-static) electric fields. To account for the coupling of electric and mechanical quantities inside the piezoelectric material, let us utilize the material law for linear piezoelectricity in e-form and Voigt notation (see Sect. 3.3)

$$\mathbf{T} = \left[\mathbf{c}^E\right]\mathbf{S} - [\mathbf{e}]^t\,\mathbf{E} \tag{4.89}$$

$$\mathbf{D} = [\mathbf{e}]\,\mathbf{S} + \left[\varepsilon^S\right]\mathbf{E} \tag{4.90}$$

with the mechanical stress $\mathbf{T}$, the mechanical strain $\mathbf{S}$, the electric flux density $\mathbf{D}$, and the electric field intensity $\mathbf{E}$. The tensors $\left[\mathbf{c}^E\right]$, $[\mathbf{e}]$, and $\left[\varepsilon^S\right]$ contain the elastic stiffness constants for constant electric field intensity, the piezoelectric stress constants and the electric permittivities for constant mechanical strain, respectively. In the next step, we insert this material law into Navier's equation (see (2.42, p. 18))

$$\varrho_0\ddot{\mathbf{u}} - \mathcal{B}^t\mathbf{T} = \mathbf{f}_V \tag{4.91}$$

as well as into the Law of Gauss (see (2.9, p. 10))

$$\nabla \cdot \mathbf{D} = q_e\,. \tag{4.92}$$

Here, $\mathbf{u}$, ϱ_0, $\mathbf{f}_V$, and q_e stand for the mechanical displacement, the material density, the volume force, and the volume charge density, respectively. Since piezoelectric

Fig. 4.16 Plate-shaped
piezoelectric material
covered with infinitely thin
electrodes on bottom
area Γ_G and top area Γ_L

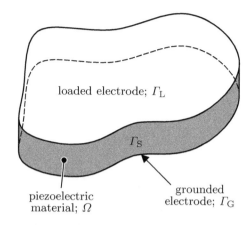

loaded electrode; Γ_L

Γ_S

piezoelectric
material; Ω

grounded
electrode; Γ_G

materials are electrically insulating, they do not contain any free volume charges,
i.e., $q_e = 0$. Under consideration of this fact and by applying the relations

$$\mathbf{S} = \mathcal{B}\mathbf{u} \quad \text{and} \quad \mathbf{E} = -\nabla V_e \,, \tag{4.93}$$

we arrive at coupled partial differential equations for $\mathbf{u}$ and the electric potential V_e

$$\varrho_0 \ddot{\mathbf{u}} - \mathcal{B}^t\left(\left[\mathbf{c}^E\right] \mathcal{B}\mathbf{u} + [\mathbf{e}]^t \nabla V_e\right) = \mathbf{f}_V \tag{4.94}$$

$$\nabla \cdot \left([\mathbf{e}] \mathcal{B}\mathbf{u} - \left[\varepsilon^S\right] \nabla V_e\right) = 0 \,, \tag{4.95}$$

which are applicable for FE simulations of piezoelectric materials. The consideration
of distinct boundary conditions for the mechanical field as well as for the electric
field requires the fundamental equations (4.91) and (4.92) in addition.

FE Method for Piezoelectric Coupling

Each PDE in (4.94) and (4.95) contains both aimed quantities, i.e., $\mathbf{u}$ and V_e. Therefore, one has to handle a coupled problem within the FE method for piezoelectricity.
Let us show the basic steps by means of a plate-shaped piezoelectric material, which is
shown in Fig. 4.16. The top area (loaded electrode; Γ_L) and the bottom area (grounded
electrode; Γ_G) of the plate are completely covered with electrodes that are assumed
to be infinitely thin. Due to the electrodes, the electric potential V_e is equal on Γ_L
as well as on Γ_G, respectively. With a view to simplifying the FE procedure, we set
the boundary conditions for the electric and the mechanical fields to (boundary $\partial\Omega$
of Ω)

$$
\begin{aligned}
\mathbf{BC} \qquad\qquad & V_e = V_0 \quad \text{on} \quad \Gamma_L \times [0, T] \\
& V_e = 0 \quad \text{on} \quad \Gamma_G \times [0, T] \\
& \mathbf{n} \cdot \mathbf{D} = 0 \quad \text{on} \quad \Gamma_S \times [0, T] \\
& \mathbf{u} = 0 \quad \text{on} \quad \Gamma_G \times [0, T] \\
& [\mathbf{T}]^t \mathbf{n} = 0 \quad \text{on} \quad \Gamma_L \times [0, T] \\
& [\mathbf{T}]^t \mathbf{n} = 0 \quad \text{on} \quad \Gamma_S \times [0, T] \\
& \Gamma_L \cup \Gamma_S \cup \Gamma_G = \partial\Omega .
\end{aligned}
$$

Thus, the electrodes feature prescribed electric potentials, the mechanical displacements at Γ_G are fixed, and the piezoelectric material is not mechanically clamped. Moreover, gravity forces within the body are neglected, i.e., $\mathbf{f}_V = 0$. By applying $\mathbf{w}(\mathbf{r})$ (vector quantity) as test function for the displacement $\mathbf{u}$ and $w(\mathbf{r})$ (scalar quantity) for the electric potential V_e, the weak form of (4.94) results in

$$
\int_\Omega \varrho_0 \mathbf{w} \cdot \ddot{\mathbf{u}} \, d\Omega + \int_\Omega (\mathcal{B}\mathbf{w})^t [\mathbf{c}^E] \, \mathcal{B}\mathbf{u} \, d\Omega + \int_\Omega (\mathcal{B}\mathbf{w})^t [\mathbf{e}]^t \, \nabla V_e \, d\Omega = 0 \qquad (4.96)
$$

$$
\int_\Omega (\nabla w)^t [\mathbf{e}] \, \mathcal{B}\mathbf{u} \, d\Omega - \int_\Omega (\nabla w)^t [\varepsilon^S] \, \nabla V_e \, d\Omega = 0 . \qquad (4.97)
$$

In the next step, we introduce Lagrange ansatz functions for $\mathbf{u}$, $\mathbf{w}$, V_e, and w (see Sects. 4.2 and 4.3). Finally, this procedure leads to the algebraic system of equations in matrix form

$$
\begin{bmatrix} \mathbf{M_u} & 0 \\ 0 & 0 \end{bmatrix} \begin{bmatrix} \ddot{\mathbf{u}} \\ \ddot{\mathbf{v}}_e \end{bmatrix} + \begin{bmatrix} \mathbf{K_u} & \mathbf{K}_{uV_e} \\ \mathbf{K}_{uV_e}^t & -\mathbf{K}_{V_e} \end{bmatrix} \begin{bmatrix} \mathbf{u} \\ \mathbf{v}_e \end{bmatrix} = \begin{bmatrix} 0 \\ 0 \end{bmatrix} \qquad (4.98)
$$

with the vectors $\mathbf{u}$ and $\mathbf{v}_e$ representing approximated values for $\mathbf{u}$ and V_e at the nodes of the computational domain Ω, respectively. The matrices $\mathbf{M_u}$, $\mathbf{K_u}$ and $\mathbf{K}_{V_e}$ as well as the right-hand side $\mathbf{f}_{V_e}$ are assembled according to Sects. 4.2 and 4.3. The additional matrix $\mathbf{K}_{uV_e}$ is a consequence of piezoelectric coupling and computes as (number n_{elem} of elements within Ω)

$$
\mathbf{K}_{uV_e} = \bigwedge_{l=1}^{n_{elem}} \mathbf{K}^l ; \qquad \mathbf{K}^l = \begin{bmatrix} \mathbf{k}_{ij}^l \end{bmatrix} ; \qquad \mathbf{k}_{ij}^l = \int_{\Omega^l} \mathcal{B}_i^t [\mathbf{e}]^t \, \tilde{\mathcal{B}}_j \, d\Omega \qquad (4.99)
$$

with

$$
\tilde{\mathcal{B}}_j = \left[\frac{\partial N_j}{\partial x}, \frac{\partial N_j}{\partial y}, \frac{\partial N_j}{\partial z} \right]^t . \qquad (4.100)
$$

The FE method also enables for piezoelectric coupling different types of analysis, i.e., static, transient, harmonic as well as eigenfrequency analysis. For the static analysis, derivatives with respect to time are omitted in the algebraic system of equations. In the same manner as for mechanical and acoustic fields, (4.98) has to be transformed into the complex frequency domain in case of harmonic and eigenfrequency analysis. To conduct a transient analysis, one can apply the Newmark scheme (see Sect. 4.1.4).

At this point, it should be mentioned that the coupled algebraic system of equations in (4.98) does not contain any attenuation. In fact, attenuation occurs in every real system and, thus, has also to be regarded here. We can easily incorporate attenuation in the FE method for piezoelectricity if complex-valued material parameters are used (see Sect. 4.3.2).

There exist extended versions of the conventional FE method for piezoelectricity since piezoelectric devices demand external electronic components for electrical excitation or readout. This does not only refer to piezoelectric sensors and actuators but also to vibration-based energy harvesting devices that exploit piezoelectric materials. For such energy harvesting devices, we mostly need a special electric matching network, which converts AC voltage to DC voltage [7, 9]. Due to piezoelectric coupling, the electric matching network has a certain retroactive effect on the harvesting devices, which changes the output of the entire system. One can realistically analyze the behavior of the entire system if the mutual coupling between energy harvesting device and electric matching network is considered in the FE method. Possible work-arounds and solutions for this task can be found in [8, 10, 25].

Example

Various piezoelectric sensors and actuators are based on piezoceramic disks. On this account, let us study the behavior of such a piezoceramic disk by means of the FE method. The considered disk (diameter $d_S = 30$ mm; thickness $t_S = 2$ mm) is made of the piezoceramic material PZT-5A and polarized in thickness direction. The decisive material parameters can be found in Table 3.5 on p. 67. Both the top and bottom surfaces of the disk are completely covered by infinitely thin electrodes. Furthermore, we assume free mechanical vibrations, which means that the disk is not clamped and there do not act external forces.

Due to the symmetry of the disk, it makes sense to restrict the computational domain Ω to a rotationally symmetric configuration. Figure 4.17a shows the 2-D geometric arrangement of the utilized FE model. Along the rotation axis (z-axis), the mechanical displacements u_ρ in radial direction have to be zero. At the remaining boundaries of Ω (i.e., Γ_G, Γ_S, and Γ_L), the normal component of the mechanical stresses was set to zero because of free mechanical vibrations. Figure 4.17b displays the used structured computational grid, which consists of 480 square elements with an edge length of 0.25 mm. Again, the FE simulations were performed with quadratic Lagrange functions, i.e., h-FEM with $p_d = 2$.

At the beginning, we will take a look at the complex-valued electrical impedance $\underline{Z}_T(f)$ of the piezoceramic disk since this frequency-resolved quantity is often essential for practical applications. In the complex domain, $\underline{Z}_T(f)$ reads as (frequency f)

Fig. 4.17 a Rotationally symmetric configuration of computational domain Ω for piezoceramic disk with diameter $d_S = 30$ mm and thickness $t_S = 2$ mm; bottom electrode Γ_G; top electrode Γ_L; b 2-D computational grid comprising 480 squares

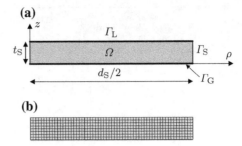

$$\underline{Z}_T(f) = \frac{\underline{u}_T(f)}{\underline{i}_T(f)} = \frac{\hat{u}_T \cdot e^{j(2\pi ft + \varphi_u)}}{\hat{i}_T \cdot e^{j(2\pi ft + \varphi_i)}} = \frac{\hat{u}_T}{\hat{i}_T} e^{j(\varphi_u - \varphi_i)} \qquad (4.101)$$

with the complex representations of the electric potential $\underline{u}_T$ between top and bottom electrodes and the electric current $\underline{i}_T$. The expressions $\hat{u}_T$ and $\hat{i}_T$ stand for the amplitudes of both quantities, whereas φ_u and φ_i indicate the phase angles. One way to calculate $\underline{Z}_T(f)$ is based on prescribing $\underline{u}_T(f) = \underline{V}_e(f)$ at the top electrode Γ_L, while the bottom electrode Γ_G is set to ground. In doing so, we require the electric current $\underline{i}_T(f)$, which results from the electric charge $\underline{Q}_T(f)$ on Γ_L through

$$\underline{i}_T(f) = j2\pi f \underline{Q}_T(f) = j2\pi f \int_{\Gamma_L} \underline{D}(f) \cdot \mathbf{n} \, d\Gamma$$

$$= j2\pi f \int_{\Gamma_L} \left\{ [\mathbf{e}] \mathcal{B}\underline{u}(f) - [\varepsilon^S] \nabla \underline{V}_e(f) \right\} d\Gamma . \qquad (4.102)$$

Here, the electric flux density $\mathbf{D}$ has been replaced by the term in the brackets of (4.95). $\underline{u}(f)$ and $\underline{V}_e(f)$ denote complex representations of the mechanical displacement $\mathbf{u}$ and electric potential V_e, respectively. By evaluating (4.101), we are, thus, able to compute $\underline{Z}_T(f)$. For FE simulations, this means that one has to sum the resulting electric charges along Γ_L. It is recommended to perform a harmonic FE analysis for reasons of efficiency. Alternatively to prescribing the electric potential, one can prescribe the charge on the top electrode. The electric impedance results then from determining the electric potential on Γ_L.

Figure 4.18 depicts the calculated magnitude $|\underline{Z}_T(f)|$ of the frequency-resolved electrical impedance for the considered piezoceramic disk. It can be clearly seen that the impedance curve contains pronounced local extrema. While local minima in $|\underline{Z}_T(f)|$ indicate resonances of mechanical vibrations, local maxima are related to antiresonances. The reason that resonances as well as antiresonances of mechanical vibrations become visible in the impedance curve lies in piezoelectric coupling. The combination of local minimum and maximum in the frequency range 60–90 kHz refers to mechanical vibrations in radial direction of the disk. At higher frequencies, there exist further combinations, which arise due to the overtones of these vibrations.

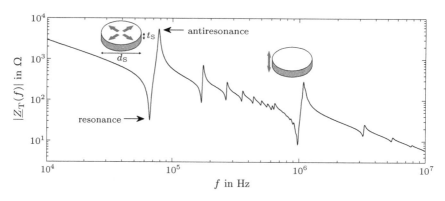

Fig. 4.18 Frequency-resolved electrical impedance $\left|\underline{Z}_T(f)\right|$ (magnitude) of considered piezoceramic disk with diameter $d_S = 30\,\text{mm}$ and thickness $t_S = 2\,\text{mm}$; piezoceramic material PZT-5A

However, the pronounced combination in the frequency range $800\,\text{kHz}$–$1.3\,\text{MHz}$ refers to mechanical vibrations in thickness direction. In Chap. 5, we will address the significance of impedance curves for material characterization.

Now, let us discuss the mechanical displacement u of the piezoceramic disk as a further simulation result. As a matter of principle, the FE method provides this quantity in each point of the computational domain. Figure 4.19 shows the normalized displacement amplitudes $\hat{u}_\rho(\rho, z)$ and $\hat{u}_z(\rho, z)$ in radial and thickness direction, respectively. Thereby, three different excitation frequencies f were selected, namely $10\,\text{kHz}$, $70\,\text{kHz}$, and $1\,\text{MHz}$. The frequency $70\,\text{kHz}$ lies in the range of the vibration resonance in radial direction, whereas $1\,\text{MHz}$ is close to the vibration resonance in thickness direction. For low excitation frequencies, the disk seems to vibrate uniformly in both directions. In contrast, high excitation frequencies cause a superposition of different vibration modes, which becomes apparent in strong local variations of $\hat{u}_\rho(\rho, z)$ and $\hat{u}_z(\rho, z)$.

Figure 4.20 depicts $\hat{u}_z(\rho)$ at the top surface Γ_L of the piezoceramic disk for different excitation frequencies. The displacement amplitude values were normalized to the amplitude $\hat{V}_e$ of the applied excitation voltage. In accordance with Fig. 4.19, $\hat{u}_z(\rho)$ remains almost constant at low frequencies and varies strongly at high frequencies. Besides, Fig. 4.20 reveals that electric excitations close to the vibration resonance in thickness direction generate extremely high values of $\hat{u}_z(\rho)$. The maximum value for $1\,\text{MHz}$ is more than ten times greater than that for $10\,\text{kHz}$. Because large displacements at high frequencies imply high surface velocities, ultrasonic transducers, which are based on thickness vibrations of piezoceramic disks, should be operated close to resonance.

$\hat{u}_\rho(\rho, z)$ $\hat{u}_z(\rho, z)$

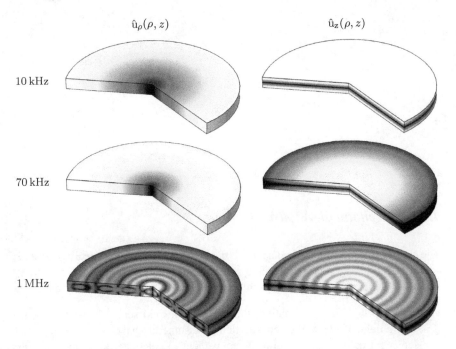

Fig. 4.19 Normalized displacement amplitudes of piezoceramic disk for different excitation frequencies f; (left) displacement amplitudes $\hat{u}_\rho(\rho, z)$ in radial direction; (right) displacement amplitudes $\hat{u}_z(\rho, z)$ in thickness direction; bright and dark colors indicate large and small amplitude values, respectively

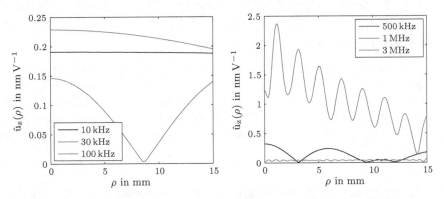

Fig. 4.20 Displacement amplitudes $\hat{u}_z(\rho)$ at top surface Γ_L of piezoceramic disk for different excitation frequencies; normalization with respect to amplitude $\hat{V}_e$ of excitation voltage

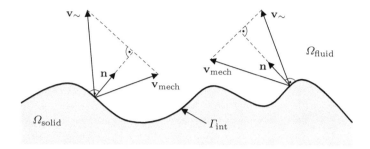

Fig. 4.21 Interface Γ_{int} between solid and fluid; normal components of mechanical velocity $\mathbf{v}_{mech}$ and acoustic particle velocity $\mathbf{v}_{\sim}$ coincide at Γ_{int}

4.5.2 Mechanical–Acoustic Coupling

The coupling of mechanical and acoustic fields in the FE method is decisive to study sound radiation of sources and sound reception of receivers, e.g., ultrasonic transducers. In general, we can distinguish between weak coupling and strong coupling.

- *Weak coupling:* In this situation, the mechanical field serves as source for the acoustic field. Thereby, the acoustic field is assumed to have no influence on the mechanical field. Weak coupling is convenient to calculate the sound radiation pattern of an ultrasonic transducer in air.
- *Strong coupling:* The mechanical field couples to the acoustic field and vice versa, i.e., coupling in both directions. As a result, the acoustic field alters the mechanical field. Strong coupling has to be considered for ultrasonic transducers operating in water.

Solid–Fluid Interface

To study mechanical–acoustic coupling, let us take a look at an interface Γ_{int} of a solid (elastic body) and a fluid (nonviscous liquid or gas), which is shown in Fig. 4.21. At each point of this interface, the normal components of both the mechanical velocity $\mathbf{v}_{mech}$ in the solid and the acoustic particle velocity $\mathbf{v}_{\sim}$ in the fluid have to coincide. With the normal vector $\mathbf{n}$, this continuity relation is given by

$$\text{condition I:} \quad \mathbf{n} \cdot (\mathbf{v}_{mech} - \mathbf{v}_{\sim}) = 0 . \tag{4.103}$$

Moreover, the fluid causes a certain pressure load on the solid at the interface Γ_{int}. This pressure load $\mathbf{f}_\Gamma$ corresponds to the mechanical stress $\mathbf{T_n}$ acting perpendicular to the surface of the solid and, consequently, computes as

$$\text{condition II:} \quad \mathbf{f}_\Gamma = \mathbf{T_n} = -\mathbf{n}p_{\sim} . \tag{4.104}$$

Note that both conditions have to be applied for strong coupling. In contrast, weak coupling is solely based on condition I.

By using the basic relations for linear continuum mechanics and acoustics

Table 4.3 Coupling conditions (condition I and II) at the solid–fluid interface Γ_{int} for pressure and potential formulation; fluid density $\varrho_{0\mathrm{f}}$ in the equilibrium state

	Pressure formulation	Potential formulation
Condition I	$\mathbf{n} \cdot \dfrac{\partial^2 \mathbf{u}}{\partial t^2} = -\dfrac{1}{\varrho_{0\mathrm{f}}} \dfrac{\partial p_\sim}{\partial \mathbf{n}}$	$\mathbf{n} \cdot \dfrac{\partial \mathbf{u}}{\partial t} = -\dfrac{\partial \Psi}{\partial \mathbf{n}}$
Condition II	$\mathbf{T_n} = -\mathbf{n} p_\sim$	$\mathbf{T_n} = -\mathbf{n} \varrho_{0\mathrm{f}} \dfrac{\partial \Psi}{\partial t}$

$$\mathbf{v}_{\mathrm{mech}} = \frac{\partial \mathbf{u}}{\partial t}, \quad \mathbf{v}_\sim = -\nabla \Psi \quad \text{and} \quad p_\sim = \varrho_{0\mathrm{f}} \frac{\partial \Psi}{\partial t}, \tag{4.105}$$

we can derive the two coupling conditions (condition I and II) at the solid–fluid interface for the acoustic pressure $p_\sim$ and the acoustic velocity potential Ψ, respectively. Table 4.3 contains the resulting equations for the different formulations.

FE Method for Mechanical–Acoustic Coupling

The numerical simulation of coupled mechanical–acoustic problems demands consideration of the coupling conditions at the interface Γ_{int}. In doing so, these conditions are incorporated as appropriate boundary conditions in the PDEs. Without limiting the generality, let us assume that deformations of the mechanical system are the only source for the acoustic field and the boundary conditions at the outer boundary of Ω_{fluid} are zero. Then, the weak forms for the mechanical domain Ω_{solid} and acoustic domain Ω_{fluid} in potential formulation become (density $\varrho_{0\mathrm{s}}$ of the solid)

$$\int_{\Omega_{\mathrm{solid}}} \varrho_{0\mathrm{s}} \mathbf{w} \cdot \ddot{\mathbf{u}} \, d\Omega + \int_{\Omega_{\mathrm{solid}}} (\mathcal{B}\mathbf{w})^{\mathrm{t}} [\mathbf{c}] \, \mathcal{B}\mathbf{u} \, d\Omega - \int_{\Gamma_{\mathrm{int}}} \mathbf{w} \cdot \mathbf{T_n} \, d\Gamma = \int_{\Omega_{\mathrm{solid}}} \mathbf{w} \cdot \mathbf{f_V} \, d\Omega \tag{4.106}$$

$$\int_{\Omega_{\mathrm{fluid}}} \frac{1}{c_0^2} w \, \ddot{\Psi} \, d\Omega + \int_{\Omega_{\mathrm{fluid}}} \nabla w \cdot \nabla \Psi \, d\Omega + \int_{\Gamma_{\mathrm{int}}} w \, \mathbf{n} \cdot \nabla \Psi \, d\Gamma = 0 \tag{4.107}$$

with the test function $\mathbf{w}(\mathbf{r})$ (vector quantity) for the mechanical field and $w(\mathbf{r})$ (scalar quantity) for the acoustic field. While for strong coupling, the interface integrals $\int_{\Gamma_{\mathrm{int}}}$ are required for both fields, weak coupling is solely based on the interface integral in (4.107).

Now, we can insert in (4.106) and (4.107) the coupling conditions from Table 4.3, which yields

$$\int_{\Omega_{\mathrm{solid}}} \varrho_{0\mathrm{s}} \mathbf{w} \cdot \ddot{\mathbf{u}} \, d\Omega + \int_{\Omega_{\mathrm{solid}}} (\mathcal{B}\mathbf{w})^{\mathrm{t}} [\mathbf{c}] \, \mathcal{B}\mathbf{u} \, d\Omega + \int_{\Gamma_{\mathrm{int}}} \varrho_{0\mathrm{f}} \mathbf{w} \cdot \mathbf{n} \frac{\partial \Psi}{\partial t} \, d\Gamma = \int_{\Omega_{\mathrm{solid}}} \mathbf{w} \cdot \mathbf{f_V} \, d\Omega \tag{4.108}$$

$$\int_{\Omega_{\mathrm{fluid}}} \frac{1}{c_0^2} w \, \ddot{\Psi} \, d\Omega + \int_{\Omega_{\mathrm{fluid}}} \nabla w \cdot \nabla \Psi \, d\Omega - \int_{\Gamma_{\mathrm{int}}} w \, \mathbf{n} \cdot \frac{\partial \mathbf{u}}{\partial t} \, d\Gamma = 0. \tag{4.109}$$

Subsequently to introducing Lagrange ansatz functions for $\mathbf{u}$, $\mathbf{w}$, Ψ, and w (see Sects. 4.3 and 4.4), we end up with a symmetric algebraic system of equations in matrix form

$$
\begin{bmatrix} \mathbf{M_u} & 0 \\ 0 & -\varrho_{0f}\mathbf{M}_\Psi \end{bmatrix} \begin{bmatrix} \ddot{\mathbf{u}} \\ \ddot{\boldsymbol{\Psi}} \end{bmatrix} + \underbrace{\begin{bmatrix} 0 & \mathbf{C}_{u\Psi} \\ \mathbf{C}_{u\Psi}^t & 0 \end{bmatrix}}_{\text{coupling}} \begin{bmatrix} \dot{\mathbf{u}} \\ \dot{\boldsymbol{\Psi}} \end{bmatrix} + \begin{bmatrix} \mathbf{K_u} & 0 \\ 0 & -\varrho_{0f}\mathbf{K}_\Psi \end{bmatrix} \begin{bmatrix} \mathbf{u} \\ \boldsymbol{\Psi} \end{bmatrix} = \begin{bmatrix} \mathbf{f_u} \\ 0 \end{bmatrix} .
$$

$$(4.110)$$

The vector $\mathbf{u}$ contains approximated values of the mechanical displacement at the nodes of Ω_{solid} and $\boldsymbol{\Psi}$ those of the acoustic velocity potential in Ω_{fluid}. The matrices $\mathbf{M_u}$, $\mathbf{M}_\Psi$, $\mathbf{K_u}$, $\mathbf{K}_\Psi$, and the right-hand side $\mathbf{f_u}$ are assembled as discussed in Sects. 4.3 and 4.4. By means of the matrix $\mathbf{C}_{u\Psi}$, we perform the coupling of the mechanical and acoustic fields. $\mathbf{C}_{u\Psi}$ is composed as

$$
\mathbf{C}_{u\Psi} = \sum_{l=1}^{n_{\text{int}}} \mathbf{C}_{u\Psi}^l ; \quad \mathbf{C}_{u\Psi}^l = \left[c_{ij}^l \right] ; \quad c_{ij}^l = \int_{\Gamma_e} \varrho_{0f}\left(\mathbf{N}_i N_j \right) \cdot \mathbf{n} \, d\Gamma \quad (4.111)
$$

with the part Γ_e of Γ_{int} and the number n_{int} of finite elements along the interface. Note that in case of weak coupling, $\mathbf{C}_{u\Psi}$ is omitted in the first line of (4.110). Therefore, the mechanical system can be computed directly without considering the acoustic field. The calculation of the acoustic field demands mechanical quantities at the interface Γ_{int} of Ω_{solid} and Ω_{fluid}.

Alternatively to the potential formulation, one is able to investigate mechanical–acoustic coupling with the pressure formulation. However, the resulting algebraic system of equations in matrix form is not symmetric anymore, which involves increasing computational effort.

As for the mechanical and acoustic fields, different types of analysis are possible here. A static analysis makes no sense because the acoustic field is based on alternating quantities. For the harmonic and eigenfrequency analysis, we transform (4.110) into the complex frequency domain. The transient analysis can be carried out again according to the Newmark scheme (see Sect. 4.1.4).

Although $\mathbf{C}_{u\Psi}$ refers to the first derivative with respect to time, it should not be confused with attenuation. In order to take attenuation into account within the coupled mechanical–acoustic system, we can use appropriate complex-valued material parameters.

The conventional mechanical–acoustic coupling demands computational grids that coincide at the interface Γ_{int} of mechanical domain Ω_{solid} and acoustic domain Ω_{fluid}. However, in many practical situations, we want to conduct independent spatial discretizations in both domains. This is especially important because the mechanical domain often calls for a finer computational grid (e.g., due to complicated structures) than the acoustic domain. An approach to get rid of the limitation at Γ_{int} is called nonconforming grids [14]. The idea of this approach lies in appropriately modifying the coupling matrix $\mathbf{C}_{\mathbf{u}\psi}$ in (4.111). Note that similar approaches also exist for other fields, e.g., electromagnetics.

In many cases, the acoustic domain Ω_{fluid} in mechanical–acoustic coupling is rather large. Owing to the comparatively low sound velocities of fluids (e.g., air), the acoustic wavelength takes small values, which implies a fine computational grid. Even though Ω_{fluid} is homogeneous, the required computational grid causes an excessive computational effort, especially for transient FE analysis. A potential remedy for the calculation of the pulse-echo behavior of ultrasonic transducers was suggested by Lerch et al. [16]. They model the piezoelectric ultrasonic transducer as well as a thin fluid layer with the FE method. The wave propagation inside the remaining fluid is described by the Helmholtz integral. Close to the transducer, the FE method gets coupled to the Helmholtz integral. A similar approach was recently presented in [18, 21]. Instead of the Helmholtz integral, the wave propagation in the fluid is described by the so-called spatial impulse response (SIR; see Sect. 7.1.2) of the ultrasonic transducer. Since there exist piecewise continuous solutions of the SIR for some shapes of the active transducer surface (e.g., piston-type transducer), this approach enables a highly efficient calculation of the sound field in the fluid. The method was successfully exploited to determine transient output signals of an acoustic microscope, which is based on a high-frequency ultrasonic transducer.

References

1. ANSYS: Software Package for Finite Element Method (2018). http://www.ansys.com
2. Babuška, I., Suri, M.: p and h-p versions of the finite element method, basic principles and properties. SIAM Rev. **36**(4), 578–632 (1994)
3. Bathe, K.J.: Finite Element Procedures. Prentice Hall, Upper Saddle River (1996)
4. Brebbia, C.A., Dominguez, J.: Boundary Elements: An Introductory Course, 2nd edn. WIT Press, Southampton (1996)
5. Briggs, W.L., Van Emden, H., McCormick, S.F.: A Multigrid Tutorial. Society for Industrial and Applied Mathematics (SIAM), Philadelphia (2000)
6. COMSOL Multiphysics: Software Package for Finite Element Method (2018). https://www.comsol.com
7. Dorsch, P., Gedeon, D., Weiß, M., Rupitsch, S.J.: Design and optimization of a piezoelectric energy harvesting system for asset tracking applications. Tech. Messen (2017). https://doi.org/10.1515/teme-2017-0102. (in press)
8. Elvin, N.G., Elvin, A.A.: A coupled finite element circuit simulation model for analyzing piezoelectric energy generators. J. Intell. Mater. Syst. Struct. **20**(5), 587–595 (2009)
9. Erturk, A., Inman, D.J.: Piezoelectric Energy Harvesting. Wiley, New York (2011)
10. Gedeon, D., Rupitsch, S.J.: Finite Element based system simulation for piezoelectric energy harvesting devices. J. Intell. Mater. Syst. Struct. **29**(7), 1333–1347 (2018)
11. Gui, W., Babuška, I.: The h, p and h-p versions of the finite element method in 1 dimension - part i. The error analysis of the p-version. Numer. Math. **49**(6), 577–612 (1986)
12. Hughes, T.J.R.: Finite Element Method: Linear Static and Dynamic Finite Element Analysis. Prentice Hall, Upper Saddle River (1987)
13. Hüppe, A., Kaltenbacher, M.: Stable matched layer for the acoustic conservation equations in the time domain. J. Comput. Acoust. **20**(1) (2012)
14. Kaltenbacher, M.: Numerical Simulation of Mechatronic Sensors and Actuators - Finite Elements for Computational Multiphysics, 3rd edn. Springer, Berlin (2015)

15. Lenk, A., Ballas, R.G., Werthschutzky, R., Pfeiefer, G.: Electromechanical Systems in Microtechnology and Mechatronics: Electrical, Mechanical and Acoustic Networks, their Interactions and Applications. Springer, Berlin (2010)
16. Lerch, R., Landes, H., Kaarmann, H.T.: Finite element modeling of the pulse-echo behavior of ultrasound transducers. In: Proceedings of International IEEE Ultrasonics Symposium (IUS), pp. 1021–1025 (1994)
17. LeVeque, R.J.: Finite Difference Methods for Ordinary and Partial Differential Equations. Society for Industrial and Applied Mathematics (SIAM), Philadelphia (2007)
18. Nierla, M., Rupitsch, S.J.: Hybrid seminumerical simulation scheme to predict transducer outputs of acoustic microscopes. IEEE Trans. Ultrason. Ferroelectr. Freq. Control **63**(2), 275–289 (2016)
19. PZFlex: Software Package for Finite Element Method (2018). https://pzflex.com
20. Rao, S.: The Finite Element Method in Engineering, 5th edn. Butterworth-Heinemann, Oxford (2010)
21. Rupitsch, S.J., Nierla, M.: Efficient numerical simulation of transducer outputs for acoustic microscopes. In: Proceedings of IEEE Sensors, pp. 1656–1659 (2014)
22. SIMetris GmbH: NACS Finite Element Analysis (2018). http://www.simetris.de
23. Szabo, B., Babuška, I.: Finite Element Analysis. Wiley, New York (1991)
24. Szabo, I.: Höhere Technische Mechanik. Springer, Berlin (2001)
25. Wu, P.H., Shu, Y.C.: Finite element modeling of electrically rectified piezoelectric energy harvesters. Smart Mater. Struct. **24**(9) (2015)
26. Ziegler, F.: Mech. Solids Fluids, 2nd edn. Springer, Berlin (1995)
27. Zienkiewicz, O., Tayler, R., Zhu, J.Z.: The Finite Element Method: Its Basis and Fundamentals, 7th edn. Butterworth-Heinemann, Oxford (2013)

Chapter 5
Characterization of Sensor and Actuator Materials

The design, comparison, and optimization of piezoelectric sensor as well as actuator devices demand reliable information (i.e., material data) about the materials that are used in the devices. For example, such information will be especially important if we want to predict the behavior of piezoelectric ultrasonic transducers by means of numerical simulations. Apart from the electrical control elements and read-out units, one can subdivide the decisive components of piezoelectric sensor and actuator devices into (i) *active* and (ii) *passive materials*:

- Active materials denote piezoelectric materials such as piezoceramics, which provide a pronounced mutual coupling of mechanical and electrical quantities.
- Passive materials refer to the remaining components of the devices, e.g., matching layers, damping elements, and housing in case of piezoelectric ultrasonic transducers.

In this chapter, the focus lies on characterizing the small-signal behavior of both active and passive materials. We are interested in precise material parameters, which are required in linearized mathematical relations like the material law for linear piezoelectricity (see Sect. 3.3) to describe the underlying material behavior. Because standard approaches for characterizing active and passive materials exhibit several disadvantages and inherent problems, various alternative characterization approaches have been suggested in the literature. Here, we will present a simulation-based approach, which is named *inverse method*. This approach has been mainly developed in the framework of the *Collaborative Research Center TRR39: Production Technologies for Light Metal and Fiber-Reinforced Composite-based Components with Integrated Piezoceramic Sensors and Actuators (PT-PIESA)* [10] and the *Research Unit FOR 894: Fundamental Flow Analysis of the Human Voice* [49]. Both projects were supported by the German Research Foundation DFG. The achieved research results have been recently published in the doctoral theses of Weiß [68] and Ilg [18] as well as in many further publications, e.g., [52, 56, 57, 69].

© Springer-Verlag GmbH Germany, part of Springer Nature 2019
S. J. Rupitsch, *Piezoelectric Sensors and Actuators*, Topics in Mining, Metallurgy and Materials Engineering, https://doi.org/10.1007/978-3-662-57534-5_5

The chapter starts with standard approaches for characterizing active and passive materials. The fundamentals of the inverse method regarding material characterization are detailed in Sect. 5.2. In Sects. 5.3 and 5.4, the inverse method will be used to identify the complete data set of piezoceramic materials and the dynamic mechanical behavior of homogenous passive materials such as thermoplastics.

5.1 Standard Approaches for Characterization

In this section, we will discuss standard approaches for characterizing linearized properties of piezoelectric materials (e.g., piezoceramics) and passive materials like metals as well as plastics. At first, the IEEE/CENELEC Standard on Piezoelectricity is outlined which allows determining the entire data set of a piezoceramic material. Section 5.1.2 deals with standard characterization approaches for passive materials, e.g., tensile and compression tests.

5.1.1 IEEE/CENELEC Standard on Piezoelectricity

Both the *IEEE Standard on Piezoelectricity 176–1987* [21] and the *CENELEC European Standard EN 50324-2* [13] suggest similar approaches for characterizing piezoelectric properties of ceramic materials and components. Here, we will address the underlying idea to determine the complete parameter set for describing the small-signal behavior of piezoceramic materials. This does not only include the identification procedure but also disadvantages and inherent problems of the standard approaches.[1]

Principle

Basically, the IEEE/CENELEC Standard exploits fundamental modes of mechanical vibrations within various test samples. Due to the electromechanical coupling of piezoelectric materials, the fundamental vibration modes also become visible in the frequency-resolved electrical impedance $\underline{Z}_T(f)$ (impedance curve) of the test samples.[2] In order to demonstrate this fact, let us consider a piezoceramic test sample of cylindrical shape (diameter $d_S = 6\,\text{mm}$; length $l_S = 30\,\text{mm}$) that is polarized in longitudinal direction and covered with electrodes on top as well as bottom surface. Figure 5.1 depicts results of a FE simulation for both magnitude $|\underline{Z}_T(f)|$ and phase $\arg\{\underline{Z}_T(f)\}$ of the frequency-resolved impedance between the electrodes. The impedance curve contains the fundamental vibration mode at $\approx 40\,\text{kHz}$ as well as

---

[1]The standard approaches for characterizing piezoceramic materials are abbreviated as IEEE/CENELEC Standard.

[2]The underscore denotes a complex-valued quantity, which is represented either by real and imaginary part or by magnitude and phase (see Chap. 2).

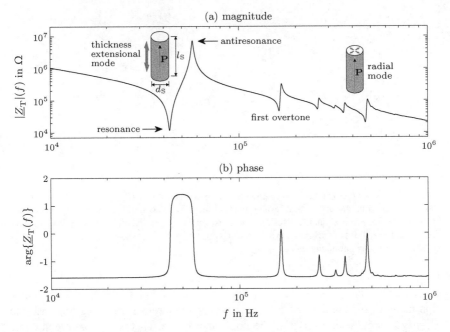

Fig. 5.1 **a** Magnitude and **b** phase of simulated frequency-resolved electrical impedance $\underline{Z}_T(f)$; test sample of cylindrical shape ($d_S = 6$ mm, $l_S = 30$ mm; piezoceramic material PIC255); material parameters from manufacturer (see Table 5.3)

its overtones, e.g., the first one at ≈ 160 kHz. Each vibration mode consists of a resonance–antiresonance pair, respectively. While resonance features a small value of $\left|\underline{Z}_T(f)\right|$, the impedance is comparatively large for antiresonance. These modes relate to mechanical vibrations in longitudinal direction of the cylindrical test sample. Besides, we can, however, observe further modes that originate from mechanical vibrations in radial direction.

The idea of the IEEE/CENELEC Standard lies in separating the fundamental vibration mode from other vibration modes within the investigated test samples. For the cylindrical sample, this means that its length should be much greater than its diameter, i.e., $l_S \gg d_S$. In such a case, we may assume monomodal mechanical vibrations for the fundamental mode. The dominating mechanical waves inside the sample can then be approximated by rather simple one-dimensional analytical relations depending solely on the coordinate in longitudinal direction. On the one hand, the analytical relations lead to the frequencies for resonance and antiresonance of the fundamental vibration mode. When these frequencies as well as the sample dimension l_S and material density ϱ_0 are known, it will be possible, on the other hand, to determine several material parameters of the piezoelectric material. For demonstration purpose, let us vary selected material parameters, namely the elastic compliance constant s_{33}^E, the electric permittivity ε_{33}^T, the piezoelectric strain constant d_{33}, and the damping coefficient α_d. As can be clearly seen in Fig. 5.2, each of these parameters remarkably

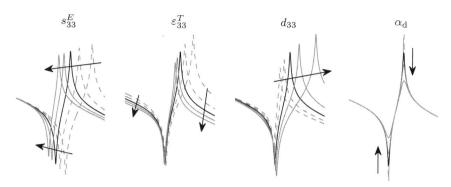

Fig. 5.2   Effect of distinct parameter variations on impedance curve $|\underline{Z}_T(f)|$ (magnitude) of fundamental vibration mode; test sample of cylindrical shape ($d_S = 6\,\mathrm{mm}$, $l_S = 30\,\mathrm{mm}$; piezoceramic material PIC255); initial data set (solid black lines) refers to manufacturer's data (see Table 5.3); dashed and solid lines in gray relate to negative and positive parameter changes, respectively; arrows indicate increasing parameter values

alters the frequency-resolved electrical impedance $\underline{Z}_T(f)$ for the fundamental vibration mode of the cylindrical sample. However, numerous parameters (e.g., s_{11}^E) do not have a significant impact on the impedance curve. That is why we are not able to determine the complete parameter set through the fundamental vibration mode of the cylindrical sample. In other words, additional test samples are required for identifying further material parameters.

Identification Procedure

In the following, we will detail the application of the IEEE/CENELEC Standard for characterizing piezoceramic materials of crystal class 6mm. For such materials, the three material tensors in d-form $\left[\mathbf{s}^E\right]$, $\left[\boldsymbol{\varepsilon}^T\right]$, and $[\mathbf{d}]$ consist altogether of ten independent parameters (see (3.32, p. 52)), i.e.,

- Elastic compliance constants: $s_{11}^E, s_{12}^E, s_{13}^E, s_{33}^E, s_{44}^E$.
- Electric permittivities: $\varepsilon_{11}^T, \varepsilon_{33}^T$.
- Piezoelectric strain constants: d_{31}, d_{33}, d_{15}.

To identify these material parameters by means of standard approaches, one has to analyze the frequency-resolved electrical impedance of at least four test samples offering five individual fundamental vibration modes. Apart from the (a) longitudinal length mode of a cylindrical test sample, the fundamental vibration modes relate to the (b) transverse length mode of a bar, the (c) radial as well as (d) thickness extensional mode of a disk, and the (e) thickness shear mode of a bar (see Fig. 5.3). The four test samples have to exhibit specific proportions of their geometric dimensions and distinct directions of electric polarization. The impedance measurements should be performed at low electrical excitation signals (e.g., $< 1\,\mathrm{V_{pp}}$) without any mechanical load; i.e., the test samples can vibrate freely.

The parameter identification requires the geometric samples dimensions as well as material density, but also characteristic frequencies at which resonance and

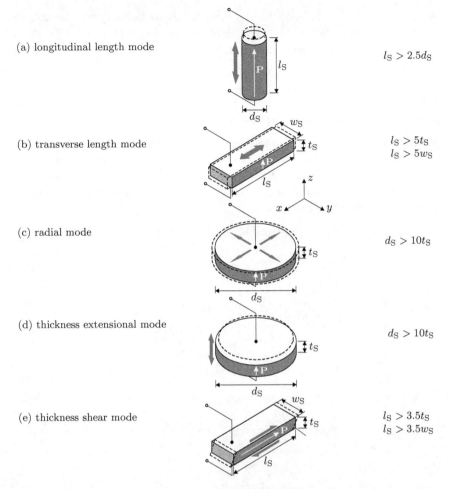

(a) longitudinal length mode

$l_S > 2.5d_S$

(b) transverse length mode

$l_S > 5t_S$
$l_S > 5w_S$

(c) radial mode

$d_S > 10t_S$

(d) thickness extensional mode

$d_S > 10t_S$

(e) thickness shear mode

$l_S > 3.5t_S$
$l_S > 3.5w_S$

Fig. 5.3 Fundamental vibration modes of piezoceramic test samples (i.e., cylinder, bar, and disk) utilized within IEEE/CENELEC Standard; electrodes with area A_S on top and bottom surface of test samples; **P** specifies direction of electric polarization; red arrows show dominating vibrations; aspect ratios (e.g., $l_S > 2.5d_S$) on right-hand side indicate recommended requirements for geometric sample dimensions

antiresonance occur, e.g., f_r. The characteristic frequencies correspond to local minima and maxima in the frequency-resolved impedance $\underline{Z}_T(f)$ and admittance $\underline{Y}_T(f)$ $= 1/\underline{Z}_T(f)$, which are given by

$$\underline{Z}_T(f) = R_T(f) + j X_T(f) \tag{5.1}$$
$$\underline{Y}_T(f) = G_T(f) + j B_T(f) \tag{5.2}$$

with R_T, X_T, G_T, and B_T being resistance, reactance, conductance, and susceptance, respectively. Before the parameter identification procedure is presented, let us define the relevant frequencies, namely (cf. Fig. 5.4c)

- Antiresonance frequency f_a: $X_T = 0$ and $\underline{Y}_T$ is small.
- Parallel resonance frequency f_p: maximum of R_T.
- Resonance frequency f_r: $X_T = 0$ and $\underline{Z}_T$ is small.
- Motional (series) resonance frequency f_s: maximum of G_T.

In addition to these frequencies, we need the capacity values C^T of the test samples at a frequency of $\approx 1\,\mathrm{kHz}$. When all information is available (i.e., characteristic frequencies, geometric sample dimensions, C^T and ϱ_0), the complete parameter set for a piezoceramic material of crystal class 6mm can be determined through the following mathematical formulas.

(a) *Longitudinal length mode* (see Fig. 5.3a)

$$\varepsilon_{33}^T = C^T \frac{l_S}{A_S} \tag{5.3}$$

$$s_{33}^D = \left[4\,\varrho_0\,f_p^2\,l_S^2\right]^{-1} \tag{5.4}$$

$$k_{33}^2 = \frac{\pi}{2}\frac{f_r}{f_a}\cot\left(\frac{\pi}{2}\frac{f_r}{f_a}\right) \tag{5.5}$$

$$s_{33}^E = s_{33}^D\left[1 - k_{33}^2\right]^{-1} \tag{5.6}$$

$$d_{33} = k_{33}\sqrt{\varepsilon_{33}^T\,s_{33}^E} \tag{5.7}$$

resulting parameters: ε_{33}^T, s_{33}^E, d_{33} (s_{33}^D, k_{33})

(b) *Transverse length mode* (see Fig. 5.3b)

$$\varepsilon_{33}^T = C^T \frac{t_S}{A_S} \tag{5.8}$$

$$s_{11}^E = \left[4\,\varrho_0\,f_s^2\,l_S^2\right]^{-1} \tag{5.9}$$

$$k_{31}^2 = \frac{\pi}{2}\frac{f_a}{f_r}\left[\frac{\pi}{2}\frac{f_a}{f_r} - \tan\left(\frac{\pi}{2}\frac{f_a}{f_r}\right)\right]^{-1} \tag{5.10}$$

$$d_{31} = k_{31}\sqrt{\varepsilon_{33}^T\,s_{11}^E} \tag{5.11}$$

resulting parameters: ε_{33}^T, s_{11}^E, d_{31} (k_{31})

(c) *Radial mode* (see Fig. 5.3c) required quantities: s_{11}^E, k_{31}

$$\varepsilon_{33}^T = C^T \frac{t_S}{A_S} \tag{5.12}$$

$$k_p = \sqrt{\frac{J_{mod}(\zeta_1) + \nu_P^p - 1}{J_{mod}(\zeta_1) - 2}} \tag{5.13}$$

The expression $J_{mod}(\zeta_1) = \zeta_1 J_0(\zeta_1) / J_1(\zeta_1)$ stands for the modified Bessel function of first order.[3] The so-called planar Poisson's ratio ν_P^p and the argument ζ_1 are given by

$$\nu_P^p = -\frac{s_{12}^E}{s_{11}^E} = 1 - \frac{2k_{31}^2}{k_p^2} \quad \text{and} \quad \zeta_1 = \eta_1 \frac{f_p}{f_s} \tag{5.14}$$

with η_1 being the solution to a transcendental equation.
resulting parameters: ε_{33}^T, s_{12}^E (k_p)

(d) *Thickness extensional mode* (Fig. 5.3d) required quantities: s_{11}^E, s_{12}^E, s_{33}^E

$$c_{33}^D = 4 \varrho_0 f_p^2 t_S^2 \tag{5.15}$$

$$k_t^2 = \frac{\pi}{2} \frac{f_r}{f_a} \cot\left(\frac{\pi}{2} \frac{f_r}{f_a}\right) \tag{5.16}$$

$$c_{33}^E = c_{33}^D \left(1 - k_t^2\right) \tag{5.17}$$

$$s_{13}^E = \sqrt{\frac{1}{2}\left[s_{33}^E \left(s_{11}^E + s_{12}^E \right) - \frac{s_{11}^E + s_{12}^E}{c_{33}^E} \right]} \tag{5.18}$$

resulting parameters: s_{13}^E (c_{33}^D, k_t, c_{33}^E)

(e) *Thickness shear mode* (see Fig. 5.3e)

$$\varepsilon_{11}^T = C^T \frac{t_S}{A_S} \tag{5.19}$$

$$s_{55}^D = \left[4 \varrho_0 f_p^2 t_S^2\right]^{-1} \tag{5.20}$$

$$k_{15}^2 = \frac{\pi}{2} \frac{f_r}{f_a} \cot\left(\frac{\pi}{2} \frac{f_r}{f_a}\right) \tag{5.21}$$

$$s_{55}^E = s_{55}^D \left[1 - k_{15}^2\right]^{-1} \tag{5.22}$$

$$d_{15} = k_{15} \sqrt{\varepsilon_{11}^T s_{55}^E} \tag{5.23}$$

resulting parameters: ε_{11}^T, $s_{55}^E = s_{44}^E$, d_{15} (s_{55}^D, k_{15})

---

[3] Bessel function $J_0(\zeta_1)$ of first kind and zero order; Bessel function $J_1(\zeta_1)$ of first kind and first order.

From the underlying characterization procedure, one can draw the following conclusions:

- Besides the ten decisive material parameters for crystal class 6mm, the standard procedure provides other useful quantities such as the electromechanical coupling factors k_{pq} (e.g., k_{31}), the so-called planar coupling factor k_p and thickness coupling factor k_t.
- The material parameter ε_{33}^T results from various test samples and, therefore, can be verified in several ways.
- Parameter computations in case of the fundamental vibration modes (a), (b), and (e) are carried out separately, while calculations for the (c) radial mode as well as (d) thickness extensional mode of the disk require parameters from other modes. It seems only natural that this fact may lead to strong errors in the final material parameter set if there exist already deviations for the required parameters, e.g., s_{11}^E.

Relevant Frequencies of Fundamental Vibration Modes

The characterization according to the IEEE/CENELEC Standard is based on the knowledge of various frequencies concerning resonance and antiresonance of fundamental vibration modes within appropriate test samples. To study the relevant frequencies in more detail, let us introduce a simple equivalent electrical circuit (Butterworth–Van Dyke equivalent circuit) that consists only of a few lumped elements, namely the series resonant circuit $R_1 L_1 C_1$ and the parallel capacitance C_0 (see Fig. 5.4a) [66]. Because this can be done for each fundamental vibration mode as well as its overtones, such equivalent electrical circuits are oftentimes used for describing the electrical behavior of piezoelectric materials. Figure 5.4b shows a characteristic curve for the frequency-resolved impedance magnitude $|\underline{Z}_{RC}(f)|$ of the electrical circuit with the minimum impedance value Z_m at frequency f_m and the maximum impedance value Z_n at f_n. In Fig. 5.4c, one can see the corresponding Nyquist plot with the axes $\Re\{\underline{Z}_{RC}\}$ and $\Im\{\underline{Z}_{RC}\}$. Contrary to the curve $|\underline{Z}_{RC}(f)|$, the Nyquist plot does not only provide information about f_m and f_n but also includes the relevant frequencies f_a, f_p, f_r as well as f_s that are utilized during parameter identification.

By simply approximating and adjusting the component values (i.e., R_1, L_1, C_1, and C_0) of the equivalent electrical circuit, we can emulate the electrical behavior in Fig. 5.1 for the fundamental vibration mode of the cylindrical test sample. Figure 5.5 illustrates the resulting frequency-resolved magnitudes of both impedance $\underline{Z}_{RC}(f)$ and admittance $\underline{Y}_{RC}(f)$ as well as their phases. From these curves, one can now determine the relevant frequencies for resonance and antiresonance, which take the values

- Resonance: $f_r = 42.98\,\text{kHz}$, $f_s = 42.96\,\text{kHz}$, $f_m = 42.94\,\text{kHz}$.
- Antiresonance: $f_a = 56.80\,\text{kHz}$, $f_p = 56.83\,\text{kHz}$, $f_n = 56.85\,\text{kHz}$.

It is obvious that the different relevant frequencies for resonance and antiresonance coincide very well, respectively. We may, thus, perform parameter identification through the IEEE/CENELEC Standard by exclusively considering the frequencies f_m

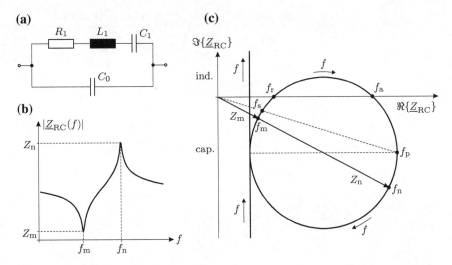

Fig. 5.4 **a** Simple equivalent electrical circuit to emulate electrical behavior for fundamental vibration mode of test samples; **b** impedance curve $\left|\underline{Z}_{RC}(f)\right|$ of equivalent circuit with minimum and maximum values at f_{m} and f_{n}, respectively; **c** Nyquist plot showing real and imaginary part of $\underline{Z}_{RC}(f)$ simultaneously as a function of f; frequencies f_{s}, f_{r}, f_{a} as well as f_{p} represent important quantities for parameter identification

and f_{n}, which can be easily figured out from $\left|\underline{Z}_{RC}(f)\right|$. Nevertheless, in case of piezoceramic materials with large attenuation, the relevant frequencies differ remarkably and, consequently, this simplification implies significant deviations in the resulting material parameters.

Disadvantages and Inherent Problems

The application of the IEEE/CENELEC Standard for characterizing piezoceramic materials is accompanied by certain disadvantages and inherent problems. There are two main points that need to be discussed. Firstly, one requires various test samples of the piezoceramic material differing in geometric shape, i.e., cylinder, bar, and disk. As already stated, the test samples have to exhibit specific proportions of their geometric dimensions as well as distinct directions of electric polarization. Therefore, the underlying identification procedure goes hand in hand with a time-consuming and expensive sample preparation.

Secondly, we have to assume monomodal mechanical vibrations within the test samples for the fundamental modes. This assumption will be, strictly speaking, only permitted if the test samples possess extreme aspect ratios [7], e.g., a very long cylinder with small diameter. Let us explain this inherent problem by means of the thickness extensional mode of a piezoceramic disk (diameter d_{S}, thickness $t_{S} = 3$ mm; see Fig. 5.3d), which is employed to determine the material parameter s_{13}^{E}. Figure 5.6 displays simulation results for the frequency-resolved electrical impedance $\left|\underline{Z}_{T}(f)\right|$ of the disk with respect to the aspect ratio d_{S}/t_{S} by varying d_{S}. To enhance comparability, $\left|\underline{Z}_{T}(f)\right|$ was multiplied by the disk area $A_{S} = d_{S}^{2}\pi/4$. In case of $d_{S} = 5t_{S}$, we

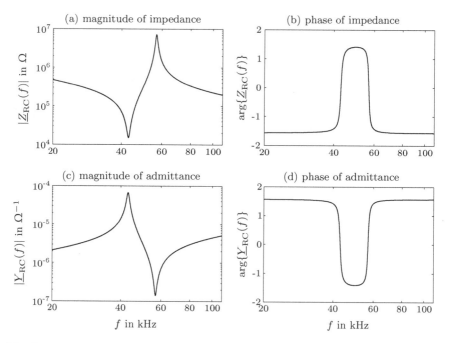

Fig. 5.5 Result of emulating fundamental vibration mode of cylindrical test sample in electrical behavior (see Fig. 5.1) through equivalent electrical circuit (see Fig. 5.4); **a** and **b** magnitude and phase of frequency-resolved impedance $\underline{Z}_{RC}(f)$; **c** and **d** magnitude and phase of frequency-resolved admittance $\underline{Y}_{RC}(f)$; component values: $R_1 = 15.0\,\text{k}\Omega$, $L_1 = 2.1\,\text{H}$, $C_1 = 6.4\,\text{pF}$, and $C_0 = 8.6\,\text{pF}$

are not able to figure out the relevant frequencies $f_r \approx f_m$ and $f_a \approx f_n$ (see (5.16)) since overtones of the radial mode strongly couple to the thickness extensional mode. Even though the disk's aspect ratio (e.g., $d_S = 20t_S$) complies the requirement $d_S > 10t_S$ of the IEEE/CENELEC Standard, this coupling may still be well pronounced. Consequently, f_m as well as f_n get shifted and the resulting material parameter s_{13}^E will be wrongly calculated. Due to decreasing magnitudes for increasing order of overtones and attenuation within the piezoceramic material, the coupling effects almost disappear for greater aspect ratios, e.g., $d_S = 50t_S$. It is, however, impossible to fabricate such a disk from the practical point of view.

5.1.2 Characterization Approaches for Passive Materials

As a matter of principle, most approaches for characterizing passive materials (e.g., plastics) are based on introducing defined mechanical excitations into a test sample. The mechanical response of the test sample to these external excitations naturally depends on both the geometric sample dimensions and the decisive mechanical material properties. In case of well-known sample dimensions, a test sample can be

Fig. 5.6 Simulated influence of radial mode of disk (piezoceramic material PIC255; thickness $t_S = 3$ mm) on its electrical behavior $|\underline{Z}_T(f)|$ for thickness extensional mode; variation of aspect ratio d_S/t_S by altering d_S; disk area A_S

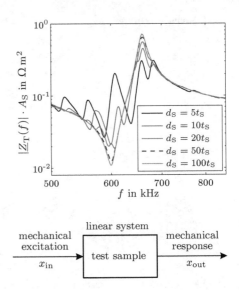

Fig. 5.7 Linear system for test sample with mechanical excitation x_{in} as input and mechanical response x_{out} as output

treated as a system with mechanical excitation and response denoting input x_{in} and output x_{out}, respectively (see Fig. 5.7). When x_{in} as well as x_{out} are sufficiently small, one will be able to additionally linearize the system by its transfer behavior $g(t, \mathbf{p})$, which exclusively depends on time t and the material parameters $\mathbf{p}$ [48]. Mechanical stresses T and mechanical strains S represent possible input and output quantities of the system, i.e.,

$$T_{in}(t), \ S_{in}(t) \ \xrightarrow{g(t,\mathbf{p})} \ T_{out}(t), \ S_{out}(t) \ . \tag{5.24}$$

Apart from the time domain, the transfer behavior of the test sample can also be described in the frequency domain. The corresponding transfer function $G(f, \mathbf{p})$ of the system depends on frequency f and the material parameters $\mathbf{p}$ again. Especially for harmonic mechanical excitations (i.e., $T_{in}(f)$), the frequency domain usually facilitates analyzing system behavior and, thus, should be preferred over the time domain.

Here, we will focus exclusively on two fundamental mechanical properties in the linear case (i.e., small-signal behavior), namely Young's modulus E_M and Poisson's ratio ν_P. Figure 5.8 shows common characterization approaches for passive materials. In the following, a brief explanation of these approaches is given.

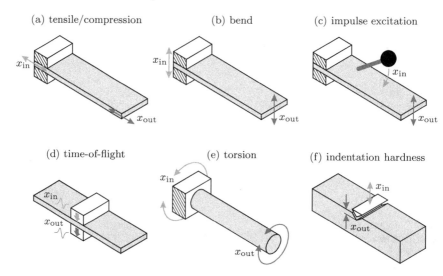

Fig. 5.8 Common characterization approaches for passive materials [18]; **a** tensile and compression test; **b** bend testing; **c** impulse excitation; **d** time-of-flight measurement; **e** torsion test; **f** indentation hardness test

Tensile and Compression Tests

Tensile and compression tests are the most common approaches for characterizing mechanical properties of materials [11]. The tests can be categorized in approaches for double-sided clamped test samples such as the classical tensile test and approaches for one-sided clamped samples that get dynamically excited. For a thin test sample (e.g., long cylinder) that is double-sided clamped, Hooke's law (see Sect. 2.2.3) yields Young's modulus through $E_M = T_{in}/S_{out}$. Thereby, T_{in} stands for the mechanical stress within the test sample and S_{out} indicates its resulting deformation due to an applied mechanical force in longitudinal direction. However, in case of more complex geometries, such simple mathematical relation may not exist and we need numerical simulations as well as additional measurement variables to figure out relevant material parameters (e.g., [72]).

A major advantage of the tensile and compression test lies in the simple experimental setup. Furthermore, it is possible to perform dynamic material characterization up to 1 kHz or even more by means of applying alternating forces.

Bend Testing

Many characterization approaches are based on harmonic or impulse excitations of a beam-shaped test sample (see *ASTM Standard E1876* [1]), which lead to mechanical vibrations of the beam. While harmonic excitations cause enforced vibrations, impulse excitations yield free vibrations of the test sample. The resulting bending does not only depend on the geometric sample dimensions but also on the material parameters. Through a comparison of measured bending resonances[4] to analytical

---

[4]Bending resonance means that the beam deflection is high at a certain frequency, which represents the so-called bending resonance frequency.

relations, it is possible to determine the Young's modulus of the investigated material. Primarily, we can find two theories for deducing appropriate relations: (i) the *Euler–Bernoulli beam theory* and (ii) the *Timoshenko beam theory* [43, 75]. The Euler–Bernoulli beam theory assumes that there arise exclusively tensile and compressive stresses above and below the neutral axis of the beam. This assumption is reasonable for thin beams, but in case of thicker beams, the computed bending will exhibit large deviations from reality. In contrast to the Euler–Bernoulli beam theory, the Timoshenko beam theory additionally considers shear stresses, which are especially important for higher vibrations modes of the beam.

In general, bend testing is able to provide frequency-dependent values for Young's modulus since one can determine E_M for each bending resonance of a beam, i.e., also for the overtones. However, due to the required beams of low thickness as well as low width, the number of suitable sample shapes is very limited.

Impulse Excitation

Besides the previously mentioned determination of Young's modulus, impulse excitation of a beam is oftentimes utilized to characterize attenuation behavior of a material, i.e., the damping ratio ξ_d [76]. This can be done by evaluating the decay of the time-dependent bending amplitude $\hat{x}$ for natural mechanical vibrations after impulse excitation. The damping ratio results from

$$\xi_d = \sqrt{\frac{D_d^2}{D_d^2 + 4\pi^2}} \quad \text{with} \quad D_d = \ln\left(\frac{\hat{x}_{i+1}}{\hat{x}_i}\right). \tag{5.25}$$

D_d stands for the logarithmic decrement of two timely subsequent amplitudes $\hat{x}_i$ and $\hat{x}_{i+1}$. Strictly speaking, impulse excitation provides a damping ratio that is only valid for the first bending resonance of the beam. Therefore, we will have to alter the geometric sample dimensions if the frequency-dependent behavior of ξ_d is needed.

Time-of-flight Measurements

Another widely used characterization approach is the so-called time-of-flight measurement (see *ASTM Standard E1875* [2]), which exploits the difference in propagation velocities of longitudinal and transverse waves within solid media. Commonly, such waves are excited and received by appropriate piezoelectric ultrasonic transducers (see Chap. 7) that have to be attached to the test sample. If the length of the propagation path l_w as well as the time-of-flight T_w for both wave types are known, we can easily determine the corresponding wave propagation velocities c_l and c_t with $c = l_w/T_w$. Because the wave propagation velocities are also given by the relations (material density ϱ_0; cf. (2.79) and (2.81), p. 25)

$$c_l = \sqrt{\frac{E_M(1 - \nu_P)}{\varrho_0(1 - 2\nu_P)(1 + \nu_P)}} \quad \text{and} \quad c_t = \sqrt{\frac{E_M}{2(1 + \nu_P)\varrho_0}}, \tag{5.26}$$

it is possible to determine Young's modulus E_M and Poisson's ratio ν_P of the investigated material.

The coupling of ultrasonic transducers to the test sample is a crucial point for time-of-flight measurements. Glycerin can be used for transmitting longitudinal waves between transducer and test sample, while honey is well suited for transmitting transverse waves. However, when a test sample exhibits small dimensions in wave propagation direction, the thickness of the coupling layers may remarkably influence the obtained results for E_M and ν_P since these layers are usually not considered in calculation. Moreover, a strongly attenuating material may damp and distort incoming waves at the receiving transducer in a manner that evaluation of time-of-flight gets impossible. Finally, the identified material parameters relate solely to the frequency f of propagating waves. If one is interested in frequency-dependent values for E_M and ν_P, f need to be changed which will demand several ultrasonic transducers.

Torsion and Indentation Hardness Tests

Torsion tests are particularly applied for very soft solids as well as biological tissue. This test can be carried out by a so-called rotational rheometer, which delivers frequency-dependent material parameters [36]. In order to introduce torsion, the investigated material sample is clamped in between two plates that are twisted against each other. Even though such rheometer can be used over a wide frequency range, it is oftentimes difficult to ensure perfect adhesion of material sample and plates. As a result, the obtained material parameters may show rather large deviations.

The indentation hardness test also represents a common approach to identify material parameters of soft as well as hard solids [11]. Thereby, a tip of defined shape is indented into the surface of a material sample. The applied tip force and resulting penetration depth or remaining impression yield characteristic parameters of the investigated material. However, indentation hardness tests are hardly suitable for dynamic characterization tasks.

Summing up, it can be stated that the standard approaches for characterizing linearized properties of piezoceramic materials and passive materials are accompanied by significant drawbacks and limitations. On the one hand, the standard approaches sometimes exploit improper simplifications (e.g., monomodal mechanical vibrations) that may lead to remarkable deviations in the identified data set. Several standard approaches do not allow determination of the aimed properties like frequency-dependent mechanical parameters of passive materials, on the other hand. Moreover, the approaches are in many cases very sophisticated regarding measurement setup and require a considerable amount of test samples. For these reasons, there exists a great demand for alternative characterization techniques. The inverse method offers many advantages over standard approaches and, thus, represents such an alternative characterization technique.

5.2 Fundamentals of Inverse Method

Before the so-called *inverse method* is applied for material characterization, let us discuss fundamentals of this method. Because the inverse method usually implies an ill-posed inverse problem, we will start with the mathematical definition of inverse problems as well as ill-posedness. Subsequently, the idea of the inverse method concerning material characterization is outlined. In Sect. 5.2.3, a special regularization approach will be shown that allows stabilization of ill-posed inverse problems. Finally, we introduce the iteratively regularized Gauss–Newton method, which is exploited in Sects. 5.3 and 5.4 to figure out a unique solution for the aimed parameter set.

5.2.1 Definition of Inverse Problems

A *direct problem* means that causes determined effects, whereas an *inverse problem* will arise when we have to identify causes from effects. That is the reason why direct and inverse problems are sometimes referred to as forward and backward calculations, respectively. From the mathematical point of view, one can formulate direct and inverse problems according to the following definition [50].

Definition 1 Let us assume a mathematical model $A : X \rightarrow Y$, which maps the set of causes X to the set of effects Y. In case of a direct problem, the effect $y \in Y$ is calculated from the cause, i.e., $y = Ax$ for $x \in X$. In case of an inverse problem, the cause results from the effect, i.e., for the effect $y \in Y$, one has to determine the cause $x \in X$ that fulfills $Ax = y$.

Various technical issues in engineering science such as tomographic imaging, system characterization, and parameter identification imply inverse problems (e.g., [22, 25, 53]). Usually, the resulting inverse problems are ill-posed, which represents the opposite of well-posed. Hadamard [16] introduced the term *well-posed problems* in 1923 by means of the definition given below.

Definition 2 Let $A : X \rightarrow Y$ be a mapping between the topological spaces X and Y. A problem (A, X, Y) will be called well-posed if three conditions are fulfilled:

(a) For any $y \in Y$, there exists a solution $x \in X$ satisfying $Ax = y$.
(b) The solution $x \in X$ is unique.
(c) The inverse mapping $A^{-1} : Y \rightarrow X$ is continuous; i.e., small changes in $y \in Y$ result in small changes in $x \in X$.

When at least one of these conditions is violated, the underlying mathematical problem will be named ill-posed.

Both the existence of a solution (Definition 2a) and its uniqueness (Definition 2b) are commonly fulfilled for inverse problems in practical situations. However, because

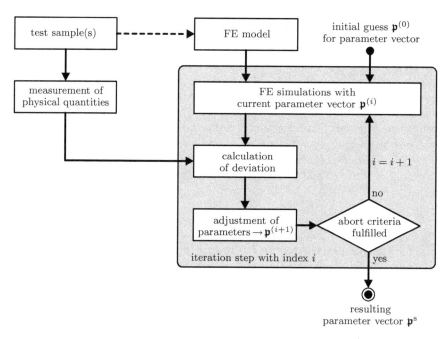

Fig. 5.9 Main steps and procedure of inverse method for material characterization [51]

inverse problems oftentimes lack stability (Definition 2c) due to certain data errors, we are faced with ill-posed inverse problems [25, 50]. In case of simulation-based material characterization, such data errors originate from imperfections in numerical modeling of the considered physical fields as well as from noisy measurements. To handle the missing stability, special regularization techniques are applied which transform the original ill-posed inverse problem into a well-posed problem (see Sect. 5.2.3).

5.2.2 Inverse Method for Material Characterization

The so-called inverse method for material characterization usually represents an ill-posed inverse problem. Basically, the idea of the inverse method lies in combining finite element (FE) simulations with measurements [33, 54, 57]. Figure 5.9 illustrates the main steps as well as the procedure of the method, which will be explained below.

Since the aim is determining decisive parameters of a material (e.g., piezoceramics), we have to investigate at least one or more test samples that are made of the investigated material. In a first step, appropriate FE models (i.e., computational grids) are required enabling precise FE simulations for the test sample(s). The numerical simulations demand an initial guess $\mathbf{p}^{(0)}$ for the desired parameter vector. Besides

FE simulations, the inverse method calls for measurements of physical quantities on the test sample(s) like the frequency-resolved electrical impedance. During the inverse method, deviations between measurements and simulations get iteratively reduced by adjusting the parameter vector $\mathbf{p}^{(i)}$ (iteration index i) on basis of these deviations. Starting from iteration step i, the adjusted parameter vector $\mathbf{p}^{(i+1)}$ is used to perform improved FE simulations at the subsequent iteration step $i + 1$, which are then compared again to measurements. This iterative procedure is repeated until one of the predefined abort criteria is fulfilled; e.g., the deviation between measurements and simulations falls below a certain limit. As a result of the inverse method, we obtain the parameter vector $\mathbf{p}^{s}$ containing characteristic quantities of the investigated material. Note that such quantities can be either explicit values for material properties or coefficients of mathematical relations, which rate the influence of additional quantities (e.g., excitation frequency) on material parameters [56, 69].

The choice of both the employed test sample(s) and the considered quantities for comparing measurements with simulations represents a crucial point for the inverse method. Actually, special attention should be paid to possible and available measurement quantities of the test sample(s). For example, if a desired material parameter does not affect any measured values, we will not be able to identify this parameter through the inverse method. Moreover, when variations of two material parameters cause similar or even identical changes in all measurement quantities, it will be impossible to identify those parameters uniquely. To avoid the mentioned problems, one should conduct parameter studies for the test sample(s) in advance [19, 52, 56]. Such parameter studies substantially support the design of appropriate test sample(s) concerning their shape and geometric dimensions. They can also provide quantitative information about parameter tolerances due to uncertainties of sample dimensions and geometrical variations of the utilized measuring points [18].

In the following two subsections, we will detail the mathematical background for iteratively adjusting the parameter vector.

5.2.3 Tikhonov Regularization

Let us assume that the simulated and measured quantities are collected in the vectors $\mathbf{q}_{\text{sim}}(\mathbf{p})$ and $\mathbf{q}_{\text{meas}}$ of dimension N_{I}, respectively. In the course of the inverse method, it is advisable to minimize the quadratic deviation[5] between $\mathbf{q}_{\text{sim}}(\mathbf{p})$ and $\mathbf{q}_{\text{meas}}$ with respect to the parameter vector $\mathbf{p}$ containing $N_{\mathbf{p}}$ entries. This so-called *least squares method* [26] yields the least squares functional $\Psi_{\text{L}}(\mathbf{p})$, which has to be minimized, i.e.,

$$\min_{\mathbf{p}} \Psi_{\text{L}}(\mathbf{p}) = \min_{\mathbf{p}} \left\| \mathbf{q}_{\text{sim}}(\mathbf{p}) - \mathbf{q}_{\text{meas}} \right\|_{2}^{2} . \tag{5.27}$$

---

[5]The quadratic deviation corresponds to the squared L_2 norm of the deviation.

However, in case of ill-posed inverse problems, the least squares method will be accompanied by an unrealistic solution $\mathbf{p}^s$ because stability is mostly violated during minimizing $\Psi_L(\mathbf{p})$ due to data errors, e.g., noisy measurements. To get rid of such problems, one should apply special regularization approaches like *Tikhonov regularization* [12, 25, 53]. In doing so, the original ill-posed problem gets replaced by a neighboring but well-posed regularized problem. The resulting Tikhonov functional $\Psi_T(\mathbf{p})$ that need to be minimized reads as

$$
\begin{aligned}
\min_{\mathbf{p}} \Psi_T(\mathbf{p}) = \min_{\mathbf{p}} &\left\{ \left\| \mathbf{q}_{sim}(\mathbf{p}) - \mathbf{q}_{meas} \right\|_2^2 + \zeta_R \left\| \mathbf{p} - \mathbf{p}^{(0)} \right\|_2^2 \right\} \\
= \min_{\mathbf{p}} &\left\{ \left[\mathbf{q}_{sim}(\mathbf{p}) - \mathbf{q}_{meas} \right]^t \left[\mathbf{q}_{sim}(\mathbf{p}) - \mathbf{q}_{meas} \right] \right. \\
&\left. + \zeta_R \left(\mathbf{p} - \mathbf{p}^{(0)} \right)^t \left(\mathbf{p} - \mathbf{p}^{(0)} \right) \right\}
\end{aligned}
\tag{5.28}
$$

with the so-called *regularization parameter* ζ_R being a positive number and the initial guess $\mathbf{p}^{(0)}$ of the aimed parameter vector. The combination of ζ_R and the expression $\| \mathbf{p} - \mathbf{p}^{(0)} \|$ leads to a certain penalization of unwanted oscillations in $\mathbf{p}$. That is the reason why the additional expression in (5.28) is oftentimes referred to as penalty term. In general, ζ_R determines the compromise between approximating the original ill-posed problem and stability of the regularized problem. Small values of ζ_R result in good approximations at the cost of decreased stability, while the regularized problem may be far away from the original one in case of large ζ_R-values. Therefore, the key in regularization is to choose this value appropriately.

The minimum of the Tikhonov functional $\Psi_T(\mathbf{p})$ follows from the first-order derivative of $\Psi_T(\mathbf{p})$ with respect to the parameter vector $\mathbf{p}$. At the point $\mathbf{p}^s$, the derivative becomes

$$
\begin{aligned}
\left. \frac{\partial \Psi_T(\mathbf{p})}{\partial \mathbf{p}} \right|_{\mathbf{p}=\mathbf{p}^s} = &\left[\left. \frac{\partial \mathbf{q}_{sim}(\mathbf{p})}{\partial \mathbf{p}} \right|_{\mathbf{p}=\mathbf{p}^s} \right]^t \left[\mathbf{q}_{sim}(\mathbf{p}^s) - \mathbf{q}_{meas} \right] \\
&+ \left[\mathbf{q}_{sim}(\mathbf{p}^s) - \mathbf{q}_{meas} \right]^t \left[\left. \frac{\partial \mathbf{q}_{sim}(\mathbf{p})}{\partial \mathbf{p}} \right|_{\mathbf{p}=\mathbf{p}^s} \right] + 2\zeta_R \left([\mathbf{p}]^s - \mathbf{p}^{(0)} \right) .
\end{aligned}
\tag{5.29}
$$

If $\mathbf{p}^s$ corresponds to the aimed solution and, thus, to the minimum of $\Psi_T(\mathbf{p})$, this derivative will be zero, i.e.,

$$
\left[\left. \frac{\partial \mathbf{q}_{sim}(\mathbf{p})}{\partial \mathbf{p}} \right|_{\mathbf{p}=\mathbf{p}^s} \right]^t \left[\mathbf{q}_{sim}(\mathbf{p}^s) - \mathbf{q}_{meas} \right] + \zeta_R \left(\mathbf{p}^s - \mathbf{p}^{(0)} \right) = 0
\tag{5.30}
$$

has to be fulfilled. To check whether $\mathbf{p}^s$ relates to a local minimum or a local maximum also satisfying (5.30), evaluation of the second-order derivative would be required in addition. Since we will apply an iteratively regularized Gauss–Newton method, which minimizes $\Psi_T(\mathbf{p})$ in each iteration step, it is, however, impossible that $\mathbf{p}^s$ leads to a local maximum of $\Psi_T(\mathbf{p})$.

5.2.4 Iteratively Regularized Gauss–Newton Method

With a view to explaining the iteratively regularized Gauss–Newton method, let us consider a mathematical function $\Gamma(\mathbf{p})$ depending on the parameter vector $\mathbf{p}$. For this function, we seek the solution $\mathbf{p}^s$ that fulfills the equation

$$\Gamma(\mathbf{p}^s) \equiv 0 , \tag{5.31}$$

which can be interpreted as an optimization problem. If $\Gamma(\mathbf{p})$ denotes a nonlinear function, *Newton's method* [8] will represent a common approach to solve (5.31). Thereby, the nonlinear function gets iteratively approximated through a series of linear functions. In the following, the starting point of Newton's method shall be the parameter vector $\mathbf{p}^{(i)}$ for the ith iteration step. To compute the parameter vector $\mathbf{p}^{(i+1)}$ for the subsequent iteration step, one has to perform a Taylor series expansion of $\Gamma(\mathbf{p})$ at the point $\mathbf{p}^{(i)}$, which takes the form

$$\Gamma(\mathbf{p}^{(i)}) + \left.\frac{\partial\Gamma(\mathbf{p})}{\partial\mathbf{p}}\right|_{\mathbf{p}=\mathbf{p}^{(i)}} \cdot \left(\mathbf{p}^{(i+1)} - \mathbf{p}^{(i)}\right) = 0 . \tag{5.32}$$

Rearranging (5.32) yields then the updated solution

$$\mathbf{p}^{(i+1)} = \mathbf{p}^{(i)} - \left[\left.\frac{\partial\Gamma(\mathbf{p})}{\partial\mathbf{p}}\right|_{\mathbf{p}=\mathbf{p}^{(i)}}\right]^{-1} \Gamma(\mathbf{p}^{(i)}) \tag{5.33}$$

to the nonlinear problem. Newton's method provides approximations that converge quadratically (i.e., quite fast) to the aimed solution $\mathbf{p}^s$. Because this method exhibits, however, only local convergence, the initial guess $\mathbf{p}^{(0)}$ of the parameter vector needs to be sufficiently close to $\mathbf{p}^s$. Otherwise, Newton's method will end up at a local minimum, which might be far away from the global minimum, i.e., from $\mathbf{p}^s$.

As mentioned in the previous subsection, the aimed solution has to satisfy (5.30). We are, thus, able to interpret this equation as a nonlinear function $\Gamma(\mathbf{p}^{(i)})$ for iteration index i, i.e.,

$$\Gamma(\mathbf{p}^{(i)}) = \left[\left.\frac{\partial\mathbf{q}_{sim}(\mathbf{p})}{\partial\mathbf{p}}\right|_{\mathbf{p}=\mathbf{p}^{(i)}}\right]^t \left[\mathbf{q}_{sim}(\mathbf{p}^{(i)}) - \mathbf{q}_{meas}\right] + \zeta_R\left(\mathbf{p}^{(i)} - \mathbf{p}^{(0)}\right) . \tag{5.34}$$

According to (5.33), Newton's method demands the first-order derivative $\partial\Gamma(\mathbf{p})/\partial\mathbf{p}$ at $\mathbf{p}^{(i)}$, which becomes

$$\left.\frac{\partial \Gamma(\mathbf{p})}{\partial \mathbf{p}}\right|_{\mathbf{p}=\mathbf{p}^{(i)}} = \underbrace{\left[\left.\frac{\partial^2 \mathbf{q}_{\text{sim}}(\mathbf{p})}{\partial \mathbf{p}^2}\right|_{\mathbf{p}=\mathbf{p}^{(i)}}\right]^{t}}_{\text{Hessian matrix}} \left[\mathbf{q}_{\text{sim}}(\mathbf{p}^{(i)}) - \mathbf{q}_{\text{meas}}\right] \tag{5.35}$$

$$+ \left[\left.\frac{\partial \mathbf{q}_{\text{sim}}(\mathbf{p})}{\partial \mathbf{p}}\right|_{\mathbf{p}=\mathbf{p}^{(i)}}\right]^{t} \left[\left.\frac{\partial \mathbf{q}_{\text{sim}}(\mathbf{p})}{\partial \mathbf{p}}\right|_{\mathbf{p}=\mathbf{p}^{(i)}}\right] + \zeta_{\text{R}}\mathbf{I}$$

with the identity matrix $\mathbf{I}$ of dimension $N_{\text{I}} \times N_{\text{I}}$. In contrast to Newton's method, the second-order derivative of $\mathbf{q}_{\text{sim}}(\mathbf{p})$ with respect to $\mathbf{p}$ (*Hessian matrix*) is neglected within the framework of the so-called *Gauss–Newton method*. In doing so, (5.35) reads as

$$\left.\frac{\partial \Gamma(\mathbf{p})}{\partial \mathbf{p}}\right|_{\mathbf{p}=\mathbf{p}^{(i)}} \approx \left[\left.\frac{\partial \mathbf{q}_{\text{sim}}(\mathbf{p})}{\partial \mathbf{p}}\right|_{\mathbf{p}=\mathbf{p}^{(i)}}\right]^{t} \left[\left.\frac{\partial \mathbf{q}_{\text{sim}}(\mathbf{p})}{\partial \mathbf{p}}\right|_{\mathbf{p}=\mathbf{p}^{(i)}}\right] + \zeta_{\text{R}}\mathbf{I} . \tag{5.36}$$

This simplification is permitted since the deviation $\mathbf{q}_{\text{sim}}(\mathbf{p}^{(i)}) - \mathbf{q}_{\text{meas}}$ gets minimized during iterative parameter identification, whereby the influence of the Hessian matrix also decreases. When (5.34) and (5.36) are inserted in (5.33), we can calculate the corrected parameter vector $\mathbf{p}^{(i+1)}$.

Up to now, the question concerning the optimal choice of the regularization parameter ζ_{R} still remains open. Without knowledge of the solution $\mathbf{p}^{\text{s}}$, it is hardly possible to explicitly determine ζ_{R} in advance. Therefore, ζ_{R} should be adjusted iteratively during parameter identification. The *iteratively regularized Gauss–Newton method* [3, 6] pursued exactly that strategy by sequentially reducing ζ_{R} from one iteration step to the next, i.e., ζ_{R} tends to zero. Strong reductions of ζ_{R} accelerate the identification procedure but may cause instabilities in the obtained parameter vectors. As stability constitutes the prime objective of the identification procedure, slight reductions of ζ_{R} (e.g., $\zeta_{\text{R}}^{(i+1)} = 0.8\zeta_{\text{R}}^{(i)}$) should, thus, be favored against strong ones.

To achieve a compact form of the iteratively regularized Gauss–Newton method, let us introduce abbreviations for the deviation of simulations and measurements

$$\mathbf{d}_{\text{I}}(\mathbf{p}^{(i)}) = \mathbf{q}_{\text{sim}}(\mathbf{p}^{(i)}) - \mathbf{q}_{\text{meas}} \tag{5.37}$$

as well as for the first-order derivative $\partial \mathbf{q}_{\text{sim}}(\mathbf{p})/\partial \mathbf{p}$

$$\mathbf{J}(\mathbf{p}^{(i)}) = \left.\frac{\partial \mathbf{q}_{\text{sim}}(\mathbf{p})}{\partial \mathbf{p}}\right|_{\mathbf{p}=\mathbf{p}^{(i)}} = \left.\frac{\partial \mathbf{d}_{\text{I}}(\mathbf{p})}{\partial \mathbf{p}}\right|_{\mathbf{p}=\mathbf{p}^{(i)}} \tag{5.38}$$

that is usually named *Jacobian matrix* (dimension $N_{\text{I}} \times N_{\text{I}}$). The correction vector $\mathbf{c}^{(i)}$ (dimension $N_{\mathbf{p}}$) to update the parameter vector by means of

$$\mathbf{p}^{(i+1)} = \mathbf{p}^{(i)} + \mathbf{c}^{(i)} \tag{5.39}$$

results then in

$$\mathbf{c}^{(i)} = -\left[\mathbf{J}(\mathbf{p}^{(i)})^{t}\,\mathbf{J}(\mathbf{p}^{(i)}) + \zeta_{R}^{(i)}\mathbf{I}\right]^{-1}$$
$$\cdot\left[\mathbf{J}(\mathbf{p}^{(i)})^{t}\cdot\mathbf{d}_{I}(\mathbf{p}^{(i)}) + \zeta_{R}^{(i)}\big(\mathbf{p}^{(i)} - \mathbf{p}^{(0)}\big)\right]\,. \qquad (5.40)$$

Finally, we have to answer the question when the iteration procedure should be terminated, i.e., at which iteration index i. For this task, Morozov [39] suggested the so-called *discrepancy principle*, which states that one should continue iteration until the L_2 norm of $\mathbf{d}_I(\mathbf{p}^{(i)})$ is greater than the data errors due to modeling imperfections as well as noisy measurements. Both error sources are, however, extremely difficult to be quantified in practical situations. On account of this fact, it is recommended to apply an empirical abort criterion instead. In case of the presented characterization of piezoceramic materials (see Sect. 5.3) and passive materials (see Sect. 5.4), a converging parameter vector $\mathbf{p}$ turned out to be such a criterion (e.g., [52, 56]).

There exist various other approaches, which allow solving nonlinear and ill-posed inverse problems. For example, the nonlinear Landweber iteration, the Levenberg–Marquardt method as well as the steepest descent and minimal error method represent such approaches [25, 32]. Generally speaking, the approaches mainly differ in the applied regularization and the resulting convergence rates. We exclusively exploit the iteratively regularized Gauss–Newton method for material characterizations presented hereinafter. Nevertheless, the other approaches can be used in a similar manner for this task.

5.3 Inverse Method for Piezoceramic Materials

The material parameters of piezoceramic materials are commonly identified by means of the IEEE/CENELEC Standard (see Sect. 5.1.1). Thereby, the frequency-resolved electrical impedances of various test samples (e.g., disk and bar) need to be acquired. Practical test samples commonly do not, however, meet the requirements of the IEEE/CENELEC Standard resulting in a certain coupling of different mechanical vibration modes. Consequently, the assumption of monomodal mechanical vibrations does not hold anymore and the identified material parameters are not appropriate for precise numerical simulations. That was the reason why several research groups have developed alternative approaches concerning characterization of piezoceramics, e.g., [24, 31, 33, 42]. Let us discuss some of the alternative approaches in more detail.

Sherrit et al. [61] presented in 2011 an experimental procedure that provides the complete data set including damping factors. In a first step, they evaluate the electrical impedance for radial and thickness extensional mode of a single piezoceramic disk. Subsequently, a piezoceramic bar is cut from the disk and analyzed according to the IEEE/CENELEC Standard. Pérez et al. [44] and in similar form Jonsson et al. [23] suggested identification approaches allowing precise FE simulations of piezoceramic disks. Their approaches are based on adjustments of FE simulations to measured impedance curves for the investigated disk. The complete set of material parameters including attenuation is determined by considering significant vibration modes (i.e., radial, edge, and thickness extension) as well as overtones. For validation purpose, the wave propagation velocity of longitudinal waves is measured and compared with analytical results. In [35], impedance measurements and acquired propagation velocities of longitudinal as well as shear waves are combined with FE simulations to completely characterize piezoceramics. Cappon and Keesman [9] described in 2012 a procedure that is also based on comparisons of FE simulations and measurements for the electrical impedance. They analyze a piezoceramic plate and restrict characterization to those parameters, which are not provided by the manufacturer. Recently, Kulshreshtha et al. [30] published a simulation-based procedure exploiting ring-shaped electrodes on a piezoceramic disk. The special electrode structures increase sensitivity of several material parameters on the frequency-resolved electrical impedance, and therefore, facilitate parameter identification.

It can be concluded that there exist various approaches addressing parameter identification for piezoceramic materials. Several characterization techniques aim at highly precise simulations for a particular sample shape and, thus, do not consider the applicability of the identified data set for different shapes. In contrast, various other approaches enable the complete characterization of piezoceramics by conducting extensive measurements and/or sample preparations. But only to some extent, these approaches are feasible to determine temperature-dependent material parameters, which might be especially important concerning piezoceramics in sensor and actuator devices. Because of the just mentioned reasons, there is a need for a characterization procedure that mainly copes with three aspects [52]: (i) Only a few test samples and measurements are required, (ii) the entire data set of the piezoceramic material is provided with respect to temperature, and (iii) the identified data set is applicable for sufficiently precise FE simulations of various sample shapes and configurations.

Here, let us exploit the inverse method to identify the complete parameter set of piezoceramic materials. In particular, the identification will be applied to characterize two ferroelectrically soft materials of crystal class 6mm, namely PIC255 and PIC155 from the company PI Ceramic GmbH [45]. Without limiting the generality, the inverse method is carried out for the e-form. In Sect. 5.3.1, we will introduce a simple phenomenological approach to account for attenuation within piezoceramic materials. Section 5.3.2 deals with measurable quantities (e.g., electrical impedance) that can serve as input for the inverse method. Afterward, an explanation of the two different block-shaped test samples including their numerical modeling is given. Moreover, we will justify the necessity of both test samples for parameter identification. Section 5.3.4 briefly addresses the underlying mathematical procedure and contains various approaches to figure out a proper initial guess for parameter identification of piezoceramics. Before temperature-dependent material parameters and coupling factors for PIC255 and PIC155 are presented in Sect. 5.3.6, a strategy to efficiently conduct the inverse method will be shown.

5.3.1 Material Parameters and Modeling of Attenuation

Basically, one may conduct the inverse method for characterizing piezoceramic materials in each form of the material law for linear piezoelectricity, e.g., d-form. Since the e-form can be, however, directly implemented in FE simulations of piezoceramic materials (see Sect. 4.5.1), let us utilize this form in the framework of the inverse method. For the crystal class 6mm, the tensors of elastic stiffness constants $[\mathbf{c}^E]$, of electric permittivities $[\varepsilon^S]$, and of piezoelectric stress constants $[\mathbf{e}]$ become (cf. (3.32, p. 52))

$$[\mathbf{c}^E] = \begin{bmatrix} c_{11}^E & c_{12}^E & c_{13}^E & 0 & 0 & 0 \\ c_{12}^E & c_{11}^E & c_{13}^E & 0 & 0 & 0 \\ c_{13}^E & c_{13}^E & c_{33}^E & 0 & 0 & 0 \\ 0 & 0 & 0 & c_{44}^E & 0 & 0 \\ 0 & 0 & 0 & 0 & c_{44}^E & 0 \\ 0 & 0 & 0 & 0 & 0 & \left(c_{11}^E - c_{12}^E\right)/2 \end{bmatrix}, \tag{5.41}$$

$$[\varepsilon^S] = \begin{bmatrix} \varepsilon_{11}^S & 0 & 0 \\ 0 & \varepsilon_{11}^S & 0 \\ 0 & 0 & \varepsilon_{33}^S \end{bmatrix}, \quad [\mathbf{e}] = \begin{bmatrix} 0 & 0 & 0 & 0 & e_{15} & 0 \\ 0 & 0 & 0 & e_{15} & 0 & 0 \\ e_{31} & e_{31} & e_{33} & 0 & 0 & 0 \end{bmatrix}. \tag{5.42}$$

The three tensors contain altogether ten independent entries, namely

- $c_{11}^E, c_{12}^E, c_{13}^E, c_{33}^E, c_{44}^E$
- $\varepsilon_{11}^S, \varepsilon_{33}^S$
- e_{31}, e_{33}, e_{15}.

Just as for the IEEE/CENELEC Standard, the material parameters are supposed to exhibit a negligible dependence on frequency for typical operating frequencies f of piezoceramic sensors and actuators, i.e., $f < 10\,\mathrm{MHz}$.

To account for attenuation within piezoceramic materials in case of harmonic excitation, it is common to introduce appropriate damping coefficients resulting in imaginary parts of the material tensors. The damping coefficients might show a complicated frequency dependence, but for phenomenological modeling of attenuation, they are usually assumed to be constant (e.g., [17, 60]). Strictly speaking, each material parameter has its individual damping coefficient due to different physical mechanisms. As a consequence, one has to characterize 10 parameters in addition, which leads to 20 sought-after quantities. It seems only natural that the identification of such an amount of parameters may cause certain problems; e.g., the resulting data set is not unique anymore. This can be mainly ascribed to the small impact of several parameters on the measured quantity. To get rid of such problems, let us apply a highly simplified phenomenological damping model instead [52, 55]. We consider attenuation within piezoceramics by a single damping coefficient α_{d}, which relates the imaginary values (index $\Im$) to the real parts (index $\Re$) of the tensors (cf. (4.67, p. 105))

$$\left[\underline{\mathbf{c}}^E\right] = \left[\mathbf{c}^E\right]_{\Re} + j\left[\mathbf{c}^E\right]_{\Im} = \left[\mathbf{c}^E\right]_{\Re}\left[1 + j\alpha_d\right] \tag{5.43}$$

$$\left[\underline{\varepsilon}^S\right] = \left[\varepsilon^S\right]_{\Re} + j\left[\varepsilon^S\right]_{\Im} = \left[\varepsilon^S\right]_{\Re}\left[1 + j\alpha_d\right] \tag{5.44}$$

$$\left[\underline{\mathbf{e}}\right] = \left[\mathbf{e}\right]_{\Re} + j\left[\mathbf{e}\right]_{\Im} \quad\;\; = \left[\mathbf{e}\right]_{\Re}\left[1 + j\alpha_d\right] \;. \tag{5.45}$$

Therewith, attenuation is assumed to be equal for the tensors, and, moreover, does not depend on excitation frequency. Although a rather simple damping model is applied, the experimental results in Sect. 5.3.6 clearly point out that a single damping coefficient yields precise FE simulations for various piezoceramic sample shapes.

5.3.2 Feasible Input Quantities

To identify the material parameters of piezoceramic materials, one has to compare simulated quantities with measured quantities. Both the electrical impedance and the surface normal velocity of the piezoceramic test samples are feasible quantities for this comparison (see Fig. 5.10). While velocities provide spatially as well as frequency-resolved information, impedance curves[6] capture only the frequency-resolved global behavior of the samples [54, 57]. Hence, velocities contain more information, which can be especially useful for parameter identification purpose. The acquisition of the spatially resolved surface normal velocity requires, however, an expensive measurement setup (e.g., laser scanning vibrometer Polytec PSV-300 [46]) that is rarely available in production facilities of piezoceramics. Moreover, the acquisition is an error-prone procedure as mainly small velocities arise outside the resonance frequency of the piezoceramic sample. With a view to achieving measurable velocity magnitudes, the electrical excitation voltage has to be remarkably increased. In doing so, the piezoceramic material may be heated up, which alters its behavior and, consequently, the material parameters [52]. Comparisons of simulations and measurements in the course of parameter identification should, therefore, be conducted exclusively with the frequency-resolved impedance that can be easily measured by an impedance analyzer, e.g., Keysight 4194A [27].

5.3.3 Test Samples

For completely characterizing piezoceramic materials through the inverse method, it is recommended to employ two different block-shaped test samples (T1 and T2) of the material [52, 55, 58]. In the following, the main specifications of these test samples are mentioned. The top and bottom area of the samples should be covered with thin-film electrodes, e.g., CuNi alloy of 1 μm thickness. Both test samples should exhibit identical geometric dimensions $l_S \times w_S \times t_S$ but different directions of electric

---

[6]An impedance curve corresponds to the frequency-resolved electrical impedance.

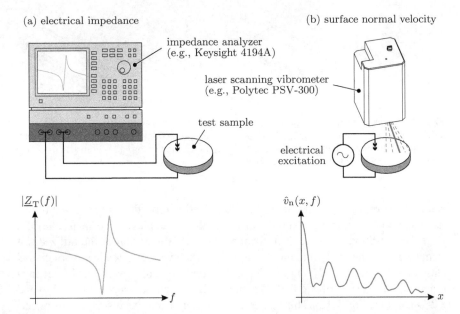

(a) electrical impedance   (b) surface normal velocity

impedance analyzer
(e.g., Keysight 4194A)

laser scanning vibrometer
(e.g., Polytec PSV-300)

test sample

electrical
excitation

$|\underline{Z}_T(f)|$

$\hat{v}_n(x, f)$

Fig. 5.10 **a** Measurement of frequency-resolved electrical impedance $|\underline{Z}_T(f)|$ of test samples with impedance analyzer; **b** measurement of spatially as well as frequency-resolved surface normal velocity $\hat{v}_n(x, f)$ of electrically excited test samples with laser scanning vibrometer; piezoceramic test samples can vibrate freely in both measurements; $|\underline{Z}_T(f)|$ and $\hat{v}_n(x, f)$ indicate magnitudes

polarization **P**. While test sample T1 is polarized in t_S-direction, the electric polarization of T2 points in w_S-direction. Thus, T1 is prevalently efficient for mechanical vibrations in thickness direction, and T2 mainly vibrates in thickness shear mode. Figure 5.11 schematically depicts the four significant vibration modes (T1-L, T1-W, T1-T, and T2) of the block-shaped test samples. Depending on the geometric dimensions, there always occur certain superpositions of different vibration modes within the test samples that can be observed in the impedance curve (see, e.g., T1-W; Fig. 5.12a). Such superpositions are considered in the framework of FE simulations for the test samples. Below, we will discuss an approach toward efficient FE simulations and present results of a parameter study for the utilized test samples.

Numerical Simulation

The presented characterization of piezoceramic materials is based on an iterative matching of numerical simulations to measurements. It is obvious that the accuracy of FE simulations always greatly depends on the spatial discretization of the computational domain. Consequently, the precision of the identified material parameters also depends on the spatial discretization of the FE model for the test samples T1 and T2. However, the finer the spatial discretization, the longer numerical simulations will take. Because the inverse method is an iterative approach requiring a considerable amount of simulation steps, the time consumption of a single step

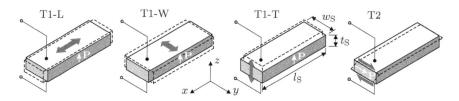

Fig. 5.11 Block-shaped test samples T1 and T2 (geometric dimensions $l_S \times w_S \times t_S$) for charac-terizing piezoceramic materials [51, 52]; T1 polarized in t_S-direction; T2 polarized in w_S-direction; T1-L, T1-W, T1-T as well as T2 refer to significant vibration modes; top and bottom area ($l_S \times w_S$) completely covered with thin-film electrodes

actually determines applicability of the identification procedure. On account of this fact, time-efficient numerical simulations for both test samples are indispensable [52].

If the condition $l_S \gg w_S \gg t_S$ for the geometric dimension of the utilized test samples is fulfilled, FE simulations can be remarkably accelerated in case of crystal class 6mm. Instead of solving a 3-D problem for the whole investigated frequency range, we are able to divide the FE simulations of the test samples into a 3-D and a 2-D part (see Fig. 5.12). This can be done as there occur clear separations of cer-tain resonance–antiresonance pairs in the frequency-resolved impedance resulting from mechanical vibrations in different directions. To compute impedance curves for mechanical vibrations in x- and y-direction of T1 (i.e., T1-L and T1-W), we have to perform 3-D simulations, which, however, require only a rough spatial discretiza-tion of the sample thickness. Model reduction is achieved by considering solely a quarter of the block-shaped samples. The simulation model of T1 features, thus, the dimensions $l_S/2 \times w_S/2 \times t_S$. For the vibration modes in z-direction (i.e., T1-T, and T2), a 2-D model with the dimensions $w_S \times t_S$ is sufficient to calculate the impedance curves. Nevertheless, in general, one should conduct comparative cal-culations of original 3-D problem and reduced 2-D problem to verify whether the model reduction is reasonable or not.

Compared with the geometric dimensions of typical test samples, the thickness of the thin-film electrodes is negligible. Therefore, the electrodes simply determine boundary conditions but do not contribute to the computational grid in the FE model. Table 5.1 contains appropriate spatial discretizations for quadratic Lagrange ansatz functions (polynomial degree $p_d = 2$; see Table 4.1) and the sample dimension $l_S \times w_S \times t_S = 30\,\text{mm} \times 10\,\text{mm} \times 2\,\text{mm}$.

Parameter Study

To answer the question why two block-shaped test samples should be utilized for characterization purpose, let us now take a closer look at the electrical behavior of those samples. Figure 5.13 depicts simulation results of the frequency-resolved impedance for distinct parameter variations in the range $\pm 10\%$ for c_{xy}^E and $\pm 50\%$ for ε_{xx}^S as well as e_{xy}. The panels refer to dominating vibration modes (i.e., T1-L, T1-W, T1-T and T2) of the test samples, respectively. Note that each horizontal line represents the impedance curve for a defined parameter configuration. While one

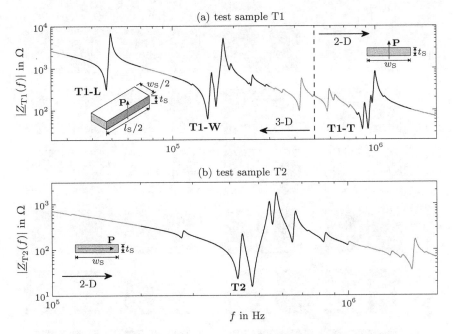

Fig. 5.12 FE simulations (3-D and 2-D) for frequency-resolved electrical impedance $|\underline{Z}_T(f)|$ (magnitude) of **a** test sample T1 and **b** test sample T2 ($l_S = 30$ mm, $w_S = 10$ mm, $t_S = 2$ mm; PIC255); material parameters from manufacturer (see Table 5.3); spatial discretization with quadratic ansatz functions (see Table 5.1); black parts of impedance curves represent frequency bands of dominating vibration modes considered within identification procedure

Table 5.1 Frequency bands and spatial discretizations (quadratic ansatz functions) used within inverse method for vibration modes of T1 and T2

Vibration mode	f_{start} in kHz	f_{stop} in kHz	Dimension	Spatial discretization
T1-L	30	70	3-D (quarter)	$15 \times 5 \times 1$
T1-W	100	300	3-D (quarter)	$15 \times 5 \times 1$
T1-T	800	1500	2-D	50×10
T2	200	1000	2-D	60×20

parameter was modified, the others remain constant. The most important findings of the parameter study are listed below [52].

- c_{11}^E, c_{12}^E, ε_{33}^S, e_{31} as well as e_{33} considerably alter the impedance curves for T1 but hardly influence the electrical behavior of T2.
- c_{13}^E and c_{33}^E cause significant variations of the impedance curves for both test samples.
- Compared with T1, the impedance curve of T2 strongly depends on c_{44}^E, ε_{11}^S, and e_{15}.

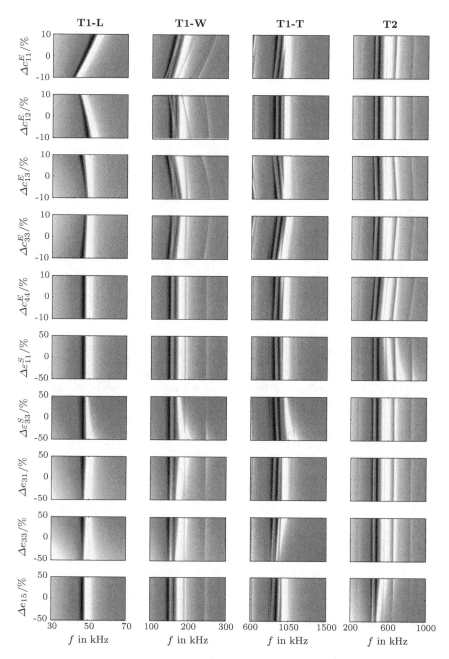

Fig. 5.13 Influence of distinct parameter variations on impedance curves $\left|\underline{Z}_\mathrm{T}(f)\right|$ of test samples T1 and T2 ($l_\mathrm{S} = 30\,\mathrm{mm}$, $w_\mathrm{S} = 10\,\mathrm{mm}$, $t_\mathrm{S} = 2\,\mathrm{mm}$; PIC255); each horizontal line in a panel refers to one impedance curve; bright and dark colors indicate large and small impedance values, respectively

Owing to these findings, it seems only natural that the reliable determination of the complete parameter set demands the consideration of both block-shaped test samples within the inverse method. In case of a piezoelectric material of another crystal class, the tensors $[\mathbf{c}^E]$, $[\varepsilon^S]$, and $[\mathbf{e}]$ as well as the amount of independent material parameters may differ. For instance, crystal class 4mm requires the elastic stiffness constant c_{66}^E in addition. Through various parameter studies and further investigations (e.g., robustness analysis) for the most relevant crystal classes showing piezoelectric coupling, the general applicability of the inverse method could be proven if we utilized both block-shaped test samples.

5.3.4   Mathematical Procedure

Overall, 11 parameters have to be identified for piezoceramic materials of crystal class 6mm. These parameters are collected in the parameter vector $\mathbf{p}$, which is for the e-form given by

$$\mathbf{p} = \left[c_{11}^E, c_{12}^E, c_{13}^E, c_{33}^E, c_{44}^E, \varepsilon_{11}^S, \varepsilon_{33}^S, e_{31}, e_{33}, e_{15}, \alpha_{\mathrm{d}}\right]^{\mathrm{t}} . \tag{5.46}$$

During the identification approach, $\mathbf{p}$ is updated iteratively. The update is based on the comparison of numerical simulations with measurements for the frequency-resolved electrical impedance, in particular its magnitude $\left|\underline{Z}_{\mathrm{T}}(f)\right|$. Let us assume that $\left|\underline{Z}_{\mathrm{T}}(f)\right|$ for each vibration mode of a piezoceramic specimen is logarithmically sampled at N_{I} discrete frequencies $(f_1, f_2, \ldots, f_{N_{\mathrm{I}}})$ in case of measurements and simulations, respectively. Therewith, the vector $\mathbf{Z}_{\mathrm{T}}$ of sampled impedance values becomes

$$\mathbf{Z}_{\mathrm{T}} = \left[\left|\underline{Z}_{\mathrm{T}}(f_1)\right|, \left|\underline{Z}_{\mathrm{T}}(f_2)\right|, \ldots, \left|\underline{Z}_{\mathrm{T}}(f_{N_{\mathrm{I}}})\right|\right]^{\mathrm{t}} . \tag{5.47}$$

Combining the impedance vectors for the significant vibrations mode (i.e., T1-L, T1-W, T1-T, and T2) of both block-shaped test samples leads to

$$\mathbf{q}_{\mathrm{meas}} = \left[\mathbf{Z}_{\mathrm{T1-L}}^{\mathrm{t}}, \mathbf{Z}_{\mathrm{T1-W}}^{\mathrm{t}}, \mathbf{Z}_{\mathrm{T1-W}}^{\mathrm{t}}, \mathbf{Z}_{\mathrm{T2}}^{\mathrm{t}}\right]^{\mathrm{t}} , \tag{5.48}$$

which includes all relevant measured quantities and, thus, features the dimension $4N_{\mathrm{I}}$. The vector $\mathbf{q}_{\mathrm{sim}}(\mathbf{p})$ containing the simulated impedance values for the parameter vector $\mathbf{p}$ is composed in the same way. So, the deviation $\mathbf{d}_{\mathrm{I}}(\mathbf{p})$ of simulations from measurements to be minimized takes the form (cf. (5.37))

$$\mathbf{d}_{\mathrm{I}}(\mathbf{p}) = \mathbf{q}_{\mathrm{sim}}(\mathbf{p}) - \mathbf{q}_{\mathrm{meas}} . \tag{5.49}$$

Since we have to estimate numerous parameters, the resulting minimization constitutes an ill-posed optimization problem yielding unstable solutions that are not unique in addition. The iteratively regularized Gauss–Newton method represents an outstanding approach to solve such ill-posed problem efficiently. Thus, it makes sense to apply exactly this approach for identifying material parameters of piezoceramics. According to the iteratively regularized Gauss–Newton method, the correction vector $\mathbf{c}^{(i)}$ to update the parameter vector $\mathbf{p}^{(i)}$ (i.e., $\mathbf{p}^{(i+1)} = \mathbf{p}^{(i)} + \mathbf{c}^{(i)}$) for iteration index i results in (cf. (5.40))

$$
\mathbf{c}^{(i)} = -\left[\mathbf{J}(\mathbf{p}^{(i)})^{\mathrm{t}} \mathbf{J}(\mathbf{p}^{(i)}) + \zeta_{\mathrm{R}}^{(i)}\mathbf{I}\right]^{-1}
$$
$$
\cdot\left[\mathbf{J}(\mathbf{p}^{(i)})^{\mathrm{t}} \cdot \mathbf{d}_{\mathrm{I}}(\mathbf{p}^{(i)}) + \zeta_{\mathrm{R}}^{(i)}\left(\mathbf{p}^{(i)} - \mathbf{p}^{(0)}\right)\right] \tag{5.50}
$$

with the Jacobian matrix $\mathbf{J}(\mathbf{p}^{(i)})$ of dimension $4N_{\mathrm{I}} \times 11$, the identity matrix $\mathbf{I}$ of dimension 11×11, and the regularization parameter $\zeta_{\mathrm{R}}^{(i)}$, respectively.

Actually, inverse methods always require an appropriate initial guess $\mathbf{p}^{(0)}$ for the aimed parameters (see Sect. 5.2). If $\mathbf{p}^{(0)}$ is too far away from the solution, the iteration procedure may converge to a local minimum, which is accompanied by unrealistic material parameters. One can distinguish between four approaches to figure out an appropriate $\mathbf{p}^{(0)}$ for characterizing piezoceramic materials [52]:

- The data set (if available) provided from the manufacturer serves as initial guess for the inverse method.
- We may exploit the identified parameters of a known piezoceramic material. Especially when test samples made of the known and unknown material feature similar electrical behavior (e.g., resonance frequency for a vibration mode), respectively, this will be an excellent approach to obtain $\mathbf{p}^{(0)}$.
- The desired material parameters represent input quantities of numerical simulations. In case of a completely unknown piezoceramic material, one should try common parameter values in a first step. These values need to be manually adjusted so that simulations roughly coincide with measurements for the test samples. It is recommended to proceed in the following order: manual adjustment of (i) ε_{xx}^{S}, (ii) c_{xy}^{E}, and (iii) e_{xy}. The adjusted parameter set can then serve as initial guess.
- As alternative to the previous approaches, one may apply the IEEE/CENELEC Standard to identify $\mathbf{p}^{(0)}$. However, this approach requires considerable work (see Sect. 5.1.1) and should, therefore, be avoided.

Another important aspect for characterizing piezoceramic materials through the inverse method concerns the value range of the desired material parameters since the individual parameters differ in several orders of magnitude (cf. Table 5.3). While elastic stiffness constants c_{xy}^{E} are in the range $10^{10}\,\mathrm{Nm}^{-2}$, electric permittivities ε_{xx}^{S} exhibit values of $\approx 10^{-8}\,\mathrm{Fm}^{-1}$. This may cause problems regarding the matrix inversion in (5.50). That is the reason why one should normalize the original parameter vector $\mathbf{p}$ to the initial guess $\mathbf{p}^{(0)}$ during identification. Consequently, the inverse method starts with a modified parameter vector, which exclusively contains entries of value 1. Besides this normalization, the L_2 norm of the deviation $\mathbf{d}_{\mathrm{I}}(\mathbf{p}^{(i)})$ should be normalized to that of the initial guess, i.e., $\mathbf{d}_{\mathrm{I}}(\mathbf{p}^{(0)})$.

Table 5.2 Relative deviations $|\Delta|$ (magnitude) in per mil of identified material parameters for different amounts of frequency points N_{I}, which are used to discretize impedance curves of each vibration mode; deviations refer to solution for $N_{\mathrm{I}} = 500$ frequency points

	c_{11}^E	c_{12}^E	c_{13}^E	c_{33}^E	c_{44}^E	ε_{11}^S	ε_{33}^S	e_{31}	e_{33}	e_{15}	α_{d}
$N_{\mathrm{I}} = 70$	12.1	19.5	13.3	1.6	3.1	9.2	2.7	25.4	0.5	3.8	80.3
$N_{\mathrm{I}} = 100$	0.6	0.6	1.0	0.6	3.2	5.0	1.0	0.3	0.9	2.5	11.3
$N_{\mathrm{I}} = 200$	0.9	1.4	0.9	0.1	0.5	0.3	0.1	1.0	0.2	0.5	0.7

5.3.5 Efficient Implementation

A decisive point toward a competitive identification procedure is the time-efficient numerical simulation for the test samples [52, 55]. As discussed in Sect. 5.3.3, this can be arranged by exploiting symmetries within the samples and rough spatial discretization. Nevertheless, several further points are crucial to obtain reliable material parameters in an acceptable time. For instance, the inverse method requires the Jacobian matrix $\mathbf{J}(\mathbf{p}^{(i)})$ at each iteration step to update the parameter vector $\mathbf{p}^{(i)}$. Because we are commonly not able to calculate $\mathbf{J}(\mathbf{p}^{(i)})$ in analytical manner, the matrix columns have to be computed numerically, e.g., by evaluating forward difference quotients.

Instead of considering the whole impedance curve of the test samples, it is reasonable to limit the identification procedure to small frequency bands comprising resonance–antiresonance pairs of the dominating vibration modes, respectively (see Table 5.1). The amount of frequency points N_{I} that are used to discretize each frequency band has a direct influence on the duration of parameter identification. The more frequency points, the longer the identification will take. However, if N_{I} is chosen too small, the determined parameters might exhibit large deviations because essential properties of the impedance curves are not captured. Table 5.2 contains the relative deviations Δ of the estimated parameters with respect to N_{I} after $M_{\mathrm{I}} = 50$ iteration steps. The deviations relate to the solution for $N_{\mathrm{I}} = 500$ frequency points. Such a fine frequency resolution enables the consideration of all relevant details in the impedance curves. Therefore, one obtains the reference solution of the inverse method. It can be clearly seen that for $N_{\mathrm{I}} = 200$, the maximum parameter deviations are much smaller than 1%, which is in any case a sufficient accuracy for material parameters.

Similar to N_{I}, the amount of iteration steps M_{I} directly affects the computation time of the identification procedure. Hence, a fast as well as robust convergence of the aimed parameter vector is desired. This is achieved by a proper choice of the regularization parameter $\zeta_{\mathrm{R}}^{(0)}$. In case of normalizing parameter vector $\mathbf{p}^{(i)}$ as well as $\mathbf{d}_{\mathrm{I}}(\mathbf{p}^{(i)})$ to the initial guess $\mathbf{p}^{(0)}$, $\zeta_{\mathrm{R}}^{(0)} = 10^3$ and a decrement of 0.5 (i.e., $\zeta_{\mathrm{R}}^{(i+1)} = 0.5\zeta_{\mathrm{R}}^{(i)}$) turned out to be a good choice. The inverse method converges after ten iterations to a solution, as shown in Fig. 5.14. By altering $\mathbf{p}^{(0)}$ up to $\pm 15\%$, it could

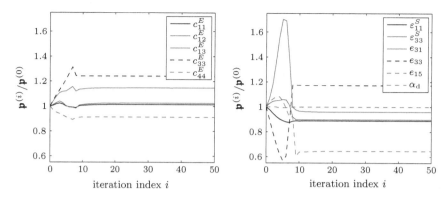

Fig. 5.14 Progress of material parameters during iterative identification (iteration index i) for PIC255; $\mathbf{p}^{(0)}$ refers to manufacturer's data

also be proven that the implemented identification procedure is robust and does not converge to a local minimum.

Besides those arrangements, the identification procedure can be accelerated if one is able to parallelize FE simulations. Especially for the determination of the Jacobian matrix $\mathbf{J}(\mathbf{p}^{(i)})$, the parallelization constitutes a remarkable advantage since each column can be computed separately. By applying all the considered points, the identification of the entire parameter set takes less than two hours on the computing server Fujitsu Celsius R920 Power with 16 cores.

5.3.6 Results for Selected Piezoceramic Materials

Now, the identified material parameters for two ferroelectrically soft materials (PIC255 and PIC155 from the company PI Ceramic GmbH [45]) of crystal class 6mm are given as well as verified. This also includes temperature dependences of both the material parameters and the resulting electromechanical coupling factors.

Before we will discuss the obtained results, let us briefly summarize the three main steps of the inverse method for piezoceramic materials:

- One has to acquire the frequency-resolved electrical impedances of the block-shaped test samples T1 and T2 for the dominating vibration modes, i.e., T1-L, T1-W, T1-T, and T2. For the investigated piezoceramic materials and geometric sample dimensions $l_\mathrm{S} \times w_\mathrm{S} \times t_\mathrm{S} = 30.0\,\mathrm{mm} \times 10.0\,\mathrm{mm} \times 2.0\,\mathrm{mm}$, this was done in the frequency bands according to Table 5.1. When another piezoceramic material should be characterized and/or the geometric sample dimensions differ, these frequency bands might require modifications.
- In addition to the measurements, appropriate FE models of T1 and T2 in the previously defined frequency bands are necessary. The FE models should enable

reliable as well as time-efficient numerical simulations of the frequency-resolved electrical impedances for all dominating vibration modes.

- By means of the iteratively regularized Gauss–Newton method, the parameter vector $\mathbf{p}$ (see (5.46)) is iteratively corrected so that simulations of the frequency-resolved electrical impedances match measurements as well as possible.

Material Parameters at Room Temperature

Table 5.3 contains both the data set provided by the manufacturer and the results of the inverse method at room temperature, i.e., $\vartheta = 25\,°\text{C}$. Note that the manufacturer exploited the IEEE/CENELEC Standard for determining the material parameters, which is currently the conventional as well as the only standardized method to completely characterize piezoceramic materials. The material parameters s_{xy}^E, ε_{xx}^T, and d_{xy} in d-form were computed by (cf. (3.33, p. 52))

$$\left[\mathbf{s}^E\right] = \left[\mathbf{c}^E\right]^{-1} \tag{5.51}$$

$$[\mathbf{d}] = [\mathbf{e}]\left[\mathbf{c}^E\right]^{-1} \tag{5.52}$$

$$\left[\varepsilon^T\right] = \left[\varepsilon^S\right] + [\mathbf{d}][\mathbf{e}]^{\mathrm{t}} . \tag{5.53}$$

To some extent, there occur deviations up to more than 25% between the different data sets for PIC255 and PIC155. On the one hand, this is a consequence of assuming monomodal mechanical vibrations within the IEEE/CENELEC Standard, which will be, strictly speaking, only fulfilled if the utilized test samples significantly exceed the geometric requirements (cf. Fig. 5.6). On the other hand, certain parameters seem to be easily identifiable by the IEEE/CENELEC Standard, e.g., ε_{33}^T. However, because such parameters partly exhibit rather large deviations, there is reason to suspect that the manufacturer analyzed test samples, which slightly differ from T1 as well as T2 in their polarization state and/or material composition.

Figure 5.15 shows a comparison of measurements and FE simulations for the frequency-resolved impedance of the test samples that were utilized to characterize PIC255. As can be clearly noticed, there will be a better match of simulations and measurements if the identified data set is used instead of material parameters provided by manufacturer. Especially for the thickness extensional mode T1-T of test sample T1 (see Fig. 5.15c), the manufacturer's data leads to remarkable deviations in the electrical behavior. Such deviations pose a problem in the simulation-assisted development of ultrasonic transducers, e.g., the precise prediction of generated sound fields.

Table 5.3 Material parameters (real parts) provided by manufacturer (MF) and those resulting from inverse method (IM) in e-form as well as d-form; relative deviations Δ in % related to manufacturer's data; piezoceramic materials PIC255 and PIC155 from PI Ceramics; material density $\varrho_0 = 7.8 \cdot 10^3$ kgm^{-3}; c_{xy}^E in 10^{10} Nm^{-2}; ε_{xx}^S in 10^{-9} Fm^{-1}; e_{xy} in NV^{-1}m^{-1}m^{-1}; s_{xy}^E in 10^{-12} m^{2}N^{-1}; ε_{xx}^T in 10^{-9} Fm^{-1}; d_{xy} in 10^{-11} mV^{-1}

e-form		c_{11}^E	c_{12}^E	c_{13}^E	c_{33}^E	c_{44}^E	ε_{11}^S	ε_{33}^S	e_{31}	e_{33}	e_{15}	α_d
PIC255	MF	12.30	7.67	7.03	9.71	2.23	8.23	7.59	-7.15	13.70	11.90	0.0200
	IM	12.46	7.86	8.06	12.06	2.04	7.31	6.81	-6.86	16.06	11.90	0.0129
	Δ	1.3	2.4	14.7	24.2	-8.7	-11.1	-10.2	-4.1	17.3	0.0	-35.4
PIC155	MF	11.08	6.32	6.89	11.08	1.91	7.75	6.34	-5.64	12.78	10.29	0.0190
	IM	12.80	8.15	8.42	11.95	1.85	6.78	5.58	-6.29	14.15	12.18	0.0127
	Δ	15.5	28.9	22.1	7.9	-3.4	-12.5	-12.0	11.5	10.8	18.4	-33.4

d-form		s_{11}^E	s_{12}^E	s_{13}^E	s_{33}^E	s_{44}^E	ε_{11}^T	ε_{33}^T	d_{31}	d_{33}	d_{15}	α_d
PIC255	MF	15.90	-5.70	-7.38	20.97	44.84	14.58	15.46	-17.40	39.29	53.36	0.0200
	IM	16.10	-5.63	-7.00	17.64	49.12	14.26	15.43	-18.41	37.93	58.43	0.0129
	Δ	1.3	-1.2	-5.2	-15.9	9.5	-2.2	-0.2	5.8	-3.5	9.5	-35.4
PIC155	MF	16.17	-4.84	-7.05	17.80	52.37	13.30	12.00	-15.40	30.70	53.90	0.0190
	IM	16.23	-5.24	-7.74	19.26	54.20	14.82	13.06	-17.87	37.00	66.02	0.0127
	Δ	0.4	8.3	9.8	8.2	3.5	11.5	8.8	16.0	20.5	22.5	-33.4

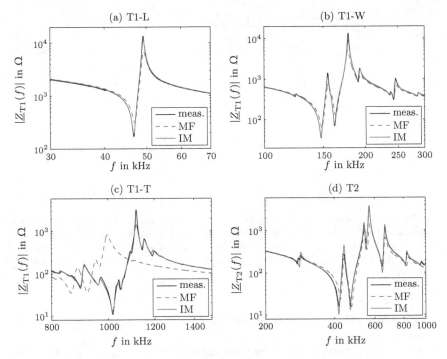

Fig. 5.15 Comparison of measurements and FE simulations for frequency-resolved impedance $\left|\underline{Z}_{\mathrm{T}}(f)\right|$ (magnitude) of test samples T1 and T2 ($l_{\mathrm{S}} = 30.0\,\text{mm}$, $w_{\mathrm{S}} = 10.0\,\text{mm}$, $t_{\mathrm{S}} = 2.0\,\text{mm}$; PIC255); panels show dominating vibration modes of test samples; FE simulations performed with material parameters from manufacturer (MF) and from inverse method (IM), respectively

Verification of Identified Parameters

With a view to verifying the results of the inverse method, various experiments are conceivable, e.g., numerical simulations for different sample geometries. Let us detail two experimental verifications [52]. First, the frequency-resolved electrical impedance $\underline{Z}_{\mathrm{T}}(f)$ and spatially resolved surface normal $v_{\mathrm{n}}(\rho)$ of a disk sample (diameter $d_{\mathrm{S}} = 30.0\,\text{mm}$; thickness $t_{\mathrm{S}} = 3.0\,\text{mm}$; PIC255) were analyzed by measurements as well as FE simulations. The velocity measurements were carried out with the laser scanning vibrometer Polytec PSV-300, which provides velocity amplitudes $\hat{v}_{\mathrm{n}}(\rho)$ with respect to the radial position ρ on the disk surface. Figure 5.16 depicts measurements as well as simulations in the frequency range of the thickness extensional mode. As the comparison demonstrates, the manufacturer's data yields unrealistic simulation results. This data set is, thus, not applicable to predict the behavior of sensor as well as actuator devices containing such piezoceramic material. In particular, the resonance frequency is shifted down remarkably. Even though neither the electrical nor mechanical behavior of the disk was considered within the

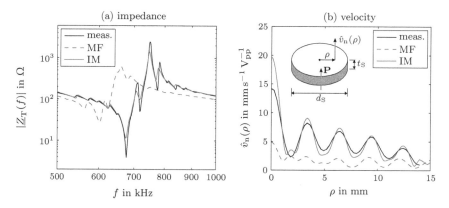

Fig. 5.16 Comparison of measurements and FE simulations for **a** frequency-resolved electrical impedance $\left|\underline{Z}_{\mathrm{T}}(f)\right|$ (magnitude) and **b** amplitudes of spatially resolved surface normal velocity $\hat{v}_{\mathrm{n}}(\rho)$ at $f = 660\,\mathrm{kHz}$ of a piezoceramic disk ($d_{\mathrm{S}} = 30.0\,\mathrm{mm}$, $t_{\mathrm{S}} = 3.0\,\mathrm{mm}$; PIC255); FE simulations performed with material parameters from manufacturer (MF) and from inverse method (IM), respectively

inverse method, simulations based on the identified data set coincide very well with measurements.

In a second experiment, the impedance curves of 5 test samples T1 and of 5 test samples T2 were acquired for PIC255 and for PIC155, respectively. Because the frequency-resolved electrical impedance of one of each test sample is used for the identification procedure, 25 different combinations exist. The inverse method was applied for each combination separately leading to 25 sets of material parameters. From this data, we are able to calculate mean value and standard deviation for the 11 desired parameters. Apart from the damping coefficient α_{d}, the relative standard deviation stays for all material parameters below 1.5%. The damping coefficient exhibits a relative standard deviation of $\approx 5\%$ for PIC255 and PIC155. According to the additional verification experiments, it is reasonable to conclude that the inverse method provides appropriate material parameters and the fabrication process of the investigated piezoceramic samples is reliable.

Temperature Dependence of Parameters

Regarding practical applications of piezoceramic materials, it is of utmost importance to know temperature dependences of the material behavior. To obtain such information, the piezoceramic test samples T1 and T2 were placed in a climatic chamber. The temperature ϑ within the climatic chamber was altered stepwise. After achieving the defined temperature and an additional holding time of 15 min, which ensures uniform temperature of the test samples [20], their frequency-resolved electrical impedances were acquired with the impedance analyzer. This procedure was repeated for at least five temperature loops in the range $-35\,°\mathrm{C}$ to $+145\,°\mathrm{C}$ for PIC255 and $-35\,°\mathrm{C}$ to $+130\,°\mathrm{C}$ for PIC155, respectively. Note that the following results always refer to the last temperature loop.

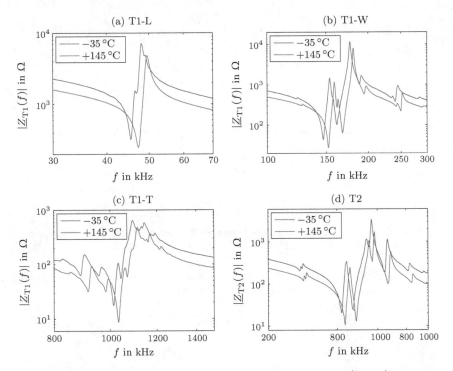

Fig. 5.17 Comparison of measured frequency-resolved electrical impedance $\left|\underline{Z}_T(f)\right|$ (magnitude) of test samples T1 and T2 ($l_S = 30.0\,\text{mm}$, $w_S = 10.0\,\text{mm}$, $t_S = 2.0\,\text{mm}$; PIC255) at two different temperatures ϑ; panels show dominating vibration modes of test samples

As Fig. 5.17 reveals, temperature highly influences the impedance curves of T1 and T2 in case of PIC255, which also holds for PIC155. It therefore stands to reason that the decisive material parameters possess considerable temperature dependences [52, 58]. To quantify these dependences, let us apply the inverse method for the investigated piezoceramic materials. Figure 5.18 depicts selected material parameters of PIC255 and PIC155 versus temperature, respectively. For better comparison, the parameters have been normalized to the values at room temperature, i.e., $\mathfrak{p}(\vartheta) / \mathfrak{p}(25\,^\circ\text{C})$. Additionally, Table 5.4 lists the relative deviations of all parameters at lowest as well as highest temperature. The elastic stiffness constants c_{xy}^E are mostly independent of ϑ, whereas the electric permittivities ε_{xx}^S exhibit a rather large temperature dependence. This fact has to be considered in the design of appropriate electronics for piezoceramic sensors and actuators. Figure 5.18 and Table 5.4 also demonstrate that PIC155 shows stronger temperature dependence than PIC255, which coincides with the rough information provided by the manufacturer. Furthermore, a certain hysteresis behavior arises that is particularly pronounced for PIC155. In other words, there occur great differences of several material parameters (e.g., ε_{11}^S; Fig. 5.6b) for increasing and decreasing temperatures. A possible explanation for this

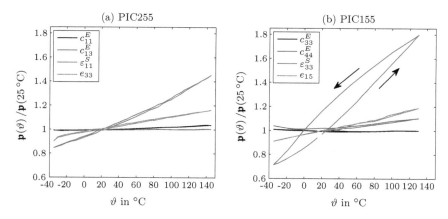

Fig. 5.18 Selected material parameters $\mathbf{p}(\vartheta)$ of **a** PIC255 and **b** PIC155 versus temperature ϑ; parameters relate to values at $\vartheta = 25\,°C$

Table 5.4 Relative variations in % of material parameters for piezoceramic materials PIC255 and PIC155 at lowest and highest temperature; variations relate to values at $\vartheta = 25\,°C$

	PIC255		PIC155	
	$\vartheta = -35\,°C$	$\vartheta = +145\,°C$	$\vartheta = -35\,°C$	$\vartheta = +130\,°C$
Δc_{11}^E	−0.7	3.7	5.0	−0.7
Δc_{12}^E	−0.2	1.6	1.8	0.2
Δc_{13}^E	−0.9	0.4	0.4	−1.7
Δc_{33}^E	−1.1	2.3	0.8	0.0
Δc_{44}^E	−1.0	7.6	3.9	10.2
$\Delta \varepsilon_{11}^S$	−15.3	45.3	−20.0	49.5
$\Delta \varepsilon_{33}^S$	−16.4	35.6	−28.6	79.6
Δe_{31}	−14.1	4.6	−23.4	14.8
Δe_{33}	−9.7	16.2	−15.0	40.0
Δe_{15}	−7.9	19.0	−9.0	18.8
$\Delta \alpha_d$	12.5	−17.6	30.9	2.8

behavior lies in intrinsic processes taking place within piezoceramic materials. Such processes may differ considerably for heating and cooling of the material.

By means of the basic relations for piezoelectricity (see Sects. 3.5 and 5.1.1)

$$k_{33}^2 = \frac{d_{33}^2}{s_{33}^E \varepsilon_{33}^T}\,, \quad k_{31}^2 = \frac{d_{31}^2}{s_{11}^E \varepsilon_{33}^T}\,, \quad k_{15}^2 = \frac{d_{15}^2}{s_{44}^E \varepsilon_{11}^T}\,, \tag{5.54}$$

one is able to compute the coupling factors k_{pq} between electrical and mechanical energy within a piezoceramic material of crystal class 6mm. Figure 5.19 displays the resulting electromechanical coupling factors for PIC255 and PIC155 with respect

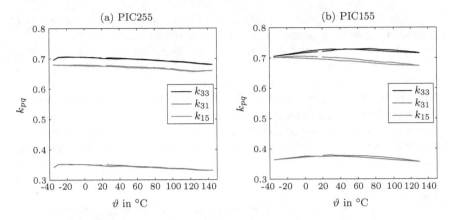

Fig. 5.19 Electromechanical coupling factors k_{33}, k_{31} as well as k_{15} of **a** PIC255 and **b** PIC155 versus temperature ϑ

to ϑ. Although the material parameters of both materials remarkably depend on temperature, k_{33}, k_{31} as well as k_{15} remain nearly constant. Thus, we can state that in the considered temperature range, the energy conversion of these piezoceramic materials is not significantly affected by the ambient temperature.

To summarize, the inverse method for piezoceramic materials provides the complete set of material parameters by analyzing the frequency-resolved electrical impedance of two block-shaped test samples. One can use the identified data set to predict both electric and mechanical small-signal behavior of arbitrarily shaped piezoceramics in a reliable way. Besides classic materials like PIC255, the inverse method was successfully applied for lead-free piezoceramics [41, 74]. Finally, it should be emphasized that the presented characterizing approach is not restricted to piezoelectric materials of crystal class 6mm. Several further numerical investigations revealed its general applicability for other crystal classes showing piezoelectric coupling, e.g., crystal class 4mm.

5.4 Inverse Method for Passive Materials

Section 5.1.2 addressed standard approaches for characterizing linearized mechanical properties (e.g., Young's modulus) of passive materials. As detailed, the standard approaches are accompanied by significant drawbacks and limitations. Owing to this fact, various alternative approaches have been developed which allow characterizing dynamic behavior of passive materials like viscoelastic solids. Several of the approaches are based on the combination of measurements and numerical simulations, e.g., [5, 28, 37, 38, 62].

Just as for piezoceramic materials, we will apply the simulation-based inverse method for identifying characteristic mechanical properties of isotropic as well as homogeneous passive materials. The focus lies on a simple and reliable method to deduce functional relations for frequency-dependent material parameters [56, 69]. Such functional relations are particularly important for precisely simulating the mechanical behavior of viscoelastic solids like plastics. In principle, the inverse method for passive materials is based on harmonically exciting an appropriate test sample by mechanical vibrations [18, 19, 68]. The resulting response (e.g., tip displacement) of the test sample depends on both its geometric dimensions and mechanical material properties. By adjusting simulated to measured responses, we are able to determine dynamic material parameters, i.e., Young's modulus as well as damping ratio.

The section starts with models for viscoelastic solids and an approach to take into account attenuation within such materials. This includes common models as well as a tailored material model. Subsequently, we briefly discuss feasible input quantities for the inverse method. In Sect. 5.4.3, the choice of test samples will be explained and justified by means of measurements as well as numerical simulations. Moreover, a detailed description of the experimental arrangement will be given. Section 5.4.4 shows the main steps toward an efficient implementation of the inverse method for passive materials. At the end, identified parameters for selected materials are listed and verified.

5.4.1 Material Model and Modeling of Attenuation

Various solid materials (e.g., plastics) show a pronounced viscoelastic behavior due to internal losses within the material. Viscoelasticity means that the relation between mechanical stresses and mechanical strains depends on time t and, consequently, on frequency f in the frequency domain. We can observe three effects for a viscoelastic material, namely (i) *stress relaxation*, (ii) *creep*, and (iii) *hysteresis* [4]. While stress relaxation refers to a decreasing mechanical stress within the material for an applied constant strain, creep indicates an increasing mechanical strain for an applied constant stress. In the context of viscoelastic materials, hysteresis occurs in the stress–strain curve because the material behavior differs for loading and unloading cycles.

Now, let us assume linear viscoelasticity, which is permitted in case of small input quantities, i.e., small mechanical strains as well as small mechanical stresses. To consider linear viscoelastic behavior of passive materials in numerical simulations, one can use frequency-dependent values for the decisive material parameters such as Young's modulus E_M and Poisson's ratio ν_P. In complex representation, the dynamic (i.e., frequency-dependent) Young's modulus $\underline{E}_M(\omega)$ takes the form (angular frequency $\omega = 2\pi f$) [56, 73]

$$\underline{E}_M(\omega) = E_\Re(\omega) + j E_\Im(\omega) \ . \tag{5.55}$$

The real part $E_\Re(\omega)$ measures energy stored and recovered per cycle. Against that, the imaginary part $E_\Im(\omega)$ characterizes energy dissipation within the viscoelastic material by internal damping. These quantities lead to the so-called loss factor $\tan \delta_d$ and damping ratio $\xi_d(\omega)$, which are defined as

$$\tan \delta_d = 2\xi_d(\omega) = \frac{E_\Im(\omega)}{E_\Re(\omega)} \,. \tag{5.56}$$

Of course, one can also introduce a dynamic Poisson's ratio $\underline{\nu}_p(\omega)$ for viscoelastic solids [48]. In a wide frequency range, Poisson's ratio mostly exhibits, however, a much less pronounced frequency dependence than Young's modulus [18, 67]. That is why we will concentrate exclusively on the dynamic Young's modulus $\underline{E}_M(\omega)$ instead of considering both quantities with respect to frequency. To incorporate the frequency dependence of Young's modulus within harmonic FE simulations, the entries of the elastic stiffness tensor [**c**] (see Sect. 2.2.3) need to be altered.

In the following, let us study common material models (e.g., Kelvin–Voigt model), which are applied for describing viscoelasticity. The models will be examined regarding the significant system property *causality*. Finally, a tailored frequency-dependent model is presented that allows reliable emulation of viscoelastic behavior for real solid materials.

Common Material Models

The linear viscoelastic behavior of solid materials is oftentimes modeled by means of mechanical analogies. In doing so, one emulates the material behavior as a linear combination of elastic springs and mechanical dashpots (see Fig. 5.20). Springs with spring rates (elastic modulus) κ_S represent the elastic part of the linear viscoelastic behavior, while the viscous part is described through dashpots with viscosities η_D. Typically, three models are listed in the literature for viscoelastic materials (e.g., [4, 34]), namely (i) the Maxwell model, (ii) the Kelvin–Voigt model, and (iii) the standard linear solid (SLS) model. Below, let us discuss the main properties of the basic models. To facilitate model comparison, the time-dependent step responses for the inputs mechanical strain $S(t)$ and mechanical stress $T(t)$ are shown in Fig. 5.21, respectively. In case of a strain input, the step responses can be interpreted as stress relaxation, whereas the step response to a stress input explains the underlying creep behavior.

- *Maxwell model*: A single spring is connected in series with a single dashpot (see Fig. 5.20a). Because spring and dashpot are subject to the same mechanical stress T, the model is also known as isostress model. The differential equation for this configuration becomes

$$\frac{T(t)}{\eta_D} + \frac{1}{\kappa_S}\frac{dT(t)}{dt} = \frac{dS(t)}{dt} \,. \tag{5.57}$$

As Fig. 5.21 depicts, the step response $T(t)$ for a strain input decays exponentially with time, which coincides with the behavior of many materials, e.g., polymers.

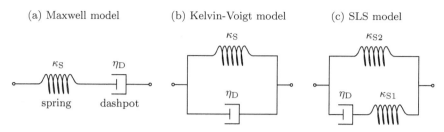

Fig. 5.20 Schematic representation of **a** Maxwell model, **b** Kelvin–Voigt model, and **c** SLS model

However, the step response $S(t)$ increases linearly with time without any bound for a stress input. Such material behavior is, of course, impossible.

- *Kelvin–Voigt model*: This model is based on a parallel connection of a single spring and a single dashpot (see Fig. 5.20b). Since both elements are subject to the same mechanical strain S, the model is also known as isostrain model. The differential equation for this configuration is given by

$$T(t) = \kappa_S S(t) + \eta_D \frac{dS(t)}{dt} .$$ (5.58)

According to the step response $S(t)$ for a stress input (see Fig. 5.21), the Kelvin–Voigt model is able to cover creep behavior of viscoelastic materials. The step response $T(t)$ for a strain input, on the other hand, does not feature the characteristic stress relaxation of such materials.

- *Standard linear solid (SLS) model*: The SLS model, which is also called Zener model, consists of one dashpot and two springs (see Fig. 5.20c). The differential equation for this configuration takes the form

$$T(t) + \tau_\varepsilon \frac{dT(t)}{dt} = \kappa_{S2} \left[S(t) + \tau_T \frac{dS(t)}{dt} \right]$$ (5.59)

with

$$\tau_\varepsilon = \frac{\eta_D}{\kappa_{S1}} \quad \text{and} \quad \tau_T = \eta_D \frac{\kappa_{S1} + \kappa_{S2}}{\kappa_{S1} \kappa_{S2}} .$$ (5.60)

In contrast to the Maxwell and Kelvin–Voigt model, the SLS model offers realistic step responses for strain and stress inputs. Therefore, it can be stated that this model is accurate for predicting both stress relaxation and creep.

Besides the mentioned basic models, there exist various other models (e.g., the fractional Zener model [47]) for modeling linear viscoelastic material behavior of solids. The models differ in complexity and may enable a highly precise imitation

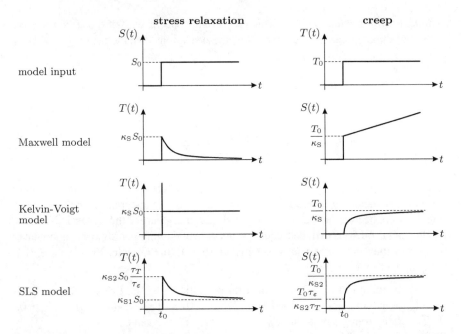

Fig. 5.21 Step responses of basic material models, i.e., Maxwell, Kelvin–Voigt as well as SLS model [4]; (left) stress relaxation $T(t)$ for strain input; (right) creep $S(t)$ for stress input

of the real material behavior. This fact implies, however, a considerable amount of model parameters that need to be identified.

Consideration of Causality

From the system point of view, causality means that current outputs of a system depend solely on its current and past inputs; i.e., an input at time t affects the system output at the earliest at time t [48]. In other words, a system cannot look to the future, which is the case for each system occurring in nature. Let us study the consequence of this fundamental system property for characterizing dynamic mechanical properties of passive materials. According to Sect. 5.1.2, we are able to treat a test sample as a linear system. If the system is excited by a mechanical strain $T(t)$, there will arise a certain mechanical stress $S(t)$ within the test sample that can be exploited to determine Young's modulus E_M. In the frequency domain, the underlying mathematical relation reads as (angular frequency ω)

$$\underline{E}_M(\omega) = \frac{\underline{T}(\omega)}{\underline{S}(\omega)} \tag{5.61}$$

with the dynamic Young's modulus consisting of real part $E_{\Re}(\omega)$ and imaginary part $E_{\Im}(\omega)$. Due to the fact that a linear system has to fulfill causality, $E_{\Re}(\omega)$ and $E_{\Im}(\omega)$ interrelate. The connection of both quantities results from the so-called

Kramers–Kronig relations [29], which represent a special case of the *Hilbert transform*. For the dynamic Young's modulus, these relations take the form

$$E_{\Re}(\omega) = E_0 + \frac{2\omega^2}{\pi} \text{CH} \int_0^\infty \frac{E_{\Im}(y)}{y\left(\omega^2 - y^2\right)} dy \tag{5.62}$$

$$E_{\Im}(\omega) = -\frac{2\omega}{\pi} \text{CH} \int_0^\infty \frac{E_{\Re}(y)}{\omega^2 - y^2} dy . \tag{5.63}$$

The expression E_0 $(\hat{=} \underline{E}_M(\omega = 0) = E_{\Re}(\omega = 0))$ denotes the static Young's modulus, and CH stands for the Cauchy principal value[7] of the integral. Since there exists a distinct connection between real part and imaginary part of the dynamic Young's modulus, it is obvious that the frequency-dependent damping ratio $\xi_d(\omega)$ cannot take arbitrary values in case of causal systems (cf. (5.56)). In principle, $\underline{E}_M(\omega)$ and $\xi_d(\omega)$ have to meet four criteria:

1. The damping ratio needs to vanish in the static case, i.e., $\xi_d(\omega = 0) = 0$. This implies automatically the conditions $E_{\Im}(\omega = 0) = 0$ and $E_{\Re}(\omega = 0) = E_0 \neq 0$.
2. The static Young's modulus E_0 has to feature a finite value.
3. Because $\xi_d(\omega)$ of a real system is always greater than zero, $E_{\Im}(\omega) \geq 0$ has to be fulfilled. From this fact and (5.62), we can follow immediately that $E_{\Re}(\omega)$ needs to increase monotonically with rising angular frequency ω, i.e., $E_{\Re}(\omega_1) < E_{\Re}(\omega_2)$ for $\omega_1 < \omega_2$.
4. The step response of a real system is always bounded. As a step (Heaviside step function) contains all frequencies, $E_{\Re}(\omega \to \infty) = E_\infty$ has to take finite values. In connection with (5.63), this also implies $E_{\Im}(\omega \to \infty) = 0$ as well as $\xi_d(\omega \to \infty) = 0$.

At this point, the question arises whether the three basic models for viscoelastic materials (i.e., Maxwell, Kelvin–Voigt, and SLS model) satisfy those criteria. To answer the question, let us take a look at $E_{\Re}(\omega)$ and $E_{\Im}(\omega)$ in Table 5.5, which result from the underlying differential equations [4]. While the Maxwell model violates criterion 1 because $E_0 = 0$, the Kelvin–Voigt model violates criterion 3 as well as 4 due to the fact that $E_{\Re}(\omega)$ remains constant and $E_{\Im}(\omega)$ increases linearly with ω. Note that violations of the criteria go hand in hand with the unrealistic step responses in Fig. 5.21. In contrast to the Maxwell and Kelvin–Voigt model, the SLS model fulfills all criteria and, thus, reflects the causal behavior of a real system.

Tailored Material Model

As discussed above, the SLS model represents the only common model for viscoelastic materials that fulfills all criteria concerning causal systems. However, the less amount of parameters adjustable in this model prevents oftentimes an accurate

---

[7]The Cauchy principal value allows solving improper integrals, which would otherwise be undefined [8].

Table 5.5 Frequency-dependent real part $E_\Re(\omega)$ and imaginary part $E_\Im(\omega)$ of dynamic Young's modulus $\underline{E}_M(\omega)$ for typical viscoelastic material models; amount of adjustable model parameters

Model	$E_\Re(\omega)$	$E_\Im(\omega)$	Parameters
Maxwell	$\dfrac{\kappa_S(\eta_D\omega)^2}{\kappa_S^2 + (\eta_D\omega)^2}$	$\dfrac{\kappa_S^2\eta_D\omega}{\kappa_S^2 + (\eta_D\omega)^2}$	2 (κ_S and η_D)
Kelvin–Voigt	κ_S	$\eta_D\omega$	2 (κ_S and η_D)
SLS	$\dfrac{\kappa_{S2}(1 + \tau_T\tau_\varepsilon\omega^2)}{1 + (\tau_\varepsilon\omega)^2}$	$\dfrac{\kappa_{S2}(\tau_T - \tau_\varepsilon)\,\omega}{1 + (\tau_\varepsilon\omega)^2}$	3 (κ_{S2}, τ_T and τ_ε)

emulation of the dynamic material behavior. That is the reason why one requires alternative models to describe viscoelastic materials. In [19, 56], a special model for the real part $E_\Re(f)$ of the dynamic Young's modulus has been suggested which consists of a constant part, a linear part, and a logarithmic part. The tailored material model is motivated by the logarithmically rising $E_\Re(f)$ with increasing frequency f in case of various isotropic plastics and reads as

$$E_\Re(f) = E_0\left[1 + \alpha_1 \cdot f + \alpha_2 \cdot \log_{10}\left(\frac{f + 1\,\text{Hz}}{\text{Hz}}\right)\right]. \tag{5.64}$$

When both α_1 and α_2 feature positive values, $E_\Re(f)$ will rise monotonically with f and, therefore, meets criterion 3 of a causal system. Moreover, (5.64) demands a static Young's modulus E_0 by which criterion 2 is automatically fulfilled. In order to determine the frequency-dependent imaginary part $E_\Im(f)$ of $\underline{E}_M(f)$ as well as the damping ratio $\xi_d(f)$, let us introduce the approximation [40]

$$E_\Im(\omega) \approx -\frac{\pi}{2}\omega\frac{\mathrm{d}E_\Re(\omega)}{\mathrm{d}\omega}, \tag{5.65}$$

which yields ($\omega = 2\pi f$)

$$E_\Im(f) = \frac{\pi}{2}E_0\left[\alpha_1 \cdot f + \frac{\alpha_2}{\ln 10}\frac{f}{f + 1\,\text{Hz}}\right]. \tag{5.66}$$

Therewith, $\xi_d(f)$ results in

$$\xi_d(f) = \frac{E_\Im(f)}{2E_\Re(f)} = \beta_d\frac{\pi\left(\alpha_1 \cdot f + \dfrac{\alpha_2}{\ln 10}\dfrac{f}{f + 1\,\text{Hz}}\right)}{4E_0\left[1 + \alpha_1 \cdot f + \alpha_2 \cdot \log_{10}(f + 1\,\text{Hz})\right]} \tag{5.67}$$

where β_d denotes an additional scaling factor enabling adjustments of the damping ratio. This factor may be useful regarding general applications of the material model for soft and hard solids.

Aside from criterion 2 and 3, the tailored model also satisfies criterion 1. Owing to the fact that $E_\Re(f)$ rises without any bound and $E_\Im(f \to \infty) \neq 0$ as well as $\xi_d(f \to \infty) \neq 0$, the model does not meet criterion 4 of causal systems. Although the model contains overall only four parameters (i.e., E_0, α_1, α_2, and β_d) and criterion 4 is violated, an appropriate choice of the parameters allows reliable emulations of the viscoelastic material behavior in a wide frequency range (e.g., [18]). The clear separation of $E_\Re(f)$ into constant part, linear part, and logarithmic part facilitates, furthermore, parameter identification. Because of the just mentioned reasons, this material model is particularly suited for characterizing dynamic behavior of passive materials.

5.4.2 Feasible Input Quantities

With a view to identifying characteristic parameters, we need both measurements and simulations of significant quantities that serve as input for the inverse method. The focus lies on the dynamic mechanical behavior of passive materials, which are assumed to feature viscoelastic properties. Therefore, it seems only natural that frequency-dependent measurements and simulations of mechanical quantities are required. A proper method to obtain significant as well as measurable quantities results from mechanically exciting a test sample at the desired frequency, i.e., a harmonic mechanical excitation. The response of the test sample to this excitation depends on its geometric dimension but also reflects mechanical material properties at the excitation frequency.

Now, let us consider a solid disk (cf. Fig. 5.22a) that gets mechanically excited at its bottom area by a sinusoidal displacement $u_1(t)$ in thickness direction. The harmonic excitation shall be defined as (time t)

$$u_1(t) = \Re\{\underline{u}_1(t)\} = \Re\{\hat{u}_1 e^{j 2\pi f t}\} = \hat{u}_1 \cos(2\pi f t) \tag{5.68}$$

with the displacement amplitude $\hat{u}_1$ and the excitation frequency f. Under the assumption of linear material behavior, the top area of the disk oscillates at the same frequency yielding the sinusoidal displacement $u_2(t)$, which becomes

$$u_2(t) = \Re\{\underline{u}_2(t)\} = \Re\{\hat{u}_2 e^{j 2\pi f t - j\varphi}\} = \hat{u}_2 \cos(2\pi f t - \varphi) \ . \tag{5.69}$$

The ratio $\hat{u}_2/\hat{u}_1$ of the displacement amplitudes and the phase angle φ depends on excitation frequency, disk geometry as well as material properties. From there, it appears possible to deduce material properties when the ratio $\hat{u}_2/\hat{u}_1$, the disk geometry, and the excitation frequency are known. In the linear case, the ratio can, thus, be interpreted as characteristic transfer behavior $H_T(f)$ of the considered test sample for displacement amplitudes. Since both the velocity $v(t)$ and the acceleration $a(t)$ directly result from the displacement $u(t) = \hat{u} \cos(2\pi f t)$ through

$$v(t) = \frac{du(t)}{dt} = \overbrace{-2\pi f \hat{u}}^{\hat{v}} \sin(2\pi f t) \tag{5.70}$$

$$a(t) = \frac{d^2u(t)}{dt^2} = \underbrace{-(2\pi f)^2 \hat{u}}_{\hat{a}} \cos(2\pi f t) , \tag{5.71}$$

$H_T(f)$ will be also equal if those quantities are used, i.e.,

$$H_T(f) = \left.\frac{\hat{u}_2}{\hat{u}_1}\right|_f = \left.\frac{\hat{v}_2}{\hat{v}_1}\right|_f = \left.\frac{\hat{a}_2}{\hat{a}_1}\right|_f . \tag{5.72}$$

In other words, it does not matter which quantities are exploited for determining $H_T(f)$. This fact can be especially important with regard to available measurement equipment, e.g., optical triangulation position sensor.

5.4.3 Test Samples

According to the above considerations, identification of dynamic material properties calls for varying frequencies f of the mechanical excitation that is applied to the test samples. Here, we will answer the question concerning appropriate test samples. In general, there does not exist a single sample shape, which allows determination of dynamic mechanical properties for soft and hard materials. By means of various investigations (e.g., [18, 56, 69]), it has become apparent that cylindrical sample shapes are feasible for soft materials, while hard materials can be characterized with the aid of beam-shaped test samples.

Soft materials demand a special sample design, which avoids mechanical deformations due to the net weight of the test sample. A flat cylinder (i.e., disk-shaped) represents such a sample shape (see Fig. 5.22a). Because mechanical clamping will cause disturbing sample deformation, it is recommended to glue the test sample on a stiff carrier plate, which is connected to the source of mechanical vibrations, e.g., an electrodynamic vibration exciter. In case of hard materials, one may utilize a long cylindrical rod (see Fig. 5.22b). Owing to the fact that hard materials are dimensionally stable, the rod can be mechanically clamped nearby its bottom area. Beam-shaped test samples allow, however, an additional direction of mechanical loading. This might be particularly important for characterizing dynamic material properties of passive materials.

Figure 5.23 depicts three different clamps (i.e., clamp C1, C2, and C3), which have proven to be effective for beam-shaped test samples [18]. Through these clamps, we can apply three individual directions of mechanical loading and, therefore, all directions that are possible for linear motion of single-sided clamped beam. Clamp C1 and C2 enable the generation of bending modes within the beam, whereas a compression–tension load is applied by means of clamp C3. In case of clamp C1

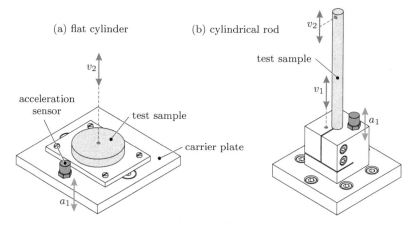

Fig. 5.22 **a** Flat cylinder serving as test sample for soft materials; cylinder is glued on carrier plate; **b** mechanically clamped cylindrical rod serving as test sample for hard materials; measurement of acceleration a_1 or velocity v_1 and velocity v_2 [18]

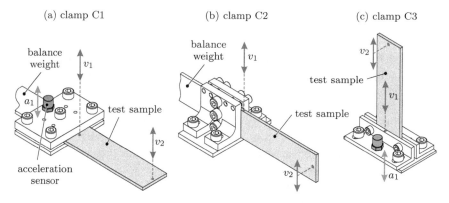

Fig. 5.23 Three different clamps for beam-shaped test samples [18]; **a** clamp C1 for bending; **b** clamp C2 for bending; **c** clamp C3 for compression–tension load; measurement of acceleration a_1 or velocity v_1 and velocity v_2

and C2, the test sample introduces an asymmetric load for the source of mechanical vibrations. To compensate the asymmetric load, one should add an identical beam that serves as balance weight.

In the following, we will discuss decisive points regarding the experimental procedure for parameter identification. This includes the mechanical excitation of the test samples as well as appropriate measuring equipment, which is required to determine the characteristic transfer function. Besides, we will study modeling approaches enabling efficient FE simulations for the test samples. Finally, a simulation-based parameter study is presented that provides useful information for parameter identification.

Mechanical Excitation of Test Samples

The identification of dynamic material properties is based on the characteristic transfer function $H_T(f)$ of an appropriate test sample for mechanical vibrations that can be provided by an electrodynamic vibration exciter or a piezoelectric stack actuator. Electrodynamic vibration exciters offer large displacements in the low-frequency region (i.e., $f < 100\,\text{Hz}$), but they are usually not suitable for operating frequencies $f > 10\,\text{kHz}$. In contrast, piezoelectric stack actuators (see Sect. 10.1) allow much higher operating frequencies. The achieved displacements of such actuators are, however, comparatively small which may lead to problems concerning determination of $H_T(f)$. Thus, one has to select the vibration source with respect to the desired frequency range of the dynamic material properties.

At this point, it should be noted that high excitation frequencies ($f > 10\,\text{kHz}$) are naturally accompanied by resonance phenomena within the clamping devices. The consideration of such phenomena during parameter identification actually poses a challenging task. For that reason, let us concentrate hereafter on determining dynamic material parameters up to 5 kHz, whereby the electrodynamic vibration exciter TIRA S 5220-120 [64] serves as source of mechanical vibrations. Nevertheless, by modifying and optimizing the clamping devices, much higher frequencies can be attained through piezoelectric stack actuators.

Measuring Equipment

In order to determine $H_T(f)$ experimentally, one has to acquire either displacements u, velocities v or accelerations a (see (5.72)) at two different positions along the test sample, e.g., at the bottom and top area of a disk. Time-dependent accelerations $a(t)$ can be easily measured by piezoelectric acceleration sensors (see Sect. 9.1.4). For example, such sensor is applicable for directly measuring mechanical excitation signals $a_1(t)$ at the carrier plate of soft cylindrical samples (see Fig. 5.22) or at the clamps of beam-shaped test samples (see Fig. 5.23). However, due to its net weight, an acceleration sensor is not suitable to acquire $a_2(t)$ since the transfer behavior $H_T(f)$ of the test sample will be strongly influenced. That is the reason why one should prefer a nonreactive measurement principle for this task.

Both optical triangulation position sensors and laser Doppler vibrometers[8] allow nonreactive as well as time-resolved measurements of mechanical movements [15, 65]. Consequently, it is possible to determine displacements, velocities as well as accelerations in a nonreactive way. Even though optical triangulation position sensors are comparatively inexpensive and can be utilized for static as well as dynamic measurements, their fixed working distance and limited displacement resolution may constitute a certain disadvantage. In contrast, laser Doppler vibrometers are rather expensive but provide outstanding displacement and velocity resolutions. Mainly,

---

[8]Laser Doppler vibrometers are also known as laser Doppler velocimeters.

Fig. 5.24 Experimental
setup for mechanical
exciting beam-shaped test
samples on basis of clamp
C2 [18]; velocity v_1
measured with out-of-plane
laser Doppler vibrometer;
velocity v_2 measured with
in-plane laser Doppler
vibrometer; electrodynamic
vibration exciter serves as
vibration source

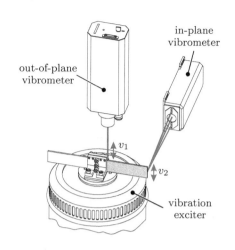

one distinguishes between *out-of-plane* and *in-plane* vibrometer. While an out-of-plane vibrometer (e.g., Polytec OFV-303 [46]) measures reflector movements along the emitted laser beam, an in-plane vibrometer (e.g., Polytec LSV-065 [46]) provides information about reflector movements perpendicular to the sensor head. Both vibrometer types will, however, work only if the test sample sufficiently reflects the incoming laser beam at the measuring points, which can be ensured by special reflecting foils. The presented experimental results for $H_T(f)$ were predominantly obtained by employing such vibrometers for measuring the velocities $v_1(t)$ and $v_2(t)$. As an example, Fig. 5.24 displays the experimental setup for a beam-shaped test sample that gets analyzed with the aid of clamp C2.

Representative Characteristic Transfer Functions

Figure 5.25 shows selected characteristic transfer functions $H_T(f)$, which were achieved with the mentioned experimental arrangements. Thereby, Fig. 5.25a and b refer to test samples of cylindrical shape and a beam-shaped test sample, respectively. The three cylindrical test samples (diameter $d_S = 50.0$ mm; height $t_S = 10.0$ mm) were made of a three component addition-cure silicone from the company Smooth-On, Inc. [63]. The mechanically soft compounds contain a two-part (A+B) Ecoflex 0030 silicon rubber and a single-part silicone thinner (T). With a view to varying mechanical stiffness of the specimen, different amounts of silicone thinner were used [19, 56, 70]. A smaller amount of thinner results in an increasing Young's modulus E_M. Each fabricated test sample consists of one part of both the component A and B, while the amount of thinner varies from zero to four parts. After the cylindrical compound has cured, it was glued on a carrier plate (cf. Fig. 5.22). As expected, $H_T(f)$ differs remarkably for the three compounds (see Fig. 5.25a); e.g.,

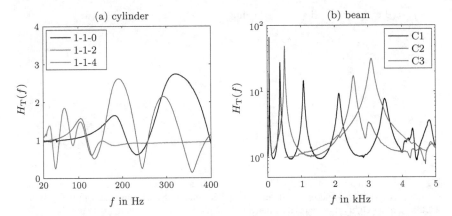

Fig. 5.25 Measured characteristic transfer functions $H_T(f)$ for different test samples; **a** cylindrical test samples of diameter $d_S = 50.0$ mm and height $t_S = 10.0$ mm differing in amount of silicone thinner, i.e., 1-1-0 and 1-1-4 mean zero and four parts silicone thinner, respectively; **b** beam-shaped PVC test sample ($w_S \times l_S \times t_S = 40.0$ mm $\times$ 150.0 mm $\times$ 4.0 mm) analyzed with clamp C1, C2 as well as C3 (see Fig. 5.23)

there occur considerable shifts of the maxima indicating resonances of mechanical vibration within the cylindrical test samples. For such mechanically very soft materials, this is an immediate consequence of Young's modulus, which takes different values for the test samples. However, it is impossible to find a constant material parameter set (i.e., independent of excitation frequency f) that describes the dynamic sample behavior in an appropriate way. The identification of frequency-dependent mechanical properties is, thus, indispensable.

In Fig. 5.25b, one can see the characteristic transfer functions $H_T(f)$ of a beam-shaped test sample that was investigated by means of the three different clamps C1, C2, and C3 (cf. Fig. 5.23). The beam was made of polyvinyl chloride (PVC) and featured the geometric dimensions $w_S \times l_S \times t_S = 40.0$ mm $\times$ 150.0 mm $\times$ 4.0 mm (cf. Fig. 5.26c). Similar to the cylindrical sample shape, resonances of mechanical vibrations are clearly visibly in $H_T(f)$ for the three clamps. Although the same beam was investigated, the vibration resonances arise at completely different frequencies; e.g., the first resonance for clamp C1 and C3 is at $f_r \approx 60$ Hz and $f_r \approx 3.1$ kHz, respectively. This fact follows from the type of mechanical excitation, which differs fundamentally for the clamps. In other words, each clamp excites other modes of mechanical vibrations within the beam. A slightly asymmetric mechanical behavior of the vibration source may, however, cause unexpected sample vibrations. For example, $H_T(f)$ for clamp C2 contains an additional peak at 3.1 kHz that originates from sample vibrations, which should, strictly speaking, exist only for clamp C3. To avoid misinterpretation and problems during parameter identification, it is, therefore, recommended to analyze the vibration behavior of the investigated beam by means of all clamps, i.e., C1, C2 as well as C3.

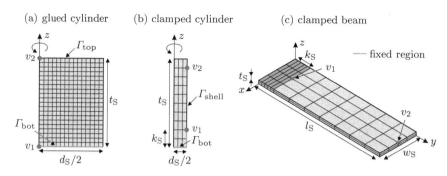

Fig. 5.26 Axisymmetric model for FE simulations of **a** glued cylinder and **b** clamped cylinder with diameter d_S and height t_S; **c** 3-D model for FE simulations of clamped beam with length l_S, width w_S, and height t_S; fixed regions indicate Dirichlet boundaries; clamping length k_S

Numerical Modeling

Apart from measurements, the inverse method for characterizing dynamic properties of passive materials demands harmonic FE simulations for the test samples. Just as in case of piezoceramic materials, the simulations should spend as little time as possible since parameter identification is based on iterative adjustments of FE simulations to measurements. For the characterization of passive materials, we utilize either cylindrical or beam-shaped test samples (see Figs. 5.22 and 5.23). Let us concentrate on efficient modeling approaches for these test samples, whereas the remaining parts (e.g., glue and clamps) of the experimental setup are not considered further.

The mechanical behavior of cylindrical test samples can be simulated efficiently by an axisymmetric model and quadratic Lagrange ansatz functions, i.e., h-FEM [56]. With a view to finding a feasible spatial discretization for the investigated cylindrical test sample, it is recommended to conduct mesh studies. The mechanical excitation v_1 at the bottom area Γ_{bot} of the cylinder represents an inhomogeneous Dirichlet boundary in z-direction (see Fig. 5.26a and b). Depending on the experimental configuration (i.e., glued or clamped samples), one has to apply additional boundary conditions in the FE model [19, 70]. For instance, homogeneous Dirichlet boundaries prevent mechanical displacements in radial directions for a glued cylinder. To obtain the characteristic transfer function $H_T(f)$ of the cylindrical test samples, we finally require a resulting velocity v_2 in z-direction. According to the experimental arrangement, this velocity refers to the center of the top area Γ_{top} or to a point along the cylinder shell Γ_{shell}, which has to be part of the computational grid.

In general, modeling of beam-shaped test samples is more complex than modeling of cylinders. Although we can exploit certain symmetries within a beam, reliable simulations of its mechanical behavior need 3-D models instead of simplified 2-D models (e.g., Kirchhoff plate) that are restricted to specific geometric ratios as well as mechanical excitations. However, conventional FE simulations for such a 3-D model may yield unrealistic results because of locking effects, especially in case of very thin beams. A potential remedy to avoid locking is the application of Legendre ansatz functions, i.e., p-FEM. As mentioned in Sect. 4.1.3, it is possible to vary the polynomial degrees for p-FEM in the different directions in space. This so-called anisotropic p-FEM allows a rough computational grid and, therefore, a less amount of finite elements for a beam-shaped test sample. An essential point toward the time-efficient application of anisotropic p-FEM is the reasonable choice of both the spatial discretization for the investigated beam and the polynomial degrees p_d in the different directions. It is recommended again to conduct mesh studies in advance before exploiting this special kind of FE simulations for iterative parameter identification.

Figure 5.26c depicts a typical computational grid, which turned out to be well suited for beams of geometric dimensions $w_S \times l_S \times t_S = 40\,\text{mm} \times 150\,\text{mm} \times 2\,\text{mm}$. Various numerical studies revealed that one should apply the highest polynomial degree (e.g., $p_d = 5$) in the direction of mechanical excitation [18], i.e., in z-direction for clamp C1. In the remaining directions, p_d can be reduced significantly. Similar to cylindrical test samples, the mechanical excitation and sample clamping are modeled by inhomogeneous and homogeneous Dirichlet boundaries at the clamping area of the beam, respectively. This area of length k_S features a comparatively fine mesh. In any case, the computational grid should cover the measuring points for the velocities v_1 as well as v_2 that are used in the experiments to determine $H_T(f)$.

Parameter Study

Now, let us take a look at simulation-based parameter studies for isotropic passive materials featuring the employed sample shapes, i.e., cylinders and beams. The focus lies on changes in the characteristic transfer behavior $H_T(f)$ due to two material parameters: (i) real part $E_{\Re}$ of Young's modulus as well as (ii) Poisson's ratio ν_P. Both parameters are here assumed to be constant with respect to excitation frequency f of the test sample. Since the damping ratio ξ_d primarily alters the height of resonance peaks in $H_T(f)$ but hardly affects resonance frequencies, this parameter is not considered within the parameter studies.

Figure 5.27 displays simulation results of the parameter study for a cylindrical test sample (diameter $d_S = 50\,\text{mm}$; height $t_S = 70\,\text{mm}$), which was uniformly excited at its bottom (cf. Fig. 5.22a). Each horizontal line in the graphs refers to the progress of $H_T(f)$ for a distinct parameter set. While one material parameter (e.g., $E_{\Re}$ in Fig. 5.27a) is varied in the range $\pm 20\%$, the other parameter (e.g., ν_P) remains constant. As expected from analytical relations, the parameter study clearly demonstrates that an increasing $E_{\Re}$ shifts resonances in $H_T(f)$ to higher values. Such distinct behavior cannot, however, be observed for variations of ν_P. Nevertheless, the significantly different modifications in $H_T(f)$ through $E_{\Re}$ and ν_P indicate that it should be possible to identify these parameters uniquely by means of appropriate cylindrical test samples.

In case of beam-shaped test samples, $E_{\Re}$ as well as ν_P were varied again in the range $\pm 20\%$. The considered beam exhibits the geometric dimensions $w_S \times l_S \times t_S = 40\,\text{mm} \times 150\,\text{mm} \times 2\,\text{mm}$. Figure 5.28 contains the results of the parameter studies for the three different clamps C1, C2, and C3, which were available in the

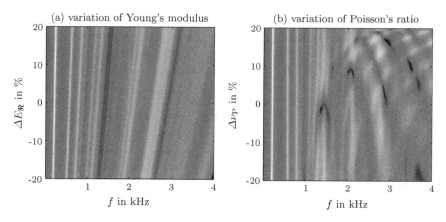

Fig. 5.27 Influence of distinct variations of **a** Young's modulus $E_\Re$ and **b** Poisson's ratio ν_P on characteristic transfer function $H_T(f)$ of cylindrical test sample ($d_S = 50$ mm; $t_S = 70$ mm); initial values: $\varrho = 1150\,\mathrm{kgm}^{-3}$, $E_\Re = 4$ MPa, $\nu_P = 0.4$, and $\xi_d = 0.025$; bright and dark colors indicate large and small values, respectively

experiments (see Fig. 5.23). Obviously, identical relative changes in $E_\Re$ and ν_P will cause similar modifications of $H_T(f)$ if the beam is investigated by means of clamp C1. Against that, such changes alter the resonance frequencies in $H_T(f)$ in opposite directions for clamp C2 as well as clamp C3. From there, we can conclude that a single clamp may not be sufficient for reliable identification of both Young's modulus and Poisson's ratio through the inverse method. It is, thus, advisable to combine the characteristic transfer function $H_T(f)$ of the beam-shaped test sample for clamp C1 with one for the other clamps, i.e., $H_T(f)$ for clamp C2 or for clamp C3. However, several investigations for real material samples revealed that the exclusive consideration of clamp C1 oftentimes leads to an accurate parameter set [18, 69].

5.4.4 Efficient Implementation

The tailored model for the dynamic Young's modulus $E_\Re(f)$ and damping ratio $\xi_d(f)$ of viscoelastic materials contains four independent parameters, namely E_0, α_1, α_2, and the additional scaling factor β_d (see (5.64) and (5.67)). Including Poisson's ratio ν_P, we will have to identify five parameters when the material density ϱ_0 is known. Consequently, the parameter vector $\mathbf{p}$ of the inverse method takes the form

$$\mathbf{p} = [E_0, \alpha_1, \alpha_2, \nu_P, \beta_d]^t \ . \tag{5.73}$$

The entries of this vector get identified iteratively by comparing FE simulations and measurement results for the characteristic transfer function $H_T(f)$ of the utilized test sample. In accordance with piezoceramic materials, let us suppose that $H_T(f)$ is

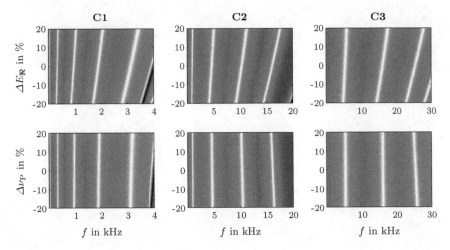

Fig. 5.28 Influence of distinct variations of Young's modulus $E_{\mathfrak{R}}$ and Poisson's ratio ν_P on characteristic transfer function $H_T(f)$ of beam-shaped test sample ($w_S \times l_S \times t_S = 40\,\text{mm} \times 150\,\text{mm} \times 2\,\text{mm}$) for clamp C1, C2, and C3; initial values: $\varrho = 1270\,\text{kgm}^{-3}$, $E_{\mathfrak{R}} = 10\,\text{MPa}$, $\nu_P = 0.4$, and $\xi_d = 0.01$; bright and dark colors indicate large and small values, respectively

sampled at N_I discrete frequencies ($f_1, f_2, \ldots, f_{N_I}$). The vectors of sampled measurements $\mathbf{q}_{\text{meas}}$ and sampled simulations $\mathbf{q}_{\text{sim}}(\mathbf{p})$ are then given by

$$\left. \begin{array}{c} \mathbf{q}_{\text{meas}} \\ \mathbf{q}_{\text{sim}}(\mathbf{p}) \end{array} \right\} = \left[H_T(f_1), H_T(f_2), \ldots, H_T(f_{N_I}) \right]^t . \tag{5.74}$$

To update the parameter vector $\mathbf{p}^{(i)}$ for iteration index i through the iteratively regularized Gauss–Newton method (see Sect. 5.2.4), one needs the correction vector $\mathbf{c}^{(i)}$ that results from the numerically evaluated Jacobian matrix $\mathbf{J}(\mathbf{p}^{(i)})$.

As it is always the case for inverse methods, we require an appropriate initial guess $\mathbf{p}^{(0)}$ for the parameter vector. Although the tailored material model offers a clear separation into constant part, linear part as well as logarithmic part, the determination of such an initial guess is a challenging task. Hereafter, an implementation will be presented that is able to cope with this problem and enables, moreover, efficient parameter identification. The underlying approach consists of three steps (see Fig. 5.29): (i) coarse adjustment through eigenfrequencies, (ii) adjustment for individual resonances, and (iii) determination of functional relations [18].

Step 1: Coarse Adjustment through Eigenfrequencies

This step will be especially useful if one does not know anything about the mechanical properties of the investigated material. Briefly, we exploit the fact that Young's modulus $E_{\mathfrak{R}}$ (real part) significantly affects resonances in the dynamic transfer behavior $H_T(f)$ of the considered test samples. Poisson's ratio ν_P also causes certain changes in $H_T(f)$, but during step 1, these changes are not taken into account and ν_P

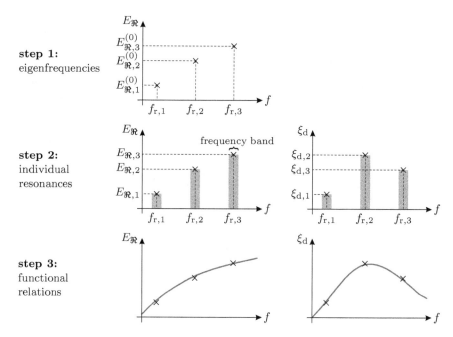

Fig. 5.29 Approach consisting of three steps to efficiently identify parameter vector $\mathfrak{p}$ (see (5.73)) for tailored material model [18]; step 1: coarse adjustment through eigenfrequencies; step 2: adjustment for individual resonances; step 3: determination of function relations

should be set to a common value, e.g., $\nu_P = 0.35$. From there, it might be possible, on the one hand, to estimate $E_\Re$ for each resonance separately by utilizing simple analytical relations (e.g., according to Euler–Bernoulli beam theory) for the test samples. Such an estimation approach is, however, only reasonable for specific test samples as well as mechanical excitations. On the other hand, we can approximate resonances in $H_T(f)$ and, consequently, $E_\Re$ for each resonance by conducting an eigenfrequency analysis based on FE simulations (see Sect. 4.3.1). The second approach is restricted neither to a particular sample shape nor mechanical excitation. That is why one should choose the eigenfrequency analysis based on FE simulations to estimate $E_\Re$ for each resonance separately.

Figure 5.30 illustrates the idea of coarse adjusting $E_\Re$ for a single resonance in $H_T(f)$ through eigenfrequency analysis. We start with a rough estimate $E_{est,1}$ of the dynamic Young's modulus at the selected resonance and perform an eigenfrequency analysis based on FE simulations for the test sample. As a result, one obtains the eigenfrequency $f_{est,1}$. Subsequently, $E_{est,1}$ is altered by a distinct portion to $E_{est,2}$, which leads to the corresponding eigenfrequency $f_{est,2}$ of the test sample. By comparing both eigenfrequencies (i.e., $f_{est,1}$ and $f_{est,2}$) to the measured resonance frequency f_{meas} and conducting linear extrapolation, it is possible to estimate $E_\Re(f_{meas})$ for the regarded resonance. Although this procedure requires solely two FE simulations for each resonance, we end up with a proper initial guess for the next step.

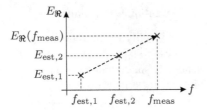

Fig. 5.30 Resonance frequencies $f_{est,1}$ and $f_{est,2}$ of test sample result from eigenfrequency analysis for Young's modulus $E_{est,1}$ and $E_{est,2}$, respectively; estimate $E_{\Re}(f_{meas})$ at measured resonance frequency f_{meas} is obtained from linear extrapolation

Step 2: Adjustment for Individual Resonances

Step 2 is based on the assumption that the relevant material parameters (i.e., $E_{\Re}$, ν_P, and ξ_d) stay almost constant in a small frequency range [19]. This also applies to small frequency bands around the resonances in the characteristic transfer function $H_T(f)$ of the test samples. With a view to identifying constant material parameters in such a frequency band, we can utilize the iteratively regularized Gauss–Newton method (see Sect. 5.2.4) for each resonance separately. In case of cylindrical test samples, the parameter vector for a single resonance becomes $\mathbf{p}_r = [E_{\Re}, \nu_P, \xi_d]^t$. As demonstrated in Fig. 5.28, the real part $E_{\Re}$ of Young's modulus and Poisson's ratio ν_P modify $H_T(f)$ of beam-shaped test samples in a similar manner. Therefore, the parameter vector for beam-shaped test samples should be limited to $\mathbf{p}_r = [E_{\Re}, \xi_d]^t$.

Step 1 provides an appropriate initial guess $E_{\Re}^{(0)}$ for each resonance. Cylindrical test samples additionally demand an initial guess $\nu_P^{(0)}$ for Poisson's ratio, which can be found in the literature (e.g., [59]) or may be determined by a common tensile test. Contrary to $E_{\Re}^{(0)}$ and $\nu_P^{(0)}$, the initial guess $\xi_d^{(0)}$ for the damping ratio is not crucial because it only scales resonance peaks in $H_T(f)$.

Due to the fact that the material parameters exhibit an enormous value range, they should be normalized to the initial guess as it was already recommended for piezoceramic materials. Finally, the questions arise how many frequency points N_I per resonance are required and which bandwidth around a single resonance needs to be considered within the inverse method. Various studies revealed that the bandwidth should cover most of the measured resonance peak and $N_I > 10$ has to be fulfilled.

After performing iterative identification, we know estimates for the decisive material parameter at the resonances f_r in $H_T(f)$ of the utilized test sample. Depending on the sample shape, this corresponds to a specific parameter combination

$$\text{cylindrical} \quad \mathbf{p}_r = \left[E_{\Re,r}, \nu_{P,r}, \xi_{d,r}\right]^t \tag{5.75}$$

$$\text{beam-shaped} \quad \mathbf{p}_r = \left[E_{\Re,r}, \xi_{d,r}\right]^t \tag{5.76}$$

for each resonance frequency f_r, respectively.

Step 3: Determination of Functional Relations

The aim of the last step lies in identifying the entire parameter set $\mathbf{p}$ (see (5.73)) of the tailored model that enables describing dynamic mechanical properties of viscoelastic materials. Instead of considering small frequency bands around the resonances, step 3 relates to the total available characteristic transfer function $H_T(f)$ of the test sample. To obtain a proper initial guess $\mathbf{p}^{(0)}$ for this final identification procedure, one should exploit the results from step 2. In doing so, the parameter $E_{\Re,r}$ for each resonance in $H_T(f)$ serves as supporting point of the underlying material model for $E_{\Re}(f)$. By conducting a least squares fit, it is possible to compute $E_0^{(0)}$, $\alpha_1^{(0)}$, and $\alpha_2^{(0)}$. Alternatively, the static Young's modulus E_0 can be determined through a tensile test and, thus, we may exclude E_0 from the identification procedure [19, 56]. While for cylindrical test samples, $\nu_P^{(0)}$ follows from the mean value of $\nu_{P,r}$ at the considered resonances, step 1 and step 2 do not provide useful information for ν_P in case of beam-shaped test samples. Hence, we have to set $\nu_P^{(0)}$ to a common value. The comparison of the supporting points $\xi_{d,r}$ with $\xi_d(f)$ in (5.67), which results from the initial parameter set (i.e., $E_0^{(0)}$, $\alpha_1^{(0)}$, and $\alpha_2^{(0)}$), yields an initial guess $\beta_d^{(0)}$ for the additional scaling factor.

It is recommended to normalize the parameter vector $\mathbf{p}^{(i)}$ during identification to its initial guess $\mathbf{p}^{(0)}$ (cf. Sect. 5.3.4). This normalization facilitates both convergence of the identification procedure and choice of a suitable regularization parameter ζ_R, which is required for the iteratively regularized Gauss–Newton method. As a matter of principle, we also have to choose the amount of frequency points N_I that are taken into account for parameter identification. Characteristic transfer functions $H_T(f)$ containing only a few resonance peaks can be treated by a less amount of frequency points, e.g., $N_I = 100$. In contrast, a broadband transfer function will demand much more frequency points (e.g., $N_I > 1000$) because each resonance peak in $H_T(f)$ needs to be covered by a sufficient amount of points.

5.4.5 Identified Parameters for Selected Materials

The presented simulation-based identification approach for passive materials is applicable for various material classes. It can be used to characterize the frequency-dependent mechanical behavior of both soft materials (e.g., elastomers and thermoplastics) and hard materials like metals as well as glass [18, 56, 68]. Here, results of selected passive materials will be detailed and verified. This is not just limited to the dynamic material behavior but also includes temperature dependence of the parameters.

Similar to piezoceramic materials, let us briefly repeat the three main steps of the inverse method for passive materials in advance:

- One has to acquire the characteristic transfer function $H_T(f)$ for an appropriate test sample that features several resonances of mechanical vibrations in the investigated

Table 5.6 Material density ϱ_0 and components of identified parameter vector $\mathfrak{p}$ for selected materials; materials sorted by ϱ_0; polypropylene PP; poly(methyl methacrylate) PMMA; polyvinyl chloride PVC; polytetrafluoroethylene PTFE

Material	ϱ_0 kgm^{-3}	E_0 Nm^{-2}	α_1 Hz^{-1}	α_2	ν_P	β_d
PP	912	$1.75 \cdot 10^9$	$9.10 \cdot 10^{-6}$	$1.39 \cdot 10^{-1}$	0.42	1.43
Silicone	1135	$6.41 \cdot 10^6$	$6.70 \cdot 10^{-6}$	$1.16 \cdot 10^{-1}$	0.49	1.34
PMMA	1183	$3.85 \cdot 10^9$	$1.09 \cdot 10^{-6}$	$7.67 \cdot 10^{-2}$	0.41	1.13
PVC	1450	$2.96 \cdot 10^9$	$8.52 \cdot 10^{-6}$	$8.64 \cdot 10^{-2}$	0.37	0.68
PTFE	2181	$1.33 \cdot 10^9$	$1.00 \cdot 10^{-8}$	$4.61 \cdot 10^{-2}$	0.47	1.52
Aluminum	2701	$6.46 \cdot 10^{10}$	$8.71 \cdot 10^{-6}$	$1.31 \cdot 10^{-5}$	0.43	0.11

frequency range. Cylinders turned out to be well suited for soft materials (e.g., silicone), whereas beam-shaped test samples should be used for hard materials.

- Since the inverse method is an iterative procedure, the FE model of the considered test sample should enable reliable as well as time-efficient simulations of $H_T(f)$.
- After determining a suitable initial guess $\mathfrak{p}^{(0)}$ according to Sect. 5.4.4, the iteratively regularized Gauss–Newton method leads to the aimed parameter vector $\mathfrak{p}$ (see (5.73)). This vector contains components that directly relate to the decisive material quantities such as the dynamic Young's modulus.

Material Behavior at Room Temperature

Table 5.6 shows the obtained parameters for several materials, which were analyzed at room temperature. The table entries are sorted by the material density ϱ_0 that was experimentally determined. As expected, many entries differ remarkably from each other. For instance, the static Young's modulus E_0 of aluminum exceeds that of silicone by four orders of magnitude. Besides, the dynamic mechanical behavior

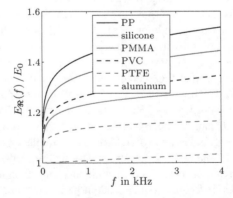

Fig. 5.31 Resulting real parts $E_{\Re}(f)$ of dynamic Young's modulus with respect to frequency f for materials listed in Table 5.6; curves normalized to static Young's modulus E_0

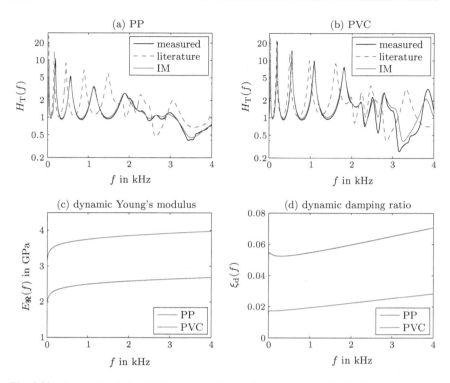

Fig. 5.32 Measured and simulated characteristic transfer functions $H_T(f)$ for beam-shaped test samples (40.0 mm × 150.0 mm × 2.0 mm) that were excited through clamp C1; **a** PP beam; **b** PVC beam; FE simulations performed with typical frequency-independent material parameters from the literature and data set identified by inverse method (IM, see Table 5.6), respectively; **c** resulting real part $E_\Re(f)$ of dynamic Young's modulus; **d** resulting dynamic damping ratio $\xi_d(f)$

of the materials is quite different; e.g., the linear part α_1 seems to be dominating for aluminum, while the logarithmic part α_2 primarily determines the behavior of the other materials. This fact is confirmed in Fig. 5.31, which displays the normalized real part $E_\Re(f)$ of Young's modulus in the frequency range from 0 Hz to 4 kHz. According to the curve progressions, there is also a reason to assume that the material density strongly influences the development of dynamic material behavior.

Figure 5.32a and b depict the characteristic transfer function $H_T(f)$ of beam-shaped test samples, which were made of PP and PVC, respectively. The beams of geometric dimensions 40.0 mm × 150.0 mm × 2.0 mm were harmonically excited through clamp C1. It can be clearly observed that the identified dynamic material properties (see Fig. 5.32c and d) yield realistic simulation results. Contrary to that, constant material parameters (i.e., frequency-independent) from the literature cause substantial deviations between measurements and FE simulations, especially at higher frequencies. If frequency-independent parameters are applied for numerical simulations, we will not be able to precisely predict the dynamic behavior of devices (e.g., sensors and actuators) containing such passive materials.

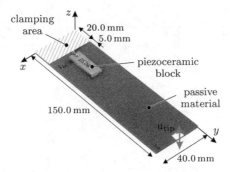

Fig. 5.33 Picture showing beam made of passive material (PMMA or PVC) equipped with piezo-ceramic block made of PIC255 [18]; beam dimensions 40.0 mm × 150.0 mm × 2.0 mm; block dimensions 10.0 mm × 30.0 mm × 2.0 mm; clamping length $k_S = 20.0$ mm of beam; electrically excited piezoceramic block causes beam deflections; measurement of resulting tip displacement u_{tip}

Verification

Several experiments were conducted to verify the identified dynamic behavior of passive materials. For example, the material parameters of silicones have been verified on basis of sample shapes that were not utilized during characterization [56]. Moreover, the resulting parameters for test samples of different geometric dimensions have been compared [18]. In the following, we will discuss a verification concerning active as well as passive materials. Figure 5.33 shows the investigated specimen, which consists of both a PMMA or PVC beam acting as passive material and a piezoceramic block made of PIC255 representing the active material. The beam of geometric dimensions 40.0 mm × 150.0 mm × 2.0 mm was clamped at one side along the clamping length $k_S = 20.0$ mm. The piezoceramic block (10.0 mm × 30.0 mm × 2.0 mm) was polarized in thickness direction (see test sample T1 in Fig. 5.11) and glued onto the beam by means of a conductive adhesive. By applying an electrical excitation between the electrodes of the piezoelectric material, the block gets deformed and, consequently, a bending moment is introduced into the beam. This bending moment causes deflections of the beam, which were measured at its free end with a laser Doppler vibrometer.

Figure 5.34a and b illustrate measurement as well as simulation results for the arising tip displacements $\hat{u}_{tip}(f)$ (amplitude) normalized to the applied electric voltage in case of the PMMA and PVC beam, respectively. Thereby, the FE simulations were carried out with three different sets of material parameters, namely (i) frequency-independent data for beam material (PMMA or PVC) and manufacturer data for PIC255, (ii) frequency-independent data for beam material and identified data set for PIC255, and (iii) identified dynamic behavior for passive material and identified data set for PIC255. The identified material parameters $\mathbf{p}$ for PMMA, PVC, and PIC255 at room temperature can be found in Tables 5.6 and 5.3. Not surprisingly, the simulation results for case (i) remarkably deviate from the measured tip

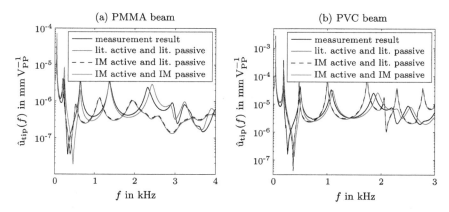

Fig. 5.34 Measured and simulated tip displacement $\hat{u}_{tip}(f)$ against excitation frequency f of beam equipped with piezoceramic block (see Fig. 5.33); **a** PMMA beam; **b** PVC beam; $\hat{u}_{tip}(f)$ normalized to applied excitation voltage of piezoceramic block; FE simulations with typical material parameters from the literature and data set identified by inverse method (IM), respectively

displacements. In contrast, simulations will coincide very well with measurements for both beam materials when the data set of case (iii) is exploited. The remaining deviations between simulations and measurements stem from irregularities in the thin adhesive layer (e.g., air pockets) between beam and piezoceramic block, which can hardly be taken into account within the FE simulation.

Apart from the just mentioned findings, Fig. 5.34a and b reveal that the deviations between simulation results for case (i) and (ii) are negligible. In other words, the material parameters of PIC255 apparently do not alter FE simulations of the ana-lyzed beam-shaped structure. This is a consequence of the strongly differing geo-metric dimensions and material properties of beam and piezoceramic block. Because the investigated frequency range contains exclusively resonances of the beam, the resonances of the piezoceramic block (cf. Fig. 5.12a) do not influence resonances in the tip displacement of the whole structure, i.e., beam equipped with piezoceramic block. When the frequency range approaches resonances of the block, the data set of the piezoceramic material will, however, become increasingly important for the overall behavior of the beam-shaped structure. Nevertheless, the studied example reveals again the significance of dynamic material properties for passive materials in order to obtain reliable simulation results.

Temperature Dependence of Material Behavior

As a matter of course, the dynamic mechanical behavior of passive materials depends on temperature ϑ. With a view to quantifying this dependence, the investigated test samples were placed in a climatic chamber. In Fig. 5.35, one can see the char-acteristic transfer function $H_T(f)$ of a PP beam (geometric dimensions 40.0 mm × 150.0 mm × 2.0 mm; clamp C1) for three different temperatures ϑ, namely −60 °C, +20 °C, and +100 °C. Since $H_T(f)$ greatly changes with ϑ, it seems only natural that the decisive parameters of a passive material also have to exhibit a certain tem-perature dependence. Figure 5.36a and b display the identified real part $E_{\Re}(f)$ of

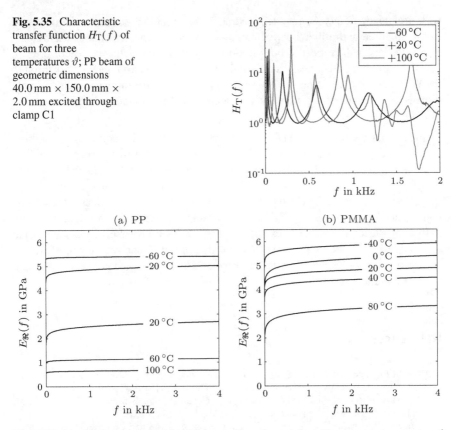

Fig. 5.35 Characteristic transfer function $H_T(f)$ of beam for three temperatures ϑ; PP beam of geometric dimensions 40.0 mm × 150.0 mm × 2.0 mm excited through clamp C1

Fig. 5.36 Identified real part $E_{\Re}(f)$ of dynamic Young's modulus with respect to temperature ϑ for **a** PP and **b** PMMA

the dynamic Young's modulus with respect to ϑ for PP and PMMA, respectively. Not surprisingly, there occurs a remarkable temperature dependence of $E_{\Re}(f)$ for both materials. If piezoelectric sensor and actuator devices are exposed to varying temperature, reliable numerical simulations for practical applications will demand the consideration of the temperature-dependent material behavior. This refers to the piezoelectric as well as passive materials within the device.

Note that there exists a unique connection between frequency dependence and temperature dependence of the mechanical material behavior, especially for viscoelastic materials such as several polymers. The underlying nonlinear mathematical relation is the so-called *Williams–Landel–Ferry equation* (WLF equation), which stands for an empirical formula associated with the time–temperature superposition [14, 71]. In the framework of the *dynamic mechanical thermal analysis* (DMTA), the mechanical material properties of a test sample are characterized in a narrow frequency band (e.g., 0.3–30 Hz) over a large temperature range [72]. The WLF equation is then exploited to specify the dynamic mechanical properties over a rather wide frequency range. As just demonstrated, the combination of simulation-based inverse method and utilized experimental arrangement enables characterizing passive materials with respect to both frequency f and temperature ϑ. Therefore, we can also extend the frequency range for the mechanical material parameters through the WLF equation. In doing so, the experimental setup concerning geometric sample dimensions, sample clamping as well as mechanical excitation does not need to be modified.

In summary, it can be stated that the inverse method yields parameters describing the linearized mechanical behavior of passive materials with respect to frequency. The comparison of numerical simulations and measurements is performed on basis of the transfer function for mechanical vibrations within an appropriate test sample. As the presented results demonstrate, the characterization approach is applicable for various homogeneous solids (e.g., PVC) that feature isotropic material behavior. According to Ilg [18], the approach can be extended for analyzing transversely isotropic passive materials like fiber-reinforced plastics. For this purpose, one requires, however, at least two test samples, which differ in the orientation of the plane of symmetry, e.g., in the orientation of the fibers.

References

1. ASTM International: Standard Test Method for Dynamic Young's Modulus, Shear Modulus, and Poisson's Ratio by Impulse Excitation of Vibration. ASTM E1875-09 (2009)
2. ASTM International: Standard Test Method for Dynamic Young's Modulus, Shear Modulus, and Poisson's Ratio by Sonic Resonance. ASTM E1875-13 (2013)
3. Bakushinskii, A.B.: The problem of the convergence of the iteratively regularized Gauss-Newton method. Comput. Math. Math. Phys. **32**(9), 1353–1359 (1992)
4. Banks, H.T., Hu, S., Kenz, Z.R.: A brief review of elasticity and viscoelasticity for solids. Adv. Appl. Math. Mech. **3**(1), 1–51 (2011)
5. Barkanov, E., Skukis, E., Petitjean, B.: Characterisation of viscoelastic layers in sandwich panels via an inverse technique. J. Sound Vib. **327**(3–5), 402–412 (2009)
6. Blaschke, B., Neubauer, A., Scherzer, O.: On convergence rates for the iteratively regularized Gauss-Newton method. IMA J. Numer. Anal. **17**(3), 421–436 (1997)
7. Brissaud, M.: Three-dimensional modeling of piezoelectric materials. IEEE Trans. Ultrason. Ferroelectr. Freq. Control **57**(9), 2051–2065 (2010)
8. Bronstein, I.N., Semendjajew, K.A., Musiol, G., Mühlig, H.: Handbook of Mathematics, 6th edn. Springer, Berlin (2015)
9. Cappon, H., Keesman, K.J.: Numerical modeling of piezoelectric transducers using physical parameters. IEEE Trans. Ultrason. Ferroelectr. Freq. Control **59**(5), 1023–1032 (2012)
10. Collaborative Research Center TRR 39: Production Technologies for Lightmetal and Fiber Reinforced Composite based Components with Integrated Piezoceramic Sensors and Actuators (PT-PIESA) (2018). http://www.pt-piesa.tu-chemnitz.de
11. Czichos, H., Seito, T., Smith, L.E.: Springer Handbook of Metrology and Testing, 2nd edn. Springer, Berlin (2011)
12. Engl, H.W., Hanke, M., Neubauer, A.: Regularization of Inverse Problems. Kluwer Academic Publishers, Dordrecht (1996)
13. European Committe for Electrotechnical Standardization (CENELEC): Piezoelectric properties of ceramic materials and components – Part 2: Methods of measurement – Low power. EN 50324-2 (2002)

14. Ferry, J.: Viscoelastic Properties of Polymers. Wiley-Interscience, Chichester (1980)
15. Göpel, W., Hesse, J., Zemel, J.N.: Sensors Volume 6 - Optical Sensors. VCH, Weinheim, New York (1992)
16. Hadamard, J.: Lectures on the Cauchy Problem in Linear Partial Differential Equations. Yale University Press, New Haven (1923)
17. Holland, R.: Representation of dielectric, elastic, and piezoelectric losses by complex coefficients. IEEE Trans. Sonics and Ultrason. **14**(1), 18–20 (1967)
18. Ilg, J.: Bestimmung, Verifikation und Anwendung frequenzabhängiger mechanischer Materialkennwerte. Ph.D. thesis, Friedrich-Alexander-University Erlangen-Nuremberg (2015)
19. Ilg, J., Rupitsch, S.J., Sutor, A., Lerch, R.: Determination of dynamic material properties of silicone rubber using one-point measurements and finite element simulations. IEEE Trans. Instrum. Meas. **61**(11), 3031–3038 (2012)
20. Ilg, J., Rupitsch, S.J., Lerch, R.: Impedance-based temperature sensing with piezoceramic devices. IEEE Sens. J. **13**(6), 2442–2449 (2013)
21. Institute of Electrical and Electronics Engineers (IEEE): IEEE Standard on Piezoelectricity. ANSI-IEEE Std. 176-1987 (1987)
22. Isakov, V.: Inverse Problems for Partial Differential Equations. Springer, Berlin (1998)
23. Jonsson, U.G., Andersson, B.M., Lindahl, O.A.: A FEM-based method using harmonic overtones to determine the effective elastic, dielectric, and piezoelectric parameters of freely vibrating thick piezoelectric disks. IEEE Trans. Ultrason. Ferroelectr. Freq. Control **60**(1), 243–255 (2013)
24. Joo, H.W., Lee, C.H., Rho, J.S., Jung, H.K.: Identification of material constants for piezoelectric transformers by three-dimensional, finite-element method and a design-sensitivity method. IEEE Trans. Ultrason. Ferroelectr. Freq. Control **50**(8), 965–971 (2003)
25. Kaltenbacher, B., Neubauer, A., Scherzer, O.: Iterative Regularization Methods for Nonlinear Ill-Posed Problems. Walter de Gruyter, Berlin (2009)
26. Kay, S.M.: Fundamentals of Statistical Signal Processing – Estimation Theory. Prentice Hall, Englewood Cliffs (1993)
27. Keysight Technologies Inc.: Product portfolio (2018). http://www.keysight.com
28. Kim, S.Y., Lee, D.H.: Identification of fractional-derivative-model parameters of viscoelastic materials from measured FRFs. J. Sound Vib. **324**(3–5), 570–586 (2009)
29. Kronig, R.d.L.: On the theory of dispersion of X-Rays. J. Opt. Soc. Am. **12**(6), 547–557 (1926)
30. Kulshreshtha, K., Jurgelucks, B., Bause, F., Rautenberg, J., Unverzagt, C.: Increasing the sensitivity of electrical impedance to piezoelectric material parameters with non-uniform electrical excitation. J. Sens. Sens. Syst. **4**(1), 217–227 (2015)
31. Kwok, K.W., Lai, H., Chan, W., Choy, C.L.: Evaluation of the material parameters of piezoelectric materials by various methods. IEEE Trans. Ultrason. Ferroelectr. Freq. Control **44**(4), 733–742 (1997)
32. Lahmer, T.: Forward and inverse problems in piezoelectricity. Ph.D. thesis, Friedrich-Alexander-University Erlangen-Nuremberg (2014)
33. Lahmer, T., Kaltenbacher, M., Kaltenbacher, B., Lerch, R., Leder, E.: FEM-based determination of real and complex elastic, dielectric, and piezoelectric moduli in piezoceramic materials. IEEE Trans. Ultrason. Ferroelectr. Freq. Control **55**(2), 465–475 (2008)
34. Lakes, R.S.: Viscoelastic Solids. CRC Press, Boca Raton (1998)
35. Li, S., Zheng, L., Jiang, W., Sahul, R., Gopalan, V., Cao, W.: Characterization of full set material constants of piezoelectric materials based on ultrasonic method and inverse impedance spectroscopy using only one sample. J. Appl. Phys. **114**(10), 104505 (2013)
36. Malkin, A., Isayev, A.: Rheology: Concepts, Methods, and Applications, 2nd edn. Elsevier, Oxford (2012)
37. Martinez-Agirre, M., Elejabarrieta, M.J.: Dynamic characterization of high damping viscoelastic materials from vibration test data. J. Sound Vib. **330**(16), 3930–3943 (2011)
38. Matter, M., Gmür, T., Cugnoni, J., Schorderet, A.: Numerical-experimental identification of the elastic and damping properties in composite plates. Compos. Struct. **90**(2), 180–187 (2009)

39. Morozov, V.A.: On the solution of functional equations by the method of regularization. Sov. Math. Dokl. **7**, 414–417 (1966)
40. O'Donnell, M., Jaynes, E.T., Miller, J.G.: Kramers-Kronig relationship between ultrasonic attenuation and wave velocity. J. Acoust. Soc. Am. **69**(3), 696–701 (1981)
41. Ogo, K., Kakimoto, K.I., Weiß, M., Rupitsch, S.J., Lerch, R.: Determination of temperature dependency of material parameters for lead-free alkali niobate piezoceramics by the inverse method. AIP Adv. **6**, 065,101–1–065,101–9 (2016)
42. Pardo, L., Algueró, M., Brebol, K.: A non-standard shear resonator for the matrix characterization of piezoceramics and its validation study by finite element analysis. J. Phys. D: Appl. Phys. **40**(7), 2162–2169 (2007)
43. Park, J.: Transfer function methods to measure dynamic mechanical properties of complex structures. J. Sound Vib. **288**(1–2), 57–79 (2005)
44. Pérez, N., Andrade, M.A.B., Buiochi, F., Adamowski, J.C.: Identification of elastic, dielectric, and piezoelectric constants in piezoceramic disks. IEEE Trans. Ultrason. Ferroelectr. Freq. Control **57**(12), 2772–2783 (2010)
45. PI Ceramic GmbH: Product portfolio (2018). https://www.piceramic.com
46. Polytec GmbH: Product portfolio (2018). http://www.polytec.com
47. Pritz, T.: Analysis of four-parameter fractional derivative model of real solid materials. J. Sound Vib. **195**(1), 103–115 (1996)
48. Pritz, T.: Frequency dependences of complex moduli and complex Poisson's ratio of real solid materials. J. Sound Vib. **214**(1), 83–104 (1998)
49. Research Unit FOR 894: Fundamental Flow Analysis of the Human Voice (2018). http://gepris. dfg.de/gepris/projekt/35819142
50. Rieder, A.: Keine Probleme mit Inversen Problemen. Vieweg, Wiesbaden (2003)
51. Rupitsch, S.J.: Simulation-based characterization of piezoceramic materials. In: Proceedings of IEEE Sensors, pp. 1–3 (2016)
52. Rupitsch, S.J., Ilg, J.: Complete characterization of piezoceramic materials by means of two block-shaped test samples. IEEE Trans. Ultrason. Ferroelectr. Freq. Control **62**(7), 1403–1413 (2015)
53. Rupitsch, S.J., Kindermann, S., Zagar, B.G.: Estimation of the surface normal velocity of high frequency ultrasound transducers. IEEE Trans. Ultrason. Ferroelectr. Freq. Control **55**(1), 225–235 (2008)
54. Rupitsch, S.J., Lerch, R.: Inverse method to estimate material parameters for piezoceramic disc actuators. Appl. Phys. A: Mater. Sci. Process. **97**(4), 735–740 (2009)
55. Rupitsch, S.J., Ilg, J., Lerch, R.: Enhancement of the inverse method enabling the material parameter identification for piezoceramics. In: Proceedings of International IEEE Ultrasonics Symposium (IUS), pp. 357–360 (2011)
56. Rupitsch, S.J., Ilg, J., Sutor, A., Lerch, R., Döllinger, M.: Simulation based estimation of dynamic mechanical properties for viscoelastic materials used for vocal fold models. J. Sound Vib. **330**(18–19), 4447–4459 (2011)
57. Rupitsch, S.J., Wolf, F., Sutor, A., Lerch, R.: Reliable modeling of piezoceramic materials utilized in sensors and actuators. Acta Mech. **223**, 1809–1821 (2012)
58. Rupitsch, S.J., Ilg, J., Lerch, R.: Inverse scheme to identify the temperature dependence of electromechanical coupling factors for piezoceramics. In: Proceedings of Joint IEEE International Symposium on Applications of Ferroelectric and Workshop on Piezoresponse Force Microscopy (ISAF/PFM), pp. 183–186 (2013)
59. Schmidt, E.: Landolt-Börnstein - Zahlenwerte und Funktionen aus Naturwissenschaft und Technik - Band IV/1, 6th edn. Springer, Berlin (1955)
60. Sherrit, S., Gauthier, N., Wiederick, H.D., Mukherjee, B.K.: Accurate evaluation of the real and imaginary material constants for a piezoelectric resonator in the radial mode. Ferroelectrics **119**(1), 17–32 (1991)
61. Sherrit, S., Masys, T.J., Wiederick, H.D., Mukherjee, B.K.: Determination of the reduced matrix of the piezoelectric, dielectric, and elastic material constants for a piezoelectric material with C∞ symmetry. IEEE Trans. Ultrason. Ferroelectr. Freq. Control **58**(9), 1714–1720 (2011)

62. Shi, Y., Sol, H., Hua, H.: Material parameter identification of sandwich beams by an inverse method. J. Sound Vib. **290**(3–5), 1234–1255 (2006)

63. Smooth-On, Inc.: Product portfolio (2018). https://www.smooth-on.com

64. TIRA GmbH: Product portfolio (2018). https://www.tira-gmbh.de/en/

65. Tränkler, H.R., Reindl, L.M.: Sensortechnik - Handbuch für Praxis und Wissenschaft. Springer, Berlin (2014)

66. Van Dyke, K.S.: The piezo-electric resonator and its equivalent network. Proc. Inst. Radio Eng. **16**(6), 742–764 (1928)

67. Wada, Y., Ito, R., Ochiai, H.: Comparison between mechanical relaxations associated with volume and shear deformations in styrene-butadiene rubber. J. Phys. Soc. Jpn. **17**(1), 213–218 (1962)

68. Weiß, S.: Messverfahren zur Charakterisierung synthetischer Stimmlippen. Ph.D. thesis, Friedrich-Alexander-University Erlangen-Nuremberg (2014)

69. Weiß, M., Ilg, J., Rupitsch, S.J., Lerch, R.: Inverse method for characterizing the mechanical frequency dependence of isotropic materials. Tech. Messen **83**(3), 123–130 (2016)

70. Weiß, S., Sutor, A., Ilg, J., Rupitsch, S.J., Lerch, R.: Measurement and analysis of the material properties and oscillation characteristics of synthetic vocal folds. Acta Acust. United Acust. **102**(2), 214–229 (2016)

71. Williams, M.L., Landel, R.F., Ferry, J.D.: The temperature dependence of relaxation mechanisms in amorphous polymers and other glass-forming liquids. J. Am. Chem. Soc. **77**(14), 3701–3707 (1955)

72. Willis, R.L., Shane, T.S., Berthelot, Y.H., Madigosky, W.M.: An experimental-numerical technique for evaluating the bulk and shear dynamic moduli of viscoelastic materials. J. Am. Chem. Soc. **102**(6), 3549–3555 (1997)

73. Willis, R.L., Wu, L., Berthelot, Y.H.: Determination of the complex young and shear dynamic moduli of viscoelastic materials. J. Acoust. Soc. Am. **109**, 611–621 (2001)

74. Yoshida, K., Kakimoto, K.I., Weiß, M., Rupitsch, S.J., Lerch, R.: Determination of temperature dependences of material constants for lead-free (Na0.5K0.5)NbO3-Ba2NaNb5O15 piezoceramics by inverse method. Jpn. J. Appl. Phys. **55**(10), 10TD02 (2016)

75. Ziegler, F.: Mechanics of Solids and Fluids, 2nd edn. Springer, Berlin (1995)

76. Zörner, S., Kaltenbacher, M., Lerch, R., Sutor, A., Döllinger, M.: Measurement of the elasticity modulus of soft tissues. J. Biomech. **43**(8), 1540–1545 (2010)

Chapter 6
Phenomenological Modeling for Large-Signal Behavior of Ferroelectric Materials

As already discussed in Chap. 3, the electromechanical coupling within piezoelectric materials can be attributed to intrinsic and extrinsic effects. If piezoelectric materials show extrinsic effects, they will be frequently named ferroelectric materials. While intrinsic effects determine the small-signal behavior, extrinsic effects dominate the large-signal behavior of those materials. The small-signal behavior can simply be described through the material law of linear piezoelectricity (see linearization in Fig. 6.1 and Sect. 3.3). In contrast, the large-signal behavior of ferroelectric materials calls for special modeling treatments since it originates from altered geometric alignments of unit cells. Such altered geometric alignments will cause nonlinear as well as hysteretic material behavior (i.e., hysteresis curves) when sufficiently large electrical and/or mechanical loads are applied. Note that this fact is of utmost importance for ferroelectric actuators, which are utilized, e.g., in high-precision positioning systems (cf. Chap. 10).

Ferroelectric actuators usually operate far below their mechanical resonance frequencies. It will, therefore, be reasonable to assume a uniform mechanical displacement along their surfaces if the bottom and top surface are completely covered with electrodes (cf Fig. 4.20 on p. 121). Without limiting the generality, we exclusively consider hereafter electrical and mechanical quantities in the thickness direction (3-direction) of ferroelectric materials. Due to this fact, components in 1-direction and 2-direction as well as indices for 3-direction of the relevant physical quantities can be omitted. Figure 6.1 exemplarily depicts symmetrical hysteresis curves of a ferroelectric material for both the electric polarization $P(E)$ and the mechanical strain $S(E)$ in case of electrical excitation with the electric field intensity E. Mainly, one can distinguish between three different working areas detailed below.

- **Bipolar working area**: The ferroelectric material is alternately driven in positive as well as negative saturation leading to $P_{\text{sat}}^{\pm}$ for the electric polarization and $S_{\text{sat}}^{\pm}$ for the mechanical strain, respectively. The resulting hysteresis curves $P(E)$

© Springer-Verlag GmbH Germany, part of Springer Nature 2019
S. J. Rupitsch, *Piezoelectric Sensors and Actuators*, Topics in Mining, Metallurgy and Materials Engineering, https://doi.org/10.1007/978-3-662-57534-5_6

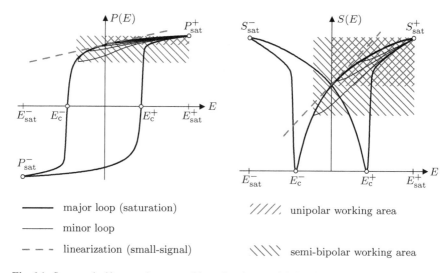

Fig. 6.1 Symmetrical hysteresis curves of ferroelectric materials for electric polarization $P(E)$ and mechanical strain $S(E)$ versus applied electric field intensity E in different working areas; electric polarization $P_{sat}^{\pm}$ and mechanical strain $S_{sat}^{\pm}$ in positive as well as negative saturation, respectively; coercive field intensity $E_c^{\pm}$; linearization relates to small-signal behavior of ferroelectric materials

and $S(E)$ (thick lines in Fig. 6.1) are known as major loops. According to its shape, $S(E)$ is also called *butterfly curve*.

- **Unipolar working area**: The ferroelectric material operates with positive or negative electric field intensities, i.e., $E \geq 0$ or $E \leq 0$. Hence, the material can be driven either in positive or negative saturation. Compared to the bipolar working area, mechanical strains and, consequently, mechanical displacements of a ferroelectric actuator get reduced remarkably. The resulting hysteresis curves $P(E)$ and $S(E)$ are referred to as minor loops (thin lines in Fig. 6.1).

- **Semi-bipolar working area**: In contrast to the unipolar working area, the ferroelectric material operates with a larger range of electric field intensities in the semi-bipolar working area. Thereby, one of the conditions $E_c^- < E \leq E_{sat}^+$ or $E_{sat}^- \leq E < E_c^+$ (coercive field intensity E_c) has to be fulfilled. As a result, the achievable mechanical strains increase but stay below values of the bipolar working area. The hysteresis curves $P(E)$ and $S(E)$ are again referred to as minor loops.

This chapter primarily deals with Preisach hysteresis modeling, which represents a phenomenological modeling approach for the large-signal behavior of ferroelectric materials in the mentioned working areas. Before we study in Sect. 6.3 alternative phenomenological modeling approaches that also focus on the macroscopic transfer behavior of ferroelectric materials, hysteresis will be mathematically defined. Moreover, an overview of material models on different length scales (e.g., atomistic scale) is given in Sect. 6.2. Contrary to phenomenological modeling approaches, those material models aim at describing the physical behavior of ferroelectric materials as accurate as possible. In Sect. 6.4, we will introduce the classical Preisach hysteresis operator $\mathcal{H}_P$, which comprises weighted elementary switching operators. Section 6.5 details different weighting procedures for the elementary switching operators. Because the classical Preisach hysteresis operator is only suitable to a limited extent for predicting hysteretic behavior of ferroelectric actuators in practical applications, a so-called generalized Preisach hysteresis model (operator $\mathcal{H}_G$) will be introduced in Sect. 6.6. This extended Preisach hysteresis model enables, e.g., the consideration of asymmetric behavior in hysteresis curves. After that, a parameter identification strategy is presented which allows reliable predictions of electrical and mechanical quantities through Preisach hysteresis modeling. To apply Preisach hysteresis modeling in practical applications of ferroelectric actuators (e.g., in high-precision positioning systems), it is of utmost importance to invert the Preisach hysteresis operator. Owing to this fact, Sect. 6.8 finally addresses an iterative inversion procedure, which enables efficient determinations of the aimed electrical excitation signals in a reasonable time. Throughout the whole chapter, piezoceramic disks made of the ferroelectrically soft materials PIC255 (manufacturer PI Ceramic GmbH [71]) as well as Pz27 and the ferroelectrically hard material Pz26 (manufacturer Meggitt Sensing Systems [65]) serve as test objects.

6.1 Mathematical Definition of Hysteresis

There exist various meanings and definitions for the term *hysteresis* in technical areas (e.g., [61, 64]). However, several similarities can be found in these definitions. Here, we especially concentrate on a transmission system with one input $x(t)$ and one output $y(t)$, both depending on time t. When the system exhibits hysteresis[1] in its transmission behavior, three properties will apply to such a system [64, 94]:

1. The output $y(t)$ is clearly defined by the progression of $x(t)$ and the initial state of the transmission system.
2. We can mathematically link $y(t)$ and $x(t)$ with the aid of nonlinear relations describing branches in the xy-plane (see, e.g., Fig. 6.7c). A change between different branches may occur at extrema of the system input $x(t)$.
3. The sequence of extrema in $x(t)$ exclusively determines the progression of the system output $y(t)$. In contrast, values in between these extrema as well as the time response of $x(t)$ do not modify the current output. For this reason, the transmission behavior is rate-independent.

Due to the fact that there always occur creep processes in ferroelectric materials, the third property is, strictly speaking, violated. Nevertheless, the superposition of a rate-independent hysteresis model with an additional approach (e.g., viscoelastic model)

---

[1]Since the system owns one scalar input and one scalar output, the hysteresis is also named *scalar hysteresis*.

can be utilized to consider creeping. Two further properties apply to the large-signal behavior of ferroelectric materials:

4. The current system output $y(t)$ is only influenced by dominating extrema[2] in $x(t)$. Past extrema of smaller magnitudes than the subsequent ones are deleted in the system history and, thus, do not alter $y(t)$.
5. Because of this *deletion property*, all hysteresis branches in the xy-plane are located within an area, which is given by the last two dominating extrema.

Apart from the listed properties, one can in general distinguish between hysteresis featuring *local memories* or *nonlocal memories* [64].

• Local memories: The upcoming path of $y(t)$ solely depends on the current value of $x(t)$.
• Nonlocal memories: In addition to the current value of the system input, past extrema of $x(t)$ affect the upcoming progression of $y(t)$.

Actually, the large-signal behavior of ferroelectric materials also depends on past extrema. Thus, we have to deal with nonlocal memories.

6.2 Modeling Approaches on Different Length Scales

Aside from the objective, we may classify modeling approaches for ferroelectric materials according to the considered length scale. Basically, five different length scales are known: (i) Atomistic, (ii) mesoscopic, (iii) microscopic, (iv) macroscopic, and (v) multiscale (see Fig. 6.2). In the following, let us briefly discuss selected modeling approaches for ferroelectric materials on these length scales.

Atomistic Scale

At the level of the atomistic scale, one considers processes taking place in the crystal lattice of a material. Thereby, common calculation methods (e.g., ab initio and *density functional theory*) from solid-state physics are frequently used. The methods yield quantitative information for lattice spacing, elastic, and stiffness tensors as well as for the spontaneous polarization within ferroelectric materials [16, 97]. Besides, so-called *core-shell models* may be applied to simulate phase transitions and motions of the domain walls [13, 25, 83]. Such models are based on electrostatic interactions among elastically supported cores and shells. Further literature concerning modeling approaches on the atomistic scale can be found in the review articles by Cohen [17] and Sepliarsky et al. [83]. In general, these modeling approaches provide valuable insight for material development. However, the required computational effort restricts their application to small volumes and short time intervals.

---

[2]A maximum/minimum will be dominant if its value is smaller/larger than the previous maximum/minimum (see Sect. 6.4).

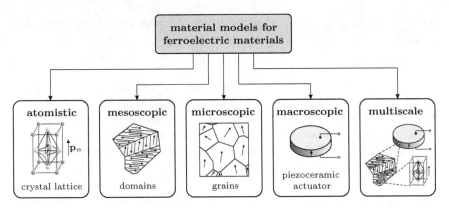

Fig. 6.2 Classification of material models for ferroelectric materials into different length scales [101]

Mesoscopic Scale

Modeling on the mesoscopic scale will be conducted if complex domain structures or lattice defects within ferroelectric materials shall be investigated. Oftentimes, the underlying modeling approaches are based on the *Landau theory*, which is understood as an extension of thermodynamic potentials by an order parameter [53]. Through this order parameter, we can explain phase transitions within materials. Note that such approaches are not only utilized on the mesoscopic scale but also on the microscopic and macroscopic scale. A well-known approach on the mesoscopic scale is the *phase field model*, where the (spontaneous) polarization serves as order parameter [39, 96]. Another method for ferroelectric materials, the *sharp interface approach* [80], is mainly based on two assumptions: (i) Each domain is a homogeneous region and (ii) material properties may jump across domain interfaces.

Microscopic Scale

The microscopic scale is very similar to the mesoscopic scale. That is why there is mostly no clear distinction between modeling approaches on these scales (e.g., [39]). One of the first approaches concerning modeling of ferroelectric materials on the microscopic scale was published by Hwang et al. [42]. They assume that the grains within the material are randomly oriented in the initial state. Since each grain has its own electric polarization, the global polarization state is neutral. By means of energy-based switching criterion depending on both electrical and mechanical excitations, the orientation of the grains is modified. Consequently, the global polarization state as well as the geometric dimension of the investigated ferroelectric material changes. Due to the simplification of equal excitations for every grain, there is a lack of accuracy. However, this modeling approach served as a basis for several further developments [39]. Huber et al. [40] suggest an alternative approach that utilizes crystal plasticity theory instead of energy-based switching criterion. Modeling approaches on the microscopic scale, which additionally consider rate-dependent behavior, can be found in [4, 9]. An overview of further computation methods is given in the review article by Kamlah [50].

Macroscopic Scale

Modeling approaches on the macroscopic scale lead to a significant reduction in computation time compared to those on the microscopic scale. Several macroscopic modeling techniques are based on the *Landau–Devonshire theory*, which is thermo-dynamically motivated and can be used to describe phase transitions within ferroelectric materials [21]. The disadvantage of this rate-independent approach lies in the restriction to monocrystalline materials and one-dimensional behavior. Bassiouny et al. [8] presented another thermodynamically consistent approach considering fer-roelectricity as well as ferroelasticity. They divided the electric polarization and mechanical strain into a reversible and an irreversible part, respectively. Alternative macroscopic modeling techniques that also exploit separate analysis of reversible and irreversible parts were developed by Kamlah and colleagues [51, 52]. Their approaches rely on phenomenological internal variables for electrical and mechanical quantities. Because electromechanical couplings within ferroelectric materials are considered in both directions, electrical and mechanical excitation can be taken into account at the same time. Further modeling approaches on the macroscopic scale are suggested by Landis [57] and Schröder et al. [81].

Multiscale Approaches

Apart from the approaches on the previously mentioned length scales, there exist various techniques that exploit simultaneous modeling on different scales. These so-called multiscale approaches aim to transfer effects on low abstraction levels to higher ones at reasonable computation time, which is usually achieved by *homoge-nization methods* within FE simulations. A multiscale approach on the atomistic and the mesoscopic scale was published by Völker et al. [97]. While they use phase field models on the mesoscopic scale, the density functional theory as well as the core-shell model are applied on the atomistic scale. Multiscale approaches combining micro-scopic and macroscopic scale with the aid of FE simulations can be found in [53, 56, 82, 96]. For instance, Keip [53] presented the *FE Square method* that is based on a microscopic *representative volume element* (abbr. RVE). At each grid point on the macroscopic scale, he deduces appropriate boundary conditions for the RVE. Aver-aging methods yield effective material parameters on the microscopic scale that can then be applied on the macroscopic scale. Contrary to multiscale approaches based on FE simulations, Smith et al. [84] developed the *homogenized energy model*, which combines mesoscopic and macroscopic scale. Similar to the Landau–Devonshire theory, they introduce a thermodynamically motivated switching criterion on the mesoscopic scale. Stochastic homogenization provides low-order macroscopic mod-els with effective parameters for ferroelectric materials. Ball et al. [7] published an extension of this approach, which enables additional consideration of mechanical stresses.

Overall, the presented approaches on the different length scales aim to model the behavior of ferroelectric materials (e.g., during poling) as accurate as possible. The resulting knowledge facilitates the research on and development of those materials. However, due to the required computational effort, most of the modeling approaches cannot be used to compensate nonlinearities of ferroelectric actuators in practical applications, e.g., positioning. Moreover, the simulation of minor loops poses a problem and the major loops are commonly of angular shape. For all these reasons, we need alternative modeling approaches that allow a sufficiently precise prediction of the actuator behavior in a reasonable computation time. The following section deals with such models for ferroelectric materials.

6.3 Phenomenological Modeling Approaches

In contrast to the approaches in Sect. 6.2, we will discuss here techniques that do not intend to model the real physical behavior of ferroelectric materials. The focus lies exclusively on the scalar transfer behavior of the fabricated transducer (e.g., piezo-ceramic actuator), i.e., input quantity as well as output quantity in predefined spatial directions. This transfer behavior is simulated through efficient phenomenological models on the macroscopic scale. Several approaches originate from plasticity theory and the research on ferromagnetic materials. For ferroelectric materials, we can classify appropriate phenomenological models into five groups [101]: (i) polynomial description, (ii) rheological models, (iii) Duhem models, (iv) fractional derivatives, and (v) switching operators. The basic principles of these groups are explained below.

Polynomial Description

There can be found many different approaches to simulate the transfer behavior of ferroelectric actuators by means of appropriate polynomials. For instance, Chonan et al. [15] describe branches in hysteresis curves of the mechanical displacement with separately parameterized polynomials for increasing and decreasing input voltage, respectively. In [93], hysteresis loops of ferromagnetic materials are modeled through piecewise linear approximations. Another technique utilizes ellipses to simulate minor loops in the transfer behavior of a piezoceramic actuator [33]. Altogether, it can be stated that polynomial descriptions will yield excellent results for the predicted output if the cycles of the input quantity are well known in advance. However, due to the fact that there is no memory of past inputs, these approaches do not meet the requirements of hysteresis models, which should be valid for general inputs (see Sect. 6.1).

Rheological Models

Basically, the term *rheology* refers to the analysis of mechanical constitutive properties for materials through the construction of ideal bodies, named *rheological models*. In doing so, we combine elementary rheological models in series and in parallel that are given by rheological state equations. Visintin [94] suggests the application of several of those elementary rheological models representing the main mechanical properties elasticity, viscosity, plasticity as well as strength. Similar to elementary rheological models, one can take advantage of lumped circuit elements from electrical engineering [58]. The parameters of the underlying state equations are derived by comparing model outputs with measurements. Reiländer et al. [76] published such a rheological model to predict the hysteretic behavior of a piezoceramic stack actuator. Further rheological models for ferroelectric actuators can be found in [74, 77].

Duhem Models

The main idea of Duhem models lies in mathematically explaining hysteretic behavior by means of differential equations and integral operators. These phenomenological models are based on the fact that one can only switch between different branches in the hysteresis curve when the derivative of the input changes its sign [61, 94]. A similar property may be attributed to rheological models. Hence, it is oftentimes difficult to distinguish between Duhem and rheological models. Very well-known representatives of generalized Duhem models are *LuGre* as well as *Dahl models* [70, 107]. Although both models can be implemented efficiently, they exhibit a number of drawbacks with regard to applications for ferroelectric materials. For instance, neither asymmetric hysteresis curves nor saturation effects can be simulated. Moreover, the fact that the input history is not considered may lead to physically impossible outputs, e.g., crossing hysteresis curves. An extended version of Duhem models for ferroelectric materials, the so-called *Bouc-Wen model* [100], is utilized for micro as well as nanopositioning (e.g., [59]). Wang et al. [98] presented a modified Bouc-Wen model to predict the hysteretic behavior of a piezoceramic stack actuator. The *Jiles–Atherton model* is a further Duhem model for ferroelectric actuators, which was originally developed for ferromagnetic materials [36, 45]. Also for this phenomenological hysteresis model, one has to cope with physically impossible outputs like unclosed hysteresis loops.

Fractional Derivatives

Another phenomenological approach for modeling hysteresis of ferroelectric materials exploits fractional derivatives. According to models of dry friction in mechanical processes, Guyomar and colleagues [34, 35] describe the electric polarization within the material through an appropriate fractional derivate. They predict the polarization for large electrical excitation with respect to excitation frequency, i.e., the dynamic behavior of ferroelectric materials. In [24], one can find an extended version to additionally consider mechanical stresses in hysteresis curves of the electric polarization. However, so far, there are not known any further approaches based on fractional derivatives, which also allow simulating hysteresis of the mechanical displacement for ferroelectric materials.

Switching Operators

Numerous phenomenological models to describe hysteretic behavior of ferromagnetic and ferroelectric materials use a weighted superposition of elementary switching operators, which are commonly named *hysterons*. Preisach [73] developed such

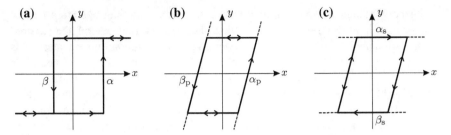

Fig. 6.3 Different elementary switching operators with input x and output y for phenomenological hysteresis models: **a** Relay operator; **b** linear play operator ($\alpha_p = -\beta_p$); **c** linear stop operator ($\alpha_s = -\beta_s$)

a model that was originally motivated by physical processes taking place within ferromagnetic materials during magnetization. With a view to applying the *Preisach hysteresis model* to various physical issues, Krasnosel'skii and Pokrovskii [55] carried out a purely mathematical examination of this approach. Moreover, they investigated three different types of elementary switching operators (see Fig. 6.3): (i) *Relay operators* that are used for Preisach hysteresis models as well as (ii) *linear play operators* (backslash operators) and (iii) *linear stop operators*. The weighted superposition of play and stop operators is commonly referred to as *Prandtl–Ishlinskii model*[3] [94].

In case of ferroelectric materials, Prandtl–Ishlinskii models are mostly based on linear play operators (e.g., [43, 75]). Al Janaideh et al. [2] presented an extension of this model to incorporate the rate-dependent behavior of smart actuators. Besides, they suggest a hyperbolic tangent function as generalized play operator, which enables the consideration of saturation effects [3]. A major problem of the play operator lies in the simulation of asymmetric hysteresis curves. Due to this fact, Dong and Tan [23] developed an asymmetric play operator. As alternative approach, Jiang et al. [44] applied especially for piezoelectric actuator systems two separate operators, one for increasing and one for decreasing inputs, respectively.

To sum up, each of the five phenomenological approaches for modeling the hysteresis of ferroelectric materials exhibits advantages and drawbacks. Many of the approaches (e.g., rheological models) have proved to be very efficient in calculation but will yield inadequate results if a precise prediction of the hysteretic behavior is required. Since both the polynomial description and the Duhem models do not use internal variables, the predicted hysteresis curves may be physically impossible, e.g., unclosed hysteresis loops as a result of the Jiles–Atherton approach. Moreover, there are not known phenomenological approaches according to fractional derivatives, which can be applied to simulate both electric polarizations and mechanical displacements of ferroelectric actuators.

---

[3] Strictly speaking, Preisach models constitute a special case of Prandtl-Ishlinkii models.

Phenomenological modeling approaches based on switching operators lead to significantly improved results than the other ones. However, in general, the required computational effort to calculate the model output for common implementations is comparatively high. Regarding asymmetric hysteresis curves as well as saturation effects, it can be stated that Preisach models are much more flexible than common Prandtl–Ishlinskii models. Nevertheless, in contrast to Preisach models, we are able to directly invert Prandtl–Ishlinskii models, which is decisive for hysteresis compensation. Besides, Preisach models require more elementary switching operators (i.e., relay operators), but the amount of parameters per a single operator is lower than for Prandtl–Ishlinskii models.

The focus in the present book lies on predicting the hysteretic behavior of ferroelectric actuators for various configurations and practical applications as precise as possible. On account of that fact, we will exclusively discuss Preisach hysteresis models in the following. This includes its efficient implementation (see Sect. 6.4.2) as well as inversion (see Sect. 6.8).

6.4   Modeling of Preisach Hysteresis Operator

In the year 1935, Preisach published a hysteresis model that is commonly known as *classical Preisach hysteresis model* [64, 73]. From the mathematical point of view, this hysteresis model belongs to the phenomenological models. It is often utilized to simulate magnetization of ferromagnetic materials as well as polarization of ferroelectric materials. Owing to the fact that we solely consider scalar inputs and outputs, the subsequent explanations refer to the scalar Preisach hysteresis model and the scalar Preisach hysteresis operator.[4] Extended versions concerning vector quantities can be found in, e.g., [48, 68, 88].

6.4.1 Preisach Hysteresis Model

To study the Preisach hysteresis model, let us assume a transmission system with the scalar input $x(t)$ and the scalar output $y(t)$, both normalized quantities depending on time t (see Fig. 6.4a). The basic idea of the Preisach hysteresis model lies in the weighted superposition of elementary switching operators $\gamma_{\alpha\beta}$. Each of them features two defined output states, i.e., -1 as well as $+1$. The switching between these two output states may occur when the operator input $x(t)$ reaches one of the changeover points α and β (see Fig. 6.4b). Mathematically, the current state of a single elementary switching operator $\gamma_{\alpha\beta,n}$ with the changeover points α_n and β_n is defined as

---

[4]For compactness, the term *scalar* is omitted in the following.

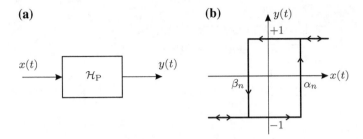

Fig. 6.4 a Preisach hysteresis operator $\mathcal{H}_P$ with input $x(t)$ and output $y(t)$, both depending on time t; **b** elementary switching operator $\gamma_{\alpha\beta,n}$ with changeover points α_n and β_n

$$\gamma_{\alpha\beta,n}[x](t) = \begin{cases} +1 & : \quad x(t) \geq \alpha_n \\ \gamma_{\alpha\beta,n}[x](t^-) : & \beta_n < x(t) < \alpha_n \\ -1 & : \quad x(t) \leq \beta_n \ . \end{cases} \tag{6.1}$$

The operator output will switch from -1 to $+1$ if the operator input $x(t)$ exceeds α_n. When $x(t)$ falls below β_n, switching from $+1$ to -1 will take place. Naturally, the operator will exclusively switch if the previous output does not coincide with the current one. Because each switching operator $\gamma_{\alpha\beta,n}$ retains its output $\left(i.e., \gamma_{\alpha\beta,n}[x](t^-)\right)$ for $\beta_n < x(t) < \alpha_n$, we are able to simulate system behavior with certain memory. Apart from that fact, the definition of the elementary switching operators implies the condition $\alpha_n \geq \beta_n$ for the changeover points.

According to the idea of Preisach hysteresis models, the weighted superposition of all possible switching operators links the input $x(t)$ to the output $y(t)$ of the transmission system. Therefore, $y(t)$ is given by (see Fig. 6.5)

$$y(t) = \mathcal{H}_P[x](t) = \iint\limits_{\alpha \geq \beta} \mu_{\mathcal{H}}(\alpha, \beta) \, \gamma_{\alpha\beta}[x](t) \, \mathrm{d}\alpha \, \mathrm{d}\beta \ . \tag{6.2}$$

Here, $\mathcal{H}_P[x](t)$ stands for the resulting Preisach hysteresis operator, which is applied to the input $x(t)$. The expression $\mu_{\mathcal{H}}(\alpha, \beta)$ individually weights the switching operators and, thus, is usually referred to as *weighting distribution*.

The value range of the changeover points α_n and β_n becomes

$$\mathcal{P} = \left\{ (\alpha_n, \beta_n) \in \mathbb{R}^2 \ : \ x_{\min} \leq \beta_n \leq x(t) \leq \alpha_n \leq x_{\max} \right\} \tag{6.3}$$

with the minimum $x_{\min}$ and the maximum $x_{\max}$ of the input. Since $\alpha \geq \beta$ has to be fulfilled, we can display this value range as triangular in the two-dimensional space with the axis α and β. Each point in this plane relates to exactly one elementary switching operator. Figure 6.6a depicts the triangular as well as three elementary switching operators. If the current outputs (-1 or $+1$) of the switching operators $\gamma_{\alpha\beta}$ are plotted in the triangular, one will obtain the so-called *Preisach plane* $\mathcal{P}(\alpha, \beta)$.

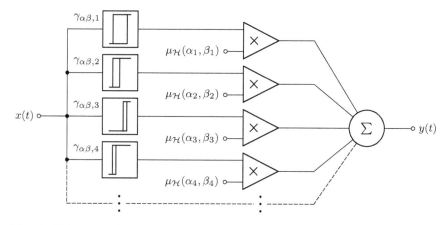

Fig. 6.5 Link of model input $x(t)$ and output $y(t)$ according to Preisach hysteresis model; elementary switching operators $\gamma_{\alpha\beta,n}$ with changeover points α_n and β_n; individual weights $\mu_{\mathcal{H}}(\alpha_n, \beta_n)$ of elementary switching operators

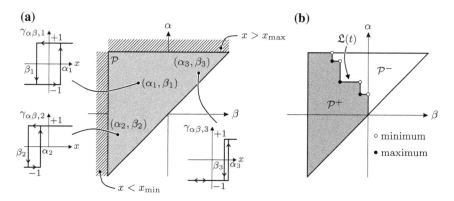

Fig. 6.6 **a** Triangular comprising value range of changeover points α_n and β_n for elementary switching operators $\gamma_{\alpha\beta,n}$; **b** Preisach plane $\mathcal{P}(\alpha, \beta)$ divided into $\mathcal{P}^+$ and $\mathcal{P}^-$, which indicate current output value ($+1$ or -1) of elementary switching operators; dividing line $\mathfrak{L}(t)$ containing maxima and minima

Moreover, due to the fact that each elementary switching operator owns his unique weighting $\mu_{\mathcal{H}}(\alpha, \beta)$, the Preisach plane can also be used to show the distribution of weights.

As discussed above, the switching operators $\gamma_{\alpha\beta}$ and, consequently, also the Preisach hysteresis operator $\mathcal{H}_P$ can only change their output if the input is altered, i.e., $\partial x/\partial t \neq 0$. The two possibilities of differential changes in the input lead to

$$\text{Ⓐ} : \qquad \frac{\partial x(t)}{\partial t} > 0 \quad \Rightarrow \quad -1 \longrightarrow +1 \quad \forall\, \gamma_{\alpha\beta,n} \;:\; \alpha_n \leq x(t) \tag{6.4}$$

$$\text{Ⓑ} : \qquad \frac{\partial x(t)}{\partial t} < 0 \quad \Rightarrow \quad +1 \longrightarrow -1 \quad \forall\, \gamma_{\alpha\beta,n} \;:\; \beta_n \geq x(t) \;, \tag{6.5}$$

which means on the one hand that for increasing inputs, solely the changeover values α_n of $\gamma_{\alpha\beta,n}$ are decisive. On the other hand, decreasing inputs relate to the changeover values β_n. As a result, we obtain at any time two interrelated areas within the Preisach plane $\mathcal{P}$, namely $\mathcal{P}^+$ and $\mathcal{P}^-$, fulfilling the property $\mathcal{P}^+ \cup \mathcal{P}^- = \mathcal{P}$. In those areas, the elementary switching operators take the output values (see Fig. 6.6b)

$$\mathcal{P}^+ = \left\{ \gamma_{\alpha\beta,n} \;:\; \gamma_{\alpha\beta,n} = +1 \right\} \quad \text{and} \quad \mathcal{P}^- = \left\{ \gamma_{\alpha\beta,n} \;:\; \gamma_{\alpha\beta,n} = -1 \right\} . \tag{6.6}$$

The dividing line between $\mathcal{P}^+$ and $\mathcal{P}^-$ is indicated with $\mathfrak{L}(t)$. Commonly, this line is a staircase-shaped curve. Depending on the current input $x(t)$ and its history, $\mathfrak{L}(t)$ is modified, e.g., the amount of steps is altered.

With a view to explaining the fundamentals of the Preisach hysteresis operator $\mathcal{H}_P$ in more detail, it is convenient to perform a graphical interpretation [36, 101]. In doing so, we choose an input signal $x(t)$ that enables us to discuss the most important characteristics of $\mathcal{H}_P$. Figure 6.7a and b depict this input signal and the current configuration of the Preisach plane $\mathcal{P}(\alpha, \beta)$ for selected instants of time, namely $t_A, \ldots, t_N$. Furthermore, the operator output $y(t)$ is plotted against the input until the considered instant of time for unweighted elementary switching operators (i.e., $\mu_{\mathcal{H}}(\alpha, \beta) = 1$; Fig. 6.7c) as well as for the weighted ones (Fig. 6.7d). Let us take a look at the different instants of time, which are parameterized with A,...,N.

- A: The input signal $x(t)$ is assumed to be zero at the beginning of the graphical interpretation. Additionally, the areas $\mathcal{P}^+$ and $\mathcal{P}^-$ should be equal. In case of a symmetric weighting distribution (i.e., $\mu_{\mathcal{H}}(\alpha, \beta) = \mu_{\mathcal{H}}(-\beta, -\alpha)$), we obtain $y(t) = 0$ as operator output.
- B: According to (6.4), the operators $\gamma_{\alpha\beta}$ will take the output value $+1$ when $x(t)$ exceeds their changeover points α. Therefore, the dividing line $\mathfrak{L}(t)$ between $\mathcal{P}^+$ and $\mathcal{P}^-$ moves upwards leading to an increase in $y(t)$.
- C: After passing through the virgin curve, $y(t)$ reaches its positive saturation. All elementary switching operators exhibit then the output value $+1$, i.e., $\mathcal{P} = \mathcal{P}^+$ and $\mathcal{P}^- = \emptyset$.
- D: Similar to B, the operators $\gamma_{\alpha\beta}$ will take the output value -1 when $x(t)$ falls below their changeover points β. As a result, the dividing line $\mathfrak{L}(t)$ between $\mathcal{P}^+$ and $\mathcal{P}^-$ moves to the left yielding a decreasing output $y(t)$.
- E...F: If the input $x(t)$ stays constant (i.e., $\partial x(t)/\partial t = 0$), $\mathfrak{L}(t)$ and, consequently, $y(t)$ will remain unchanged.
- G...H: In case of inputs outside of the defined range $x_{\min} \leq x(t) \leq x_{\max}$, $y(t)$ also remains unchanged. Contrary to the positive saturation in C, the Preisach plane becomes $\mathcal{P} = \mathcal{P}^-$ for negative saturation, i.e., $\mathcal{P}^+ = \emptyset$.

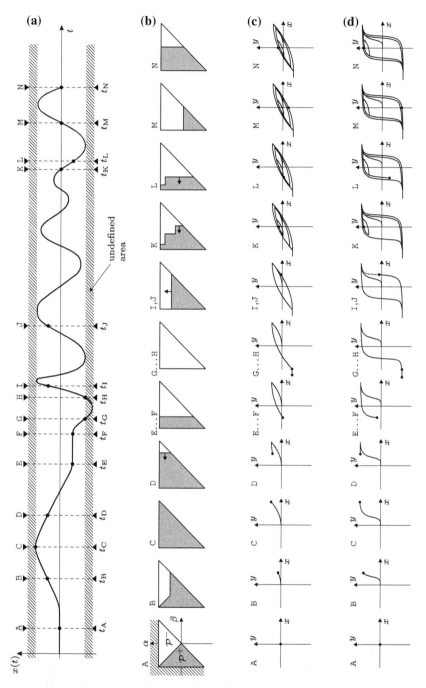

Fig. 6.7 Graphical interpretation of Preisach hysteresis operator $\mathcal{H}_P$ [101]; **a** progression of input signal $x(t)$ with respect to time t; **b** Preisach plane $\mathcal{P} = \mathcal{P}^+ \cup \mathcal{P}^-$ for A,...,N at $t_A, \ldots, t_N$; operator output y versus input x for **c** unweighted (i.e., $\mu_{\mathcal{H}}(\alpha, \beta) = 1$) and **d** weighted elementary switching operators $\gamma_{\alpha\beta}$

- I, J: Although the slope of $x(t)$ differs for t_I and t_J, the configuration of $\mathcal{P}(\alpha, \beta)$ stays constant, which leads to $y(t_I) = y(t_J)$. We are, therefore, not able to consider a rate-dependent system behavior by means of the classical Preisach hysteresis model.
- K: From t_J to t_K, there occur several successive local extrema in $x(t)$. If one of these maxima/minima is smaller/larger than the previous extrema of the same type, it will be called *dominating extremum*. Such extrema determine the corner points of $\mathcal{L}(t)$ in $\mathcal{P}(\alpha, \beta)$ and, according to the definition of the Preisach hysteresis operator, they affect the subsequent progression of $y(t)$. Due to this fact, we can state that these extrema represent the memory within the hysteresis model.
- L: Inputs that are higher in magnitude than the previous extremum cause movements of $\mathcal{L}(t)$ in horizontal and vertical direction, respectively. Since the previous extremum is deleted in $\mathcal{P}(\alpha, \beta)$, the underlying principle is commonly referred to as *wiping-out rule* or *deletion rule* of the Preisach hysteresis operator.
- M, N: If we do not know the history of $x(t)$, the current configuration of the Preisach plane will also be unknown which makes it impossible to compute subsequent states of $y(t)$. This can be avoided by driving the system into positive or negative saturation. Therewith, a defined state of $\mathcal{P}(\alpha, \beta)$ is achieved.

From the graphical interpretation, two additional key findings arise: (i) Since past extrema of the input affect the current output of $\mathcal{H}_P$, the classical Preisach hysteresis model is applicable to describe hysteresis exhibiting nonlocal memories. (ii) The comparison of unweighted and weighted elementary switching operators in Fig. 6.7c, d reveals that the distribution of weights $\mu_{\mathcal{H}}(\alpha, \beta)$ has a major influence on the shape of the hysteresis curve. It is, therefore, of utmost importance to identify an appropriate distribution for $\mu_{\mathcal{H}}(\alpha, \beta)$ because only in this way, we can reliably predict the system behavior.

6.4.2 Efficient Numerical Calculation

The Preisach hysteresis operator $\mathcal{H}_P$ and its inversion as well as the identification of the weighting distribution $\mu_{\mathcal{H}}(\alpha, \beta)$ require a large number of individual computation steps. With regard to practical applications of the hysteresis operator, the efficient numerical calculation is, thus, of utmost importance. For this purpose, a novel approach was developed at the Chair of Sensor Technology (Friedrich-Alexander-University Erlangen-Nuremberg) within the framework of the doctoral thesis by Wolf [101]. The key points of the approach are explained in the following.

Discretization

The implementation of the Preisach hysteresis operator on a computer system demands various discretizations, which are listed below.

- The continuous input $x(t)$ is converted to a discrete-time and discrete-value version by an analog-to-digital conversion since the subsequent signal processing is computer-based. Similarly, the operator output $y(t)$ is a discrete-time and discrete-value signal. Let us assume equidistant sampling with sampling time ΔT. Hence, the available input and output signal become $x(t_k = k\Delta T)$ and $y(t_k = k\Delta T)$, whereas $k \in \mathbb{N}^+$ denotes the index of the sampling point, respectively.[5] Moreover, $x(k)$ is normalized to its maximum, i.e.,

$$x(k) = \frac{X(k)}{2 \cdot \max(|X(k)|)} \quad \Rightarrow \quad x(k) \in [-0.5, 0.5] \qquad (6.7)$$

with $X(k)$ representing the original discrete-time and discrete-value input. For the changeover points α and β of the elementary switching operators $\gamma_{\alpha\beta}$, the normalization leads to the condition $-0.5 \leq \beta \leq \alpha \leq 0.5$.

- According to the definition of the Preisach hysteresis operator $\mathcal{H}_\mathrm{P}$ in (6.2), the output results from the input by analytically evaluating a double integral in the two-dimensional space. However, there does not exist an analytical solution for this integral. That is the reason why we have to perform a summation of the spatially discretized triangular instead, which contains discretized values for the changeover points α and β (see Fig. 6.8a). Without limiting the generality, the possible values of both changeover points are discretized in M equally distributed intervals leading to $\alpha(i = 1, \ldots, M)$ and $\beta(j = 1, \ldots, M)$.
- Due to the discretization of α and β, the configuration of the Preisach plane $\mathcal{P}(\alpha, \beta)$ for time step k as well as the weighting distribution $\mu_\mathcal{H}(\alpha, \beta)$ can be written as matrices, both featuring the dimension $M \times M$. They are given by (matrix elements $\mathcal{P}_{ij}(k)$ and μ_{ij})

$$\left.\begin{aligned} \mathcal{P}(k) &= \left[\mathcal{P}_{ij}(k)\right] \qquad \text{with } \mathcal{P}_{ij} \in \{-1, 1\} \\ \mu &= \left[\mu_{ij}\right] \qquad \text{with } \mu_{ij} \in \mathbb{R}_0^+ \end{aligned}\right\} \ \forall (i, j) \in \Lambda . \qquad (6.8)$$

Λ represents the definition area of the spatially discretized Preisach plane, i.e., $\Lambda = \{(i, j) : i \leq M + 1, \ j \leq M + 1 - i\}$. Note that outside of this definition area, the matrix elements $\mathcal{P}_{ij}(k)$ and μ_{ij} are zero, respectively.

The discretization M of α and β determines the resulting discretization of the operator output $y(k)$. A finer discretization leads to a higher resolution of $y(k)$. However, the greater M, the longer the computation will take for common implementations of the Preisach hysteresis operator. On this account, one has to find a compromise between output resolution and computation time. The path toward an implementation enabling both fine discretization and reasonable computation time is detailed below.

Numerical Calculation

As stated above, a weighted summation of the spatially discretized Preisach plane $\mathcal{P}(k)$ is necessary to compute the operator output $y(k)$ for time step k. The double integral in (6.2) changes into the double summation (element μ_{ij} of the weighting

---

[5]To achieve a compact notation, we use the abbreviation $x(k\Delta T) \mathrel{\widehat{=}} x(k)$.

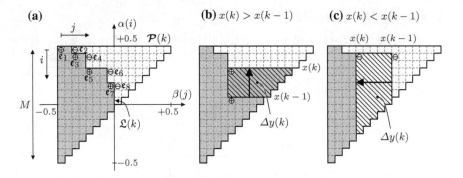

Fig. 6.8 **a** Spatially discretized Preisach plane $\mathcal{P}(k)$ for time step k; modification of Preisach plane for **b** increasing and **c** decreasing input $x(k)$ of Preisach hysteresis operator $\mathcal{H}_\mathrm{P}$; $\oplus$ and $\ominus$ indicate maxima and minima of dominating extrema, respectively

matrix μ)

$$ y(k) = \sum_{i=1}^{i_{\max}} \sum_{j=1}^{j_{\max}} \mathcal{P}_{ij}(k)\,\mu_{ij} \quad \text{with} \quad \begin{cases} i_{\max} = M + 1 - j \\ j_{\max} = M + 1 - i \ . \end{cases} \tag{6.9} $$

Note that $\mathcal{P}_{ij}(k)$ refers to the current state (i.e., $\{-1, 1\}$) of the elementary switching operator $\gamma_{\alpha\beta}$ featuring the changeover points $\alpha(i)$ and $\beta(j)$. Because this calculation is highly inefficient, a differential scheme should be applied instead, which exclusively considers modifications within the Preisach plane $\mathcal{P}(k)$ for time step k. The modifications lead to changes $\Delta y(k)$ in the operator output that are equal to the swept area in the Preisach plane (see Fig. 6.8b and c). A proper method to determine $\Delta y(k)$ is the so-called *Everett function* $\mathcal{E}$ [26]. Therewith, the current output $y(k)$ becomes [36]

$$
\begin{aligned}
y(k) &= y(k-1) + \Delta y(k) \\
&= y(k-1) + \mathcal{E}(x(k-1), x(x)) \\
&= y(k-1) + \mathrm{sign}(x(k) - x(k-1)) \cdot \iint_{\Delta y(k)} \mu_\mathcal{H}(\alpha, \beta)\,\mathrm{d}\alpha\,\mathrm{d}\beta \ .
\end{aligned}
\tag{6.10}
$$

The trapezoidal area $\Delta y(k)$ results from the difference of partial areas within the weighted Preisach plane, which are given by the successive inputs $x(k-1)$ and $x(k)$. Depending on the direction of change in the input, $\Delta y(k)$ must be added to (see Fig. 6.8b) or subtracted from (see Fig. 6.8c) the previous output $y(k-1)$. We take this fact into account by the signum function $\mathrm{sign}(\cdot)$.

In order to evaluate the Preisach hysteresis operator $\mathcal{H}_\mathrm{P}$ efficiently for each time step k, it is useful to conduct as many calculation steps as possible in advance. The optimized approach mainly comprises the three following substeps.

1. Computation of the Everett Matrix

Since the distribution of the discretized weighting distribution μ_{ij} is time-invariant, we can compute the swept areas in advance, i.e., the Everett function $\mathcal{E}$. The *Everett matrix* $\mathcal{E} = [\mathcal{E}_{ij}]$ stands for the numerically integrated weighting distribution and reads as

$$\mathcal{E}_{ij} = \sum_{r=i}^{r_{max}} \sum_{s=j}^{s_{max}} \mu_{rs} \quad \text{with} \quad \begin{cases} r_{max} = M + 1 \\ s_{max} = M + 1 - i \,. \end{cases} \tag{6.11}$$

Each component $\mathcal{E}_{i_n j_n}$ refers to the sum over a triangular $(i \geq i_n$ and $j \geq j_n)$ in the weighting matrix μ.

2. Configuration of the Preisach Plane

Now, the Everett matrix $\mathcal{E}$ can be utilized to compute the operator output $y(k)$. In doing so, we require the current dominating extrema of the input history. The extrema are located on the dividing line $\mathfrak{L}(k)$ between $\mathcal{P}^+$ and $\mathcal{P}^-$. Let us assume $m(k)$ dominating extrema at time step k named $\mathfrak{e}_1, \ldots, \mathfrak{e}_{m(k)}$. The vectors $\mathbf{e}_i(k)$ and $\mathbf{e}_j(k)$ of length $m(k)$ indicate the location of those extrema in the spatially discretized Preisach plane, i.e.,

$$\left. \begin{aligned} \mathbf{e}_i(k) &= \left[i_1(k), \ldots, i_n(k), \ldots, i_{m(k)}(k)\right]^{\mathrm{t}} \\ \mathbf{e}_j(k) &= \left[j_1(k), \ldots, j_n(k), \ldots, j_{m(k)}(k)\right]^{\mathrm{t}} \end{aligned} \right\} \text{ with } 1 \leq n \leq m(k) \leq M \,.$$
$$\tag{6.12}$$

As we have to distinguish whether the extremum represents a minimum or a maximum (see Fig. 6.8a), an additional vector $\mathbf{s}(k)$ is necessary, which contains the sign of each dominating extremum. This vector of length $m(k)$ is defined as

$$\mathbf{s}(k) = \left[\mathfrak{s}_1(k), \ldots, \mathfrak{s}_n(k), \ldots, \mathfrak{s}_{m(k)}(k)\right]^{\mathrm{t}} \tag{6.13}$$

with

$$\mathfrak{s}_n(k) = \begin{cases} -1 & : i_n = i_{n-1} \quad \text{(minimum)} \\ +1 & : j_n = j_{n-1} \quad \text{(maximum)} \,. \end{cases} \tag{6.14}$$

For the subsequent time step $k + 1$, the vectors $\mathbf{e}_i(k)$, $\mathbf{e}_j(k)$ as well as $\mathbf{s}(k)$ of the previous time step k need to be updated. Thereby, the following operations are applied:

- When the operator input increases (i.e., $x(k + 1) > x(k)$), the current value $x(k + 1)$ will be compared to the changeover points $\alpha(\mathbf{e}_i(k))$. In case of a decreasing input (i.e., $x(k + 1) < x(k)$), the comparison is carried out with respect to $\beta(\mathbf{e}_j(k))$. According to the definition of i and j (cf. Fig. 6.8a), an increasing input yields $i_{m(k+1)}(k+1) < i_{m(k)}(k)$ and a decreasing one $j_{m(k+1)}(k+1) < j_{m(k)}(k)$.

- A change of sign in the input slope leads to an additional dominating extremum, i.e., $m(k+1) = m(k) + 1$. Consequently, the length of $\mathbf{e}_i(k)$, $\mathbf{e}_j(k)$ and $\mathbf{s}(k)$ increases by one, respectively.
- If $x(k+1) > \alpha_n(k)$ is fulfilled for increasing input signals, the vectors $\mathbf{e}_i(k)$, $\mathbf{e}_j(k)$ and $\mathbf{s}(k)$ will be shortened to the length $n-1$. The same applies to decreasing inputs in case of $x(k+1) < \beta_n(k)$. Then, the nth entry of the vectors contains the location of the last dominating extremum in the spatially discretized Preisach plane as well as the sign.

3. Calculation of the Operator Output

The operator output $y(k)$ for time step k results from substep 1 and 2. We use the vectors $\mathbf{e}_i(k)$ and $\mathbf{e}_j(k)$ to select for every dominating extremum of the input $x(k)$ one entry in the Everett matrix $\mathcal{E}$. Furthermore, the selected entries are superimposed with respect to the signs of the dominating extrema, which are listed in $\mathbf{s}(k)$. Altogether, $y(k)$ computes as

$$y(k) = \frac{1}{2}\mathcal{E}_{i_1(k)j_1(k)} \cdot \mathbf{s}_1(k) + \sum_{n=2}^{m(k)} \mathcal{E}_{i_n(k)j_n(k)} \cdot \mathbf{s}_n(k) \ . \tag{6.15}$$

By means of this approach, the evaluation of the Preisach hysteresis operator is optimized. Compared to the common implementation of the Everett function (6.10), we can reduce the computation time by a factor of more than 100 [101]. Since the common implementation is mostly restricted to discretizations $M < 100$ of the Preisach plane, additional interpolation algorithms are required to achieve a reasonable resolution of the operator output [36, 76]. In contrast, the presented approach allows fine discretizations (e.g., $M = 300$) and, therefore, a high resolution without any interpolation.

6.5 Weighting Procedures for Switching Operators

As has been shown in Fig. 6.7, the weighting distribution $\mu_{\mathcal{H}}(\alpha, \beta)$ remarkably affects the output of the Preisach hysteresis operator $\mathcal{H}_P$ and, therefore, the resulting hysteresis curve. That is the reason why one can find numerous publications addressing identification as well as description of $\mu_{\mathcal{H}}(\alpha, \beta)$ for ferromagnetic and ferroelectric materials.

Before we study suitable weighting procedures and identifications, let us deduce physically motivated properties of $\mu_{\mathcal{H}}(\alpha, \beta)$. Switching processes taking place within ferroelectric materials arise from complex interactions of mechanical and electric fields on mesoscopic as well as microscopic length scales. We do not consider such interactions because the Preisach hysteresis operator represents a purely phenomenological modeling approach. However, from the macroscopic point of view, there occurs a statistical accumulation of domains, which show a specific switching property that is very well explained by selected elementary switching operators $\gamma_{\alpha\beta}$. Consequently, the weights $\mu_{\mathcal{H}}(\alpha, \beta)$ for these elementary switching operators have to possess high numerical values. In this context, four assumptions can be made for simulating the large-signal behavior of ferroelectric materials by means of Preisach hysteresis models:

1. A positive change in the electric field intensity $E(t)$ with time increases the electric polarization $P(t)$ of the ferroelectric material. In the same way, a negative change reduces $P(t)$. Therefore, the weight $\mu_{\mathcal{H}}(\alpha, \beta)$ for each elementary switching operator has to be positive, i.e.,

$$\mu_{\mathcal{H}}(\alpha, \beta) > 0 \quad \forall \, \alpha, \beta \; : \; -0.5 \leq \beta \leq \alpha \leq +0.5 \, . \qquad (6.16)$$

2. The switching behavior of unloaded domains within ferroelectric materials is assumed to be symmetrical. Elementary switching operators with the changeover points $\alpha = -\beta$ emulate this behavior. As a consequence, we can expect that the weights for those operators are rather large.
3. If ferroelectric materials are excited with a symmetric electrical signal regarding its magnitude, the resulting magnitude of $P(t)$ will be mostly symmetrical, too. To consider such material behavior, the weighting distribution $\mu_{\mathcal{H}}(\alpha, \beta)$ should be symmetrical about the axis $\alpha = -\beta$, i.e., $\mu_{\mathcal{H}}(\alpha, \beta) = \mu_{\mathcal{H}}(-\beta, -\alpha)$.
4. As hysteresis curves of the electric polarization indicate, the steepest slope is reached close to the coercive field intensity $E_{\mathrm{c}}^{\pm}$. This implies that, statistically, the majority of domains within ferroelectric materials will switch when the applied electric field intensity is similar to $E_{\mathrm{c}}^{\pm}$. In the weighting distribution, the normalized coercive field intensities $e_{\mathrm{c}}^{\pm}$ are located at the axis $\alpha = -\beta$. Hence, there also arise the maximum values of the weights.

Principally, one can distinguish between two approaches to determine weighting distributions $\mu_{\mathcal{H}}(\alpha, \beta)$ for the Preisach hysteresis model: (i) $\mu_{\mathcal{H}}(\alpha, \beta)$ is spatially discretized in elements and (ii) $\mu_{\mathcal{H}}(\alpha, \beta)$ is defined through an analytical function. We will study the main aspects of selected implementations for both approaches in Sects. 6.5.1 and 6.5.2.

6.5.1 Spatially Discretized Weighting Distribution

Here, let us concentrate on two different implementations to obtain the spatially discretized weighting distribution $\mu = \left[\mu_{ij}\right]$. While the first implementation is based on *first-order reversal curves* (FORCs), the second one minimizes deviations between appropriate measurements and simulations.

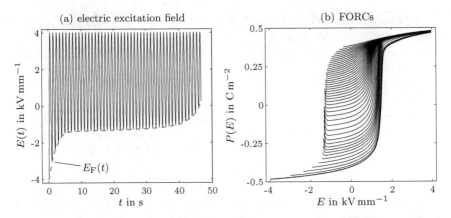

Fig. 6.9 **a** Possible input signal $E(t)$ to obtain FORCs (first-order reversal curves) for ferroelectric materials; reversal field intensity $E_F(t)$; **b** resulting FORCs for piezoceramic disk (diameter 10.0 mm; thickness 2.0 mm; material PIC255)

First-Order Reversal Curves

First-order reversal curves result from alternately loading the investigated material with a particular sequence of increasing and decreasing input signals. In case of ferroelectric materials, such a sequence starts at the electric field intensity E_{sat}^-, which leads to the negative saturation $P(E_{sat}^-)$ of the electric polarization. The electric field intensity is always increased up to positive saturation (i.e., E_{sat}^+) and, then, reduced again to a value slightly higher than the previous minimum (see Fig. 6.9a). Consequently, the local minimum $E_F(t)$ (reversal field intensity) of the input increases and its maximum remains constant with respect to time t. For the mentioned input sequence, the FORCs are defined as part of the hysteresis curve ranging from local minimum $P(E_F)$ to global maximum $P(E_{sat}^+)$, respectively (see Fig. 6.9b).

Mayergoyz [62, 64] exploited FORCs to identify spatially discretized weighting distributions for ferromagnetic materials. In doing so, he evaluated the second-order partial derivative of the acquired FORCs and performed a special coordinate transform. Some research groups (e.g., Stanco et al. [86] and Stoleriu et al. [87]) applied a similar approach for characterizing ferroelectric materials. However, the identification of the spatially discretized weighting distribution μ for those materials through FORCs exhibits various drawbacks [101]. The main drawbacks are the following:

- The slope steepness in hysteresis curves for ferroelectric materials is usually much larger than for ferromagnetic materials. Since especially at the steepest slopes, most switching processes of the unit cells take place within the ferroelectric materials, one has to change the reversal field intensity $E_F(t)$ slowly (see Fig. 6.9a). For this reason, the required measurement effort increases remarkably.
- To some extent, there occur negative entries in the identified μ for ferroelectric materials, which result from creep effects during the extensive FORCs acquisition.

Strictly speaking, such negative entries contradict the previously deduced Assumption 1 (see p. 214) for the weighting distribution.

- With a view to measuring FORCs, the investigated ferroelectric material has to reach its negative as well as positive saturation. It seems only natural that this is not always possible in practical applications of actuators, which incorporate such materials.

- The Preisach hysteresis model will be particularly well suited for predicting hysteretic behavior of materials if the input signals for identification are similar to those in the application. However, FORCs result from a predefined input sequence ranging from negative to positive saturation. Owing to this fact, we favor identification procedures allowing a flexible choice of input sequences instead.

According to the listed drawbacks, alternative approaches (see, e.g., Sect. 6.5.2) are required which yield weighting distributions for ferroelectric materials.

Adjustment of Simulations

Contrary to the previously mentioned method for the identification of μ, Kaltenbacher and Kaltenbacher [49] suggest an approach that is based on comparing measurements to outputs of the Preisach hysteresis operator $\mathcal{H}_P$. Hegewald [36, 37] firstly applied this approach for ferroelectric materials. To distinguish the resulting spatially discretized weighting distribution from the later ones, let us introduce the notation $\mu_{\mathrm{HEG}} = [\mu_{\mathrm{HEG},ij}]$, which represents the aimed quantity. The principal idea of the approach is minimizing the least squares error between normalized acquired data for the electric polarization $p_{\mathrm{meas}}(k)$ and predicted model outputs, i.e., (time step $k = 1, \ldots, k_{\mathrm{max}}$)

$$\min_{\mu_{\mathrm{HEG}}} \sum_{k=1}^{k_{\mathrm{max}}} \left[p_{\mathrm{meas}}(k) - \sum_{i=1}^{i_{\mathrm{max}}} \sum_{j=1}^{j_{\mathrm{max}}} \mathcal{P}_{ij}(k) \cdot \mu_{\mathrm{HEG},ij} \right]^2 . \tag{6.17}$$

Thereby, the matrix elements $\mu_{\mathrm{HEG},ij}$ are iteratively adjusted in a convenient way. If the changeover points α and β of the elementary switching operators $\gamma_{\alpha\beta}$ are discretized in M intervals, respectively, μ_{HEG} will contain $n_{\mathrm{HEG}} = (M^2 + M)/2$ independent entries.

In the following, we take a look at results for a piezoceramic disk (diameter 10.0 mm; thickness 2.0 mm), which is made of the ferroelectrically soft material Pz27. To experimentally determine the electric polarization P in thickness direction, a *Sawyer–Tower circuit* [79] was utilized,[6] i.e., an additional capacitor C_{ST} was connected in series to the investigated piezoceramic disk. Note that C_{ST} has to feature a high capacitance value as well as a high insulating resistance. Figure 6.10a depicts both measured hysteresis curves $P_{\mathrm{meas}}(E)$ and simulated ones $P_{\mathrm{sim}}(E)$ with respect to the applied electric field intensity E. The waveform of $E(t)$ that was utilized for exciting the piezoceramic disk is shown in Fig. 6.10e. This waveform also served as input sequence to identify μ_{HEG} through minimizing the least squares error (6.17).

---

[6]The Sawyer–Tower circuit was applied for all measurements of P in this chapter.

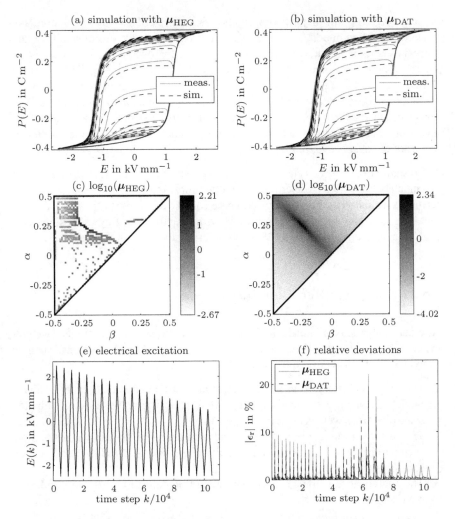

Fig. 6.10 Comparison of measurements and simulations for $P(E)$ in case of **a** μ_{HEG} and **b** μ_{DAT}; **c** and **d** spatially discretized weighting distributions for $M = 67$; **e** electrical excitation $E(k)$ for identifying μ_{HEG} and μ_{DAT}; **f** normalized relative deviations $|\epsilon_{\text{r}}|$ (magnitude) between simulations and measurements; piezoceramic disk (diameter 10.0 mm; thickness 2.0 mm; material Pz27)

As the comparison in Fig. 6.10a reveals, $P_{\text{meas}}(E)$ and $P_{\text{sim}}(E)$ coincide very well. The deviations between them can be mainly ascribed to creep processes, which we do not considered in the classical Preisach hysteresis model. Even though the excitation signal refers to a large operating range of the piezoceramic disk, the simulation procedure yields promising results.

The identified spatially discretized weighting distribution μ_{HEG} for $M = 67$ intervals (i.e., $n_{\text{HEG}} = 2278$ entries) is given in Fig. 6.10c. μ_{HEG} exhibits a wide value range and only for a few combinations of changeover points, the obtained weights are large. Actually, the majority of weights is rather small. This may lead to problems when the dividing line $\mathfrak{L}(t)$ (cf. Fig. 6.6b) crosses isolated regions of small weights since the operator output will hardly change [101]. Moreover, μ_{HEG} is strongly asymmetric, which does not agree with the deduced Assumption 3 (see p. 214). Besides, due to the fact that n_{HEG} parameters have to be modified in each iteration step, the minimizing procedure for the identification of μ_{HEG} limits M to small values. As a consequence, the Preisach plane $\mathcal{P}$ is also roughly discretized. We require, therefore, an additional interpolation method to handle operator inputs between the discretized values of the changeover points α and β.

To summarize, one can state that the adjustment of simulations to measurements is a much better option to determine spatially discretized weighting distributions for ferroelectric materials than the approach based on FORCs. From the practical point of view, the large amount of required parameters n_{HEG}, however, may cause significant problems, e.g., uniqueness as well as robustness of the identified parameters.

6.5.2 Analytical Weighting Distribution

Alternatively to determining the individual entries of the spatially discretized weighting distribution, we can describe $\mu_{\mathcal{H}}(\alpha, \beta)$ through analytical functions. Such an analytical function is desired to fulfill three properties:

1. The analytical description of the weighting distribution should enable reliable modeling for different working areas of ferroelectric actuators, i.e., unipolar, semi-bipolar as well as bipolar working areas.
2. With a view to uniquely identifying the parameters of the analytical function, each parameter should exclusively modify one property of the hysteresis curve, e.g., slope steepness.
3. The analytical function for $\mu_{\mathcal{H}}(\alpha, \beta)$ should be defined by a small number of parameters.

If an analytical function fulfills these properties, we will be able to describe and to identify weighting distributions for Preisach hysteresis models in a rather simple manner. Property 2 is especially useful to consider additional influencing factors (e.g., mechanical prestress) on hysteresis curves by means of generalized Preisach hysteresis models (see Sect. 6.6). Concerning property 3, it is essential to find a good compromise between number of parameters and desired accuracy of the model output. Although an increased number of parameters may lead to a better accuracy, the uniqueness of the parameters is remarkably reduced.

Analytical functions for defining weighting distributions do not specify the spatial discretization M of the Preisach plane $\mathcal{P}$ in advance. Consequently, the actual weight $\mu_{\mathcal{H}}(\alpha, \beta)$ can be computed for each combination of changeover points without performing any interpolation. Note that from the practical point of view, one also has to carry out spatial discretization since this is required for efficient numerical evaluations of the Preisach hysteresis operator $\mathcal{H}_P$ (see Sect. 6.4.2). Nevertheless, M for identifying the parameters of the analytical function and for utilizing $\mathcal{H}_P$ in practical applications can differ.

DAT Function

As saturation curves are quite similar to an arctangent function, Sutor et al. [89] suggest a special analytical function for describing $\mu_{\mathcal{H}}(\alpha, \beta)$, which is based on the second-order derivative of the arctangent function. The second-order derivative is attributable to the double integral within Preisach hysteresis models (see (6.2)). For this reason, the underlying function is commonly named *DAT* (derivative arc tangent) function and reads as

$$\mu_{\mathrm{DAT}}(\alpha, \beta) = \frac{B}{1 + \left\{ \left[(\alpha + \beta)\, \sigma \right]^2 + \left[(\alpha - \beta - h)\, \sigma \right]^2 \right\}^{\eta}} \qquad (6.18)$$

with four independent parameter ($n_{\mathrm{DAT}} = 4$) yielding the dimensionless parameter vector $\mathbf{p} = [B, \eta, h, \sigma]^{\mathrm{t}}$. Originally, the DAT function was intended for modeling ferromagnetic materials. Wolf et al. (e.g., [104, 106]) utilized this analytical function to predict the large-signal behavior of ferroelectric materials through the Preisach hysteresis operator. In the following, let us concentrate on an extended version of the DAT function [101], which is given by

$$\mu_{\mathrm{DAT}}(\alpha, \beta) = \frac{B}{1 + \left\{ \left[(\alpha + \beta + h_1)\, \sigma_1 \right]^2 + \left[(\alpha - \beta - h_2)\, \sigma_2 \right]^2 \right\}^{\eta}} \qquad (6.19)$$

and, thus, contains six independent parameters, i.e., $\mathbf{p} = [B, \eta, h_1, h_2, \sigma_1, \sigma_2]^{\mathrm{t}}$ and $n_{\mathrm{DAT}} = 6$. Compared to (6.18), the extended DAT function[7] is more flexible but exhibits two additional parameters. We arrive at the discretized weighting distribution $\boldsymbol{\mu}_{\mathrm{DAT}} = \left[\mu_{\mathrm{DAT},ij} \right]$ by spatially discretizing the Preisach plane in M equally distributed intervals for the changeover points α and β. Again, the discretized version $\boldsymbol{\mu}_{\mathrm{DAT}}$ contains $(M^2 + M)/2$ entries.

Figure 6.11 displays a particular weighting distribution $\mu_{\mathrm{DAT}}(\alpha, \beta)$ according to the DAT function as three- and two-dimensional representation in the Preisach plane $\mathcal{P}$, respectively. The pronounced maximum is affected by the parameters of the analytical function in a different manner. The individual impacts on the weighting distribution and on the resulting hysteresis curve (see Fig. 6.12) are as follows:

- B exclusively scales $\mu_{\mathrm{DAT}}(\alpha, \beta)$ and, therefore, modifies the magnitude of hysteresis curves. We can utilize B to compensate unwanted changes in hysteresis curves due to the parameters η, σ_1, and σ_2.
- η prevalently affects the shape of the maximum in $\mu_{\mathrm{DAT}}(\alpha, \beta)$. For instance, a large value for η causes a steep decrease of this maximum. Consequently, the slope steepness of hysteresis curves is specified.

---

[7]The extended version of the DAT function is hereinafter also called DAT function.

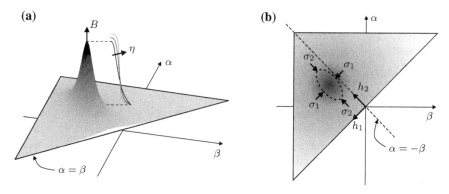

Fig. 6.11 Influence of parameters B, η, h_1, h_2, σ_1, and σ_2 on weighting distribution $\mu_{\text{DAT}}(\alpha, \beta)$ in Preisach plane $\mathcal{P}$; **a** three-dimensional and **b** two-dimensional representation

- h_1 shifts the maximum of $\mu_{\text{DAT}}(\alpha, \beta)$ in $\mathcal{P}$ along the axis $\alpha = \beta$. This parameter is especially important for modeling asymmetric hysteresis curves, which may be caused by, e.g., a bias e_{bias} in the normalized electric field intensity (see Sect. 6.6.2).
- h_2 shifts the maximum of $\mu_{\text{DAT}}(\alpha, \beta)$ in $\mathcal{P}$ along the axis $\alpha = -\beta$. In case of ferroelectric materials, we are able to vary the normalized coercive field intensity $e_{\text{c}}^{\pm}$ of hysteresis curves through h_2.
- σ_1 and σ_2 modify the maximum's width of $\mu_{\text{DAT}}(\alpha, \beta)$ in direction of the axes $\alpha = \beta$ and $\alpha = -\beta$, respectively. As a result, one alters magnitudes of hysteresis curves as well as shapes of minor loops.

In summary, the DAT function is able to specifically influence decisive properties of simulated hysteresis curves for ferroelectric materials, e.g., coercive field intensity and slope steepness. As already discussed, this fact is especially important for generalized hysteresis models based on the Preisach hysteresis operator.

Just as it is suggested by Hegewald [37] for identifying the entries of μ_{HEG}, the parameters B, η, h_1, h_2, σ_1, and σ_2 of the DAT function result from iterative adjustments of model outputs to appropriate measurements. In doing so, we have to minimize deviations between simulated and measured signals. The underlying optimization procedure represents an ill-posed problem (see Chap. 5). Hence, we require an appropriate regularization approach, which is provided by the Levenberg–Marquardt algorithm and the iteratively regularized Gauss–Newton algorithm. Both algorithms yield the aimed parameters for the DAT function in reasonable computation time but demand a proper initial guess for the parameter vector $\mathbf{p}$. Such an initial guess can be figured out by manually adjusting simulations to measurements. In [101], robustness as well as reliability of the entire parameter identification is proven through different examples.

With a view to comparing the different weighting procedures μ_{DAT} and μ_{HEG}, let us also apply the DAT function to the previously mentioned measurement data for the piezoceramic disk (diameter 10.0 mm; thickness 2.0 mm; material Pz27). Figure 6.10d depicts the identified spatially discretized weighting distribution μ_{DAT}

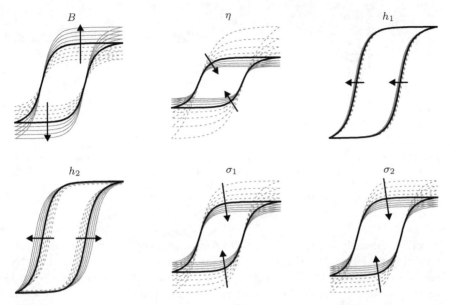

Fig. 6.12 Parameter study for DAT function $\mu_{\mathrm{DAT}}(\alpha, \beta)$ to individually rate impacts of $B, \eta, h_1, h_2,$ $\sigma_1,$ and σ_2 on hysteresis curves, e.g., $P(E)$; parameters are varied in the range $\pm[10, 20, 30, 40]\%$; dashed and solid lines refer to negative and positive parameter changes, respectively; arrows indicate increasing parameter values [101]

for $M = 67$ intervals. In contrast to μ_{HEG}, μ_{DAT} is symmetrical about the axis $\alpha = -\beta$, which coincides with Assumption 3 (see p. 214). Furthermore, there do not arise isolated regions in μ_{DAT} where the weights exhibit negligible values. This fact is a consequence of the utilized analytical function.

As the comparisons of measurements $P_{\mathrm{meas}}(E)$ and simulations $P_{\mathrm{sim}}(E)$ for the electric polarization in Fig. 6.10a and b reveal, μ_{HEG} and μ_{DAT} lead to prediction results of similar quality. Even though the DAT function is defined by a much smaller amount of independent parameters (i.e., $n_{\mathrm{DAT}} \ll n_{\mathrm{HEG}}$), which also facilitates their identification, the normalized relative deviations[8] ϵ_r between $P_{\mathrm{meas}}(E)$ and $P_{\mathrm{sim}}(E)$ for μ_{DAT} are only marginally higher than for μ_{HEG} (see Fig. 6.10f). It is, thus, reasonable to assume that μ_{DAT} and μ_{HEG} feature identical performance. However, from the physical point of view, the resulting weights μ_{DAT} are more reliable than μ_{HEG}. Because of the aforementioned arguments, the DAT function should be generally preferred to identify weighting distributions for Preisach hysteresis operators.

Gaussian Function and Lorentz Function

There exist several further analytical functions to describe the weighting distribution $\mu_{\mathcal{H}}(\alpha, \beta)$ for Preisach hysteresis models. Especially in case of ferromagnetic

---

[8]In this book, the normalized relative deviation ϵ_r usually indicates the absolute deviation related to the difference between the considered maximum and minimum.

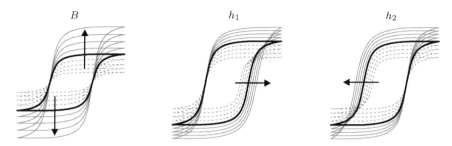

Fig. 6.13 Parameter study for Lorentz function $\mu_{\mathrm{LOR}}(\alpha, \beta)$ to individually rate impacts of B, h_1 and h_2 on hysteresis curves, e.g., $P(E)$; parameters are varied in the range $\pm[10, 20, 30, 40]\%$; dashed and solid lines refer to negative and positive parameter changes, respectively; arrows indicate increasing parameter values [101]

materials, the analytical functions are oftentimes motivated by statistical accumulations of switching processes taking place within the material. In order to consider this fact, one can use two-dimensional *Gaussian* and *Lorentz distributions* for analytical description (e.g., [5, 20, 27, 90]). As Azzerboni et al. [6] suggest in a similar manner, the Gaussian and Lorentz function (distribution) are given by

$$\mu_{\mathrm{GAUSS}}(\alpha, \beta) = B^2 \cdot \exp\left[-\frac{1}{2}\left(\frac{\alpha - \beta - 2h_1}{h_1}\sigma_1 \right)^2 - \frac{1}{2}\left(\frac{\alpha + \beta}{h_2}\sigma_2 \right)^2 \right] \quad (6.20)$$

$$\mu_{\mathrm{LOR}}(\alpha, \beta) = \frac{B}{1 + \left(\dfrac{\beta + h_1}{h_1}\sigma_1 \right)^2} \cdot \frac{B}{1 + \left(\dfrac{\alpha - h_2}{h_2}\sigma_2 \right)^2}, \quad (6.21)$$

respectively. The impacts of the parameters B, h_1, h_2, σ_1, and σ_2 on the weighting distribution approximately correspond to those of the DAT function. Again, B exclusive scales the magnitude of $\mu_{\mathrm{GAUSS}}(\alpha, \beta)$ and $\mu_{\mathrm{LOR}}(\alpha, \beta)$, which is equal to scaling hysteresis curves. Through the parameters σ_1 and σ_2, we alter slope steepness in hysteresis curves. However, in distinction from the DAT function, h_1 as well as h_2 do not allow an independent adjustment of the normalized coercive field intensity $e_c^{\pm}$ (see Figs. 6.12 and 6.13). They also change magnitudes of major loops, which may cause problems during parameter identification. On account of these facts, both the Gaussian function and the Lorentz function are not optimally suited for generalized Preisach hysteresis models.

Figure 6.14a shows measured and simulated major loops for the electric polarization $P(E)$ of a piezoceramic disk (diameter 10.0 mm; thickness 2.0 mm; material Pz27). The identified spatially discretized weighting distributions μ_{DAT}, μ_{GAUSS} as well as μ_{LOR} are depicted in Fig. 6.14b–d; Table 6.1 contains the underlying parameters. In general, μ_{DAT}, μ_{GAUSS} as well as μ_{LOR} lead to reliable model outputs of the Preisach hysteresis operator. A more detailed comparison, however, points out that the deviations between measurements and simulations with the Gaussian function

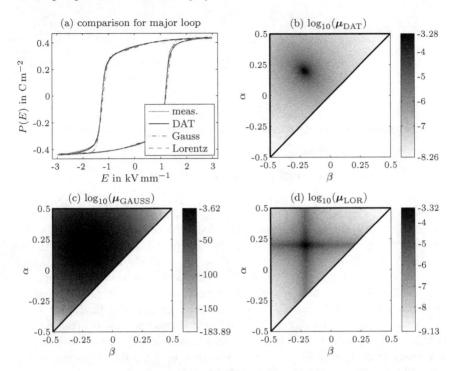

Fig. 6.14 a Comparison of measured and simulated major loops $P(E)$ for different analytical weighting distributions; **b, c,** and **d** resulting spatially discretized weighting distributions for $M = 800$; piezoceramic disk (diameter 10.0 mm; thickness 2.0 mm; material Pz27)

Table 6.1 Resulting parameters for DAT function $\mu_{DAT}(\alpha, \beta)$, Gaussian function $\mu_{GAUSS}(\alpha, \beta)$, and Lorentz function $\mu_{LOR}(\alpha, \beta)$; parameters a and b refer to reversible parts (see Sect. 6.6.1) in hysteresis curve

	B	η	h_1	h_2	σ_1	σ_2	a	b
DAT	338.105	1.467	0.010	0.411	47.12	36.30	0.052	5.34
Gaussian	0.016	–	0.206	0.119	5.06	3.20	0.053	7.30
Lorentz	0.022	–	0.200	0.211	13.20	12.08	0.051	5.05

are significantly higher than those with the DAT function as well as Lorentz function. Close to the reversal points of the applied electric field intensity E, measurements and simulations with μ_{GAUSS} differ considerably. Hence, $\mu_{DAT}(\alpha, \beta)$ and $\mu_{LOR}(\alpha, \beta)$ should be preferred as analytical function for predicting the large-signal behavior of ferroelectric materials. Nevertheless, when we are also interested in generalized Preisach hysteresis models, $\mu_{DAT}(\alpha, \beta)$ will be currently the only known analytical function providing uniqueness, accuracy as well as flexibility of the weighting distribution and, consequently, of the Preisach hysteresis operator [101].

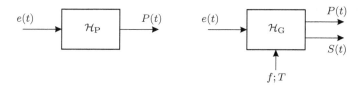

Fig. 6.15 Inputs as well as outputs of classical Preisach hysteresis operator $\mathcal{H}_P$ and generalized Preisach hysteresis operator $\mathcal{H}_G$; $e(t)$ stands for normalized electric field intensity

6.6 Generalized Preisach Hysteresis Model

The classical Preisach hysteresis operator $\mathcal{H}_P$ is only suitable to a limited extent for predicting hysteretic behavior of ferroelectric materials. For this reason, it is important to improve and modify $\mathcal{H}_P$, which leads to so-called generalized Preisach hysteresis models with the underlying operator $\mathcal{H}_G$. Several types of model generalizations were developed as well as implemented by Wolf and colleagues (e.g., [78, 104, 106]). Principally, those generalizations can be divided into three categories. The first category aims at increasing flexibility to consider reversible parts (Sect. 6.6.1) and asymmetric behavior (Sect. 6.6.2), while the second one deals with the mechanical large-signal behavior of ferroelectric materials (Sect. 6.6.3). The third category addresses extensions of $\mathcal{H}_P$ to take into account both the rate-dependent behavior of the materials (Sect. 6.6.4) and the impact of applied uniaxial mechanical stresses (Sect. 6.6.5). In contrast to $\mathcal{H}_P$, the resulting generalized Preisach hysteresis operator $\mathcal{H}_G$ features an additional output $S(t)$ for mechanical strains of ferroelectric materials (see Fig. 6.15). Moreover, $\mathcal{H}_G$ is equipped with two further inputs concerning frequency f of the electrical excitation signal and mechanical stress T within the material. Below, the model generalizations are studied separately.

6.6.1 Reversible Parts

Weighting distributions $\mu_{\mathcal{H}}(\alpha, \beta)$ of the Preisach hysteresis operator $\mathcal{H}_P$ consist in general of finite values. As a result, the gradient $\partial y(k)/\partial x(k)$ of discrete-time model output $y(k) = \mathcal{H}_P[x](k)$ with respect to discrete-time input $x(k)$ is always zero at the reversal points. This characteristic can be seen in Figs. 6.12 and 6.13. However, measured output quantities of ferroelectric materials exhibit certain saturation trends. For instance, measured hysteresis curves $P(E)$ for the electric polarization show the property $\partial P/\partial E|_{E=E_{\text{sat}}} \neq 0$. Mainly, this is due to reversible effects (intrinsic effects; see Sect. 3.4.1) taking place within ferroelectric materials. Based on the assumption that there also occurs saturation for such effects, they are oftentimes modeled through an appropriate arctangent functions. Sutor et al. [89] suggest an additional linear part c, which was primarily applied for describing the large-signal behavior of ferromagnetic materials. Therewith, the entire discrete-time model output $y(k)$ becomes

$$y(k) = \mathcal{H}_P[x](k) + y_{rev}(x(k)) \ , \tag{6.22}$$

whereas the added reversible part y_{rev} is given by

$$y_{rev}(x(k)) = a \cdot \arctan(b \cdot x(k)) + c \cdot x(k) \ . \tag{6.23}$$

Let us utilize a slightly different approach for ferroelectric materials. Reversible parts are again considered by means of an arctangent function. Instead of adding those parts to $y(k)$, we incorporate them directly in the weighting distribution $\mu_{\mathcal{H}}(\alpha, \beta)$. This will offer particular advantages when the generalized Preisach hysteresis operator $\mathcal{H}_G$ has to be inverted (see Sect. 6.8). With a view to explaining the approach, it is helpful to take a closer look at the weighting distribution in the Preisach plane $\mathcal{P}$ (see Fig. 6.16a). The incorporation of reversible parts is performed through weights on the axis $\delta = \alpha = \beta$ [64, 69]. Due to the properties of the elementary switching operators $\gamma_{\alpha\beta}$, these weights will be cumulatively summed up if the operator input $x(k)$ increases. On the other hand, in case of a decreasing input, the weights are cumulatively subtracted. These cumulative operations represent the incorporated reversible part $y_{rev}(k)$ of the modified Preisach hysteresis model. Hence, $y_{rev}(k)$ reads as

$$y_{rev}(k) = \int_{\delta_1}^{\delta_2} \mu_{\mathcal{H}}(\delta, \delta) \, \gamma_{\delta\delta}[x](k) \, d\delta \tag{6.24}$$

in analytical formulation. Here, δ_1 and δ_2 stand for the lower and upper integration limit, respectively. While δ_1 is -0.5 for increasing inputs, decreasing inputs yield $\delta_1 = 0.5$ (see Fig. 6.16b). Since (6.24) contains an integration, we have to differentiate (6.23) with respect to $x = \delta$, which leads to the reversible parts $r(\delta)$ in the Preisach plane

$$r(\delta) = r(x) = \frac{ab}{M\left[1 + b^2(x + h_1/2)^2\right]} + c \ . \tag{6.25}$$

Note that $r(\delta)$ exclusively defines the weights along the axis $\alpha = \beta$, e.g., $\mu_{\mathcal{H}}$ $(-0.3, -0.3) = r(-0.3)$. The function is composed of the dimensionless parameters a, b, c, and h_1. By conducting a normalization to M in (6.25), the resulting output $y_{rev}(k)$ is largely independent of the utilized spatial discretization.

As indicated in Fig. 6.16a, a, b, c, and h_1 modify $r(\delta)$ in a distinct way, respectively. The parameter a scales the maximum and b its extension. Through c, we can add an offset on the axis $\alpha = \beta$ causing a linear part in the model output. In accordance with the DAT function $\mu_{DAT}(\alpha, \beta)$, h_1 shifts the maximum of $r(\delta)$ along the axis $\alpha = \beta$. The reason for introducing the parameter h_1 will be discussed in

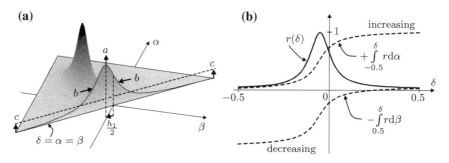

Fig. 6.16 **a** Weighting distribution $\mu_{DAT}(\alpha, \beta)$ and reversible parts $r(\delta)$ in Preisach plane $\mathcal{P}$; arrows indicate increasing parameter values; **b** normalized version of function $r(\delta)$ and its integration for increasing as well as decreasing model inputs, respectively; parameter values for $r(\delta)$ (see (6.25)): $a = 0.4, b = 15, c = 0$ and $h_1 = 0.1$ [101]

Sect. 6.6.2. On account of the fact that c can usually be omitted in (6.25), $r(\delta)$ and, consequently, $y_{rev}(k)$ are described by the two additional parameters a and b [101]. Considering also the DAT function, the analytical weighting distribution for modeling the large-signal behavior as well as reversible parts comprises eight independent parameters, i.e.,

$$\mathfrak{p} = [a, b, B, \eta, h_1, h_2, \sigma_1, \sigma_2]^t \ . \tag{6.26}$$

At this point, it should be mentioned that the reversible parts have already been applied in Figs. 6.10 and 6.14 to obtain the results by means of analytical weighting distributions. Table 6.1 additionally contains the parameters a and b for $\mu_{DAT}(\alpha, \beta)$, $\mu_{GAUSS}(\alpha, \beta)$ as well as $\mu_{LOR}(\alpha, \beta)$.

6.6.2 Asymmetric Behavior

According to the assumptions in Sect. 6.5 (see Assumptions 2 and 3 on p. 214), the switching behavior of unloaded domains inside ferroelectric materials is symmetrical to the applied electric field intensity. However, especially ferroelectrically hard materials often exhibit asymmetric hysteresis curves. This fact can be mainly ascribed to a restricted mobility of domain walls (see also Sect. 3.6.2). Such pinned or clamped domain walls originate from defects and imperfections in the crystal lattice [18]. As a result, a bias field intensity may arise which has to be compensated by the applied electric field in order to initiate domain switching processes.

The bias field intensity E_{bias} (normalized value e_{bias}) can be considered within the generalized Preisach hysteresis model by shifting an originally symmetric weighting distribution $\mu_{\mathcal{H}}(\alpha, \beta)$ along the axis $\alpha = \beta$. For this purpose, we introduced in the DAT function the parameter h_1 that shifts the maximum of $\mu_{DAT}(\alpha, \beta)$ (see Fig. 6.11b) as well as of the reversible parts $r(\delta)$ (see Fig. 6.16a) in the Preisach plane $\mathcal{P}$.

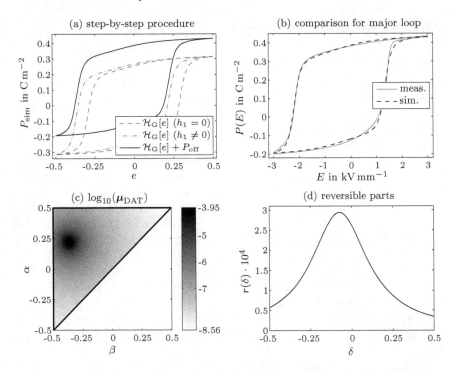

Fig. 6.17 a Step-by-step procedure to simulate asymmetric hysteresis curves through Preisach hysteresis modeling; **b** comparison of measurements and simulations for a major loop of $P(E)$; **c** spatially discretized weighting distribution μ_{DAT} for $M = 800$; **d** reversible parts $r(\delta)$ according to (6.25); piezoceramic disk (diameter 10.0 mm; thickness 2.0 mm; material Pz26)

Apart from the asymmetric shape of a hysteresis curve $y(x)$ with respect to the applied input x, the output y itself can be asymmetric in addition leading to $\max(y) \neq -\min(y)$ (see, e.g., Fig. 6.17b). But if every elementary switching operator $\gamma_{\alpha\beta}$ of the Preisach hysteresis model takes the output value -1 (i.e., $\mathcal{P} = \mathcal{P}^-$) or $+1$ (i.e., $\mathcal{P} = \mathcal{P}^+$), y will be symmetrical, which means $\max(y) = -\min(y)$. The Preisach hysteresis operator is, thus, not sufficient to model asymmetric outputs ranging from $\min(y)$ to $\max(y)$. In case of asymmetric hysteresis curves $P(E)$ for ferroelectric materials, we have to add an appropriate offset P_{off} to the electric polarization that is determined through measurements [101].

To present the applicability of the generalized Preisach hysteresis operator $\mathcal{H}_G$, let us investigate a piezoceramic disk (diameter 10.0 mm; thickness 2.0 mm) made of the ferroelectrically hard material Pz26. Figure 6.17a illustrates the step-by-step procedure to obtain an asymmetric hysteresis curve through Preisach modeling. As the comparison of measurements and simulations for a major loop in Fig. 6.17b reveals, $\mathcal{H}_G$ yields reliable results. The asymmetric behavior with respect to E is well described by the function parameter h_1. Figure 6.17c and d depict the obtained weighting distribution μ_{DAT} and $r(\delta)$ denoting reversible parts on the axis $\alpha = \beta$.

6.6.3 Mechanical Deformations

Until now, we have concentrated on Preisach modeling for the electrical behavior (i.e., $P(E)$) of ferroelectric materials. In many practical applications (e.g., high-precision positioning systems) of those materials, it is, however, of utmost importance to consider their mechanical behavior in addition. With a view to simulating the mechanical large-signal behavior of piezoceramic materials, let us briefly repeat relevant physical processes on the atomistic as well as mesoscopic scale (see Sect. 3.6.2). The spontaneous polarization $\mathbf{p}_n$ of a unit cell points in direction of its largest geometric dimension. Because each $\mathbf{p}_n$ is almost perfectly aligned in parallel to the applied electric field E in the saturation state of the piezoceramic material, the macroscopic mechanical deformation also reaches the highest value in direction of E. During changes from positive to negative saturation (and vice versa), several domains switch to the ferroelastic intermediate stage at first, which causes a negative mechanical deformation of the material [14]. When $|E|$ exceeds thereupon the coercive field intensity $\left| E_c^{\pm} \right|$, the domains will be aligned again in direction of E, i.e., $180°$ with respect to their original orientation. The macroscopic polarization state of the piezoceramic material, thus, changes its sign, while the mechanical deformations are equal for positive and negative saturation. In other words, electric polarization and mechanical deformation differ significantly in terms of the underlying large-signal behavior.

We can reasonably describe the mechanical deformation $S(E)$ of ferroelectric actuators through a generalized Preisach hysteresis model (e.g., [32, 38]) if they operate in unipolar and semi-bipolar working areas (i.e., $E > E_c^-$). The function parameter c (see (6.25)) within the presented operator $\mathcal{H}_G$ provides shifts of simulated deformations in vertical direction, which are possibly existing in these working areas. On the contrary, butterfly curves resulting for bipolar working area demand further extensions of the Preisach hysteresis operator. For this purpose, one can find two different approaches in literature:

- Kadota and Morita [47] introduced a tristable hysteron to model the ferroelastic intermediate stage of domains. As the name indicates, this hysteron features three stable states, i.e., -1, 0 as well as $+1$. Since the approach requires a four-dimensional weighting distribution, complexity of the underlying Preisach hysteresis models considerably increases.
- Due to the fact that mechanical deformations of ferroelectric materials are equal for positive and negative saturation, we may rectify the output $y = \mathcal{H}_P[x]$ of the classical hysteresis operator (e.g., [36]).

Concerning practical applications of Preisach hysteresis models for ferroelectric materials, the second approach should be preferred. That is why we will concentrate exclusively on this approach for describing mechanical deformations. Hegewald [36] conducted rectification of the operator output guided by the approximation $S \propto P^2$. To model the large-signal behavior of mechanical deformations by means of the Preisach hysteresis operator, he utilized the same weighting distribution as for the electric polarization, i.e., $\mu_{\text{HEG}}(\alpha, \beta)$. Under certain circumstances, the approximation $S \propto P^2$ yields satisfactory results, but in general, the deviations between simulations and measurements of $S(E)$ are rather high. The following findings can be deduced for computing mechanical large-signal behavior through Preisach hysteresis operators [101]:

- For several ferroelectric materials, the rectified electric polarization significantly differs from the macroscopic mechanical deformation. It is, therefore, necessary to identify individual weighting distributions for the polarization and the deformation, which are parameterized by the vectors $\mathfrak{p}_P$ and $\mathfrak{p}_S$, respectively.
- As mentioned above, domain switching processes by 180° within ferroelectric materials alter the sign of the electric polarization but do not modify their mechanical deformation. Thus, it seems reasonable to rectify the operator output y through computing its absolute value instead of squaring. In doing so, the function parameters (e.g., B) of $\mu_{\text{DAT}}(\alpha, \beta)$ influence hysteresis curves for polarization and deformation in a similar manner (see p. 219).
- To account for asymmetric large-signal behavior of mechanical deformations, we have to extend the generalized Preisach hysteresis operator $\mathcal{H}_G$.

An appropriate method to consider these findings for modeling mechanical deformations of ferroelectric materials is given by (time step k; normalized electric field intensity e)

$$S(k) = \{c_1 + |\mathcal{H}_G[e](k) + c_2| + c_3(e - 0.5)\} \cdot 100\% . \tag{6.27}$$

Hence, the three parameters c_1, c_2 and c_3 are required in addition. Overall, Preisach hysteresis modeling for mechanical deformations comprises 11 independent parameters, namely

$$\mathfrak{p}_S = [a, b, B, c_1, c_2, c_3, \eta, h_1, h_2, \sigma_1, \sigma_2]^t . \tag{6.28}$$

Five parameters refer to the DAT function $\mu_{\text{DAT}}(\alpha, \beta)$ and two parameters to reversible parts in the large-signal behavior.

In order to demonstrate this modeling approach, let us investigate a ferroelectrically hard piezoceramic disk (diameter 10.0 mm; thickness 2.0 mm; material Pz26), which usually features an asymmetric large-signal behavior for both polarization and deformation. The mechanical deformations of the piezoceramic disk were acquired with a linear variable differential transformer (abbr. LVDT [99]) that was optimized for measuring small displacements.[9] Figure 6.18a shows the basic steps in simulating a butterfly curve according to (6.27). Asymmetric behavior is incorporated in the generalized Preisach hysteresis operator $\mathcal{H}_G$ through the parameter h_1. Before computing the absolute value, we add an offset c_2 to the mechanical deformation. Therewith, one can model differences in maximum values of S, i.e., $S_{\text{max}}^- \neq S_{\text{max}}^+$. Finally, the linear equation $c_1 + c_3(e - 0.5)$ is added to consider different slope stiffnesses as well as

---

[9]The LVDT was applied for all measurements of S in this chapter.

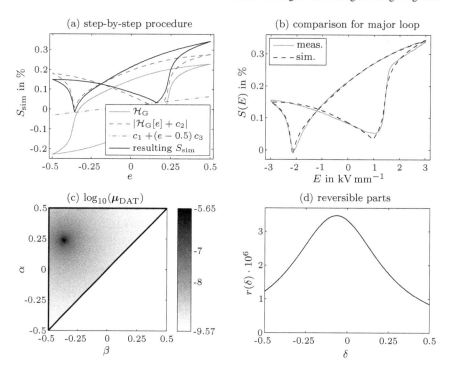

Fig. 6.18 **a** Step-by-step procedure to simulate mechanical deformations of ferroelectric materials through Preisach hysteresis modeling; **b** comparison of measured and simulated major loop (butterfly curve) $S(E)$; **c** spatially discretized weighting distribution μ_{DAT} for $M = 800$; **d** reversible parts $r(\delta)$ according to (6.25); piezoceramic disk (diameter 10.0 mm; thickness 2.0 mm; material Pz26)

the circumstance $S_{min}^{-} \neq S_{min}^{+}$. To identify the parameter vector $\mathbf{p}_S$, simulations have to be adjusted so that they match measurements best possible.

As Fig. 6.18b reveals, measured and simulated mechanical deformations of the piezoceramic disk coincide very well. It can be stated that the presented Preisach hysteresis model is ideally suited for predicting mechanical deformations of ferroelectric materials. Figure 6.18c and d depict the resulting weighting distribution μ_{DAT} and the reversible parts $r(\delta)$, respectively.

6.6.4 Rate-Dependent Behavior

Even though a piezoceramic material is macroscopically excited in a uniform manner, domain switching processes inside the material do not take place simultaneously. Depending on the alteration rate of the excitation signal, this may remarkably influence macroscopic quantities (e.g., mechanical strain) [46]. The macroscopic rate-dependent behavior of piezoceramic materials originates from their inhomogeneous inner structure, which yields locally different electric field intensities as well as mechanical stresses for the domains. In a first approximation, we can assume that a single domain switches quickly after its individual switching energy is locally provided [72]. Such switching processes and reversible as well as irreversible ion displacements alter the spatial distribution of both electric field intensity and mechanical stress inside the piezoceramic material. As a result, a large number of domains do not switch immediately but depending on previous switching processes of neighboring domains and ion displacements within them. From the macroscopic point of view, this leads to the creep behavior of electric polarization and mechanical strain (e.g., [109]).

Outputs $y(t)$ of the classical Preisach hysteresis operator $\mathcal{H}_P$ solely depend on temporal succession of the input $x(t)$ (see I and J in Fig. 6.7). Therefore, the alteration rate of $x(t)$ with respect to time (i.e., $\partial x(t)/\partial t$) does not affect $y(t)$. Since macroscopic electric polarizations and macroscopic mechanical strains of ferroelectric materials exhibit such a dependence, one has to extend $\mathcal{H}_P$, which results in so-called *dynamic* or *rate-dependent Preisach hysteresis models*. Mayergoyz [63] suggests a dynamic Preisach hysteresis model that is based on a varying weighting distribution $\mu_\mathcal{H}(\alpha, \beta)$ according to the partial time derivative of $y(t)$. Viswamurthy et al. [95] applied this approach to describe dynamic hysteresis of piezoceramic stack actuators. In several other research works (e.g., [67, 85]), the partial time derivative of $x(t)$ is used instead for modifying $\mu_\mathcal{H}(\alpha, \beta)$. As alternative to changing the weighting distribution, Bertotti [11] introduced time-dependent elementary switching operators for dynamic Preisach modeling. In contrast to common relay operators (see Fig. 6.3a), these operators can take continuous values between -1 and 1. Actually, their practical implementation is rather complicated and, thus, the operator output cannot be calculated efficiently. Füzi [31] developed a dynamic Preisach hysteresis model through applying an appropriate time lag for $x(t)$. Thereby, the resulting hysteresis operator loses its physical meaning, which poses a significant problem regarding generalization.

A different class of dynamic Preisach hysteresis models for ferroelectric materials is based on so-called *creep operators*. Such an operator is connected in series to the output of the classical Preisach hysteresis operator, i.e., $y(t) = \mathcal{H}_P[x](t)$ represents the creep operator's input. Hegewald [36] as well as Reilä̈nder [76] utilized a rheological modeling approach to achieve appropriate creep operators for ferroelectric materials. This phenomenological approach is commonly named *Kelvin–Voigt model*. A single elementary creep operator can be understood as a parallel connection of one spring and one damper element (cf. Fig. 5.20 on p. 168). By means of individually weighting several of those elementary creep operators, we are able to describe the creep behavior of ferroelectric materials in a reliable way (e.g., [36]). Although elementary creep operators can be efficiently implemented, the amount of necessary parameters for dynamic Preisach hysteresis models increases remarkably. Consequently, the uniqueness of the parameters might get lost during their identification.

With a view to practical applications of dynamic Preisach hysteresis models for ferroelectric actuators, we are primarily interested in a modeling approach that requires only a few additional parameters. Because ferroelectric actuators operate in most applications in a limited frequency range, one may simply extend the classical Preisach hysteresis operator $\mathcal{H}_P$. As presented in [67, 85], let us also alter the weighting distribution. For this purpose, a special procedure for piezoceramic actuators has been developed at the Chair of Sensor Technology (Friedrich-Alexander-University Erlangen-Nuremberg), which is based on the weighting distribution $\mu_{DAT}(\alpha, \beta)$. In short, the analytical function defining $\mu_{DAT}(\alpha, \beta)$ is modified with respect to the frequency f of the excitation signal [78, 101, 103, 104]. This leads to a dynamic weighting distribution $\mu_{DAT}(\alpha, \beta, f)$ and, therefore, we obtain a dynamic Preisach hysteresis model.

In the following, the developed procedure is illustrated on the example of a ferro-electrically soft piezoceramic disk (diameter 10.0 mm; thickness 2.0 mm; material Pz27). Figure 6.19a shows resulting hysteresis curves for the acquired electric polarization $P_{meas}(E, f)$ with respect to the excitation frequency f. Thereby, the electrical excitation was chosen so that the hysteresis curves contain major loops as well as first-order reversal curves for selected excitation frequencies. The frequencies range from 0.01 to 5 Hz and are almost logarithmically distributed in this range. While the electric polarization $P_{sat}^{\pm}$ in the saturation state and the remanent polarization $P_r^{\pm}$ stay nearly constant, the coercive field intensity $E_c^{\pm}$ exhibits a significant dependence on f. If f increases, $\left|E_c^{\pm}\right|$ will also increase which, thus, widens the hysteresis curve.

To incorporate the measured behavior of the investigated piezoceramic material in our Preisach hysteresis model, let us take a look at Fig. 6.12. The parameter study reveals that the parameter h_2 of the DAT function has a similar effect on hysteresis curves as f. For this purpose, it makes sense to exclusively alter h_2 with respect to f in order to obtain a dynamic Preisach hysteresis model. It is recommended to proceed as follows: (i) As a first step, one should identify the entire parameter set of the Preisach hysteresis model at one excitation frequency; (ii) subsequently, the dependence of h_2 on f should be evaluated, i.e., the remaining parameters (e.g., B) are not modified. In case of the investigated piezoceramic disk, the entire parameter set $\mathfrak{p}_P$ for the electric polarization was identified through an appropriate adjustment of simulations to measurements at $f = 0.1$ Hz. Note that for the other excitation frequencies, h_2 was simply changed within $\mathfrak{p}_P$. As the simulated hysteresis curves of the electric polarization $P_{sim}(E, f)$ in Fig. 6.19c point out, we can describe the frequency dependent behavior of the piezoceramic disk very well with the suggested dynamic Preisach hysteresis model. This does not only refer to the major loops but also to the minor loops.

Figure 6.19e depicts the identified values for h_2 with respect to f for the electric polarization of the investigated disk, i.e., $h_{2,P}(f)$. These values can serve as data points of a smoothing function $\psi_{smooth}(f)$ for $h_{2,P}(f)$. Due to the progression of the data points, logarithmic as well as exponential functions are appropriate smoothing functions [101]. Here, let us utilize a special exponential function, which is given by

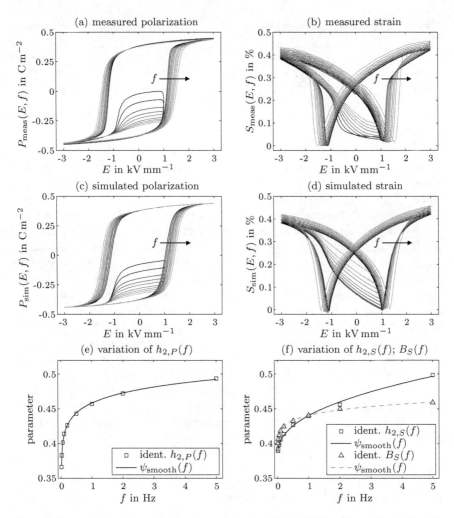

Fig. 6.19 a and **b** Measured electric polarization $P_{meas}(E, f)$ and mechanical strain $S_{meas}(E, f)$ with respect to excitation frequency f; **c** and **d** simulated curves $P_{sim}(E, f)$ and $S_{sim}(E, f)$ through Preisach hysteresis modeling; **e** and **f** resulting parameter values as well as smoothing function $\psi_{smooth}(f)$ according to (6.29); excitation frequencies $f \in \{0.01, 0.02, 0.05, 0.1, 0.5, 1, 2, 5\}$ Hz; piezoceramic disk (diameter 10.0 mm; thickness 2.0 mm; material Pz27)

$$\psi_{smooth}(f) = \varsigma_1 + \varsigma_2 \cdot f^{\varsigma_3} \qquad (6.29)$$

with the function parameters ς_1, ς_2, and ς_3. Hence, one is able to estimate values $h_{2,P}(f)$ for excitation frequencies even if measured data for that frequencies are not available. Thereby, two additional parameters are required. Table 6.2 contains the resulting parameters of the smoothing function $\psi_{smooth}(f)$ for $h_{2,P}(f)$.

Table 6.2 Parameters ς_i of smoothing function $\psi_{\text{smooth}}(f)$ in (6.29) for dynamic Preisach hysteresis model

		ς_1	ς_2	ς_3
Polarization	$h_{2,P}(f)$	−0.2121	0.6703	0.0308
Strain	$h_{2,S}(f)$	0.3819	0.0583	0.4206
	$B_S(f)$	0.3315	0.1091	0.0995

Now, we concentrate on the dynamic large-signal behavior of the mechanical strain. Figure 6.19b depicts resulting hysteresis curves for acquired mechanical strains $S_{\text{meas}}(E, f)$ of the investigated piezoceramic disk with respect to the excitation frequency f. The minima $S_{\text{min}}^{\pm}$ remain nearly constant, while the maxima $S_{\text{max}}^{\pm}$ strongly depend on f. That is the reason why apart from h_2, the parameter B of the DAT function has to be modified. However, in order to predict the dynamic large-signal behavior of the mechanical strain by means of a dynamic Preisach hysteresis model, one can perform the same steps as for the electric polarization. Again, the entire parameter set $\mathbf{p}_S$ should be identified for a certain excitation frequency f (here 0.1 Hz) and after that the dependence of h_2 as well as of B on f should be evaluated. As the comparison of measurements $S_{\text{meas}}(E, f)$ and simulations $S_{\text{sim}}(E, f)$ (see Fig. 6.19d) indicates, the presented dynamic Preisach hysteresis model is also applicable for the mechanical strain. Figure 6.19f shows the identified values for $h_{2,S}(f)$ and $B_S(f)$ with respect to f as well as the smoothing functions $\psi_{\text{smooth}}(f)$ for both parameters according to (6.29). The underlying parameters ς_1, ς_2, and ς_3 are listed in Table 6.2. In summary, dynamic Preisach hysteresis modeling for the large-signal behavior of mechanical strains requires four additional parameters.

6.6.5 Uniaxial Mechanical Stresses

In various practical applications, ferroelectric actuators are mechanically clamped or loaded causing a certain mechanical prestress within the ferroelectric material. For instance, piezoelectric stack actuators have to be mechanically prestressed in order to prevent damage during operation (see Sect. 10.1). Mechanical stresses arising within a ferroelectric material can, however, alter its electrical as well as mechanical behavior significantly [106, 109, 110]. To demonstrate this fact, we consider the large-signal behavior of a ferroelectrically soft piezoceramic disk (diameter 10.0 mm; thickness 2.0 mm; material Pz27). Figures 6.20a and 6.21a depict the acquired electric polarization $P_{\text{meas}}(E, T)$ and mechanical strain $S_{\text{meas}}(E, T)$ of the disk for varying uniaxial mechanical prestresses T, respectively. The mechanical load was applied in thickness directions (3-direction) of the disk through a tension-compression testing machine. It can be clearly seen that both the electrical and mechanical behavior strongly depend on the mechanical prestress inside the disk. The reason for this lies in switching processes of domains and in the internal structure of piezoceramic mate-

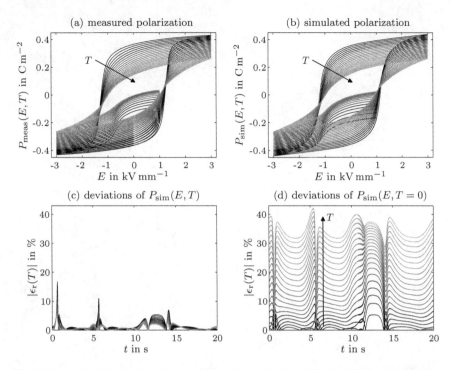

Fig. 6.20 **a** and **b** Measured and simulated electric polarization $P(E, T)$ with respect to applied mechanical prestress T; collected normalized relative deviations $|\epsilon_r(T)|$ (magnitude) in case of **c** considered prestress during Preisach hysteresis modeling and **d** without consideration of T; applied mechanical prestress $T = [0;\ 100]$ MPa in steps of 5 MPa; piezoceramic disk (diameter 10.0 mm; thickness 2.0 mm; material Pz27)

rials. The polarization direction (spontaneous polarization) within the units cells is preferably aligned in parallel to the applied electric field E. Against that due to the connection of polarization direction and largest geometric dimension of the unit cells, they are preferably aligned orthogonal to the applied mechanical stress T. Consequently, macroscopic polarizations as well as mechanical strains will be decreased if the directions of E and T coincide, which is the case for the investigated piezoceramic disk. The greater T, the more domains will stay in the ferroelastic intermediate stage during poling and can no longer be aligned in the direction of E [14]. Hence, coercive field intensity $\left|E_c^\pm\right|$, remanent polarization $\left|P_r^\pm\right|$, polarization $\left|P_{sat}^\pm\right|$ in the saturation state as well as the maximum mechanical strain $S_{max}^\pm$ of the piezoceramic material are reduced which yield smaller hysteresis curves (see Figs. 6.20a and 6.21a).

To utilize Preisach hysteresis modeling for the large-signal behavior of mechanically prestressed ferroelectric materials, one may identify the entire parameter set of the generalized model for the current situation. That will be, however, only possible if the mechanical prestress remains constant during operation. In case of time-varying mechanical loads, it makes sense to consider the resulting mechanical prestress as

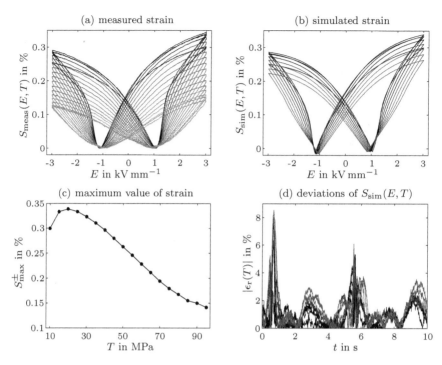

Fig. 6.21 **a** and **b** Measured and simulated mechanical strain $S(E, T)$ with respect to applied mechanical prestress T; **c** maximum value of the mechanical strain $S_{\max}^{\pm}$ versus prestress; **d** collected normalized relative deviations $|\epsilon_r(T)|$ (magnitude) in case of considered prestress during Preisach hysteresis modeling; applied mechanical prestress for measurements $T = [10; 95]$ MPa and simulations $T = [10; 50]$ MPa in steps of 5 MPa, respectively; piezoceramic disk (diameter 10.0 mm; thickness 2.0 mm; material Pz27)

additional input of a generalized Preisach hysteresis model $\mathcal{H}_G$ (see Fig. 6.15). For ferromagnetic materials, there can be found several publications concerning the incorporation of mechanical prestresses in Preisach hysteresis models. Some of the available methods are briefly described below. Adly et al. [1] suggest an approach that is based on the superposition of two Preisach hysteresis operators. While the magnetic field intensity serves as input for the first hysteresis operator, the mechanical stress is the input for the second one. Since the weighting distribution of the first hysteresis operator depends on stress and those of the second one on magnetic field intensity, a mutual coupling of magnetic and mechanical quantities is achieved. The particular problem here is the identification of appropriate weighting distributions. Bergqvist and Engdahl [10] use a single Preisach hysteresis operator with one input, which is given by combining magnetic field intensity and mechanical stress. Because of the fact that each elementary switching operator $\gamma_{\alpha\beta}$ requires an individual input resulting from this combination, model complexity increases extensively. Enhancements of both methods are mentioned in, e.g., [19, 60]. In contrast to ferromagnetic materials, the number of publications dealing with Preisach hysteresis models for ferroelectric materials under additional consideration of mechanical prestress is currently very low. Hughes and Wen [41] early recognized a need for Preisach hysteresis models with two separate inputs for electrical excitation and mechanical prestress but did not pursue this path toward a generalized approach. Freeman and Joshi [30] introduced a hysteron depending on the applied mechanical prestress. However, they only presented simulation results of the rate-independent approach and did not conduct verifications through measurements on test samples.

Due to the lack of appropriate Preisach hysteresis models enabling consideration of mechanical prestress within ferroelectric materials, an appropriate generalized Preisach hysteresis model was developed at the Chair of Sensor Technology (Friedrich-Alexander-University Erlangen-Nuremberg) [101, 105, 106]. Let us explain the underlying idea by the aforementioned large-signal behavior of the piezoceramic disk in case of uniaxial mechanical prestress. During acquisition of the electric polarization $P_{\mathrm{meas}}(E, T)$, the mechanical prestress was increased starting from 0 to 100 MPa in steps of 5 MPa. The curves in Fig. 6.20a refer to the steady state, which means that mechanical creep processes taking place within a prestressed ferroelectric material had already decayed [101]. Similar to the procedure for the rate-dependent behavior of ferroelectric materials (see Sect. 6.6.4), one can modify the weighting distribution of a classical Preisach hysteresis model with respect to the applied mechanical load. Here, we introduce the weighting distribution $\mu_{\mathrm{DAT}}(\alpha, \beta, T)$, which is, thus, also a function of the mechanical prestress T. As the comparison of $P_{\mathrm{meas}}(E, T)$ in Fig. 6.20a and the parameter study in Fig. 6.12 reveals, the function parameters B, η and h_2 should be altered according to the applied mechanical load. This can be ascribed to the fact that $\left|P_{\mathrm{sat}}^{\pm}\right|$, $\left|P_{\mathrm{r}}^{\pm}\right|$, $\left|E_{\mathrm{c}}^{\pm}\right|$ as well as the slope steepness nearby $\left|E_{\mathrm{c}}^{\pm}\right|$ change through T. Figure 6.20b displays simulated electric polarizations $P_{\mathrm{sim}}(E, T)$ for different values of T. The entire parameter set of the Preisach hysteresis operator was identified for the mechanically unloaded disk, i.e., $T = 0$. Note that in the loaded case (i.e., $T \neq 0$), we solely modified B, η as well as h_2. Figure 6.22a–c contain the resulting parameters $B_P(T)$, $\eta_P(T)$ and $h_{2,P}(T)$ with respect to T. Because these parameters feature smooth progression, they can serve as data points of the smoothing function $\psi_{\mathrm{smooth}}(T)$

$$\psi_{\mathrm{smooth}}(T) = \varsigma_1 + \varsigma_2 \mathrm{e}^{\varsigma_3 \cdot T/(1\,\mathrm{MPa})} \ . \tag{6.30}$$

Consequently, the generalized Preisach hysteresis operator $\mathcal{H}_{\mathrm{G}}$ comprises nine additional function parameters. For the investigated piezoceramic disk, these function parameters are listed in Table 6.3. Finally, the relative deviation $\epsilon_{\mathrm{r}}(T)$ between measured and simulated electric polarization is shown for two cases. While in Fig. 6.20c, the applied mechanical prestress was considered in Preisach modeling, Fig. 6.20d depicts the results if we neglect this dependence. The comparison of the figures emphasizes once again the necessity of incorporating mechanical prestress in Preisach hysteresis models for ferroelectric materials.

As a next step, let us take a closer look at the mechanical behavior of the piezoceramic disk. In Fig. 6.21a, one can see the acquired butterfly curves $S_{\mathrm{meas}}(E, T)$ in the

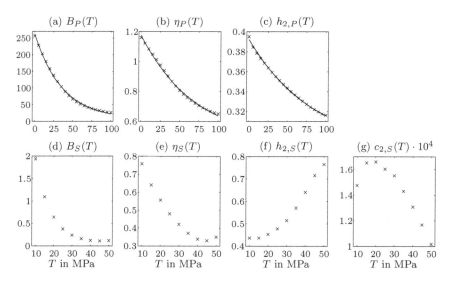

Fig. 6.22 Resulting parameter values **a–c** for electric polarization P (see Fig. 6.20) and **d–g** for mechanical strain S (see Fig. 6.21) with respect to applied mechanical prestress T, respectively; **a–c** contain smoothing functions $\psi_{\text{smooth}}(T)$ according to (6.30)

Table 6.3 Parameters ς_i of smoothing function $\psi_{\text{smooth}}(T)$ in (6.30) to consider uniaxial mechanical prestress T in generalized Preisach hysteresis model $\mathcal{H}_G$

	ς_1	ς_2	ς_3
$B_P(T)$	5.7139	255.8991	−0.0275
$\eta_P(T)$	0.4657	0.7089	−0.0142
$h_{2,P}(T)$	0.2819	0.1102	−0.0118

steady state for different mechanical prestresses, which range from 10 to 95 MPa (step size 5 MPa). It is interesting to note that the maximum mechanical strains $S^{\pm}_{\text{max}}$ of the piezoceramic disk slightly increase for small values of T (see Fig. 6.21c). This is mainly attributable to the increased mobility of domains within piezoceramic materials in the ferroelastic intermediate stage [101]. However, a further increase of the mechanical load strongly reduces $S^{\pm}_{\text{max}}$. To simulate the mechanical strain of the disk by means of Preisach hysteresis modeling, we modify the weighting distribution again with respect to the applied mechanical load, which leads to $\mu_{\text{DAT}}(\alpha, \beta, T)$. In contrast to $P_{\text{meas}}(E, T)$, parameter studies indicate that adjusting B, η, and h_2 is not sufficient to describe $S_{\text{meas}}(E, T)$ in a reliable way. The parameter c_2 (see (6.27)) has also to be varied with respect to the applied prestress. Such as for the electric polarization, the entire parameter set of the Preisach hysteresis operator was identified for the mechanically unloaded disk. In the loaded case, we only modified B, η, h_2 and c_2. Figure 6.21b depicts the simulated butterfly curves $S_{\text{sim}}(E, T)$ for the piezoceramic disk in case of mechanical prestresses ranging from 10 MPa to 50 MPa.

The underlying parameters $B_S(T)$, $\eta_S(T)$, $h_{2,S}(T)$ as well as $c_{2,S}(T)$ are depicted in Fig. 6.22d–g. Due to the fact that $S_{max}^{\pm}$ slightly increases at first, one has to utilize more complicated smoothing functions than for the electric polarization. Within a limited prestress range, it is, nevertheless, possible to conduct similar approximations as in (6.30). At the end, Fig. 6.21d displays normalized relative deviations $\epsilon_r(T)$ between measured and simulated mechanical strains of the piezoceramic disk for different mechanical prestresses. These deviations stay mostly far below 10%, which confirms once more the applicability of the presented Preisach modeling approach.

6.7 Parameter Identification for Preisach Modeling

Classical as well as generalized Preisach hysteresis modeling for ferroelectric materials requires several parameters that have to be identified. For the electric polarization P and mechanical strain S, we collect these parameters in the vectors (cf. (6.26) and (6.28))

$$\mathfrak{p}_P = \left[a_P, b_P, B_P, \eta_P, h_{1,P}, h_{2,P}, \sigma_{1,P}, \sigma_{2,P}\right]^t \tag{6.31}$$

$$\mathfrak{p}_S = \left[a_S, b_S, B_S, c_1, c_2, c_3, \eta_S, h_{1,S}, h_{2,S}, \sigma_{1,S}, \sigma_{2,S}\right]^t , \tag{6.32}$$

respectively. In Sect. 6.7.1, a identification strategy is presented allowing reliable simulations for the different working areas of ferroelectric actuators, i.e., bipolar, unipolar as well as semi-bipolar working areas. The underlying approach is then applied to a piezoceramic disk (Sect. 6.7.2), which is made of the ferroelectrically soft material PIC255.

6.7.1 Identification Strategy for Model Parameters

Just as in Chap. 5, the parameter identification represents an ill-posed inverse problem. The desired parameter vectors $\mathfrak{p}_P$ as well as $\mathfrak{p}_S$ result from comparisons of measurements and simulations, i.e., outputs of the Preisach hysteresis operator. Due to this fact, one has to acquire adequate electrical and mechanical quantities. Through iterative adjustments of the parameters, the deviations between simulations and measurements get reduced until a sufficiently good match is found. The success of the iterative adjustments mainly depends on two points: (i) The measurement signals utilized for identification and (ii) the initial guess $\mathfrak{p}_{P;S}^{(0)}$ of the parameter vectors. For Preisach hysteresis modeling, it is recommended to apply measurement signals that are close to the excitation signals actually occurring in practical applications. In other words, we should select measurement signals with respect to the working area of the ferroelectric actuator. Because Preisach hysteresis operators demand inputs in

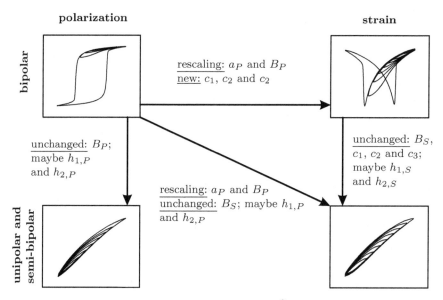

Fig. 6.23 Efficient strategy to find appropriate initial guess $\mathfrak{p}_{P;S}^{(0)}$ and to reliably identify $\mathfrak{p}_{P;S}$ for electric polarization and mechanical strain in different working areas of ferroelectric actuators [101]

the range $[-0.5, 0.5]$, the raw data has to be normalized to its maximum (see (6.7)). Such normalizations are necessary in each working area of the ferroelectric actuator.

The initial guess strongly affects convergence of the identification approach as well as its duration. To find $\mathfrak{p}_{P;S}^{(0)}$ for the working areas, a specific procedure is indispensable. Figure 6.23 depicts an entire identification strategy that has proven to be effective for piezoceramic materials [101]. The presented strategy can be divided into two parts, which are discussed below. While the first part exclusively relates to bipolar working areas, the second one deals with unipolar as well as semi-bipolar working areas.

- **Bipolar Working Area**: An appropriate initial guess $\mathfrak{p}_{P}^{(0)}$ to predict P of piezo-ceramic materials in the bipolar working area (i.e., saturation and major loops) results from manually adjusting the parameters according to Fig. 6.12. After conducting iterative parameter adjustment on basis of an optimization approach (e.g., Levenberg–Marquardt algorithm), one obtains the solution $\mathfrak{p}_{P}^{s}$. This vector serves as starting point for identifying the parameter vector $\mathfrak{p}_{S}^{s}$, which yields reliable simulations for $S(E)$ in the bipolar working area, i.e., butterfly curves. In particular, with the exception of a_P and B_P, the components of $\mathfrak{p}_{P}^{s}$ can be used directly as initial guess for $\mathfrak{p}_S$. Due to our definition of the Preisach hysteresis operator, we have to rescale a_P^s and B_P^s to achieve

Table 6.4 Resulting parameters (i.e., components of $\mathbf{p}_P^s$ and $\mathbf{p}_S^s$) for Preisach hysteresis modeling of electric polarization P and mechanical strain S in different working areas of the piezoceramic disk; bold numbers indicate parameters excluded from identification

	B	η	h_1	h_2	σ_1	σ_2
P_{bipolar}	1868.5	1.275	0.011	0.450	76.7	167.5
P_{unipolar}	**1868.5**	0.920	**0.011**	**0.450**	337.7	181.6
$P_{\text{semi-bipolar}}$	**1868.5**	0.881	0.143	**0.450**	392.1	160.8
S_{bipolar}	4.432	1.157	0.009	0.434	34.2	137.6
S_{unipolar}	**4.432**	1.089	0.675	0.041	12.2	89.3
$S_{\text{semi-bipolar}}$	**4.432**	0.420	0.090	**0.434**	5884.4	1045.2
	$a \cdot 10_3$	b	c_1	$c_2 \cdot 10^3$	$c_3 \cdot 10^3$	
P_{bipolar}	53.8	4.624	–	–	–	
P_{unipolar}	59.6	1.718	–	–	–	
$P_{\text{semi-bipolar}}$	70.5	1.563	–	–	–	
S_{bipolar}	1.5	3.641	0	−0.096	0.273	
S_{unipolar}	16.8	0.062	**0**	**−0.096**	**0.273**	
$S_{\text{semi-bipolar}}$	1.5	1.608	**0**	**−0.096**	**0.273**	

$$\left. \begin{array}{l} a_S^{(0)} = \varsigma \cdot a_P^s \\ B_S^{(0)} = \varsigma \cdot B_P^s \end{array} \right\} \quad \text{with} \quad \varsigma = \frac{2(S_{\text{max}} - S_{\text{min}}) \cdot 1\,\text{C}\,\text{m}^{-2}}{(P_{\text{max}} - P_{\text{min}}) \cdot 100\%} . \tag{6.33}$$

The initial guess for the further parameters c_1, c_2 as well as c_3 results from geometric considerations shown in Fig. 6.18a.

- **Unipolar and Semi-bipolar Working Areas**: For these working areas, $\mathbf{p}_P^s$ and $\mathbf{p}_S^s$ from the bipolar working area represent appropriate initial guesses. However, with a view to ensuring convergence of the subsequent optimization approach, the parameter B should be excluded from identification, i.e., B_P^s and B_S^s as identified for the bipolar working area are directly used. It might also be necessary to exclude h_1 and h_2 during optimization, i.e., $h_{1,P}, h_{2,P}, h_{1,S}$ as well as $h_{2,S}$. If $S(E)$ is simulated in unipolar and semi-bipolar working areas without the model extension in (6.27), we can utilize a rescaled version of $\mathbf{p}_P^s$ as suitable initial guess. Again, it is recommended to exclude B_S from identification and maybe $h_{1,S}$ as well as $h_{2,S}$ in addition.

6.7.2 Application to Piezoceramic Disk

Now, let us apply the aforementioned identification strategy to a piezoceramic disk (diameter 10.0 mm; thickness 2.0 mm), which is made of the ferroelectrically soft material PIC255. Table 6.4 contains the resulting components of $\mathbf{p}_P^s$ and $\mathbf{p}_S^s$ in the different working areas. As can be clearly seen, the identified parameters differ considerably for both the working areas and the physical quantities (i.e., electric polarization or mechanical strain). This emphasizes once again the importance of determining individual parameter vectors.

Figures 6.24 and 6.25 depict various measurements as well as simulations for the piezoceramic disk in unipolar and semi-bipolar working areas, respectively. The left panels deal with electric polarizations P of the disk and the right panels show the obtained mechanical strains S. Due to the fact that the initial polarization state of the disk is, strictly speaking, unknown in both working areas, it is not possible to determine absolute values for P and S. We exclusively quantify changes of the quantities instead, which are denoted by ΔP and ΔS. As the comparisons in the Figs. 6.24a, b and 6.25a, b reveal, Preisach hysteresis modeling yields reliable simulations for $\Delta P(E)$ and $\Delta S(E)$. This can also be seen in the Figs. 6.24c, d and 6.25c, d, which display the time signals utilized for identifying $\mathbf{p}_P^s$ and $\mathbf{p}_S^s$ (see Table 6.4). With a view to demonstrating the applicability of Preisach hysteresis modeling for piezoceramic actuators, additional comparisons were carried out by means of further time signals (Figs. 6.24e, f and 6.25e, f). Although these time signals have not been considered during parameter identification, simulations coincide very well with measurements. This is confirmed by the normalized relative deviation ϵ_r of the simulation results as shown in the Figs. 6.24g, h and 6.25g, h. In the particular cases, $|\epsilon_r|$ always stays below 6%. Summing up, it can be stated again that Preisach hysteresis modeling represents an excellent approach to predict the large-signal behavior of piezoceramic actuators, especially in unipolar and semi-bipolar working areas.

6.8 Inversion of Preisach Hysteresis Model

To conduct model-based compensation of hysteresis effects within ferroelectric actuators, we have to determine that input quantity $x_{inv}(k)$ for time step k, which yields the desired target output $y_{tar}(k)$. Under certain circumstances, it may be necessary to consider also specific boundary conditions such as applied mechanical prestress T and excitation frequency f. That is the reason why we define here input quantities $x_{inv}(k)$, target quantities $y_{tar}(k)$, and boundary conditions z_{bou} as follows (excitation voltage $u(k)$; mechanical displacement $d(k)$):

- $x_{inv}(k) \in \{E(k), u(k)\}$
- $y_{tar}(k) \in \{P(k), S(k), d(k)\}$
- $z_{bou} \in \{T, f\}$.

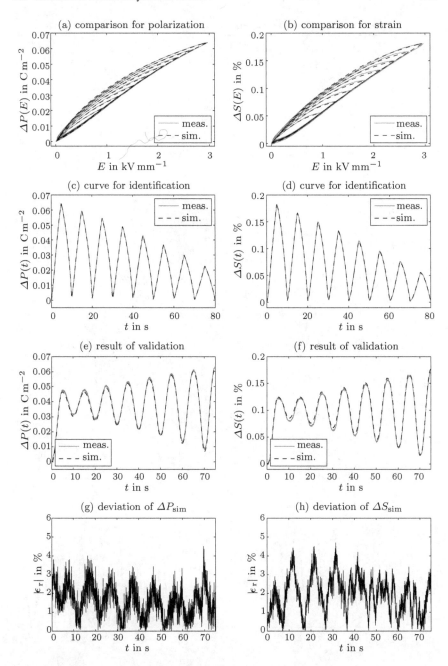

Fig. 6.24 **a** and **b** Comparison of measured and simulated hysteresis curves $\Delta P(E)$ and $\Delta S(E)$ in **unipolar** working area of the piezoceramic disk; **c** and **d** time signals for identifying weighting distributions; **e** and **f** time signals for validating Preisach hysteresis modeling; **g** and **h** resulting normalized relative deviations $|\epsilon_r|$ (magnitude) for validation signals

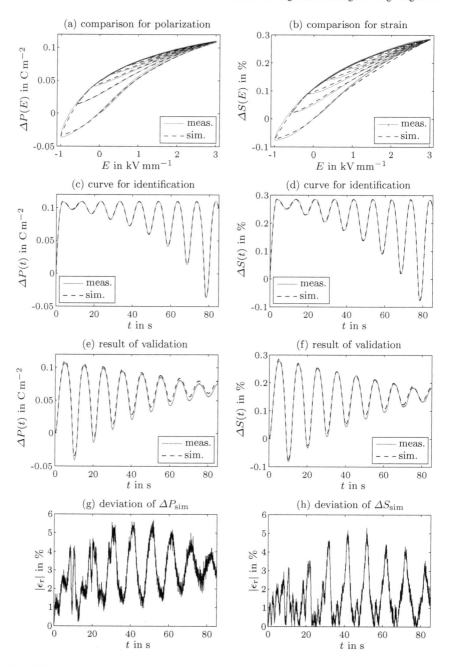

Fig. 6.25   a and **b** Comparison of measured and simulated hysteresis curves $\Delta P(E)$ and $\Delta S(E)$ in **semi-bipolar** working area of the piezoceramic disk; **c** and **d** time signals for identifying weighting distributions; **e** and **f** time signals for validating Preisach hysteresis modeling; **g** and **h** resulting normalized relative deviations $|\epsilon_\mathrm{r}|$ (magnitude) for validation signals

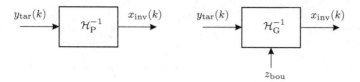

Fig. 6.26 Inverted (classical) Preisach hysteresis operator $\mathcal{H}_P^{-1}$ and inverted generalized Preisach hysteresis operator $\mathcal{H}_P^{-1}$

Generalized Preisach hysteresis modeling can be used to predict the hysteretic behavior of electric polarizations $P(k)$ and mechanical strains $S(k)$ for ferroelectric materials. Since these quantities represent target quantities, the underlying Preisach hysteresis operator $\mathcal{H}_P$ has to be inverted. In other words, the inverted generalized Preisach hysteresis operator $\mathcal{H}_G^{-1}$ and, thus, the inverted Preisach hysteresis operator $\mathcal{H}_P^{-1}$ are required for model-based compensation of hysteresis effects (see Fig. 6.26). However, owing to the fact that the elementary switching operators $\gamma_{\alpha\beta}$ exhibit discontinuities at the changeover points α and β, there does not exist a closed-form solution for this task. Consequently, $\mathcal{H}_P$ has to be inverted numerically.

One can find various approaches in literature to obtain an appropriate approximation of $\mathcal{H}_P^{-1}$. Several methods are based on iterative algorithms for locally inverting discretized Preisach hysteresis models. For instance, Mittal and Menq [66] as well as Tan and Baras [92] exploited such algorithms to compensate hysteresis of electromagnetic and magnetostrictive actuators. Viswamurthy and Ganguli [95] utilized a locally inverted Preisach hysteresis model for controlling mechanical vibrations through piezoelectric stack actuators. A different approach to achieve $\mathcal{H}_P^{-1}$ results from exchanging its input and output (e.g., [22, 91]). Thereby, the weighting distribution $\mu_{\mathcal{H}}(\alpha, \beta)$ for $\gamma_{\alpha\beta}$ has also to be inverted. With a view to ensuring positive weighting distributions, Bi et al. [12] introduced an analytical weighting distribution as well as an additional switching operator. They applied this approach for ferromagnetic materials and present convincing results. Due to exchanging input and output of $\mathcal{H}_P$, the physical meaning, however, gets lost which may cause problems regarding generalized Preisach hysteresis models.

Here, let us discuss an inverted Preisach hysteresis model that was developed by Wolf and colleagues [101, 102]. Section 6.8.1 deals with the underlying iterative inversion procedure, which is characterized in Sect. 6.8.2. Subsequently, the main steps toward an inverted generalized Preisach hysteresis model are addressed. Finally, model-based hysteresis compensation is applied to a piezoceramic disk.

6.8.1 Inversion Procedure

The computation of the sought-after input quantity $x_{\mathrm{inv}}(k)$ yielding the target quantity $y_{\mathrm{tar}}(k)$ for time step k is performed incrementally. The target quantity has to be

Fig. 6.27 Simplified flow
chart for incrementally
determining outputs of
inverted Preisach hysteresis
operator $\mathcal{H}_P^{-1}$; $y_{tar}(k)$
and $x_{inv}(k) = \mathcal{H}_P^{-1}[y_{tar}](k)$
represent desired target
quantity and sought-after
quantity for time step k,
respectively

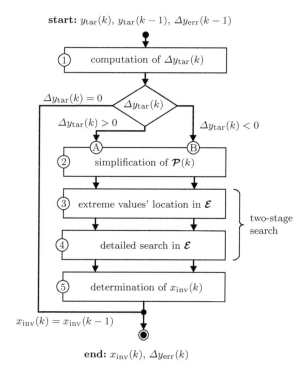

considered for the current as well as previous time step, i.e., $y_{tar}(k)$ and $y_{tar}(k-1)$.
At each time step k, we analyze and simplify the current configuration of the Preisach
plane $\mathcal{P}(k)$. Thereby, the vectors $\mathbf{e}_i(k)$, $\mathbf{e}_j(k)$ and $\mathbf{s}(k)$ are used to indicate location
as well as sign of dominating extrema in $\mathcal{P}(k)$ (see (6.12) and (6.13)). The inversion
procedure is mainly based on a two-stage evaluation of the Everett matrix $\mathcal{E} = [\mathcal{E}_{ij}]$.
Figure 6.27 shows a simplified flow chart of the entire inversion procedure comprising five steps, which are explained below.

1. Computation of the Increment $\Delta y_{tar}(k)$

In a first step, the increment $\Delta y_{tar}(k)$ is computed which represents the change
of the target output y_{tar} from time step $k-1$ to time step k, i.e.,

$$\Delta y_{tar}(k) = y_{tar}(k) - y_{tar}(k-1) \tag{6.34}$$

If $\Delta y_{tar}(k) = 0$ is fulfilled, one can directly continue with the subsequent time
step $k+1$. The resulting output of the inverted Preisach hysteresis model is
then given by $x_{inv}(k) = x_{inv}(k-1)$. This also holds for several further special
cases like saturation in the Preisach plane and an increment $\Delta y_{tar}(k)$ that is
smaller than the discretization error $\Delta y_{err}(k)$ from the previous iteration. How-

ever, when $\Delta y_{tar}(k) \neq 0$, we need to distinguish between two cases depending on its sign (see (6.4) and (6.5)):

Ⓐ increasing $y_{tar}(k)$, i.e., $\Delta y_{tar}(k) > 0 \Rightarrow$ modification of α

Ⓑ decreasing $y_{tar}(k)$, i.e., $\Delta y_{tar}(k) < 0 \Rightarrow$ modification of β.

Hence, the dividing line $\mathcal{L}(k)$ in the Preisach plane $\mathcal{P}(k)$ is modified.

2. Simplification of the Preisach Plane $\mathcal{P}(k)$

Actually, there exist various different configurations of $\mathcal{P}(k)$, e.g., number of steps in $\mathcal{L}(k)$ (see, e.g., Fig. 6.7). To standardize the subsequent inverting approach in step 3 and 4, let us simplify the configurations by reducing them to two cases, which are displayed in Fig. 6.28. For the particular configurations, the reduction implies deleting the mth entry of the vectors $\mathbf{e}_i(k)$, $\mathbf{e}_j(k)$ and $\mathbf{s}(k)$. Consequently, the output of the Preisach hysteresis operator changes by $\Delta y_{simp}(k)$ (hatched area in Fig. 6.28), which has to be included in the current increment $\Delta y_{tar}(k)$ of the target output, i.e.,

$$\Delta y'_{tar}(k) = \Delta y_{tar}(k) + \Delta y_{simp}(k) \cdot \mathbf{s}_m(k)$$
$$= \Delta y_{tar}(k) + \mathcal{E}_{i_m j_m} \cdot \mathbf{s}_m(k) . \tag{6.35}$$

The following two steps deal with an iterative search in the Everett matrix $\mathcal{E}$. While the first one represents a coarse search, the second one is a detailed search.

3. Evaluation of Extreme Values' Locations in the Everett Matrix $\mathcal{E}$

The first iterative search steps exclusively considers dominating extrema in $\mathcal{P}(k)$ that are specified through the vectors $\mathbf{e}_i(k)$, $\mathbf{e}_j(k)$, and $\mathbf{s}(k)$. If necessary, the wiping-out rule of the Preisach hysteresis operator has to be applied in addition. Principally, the first iterative search step consists of three substeps (see Fig. 6.29).

- The starting point is the dominating extremum (index m), which exhibits the smallest magnitude. From this extremum, we readout the entries of $\mathcal{E} = [\mathcal{E}_{ij}]$ in descending order according to the components of $\mathbf{e}_i(k)$ and $\mathbf{e}_j(k)$. This procedure is conducted until the condition

$$\underbrace{\left| \sum_{\nu=m}^{n+1} \mathcal{E}_{i_\nu j_\nu} \cdot \mathbf{s}_\nu(k) \right|}_{\Delta y_{ext}(k)} < \left| \Delta y'_{tar}(k) \right| < \left| \sum_{\nu=m}^{n} \mathcal{E}_{i_\nu j_\nu} \cdot \mathbf{s}_\nu(k) \right| \tag{6.36}$$

is fulfilled.

- Now, the components $m, \ldots, n+1$ of $\mathbf{e}_i(k)$, $\mathbf{e}_j(k)$ and $\mathbf{s}(k)$ are used to adjust the modified increment $\Delta y'_{tar}(k)$ of the target function by means of $\mathcal{E}_{i_\nu j_\nu}$, which leads to

$$\Delta y''_{\text{tar}}(k) = \Delta y'_{\text{tar}}(k) + \Delta y_{\text{ext}}(k)$$

$$= \Delta y'_{\text{tar}}(k) + \sum_{\nu=m}^{n+1} \mathcal{E}_{i_\nu j_\nu} \cdot \mathfrak{s}_\nu(k) . \tag{6.37}$$

- At the end of the first iterative search step, the components $m, \ldots, n+1$ of $\mathbf{e}_i(k)$, $\mathbf{e}_j(k)$ and $\mathfrak{s}(k)$ are deleted. We store the indices (i_ν, j_ν) of the extremum that was deleted at last.

  As a result, one knows the two dominating extrema between which the sought-after entry is located in the Everett matrix $\mathcal{E}$. Besides, the configuration of the Preisach plane is further simplified.

4. Detailed Search for Correct Entry in the Everett Matrix $\mathcal{E}$

The second iterative search (detailed search) can be performed in a strongly restricted region of $\mathcal{E}$. For an increasing target output $y_{\text{tar}}(k)$ (i.e., case Ⓐ), the search is done along column j_m (see Fig. 6.30). In the other case (i.e., Ⓑ), one has to search in row i_m. The procedure starts in both cases at the entry $\mathcal{E}_{i_\nu j_\nu}$ featuring the indices (i_ν, j_ν) that were stored in step 3. It is desired to find the entry $\mathcal{E}_{i_r j_s}$, which coincides with $y''_{\text{tar}}(k)$ best possible, i.e.,

$$\min\left(\left|\Delta y''_{\text{tar}}(k) - \mathcal{E}_{i_r j_s}\right|\right) \quad \text{with} \quad \begin{cases} j_s = j_m & \text{for Ⓐ} \\ i_r = i_m & \text{for Ⓑ} . \end{cases} \tag{6.38}$$

An efficient method for this task is the *divide and conquer search algorithm* [54]. Even if a fine spatial discretization (e.g., $M = 800$) of the Preisach plane is utilized, the inverting procedure will require reasonable computation time.

5. Determination of the sought-after Input $x_{\text{inv}}(k)$

The indices (i_r, j_s) from step 4 are used to update $x_{\text{inv}}(k-1)$. Depending on the progression of the target output $y_{\text{tar}}(k)$, i.e., whether it is rising or falling, we choose one of the following equations (cf. Fig. 6.31)

$$x_{\text{inv}}(k) = \frac{M - i_r}{M - 1} - 0.5 \qquad \text{for Ⓐ} \tag{6.39}$$

$$x_{\text{inv}}(k) = 0.5 - \frac{M - j_s}{M - 1} \qquad \text{for Ⓑ} . \tag{6.40}$$

Furthermore, the discretization error $\Delta y_{\text{err}}(k)$ between the increments of actually computed target quantity $\Delta y_{\text{inv}}(k)$ and of desired target output $\Delta y_{\text{tar}}(k)$ is calculated, which is, therefore, given by

$$\Delta y_{\text{err}}(k) = \Delta y_{\text{inv}}(k) - \Delta y''_{\text{tar}}(k) = \mathcal{E}_{i_r j_s} \cdot \mathfrak{s}_m(k) - \Delta y''_{\text{tar}}(k) . \tag{6.41}$$

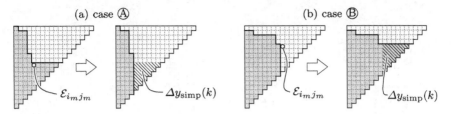

Fig. 6.28 Simplification of Preisach plane $\mathcal{P}(k)$

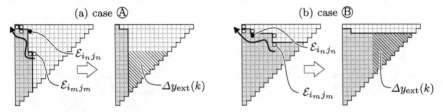

Fig. 6.29 Coarse search based on evaluation of extreme values' location in Everett matrix $\mathcal{E} = [\mathcal{E}_{ij}]$ and further simplification of Preisach plane

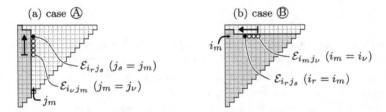

Fig. 6.30 Detailed search to figure out entry $\mathcal{E}_{i_r j_s}$ in Everett matrix

Note that $\Delta y_{\text{err}}(k)$ has to be considered in step 1. Finally, we update the vectors $\mathfrak{e}_i(k)$, $\mathfrak{e}_j(k)$, and $\mathfrak{s}(k)$ according to the current input quantity $x_{\text{inv}}(k)$. This results in $\mathfrak{e}_i(k+1)$, $\mathfrak{e}_j(k+1)$ as well as $\mathfrak{s}(k+1)$.

At the end of the whole inverting procedure, information is available which is necessary to determine $x_{\text{inv}}(k+1)$ for the subsequent time step $k+1$. In doing so, we start again with step 1 by considering the quantities $y_{\text{tar}}(k)$, $y_{\text{tar}}(k+1)$, and $\Delta y_{\text{err}}(k)$.

6.8.2 Characterization of Inversion Procedure

To characterize the inversion procedure, let us check its functionality and rate its efficiency in addition. These investigations are carried out through a serial connection of inverted Preisach hysteresis operator $\mathcal{H}_P^{-1}$ and original one, i.e., $\mathcal{H}_P$ (see Fig. 6.32) [101, 102]. We assume a target quantity $y_{\text{tar}}(k)$ that represents the input of $\mathcal{H}_P^{-1}$. The resulting output $x_{\text{inv}}(k) = \mathcal{H}_P^{-1}[y_{\text{tar}}](k)$ serves then again as input of $\mathcal{H}_P$,

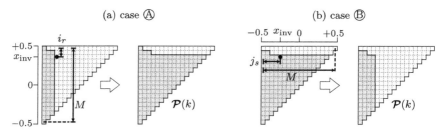

Fig. 6.31 Calculation of sought-after quantity $x_{inv}(k)$ resulting from indices (i_r, j_s); update of Preisach plane $\mathcal{P}(k)$

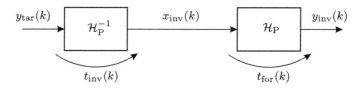

Fig. 6.32 Approach to check functionality and to rate efficiency of inversion procedure; computation time $t_{inv}(k)$ for inversion procedure; computation time $t_{for}(k)$ for evaluating $\mathcal{H}_P$ in forward direction

which leads to the final output $y_{inv}(k) = \mathcal{H}_P[x_{inv}](k)$. Hence, one is able to compare the desired target quantity $y_{tar}(k)$ with the quantity $y_{inv}(k)$ actually determined.

Figure 6.33a displays the utilized discrete-time target signal consisting of an offset and two superimposed sine waves. The sine waves feature different amplitudes as well as frequencies, respectively. For the evaluation of $\mathcal{H}_P$ and $\mathcal{H}_P^{-1}$, a spatial discretization of $M = 200$ was applied in the Preisach plane. Figure 6.33b compares the desired target quantity with the output of the serial connection in a small time window. As the comparison reveals, $y_{inv}(k)$ coincides very well with $y_{tar}(k)$. Apart from deviations due to the spatial discretization of the Preisach plane, there do not arise any further deviations. It can, thus, be stated that the inversion procedure provides reliable results.

The computation time of the inversion procedure denotes a decisive criterion with regard to practical applications. Strictly speaking, the maximum duration $t_{inv,max}$ that is required for a single time step determines the maximum sampling rate $f_{inv,max} = 1/t_{inv,max}$ for inverting the target quantity $y_{tar}(k)$. If model-based hysteresis compensation is applied in open- or closed-loop control, $x_{inv}(k)$ can be updated after the time interval $t_{inv,max}$, i.e., $t_{k+1} - t_k \geq t_{inv,max}$. In Fig. 6.33c, one can see the duration $t_{inv}(k)$ for time step k, which is required for inverting $y_{tar}(k)$ in the considered time window (cf. Fig. 6.33b). The calculations were conducted on a standard desktop PC.[10] Interestingly, $t_{inv}(k)$ takes mainly two values. The lower value results from termination conditions in step 1 of the inversion procedure, whereas the higher value $t_{inv,max}$ is a consequence of running through all steps (i.e., step 1 to step 5). Note that even if the target quantity y_{tar} exhibits an arbitrary progress, $t_{inv,max}$ will never be exceeded.

---

[10]Desktop PC: Intel Core i5 with 3.19 GHz and 4 GB RAM.

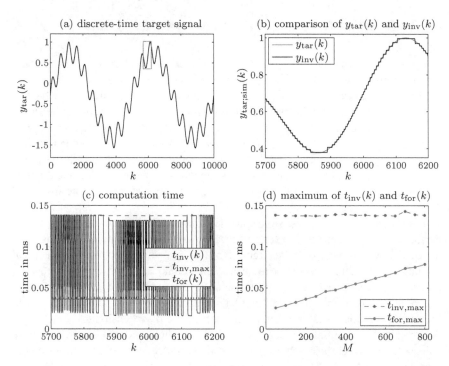

Fig. 6.33 Characterization of inversion procedure: **a** Discrete-time target signal $y_{tar}(k)$; **b** comparison of $y_{tar}(k)$ and output $y_{inv}(k)$ (cf. Fig. 6.32) for spatial discretization $M = 200$; **c** computation times for inversion procedure $t_{inv}(k)$ as well as forward calculation $t_{for}(k)$; maximum computation time $t_{inv,max}$; **d** comparison of $t_{inv,max}$ and $t_{for,max}$ with respect to M

As a result, $t_{inv,max} < 0.15$ ms is guaranteed for the spatial discretization $M = 200$, which leads to the sampling rate $f_{inv,max} = 6.67$ kHz.

Figure 6.33d depicts maximum durations $t_{inv,max}$ of the inversion procedure for different spatial discretizations M. Moreover, maximum durations $t_{for,max}$ per time step (cf. Fig. 6.32) are shown for evaluating the Preisach hysteresis operator $\mathcal{H}_P$ in forward direction. It is worth to emphasize that $t_{inv,max}$ stays almost constant in the considered range of spatial discretizations. This behavior can be ascribed to the efficient divide and conquer search algorithm in step 4 of the inversion procedure. However, contrary to $t_{inv,max}$, the duration $t_{for,max}$ increases almost along a straight line with rising M.

According to these findings, the presented inversion procedure is an efficient method for inverting Preisach hysteresis operators. Since the underlying algorithm allows time-efficient computation of the desired quantities, it can be exploited for both open- and closed-loop control of actuators exhibiting hysteretic behavior. The inversion procedure is not restricted to ferroelectric actuators but can also be used for actuators containing ferromagnetic materials.

6.8.3 Inverting Generalized Preisach Hysteresis Model

The generalized Preisach hysteresis operator $\mathcal{H}_G$ (see Sect. 6.6) for ferroelectric materials comprises reversible parts, asymmetric behavior, mechanical deformations as well as consideration of rate-dependent behavior and applied uniaxial mechanical stresses. If generalization is restricted to reversible parts and asymmetric behavior, we can evaluate the inverted generalized Preisach hysteresis operator $\mathcal{H}_G^{-1}$ in the same manner as given in Sect. 6.8.1. This can be ascribed to the fact that both generalizations directly alter the weighting distribution $\mu_\mathcal{H}(\alpha, \beta)$ for the Preisach hysteresis model. However, in case of the remaining generalizations (e.g, mechanical deformation), further important points arises during inverting $\mathcal{H}_G$, which are discussed below.

Let us start with the inversion approach for mechanical deformations S and mechanical displacements d of ferroelectric materials. In the bipolar working area, there exist two solutions of these target quantities for positive and negative electrical excitations. Thus, it is impossible to invert S and d uniquely. Ferroelectric actuators, however, usually operate in unipolar and semi-bipolar working areas. Due to this fact, we are able to describe the underlying large-signal behavior through a generalized Preisach hysteresis operator $\mathcal{H}_G$ that does not require the extension given in (6.27). As a result, the sought-after input quantities electric field intensity E and excitation voltage u of ferroelectric actuators can be determined according to the inversion procedure in Sect. 6.8.1. For instance, the target quantity d(k) for time step k serves as input of the inverted generalized Preisach hysteresis operator $\mathcal{H}_G^{-1}$, which leads to the output $u(k) = \mathcal{H}_G^{-1}[\text{d}](k)$.

To consider rate-dependent behavior and mechanical stresses for ferroelectric materials by means of $\mathcal{H}_G$, one has to take additional inputs (i.e., $z_{\text{bou}} \in \{T, f\}$) into account, respectively. The inputs modify the spatially discretized weighting distribution μ and, consequently, the Everett matrix $\mathcal{E}$ (see Sects. 6.6.4 and 6.6.5). It is of utmost importance to incorporate such modifications in the inversion procedure since only by doing so, we are able to determine reliable outputs of $\mathcal{H}_G^{-1}$. For that reason, $\mu(z_{\text{bou}})$ as well as $\mathcal{E}(z_{\text{bou}})$ should be calculated for different inputs z_{bou} in advance [101]. In practical applications of ferroelectric actuators, the task is to select an appropriate spatially discretized weighting distribution and Everett matrix. The selection depends, of course, on the boundary conditions z_{bou}, which actually occur during application.

6.8.4 Hysteresis Compensation for Piezoceramic Disk

Here, model-based compensation of hysteresis effects through an inverted generalized Preisach hysteresis operator is applied to a piezoceramic disk (diameter 10.0 mm; thickness 2.0 mm), which is made of the ferroelectrically soft material PIC255. Before the results are presented, let us discuss a particular hardware-based approach

(a)

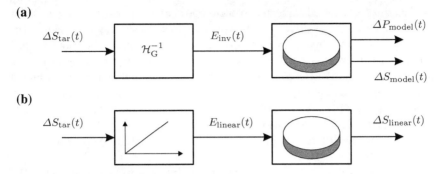

(b)

Fig. 6.34 Block diagram to achieve desired mechanical strains $\Delta S_{\text{tar}}(t)$ of the piezoceramic disk for **a** model-based hysteresis compensation and **b** uncompensated case (i.e., linearization); determined quantities: $E_{\text{inv}}(t)$ and $E_{\text{linear}}(t)$; measured quantities: $\Delta P_{\text{model}}(t)$, $\Delta S_{\text{model}}(t)$ as well as $\Delta S_{\text{linear}}(t)$

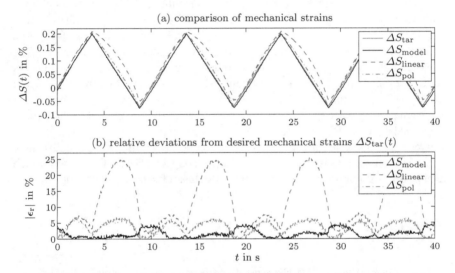

Fig. 6.35 a Comparison of desired mechanical strains $\Delta S_{\text{tar}}(t)$ and achieved quantities with respect to time t; measured quantities: $\Delta S_{\text{model}}(t)$ and $\Delta S_{\text{linear}}(t)$; computed quantity: $\Delta S_{\text{pol}}(t)$; **b** normalized relative deviations $|\epsilon_{\text{r}}|$ (magnitude) between resulting strains and desired ones; piezoceramic disk (diameter 10.0 mm; thickness 2.0 mm; material PIC255)

for compensating nonlinearities of ferroelectric actuators. Contrary to model-based compensation where we use electrical voltage as excitation signal, this hardware-based approach directly relates to the electric polarization (e.g., [28, 29, 108]). To influence the electric polarization within the ferroelectric material, electric charges Q are impressed on the actuator electrodes by means of an appropriate charge drive circuit. It is assumed that Q is directly proportional to the resulting mechanical strain S of the ferroelectric actuator, i.e., $Q \propto S$. Although a remarkable reduction of nonlinearities is achieved compared to open-loop configurations operating with voltage excitations, charge drive circuits usually exhibit substantial drawbacks. This includes limited low-frequency performance, dependence of voltage gain on capacitance of the ferroelectric material as well as time-consuming tuning procedure. Besides, when the ferroelectric actuator is driven into saturation (e.g., semi-bipolar working area), the relation of electric charge and mechanical displacement will be no longer linear [101]. As a result, the deviations of desired and actually achieved displacements of the ferroelectric actuator increase.

To compare the different types of compensations, let us desire a triangular-shaped time signal for the mechanical strain of the piezoceramic disk. The model-based compensation exploits the inverted generalized Preisach hysteresis operator $\mathcal{H}_G^{-1}$ to obtain the electrical excitation signal $E_{inv}(t)$, which is then applied to the disk sample for measurements (see Fig. 6.34a). Against that, we emulate mechanical strains $\Delta S_{pol}(t)$ of the hardware-based solution by rescaling the electric polarization $P_{model}(t)$. The expression $P_{model}(t)$ stands for the measured electric polarization actually occurring in the disk. Since the underlying rescaling does not exhibit any dependencies on electronic components, it represents the best case for charge drive circuits in open-loop configuration. Figure 6.35a depicts target strains $\Delta S_{tar}(t)$, measured strains $\Delta S_{model}(t)$ for model-based compensation of hysteresis effects as well as those for hardware-based compensation $\Delta S_{pol}(t)$. Moreover, measured strains $\Delta S_{linear}(t)$ for the uncompensated case are given meaning that the electrical excitation signal $E_{linear}(t)$ is assumed to be directly proportional to the desired mechanical strain $\Delta S_{tar}(t)$ (see Fig. 6.34b). The comparison of the different strain curves clearly indicates that $\Delta S_{model}(t)$ coincides best with $\Delta S_{tar}(t)$. In contrast, there occur normalized relative deviations of $\Delta S_{linear}(t)$ up to 25% (see Fig. 6.35b), which emphasizes the importance of considering hysteresis effects in actuator applications.

References

1. Adly, A.A., Mayergoyz, I.D., Bergqvist, A.: Preisach modeling of magnetostrictive hysteresis. J. Appl. Phys. **69**(8), 5777–5779 (1991)
2. Al Janaideh, M., Su, C.Y., Rakheja, S.: Development of the rate-dependent Prandtl-Ishlinskii model for smart actuators. Smart Mater. Struct. **17**(3), 035,026 (2008)
3. Al Janaideh, M., Rakheja, S., Su, C.Y.: A generalized Prandtl-Ishlinskii model for characterizing the hysteresis and saturation nonlinearities of smart actuators. Smart Mater. Struct. **18**(4), 045,001 (2009)
4. Arockiarajan, A., Sansour, C.: Micromechanical modeling and simulation of rate-dependent effects in ferroelectric polycrystals. Comput. Mater. Sci. **43**(4), 842–854 (2008)
5. Azzerboni, B., Cardelli, E., Finocchio, G., La Foresta, F.: Remarks about Preisach function approximation using Lorentzian function and its identification for nonoriented steels. IEEE Trans. Magn. **39**, 3028–3030 (2003)
6. Azzerboni, B., Carpentieri, M., Finocchio, G., Ipsale, M.: Super-Lorentzian Preisach function and its applicability to model scalar hysteresis. Phys. B: Condens. Matter **343**, 121–126 (2004)
7. Ball, B.L., Smith, R.C., Kim, S.J., Seelecke, S.: A stress-dependent hysteresis model for ferroelectric materials. J. Intell. Mater. Syst. Struct. **18**(1), 69–88 (2007)

8. Bassiouny, E., Ghaleb, A.F., Maugin, G.A.: Thermodynamical formulation for coupled electromechanical hysteresis effects - I. Basic equations. IEEE Trans. Ultrason. Ferroelectr. Freq. Control **26**(12), 1279–1295 (1988)

9. Belov, A.Y., Kreher, W.S.: Viscoplastic models for ferroelectric ceramics. J. Eur. Ceram. Soc. **25**(12), 2567–2571 (2005)

10. Bergqvist, A., Engdahl, G.: A stress-dependent magnetic Preisach hysteresis model. IEEE Trans. Magn. **27**(6 pt 2), 4796–4798 (1991)

11. Bertotti, G.: Dynamic generalization of the scalar Preisach model of hysteresis. IEEE Trans. Magn. **28**(5), 2599–2601 (1992)

12. Bi, S., Sutor, A., Lerch, R., Xiao, Y.: An efficient inverted hysteresis model with modified switch operator and differentiable weight function. IEEE Trans. Magn. **49**(7), 3175–3178 (2013)

13. Boddu, V., Endres, F., Steinmann, P.: Molecular dynamics study of ferroelectric domain nucleation and domain switching dynamics. Sci. Rep. **7**(1) (2017)

14. Chaplya, P.M., Carman, G.P.: Dielectric and piezoelectric response of lead zirconate-lead titanate at high electric and mechanical loads in terms of non-180° domain wall motion. J. Appl. Phys. **90**(10), 5278–5286 (2001)

15. Chonan, S., Jiang, Z., Yamamoto, T.: Nonlinear hysteresis compensation of piezoelectric ceramic actuators. J. Intell. Mater. Syst. Struct. **7**(2), 150–156 (1996)

16. Cohen, R.E.: Origin of ferroelectricity in perovskite oxides. Nature **358**(6382), 136–138 (1992)

17. Cohen, R.E.: Theory of ferroelectrics: A vision for the next decade and beyond. J. Phys. Chem. Solids **61**(2), 139–146 (2000)

18. Damjanovic, D.: Ferroelectric, dielectric and piezoelectric properties of ferroelectric thin films and ceramics. Rep. Prog. Phys **61**, 1267–1324 (1998)

19. Davino, D., Giustiniani, A., Visone, C.: Design and test of a stress-dependent compensator for magnetostrictive actuators. IEEE Trans. Magn. **2**, 646–649 (2010)

20. Della Torre, E., Vajda, F.: Parameter identification of the complete-moving-hysteresis model using major loop data. IEEE Trans. Magn. **30**(6), 4987–5000 (1994)

21. Devonshire, A.F.: Theory of ferroelectrics. Adv. Phys. **3**(10), 85–130 (1954)

22. Dlala, E., Saitz, J., Arkkio, A.: Inverted and forward Preisach models for numerical analysis of electromagnetic field problems. IEEE Trans. Magn. **42**, 1963–1973 (2006)

23. Dong, R., Tan, Y.: A modified Prandtl-Ishlinskii modeling method for hysteresis. Phys. B: Condens. Matter **404**(8–11), 1336–1342 (2009)

24. Ducharne, B., Zhang, B., Guyomar, D., Sebald, G.: Fractional derivative operators for modeling piezoceramic polarization behaviors under dynamic mechanical stress excitation. Sens. Actuators A: Phys. **189**, 74–79 (2013)

25. Endres, F., Steinmann, P.: Molecular statics simulations of head to head and tail to tail nanodomains of rhombohedral barium titanate. Comput. Mater. Sci. **97**, 20–25 (2015)

26. Everett, D.H.: A general approach to hysteresis. Part 4. An alternative formulation of the domain model. Trans. Faraday Soc. **51**, 1551–1557 (1955)

27. Finocchio, G., Carpentieri, M., Cardelli, E., Azzerboni, B.: Analytical solution of Everett integral using Lorentzian Preisach function approximation. J. Magn. Magn. Mater. **300**, 451–470 (2006)

28. Fleming, A.J.: Charge drive with active DC stabilization for linearization of piezoelectric hysteresis. EEE Trans. Ultrason. Ferroelectr. Freq. Control **60**(8), 1630–1637 (2013)

29. Fleming, A.J., Moheimani, S.O.R.: Improved current and charge amplifiers for driving piezoelectric loads, and issues in signal processing design for synthesis of shunt damping circuits. J. Intell. Mater. Syst. Struct. **15**(2), 77–92 (2004)

30. Freeman, A.R., Joshi, S.P.: Numerical modeling of PZT nonlinear electromechanical behavior. Proc. SPIE - Int. Soc. Opt. Eng. **2715**, 602–613 (1996)

31. Füzi, J.: Computationally efficient rate dependent hysteresis model. COMPEL - Int. J. Comput. Math. Electr. Electron. Eng. **18**(3), 445–457 (1999)

32. Ge, P., Jouaneh, M.: Generalized preisach model for hysteresis nonlinearity of piezoceramic actuators. J. Precis. Eng. **20**, 99–111 (1997)
33. Gu, G., Zhu, L.: Modeling of rate-dependent hysteresis in piezoelectric actuators using a family of ellipses. Sens. Actuators A: Phys. **165**(2), 303–309 (2011)
34. Guyomar, D., Ducharne, B., Sebald, G.: Dynamical hysteresis model of ferroelectric ceramics under electric field using fractional derivatives. J. Phys. D: Appl. Phys. **40**(19), 6048–6054 (2007)
35. Guyomar, D., Ducharne, B., Sebald, G., Audiger, D.: Fractional derivative operators for modeling the dynamic polarization behavior as a function of frequency and electric field amplitude. EEE Trans. Ultrason. Ferroelectr. Freq. Control **56**(3), 437–443 (2009)
36. Hegewald, T.: Modellierung des nichtlinearen Verhaltens piezokeramischer Aktoren. Ph.D. thesis, Friedrich-Alexander-University Erlangen-Nuremberg (2007)
37. Hegewald, T., Kaltenbacher, B., Kaltenbacher, M., Lerch, R.: Efficient modeling of ferroelectric behavior for the analysis of piezoceramic actuators. J. Intell. Mater. Syst. Struct. **19**(10), 1117–1129 (2008)
38. Hu, H., Ben Mrad, R.: On the classical Preisach model for hysteresis in piezoceramic actuators. Mechatronics **13**(2), 85–94 (2002)
39. Huber, J.E.: Micromechanical modelling of ferroelectrics. Curr. Opin. Solid State Mater. Sci. **9**(3), 100–106 (2005)
40. Huber, J.E., Fleck, N.A., Landis, C.M., McMeeking, R.M.: A constitutive model for ferroelectric polycrystals. J. Mech. Phys. Solids **47**, 1663–1697 (1999)
41. Hughes, D., Wen, J.T.: Preisach modeling of piezoceramic hysteresis; independent stress effect. Math. Control Smart Struct. **2442**, 328–336 (1995)
42. Hwang, S.C., Lynch, C.S., McMeeking, R.M.: Ferroelectric/ferroelastic interactions and a polarization switching model. Acta Metall. Mater. **43**(5), 2073–2084 (1995)
43. Janocha, H., Kuhnen, K.: Real-time compensation of hysteresis and creep in piezoelectric actuators. Sens. Actuators A: Phys. **79**(2), 83–89 (2000)
44. Jiang, H., Ji, H., Qiu, J., Chen, Y.: A modified Prandtl-Ishlinskii model for modeling asymmetric hysteresis of piezoelectric actuators. EEE Trans. Ultrason. Ferroelectr. Freq. Control **57**(5), 1200–1210 (2010)
45. Jiles, D.C., Atherton, D.L.: Theory of ferromagnetic hysteresis. J. Magn. Magn. Mater. **61**(1), 48–60 (1986)
46. Jung, H., Shim, J.Y., Gweon, D.: New open-loop actuating method of piezoelectric actuators for removing hysteresis and creep. Rev. Sci. Instrum. **71**(9), 3436–3440 (2000)
47. Kadota, Y., Morita, T.: Preisach modeling of electric-field-induced strain of ferroelectric material considering 90° domain switching. Jpn. J. Appl. Phys. **51**(9), 09LE08–1–09LE081–6 (2012)
48. Kahler, G.R., Della Torre, E., Cardelli, E.: Implementation of the Preisach-Stoner-Wohlfarth classical vector model. IEEE Trans. Magn. **46**(1), 21–28 (2010)
49. Kaltenbacher, B., Kaltenbacher, M.: Modeling and iterative identification of hysteresis via Preisach operators in PDEs. Lect. Adv. Comput. Methods Mech. **1**, 1–45 (2007)
50. Kamlah, M.: Ferroelectric and ferroelastic piezoceramics - modeling of electromechanical hysteresis phenomena. Contin. Mech. Thermodyn. **13**, 219–268 (2001)
51. Kamlah, M., Böhle, U.: Finite element analysis of piezoceramic components taking into account ferroelectric hysteresis behavior. Int. J. Solids Struct. **38**(4), 605–633 (2001)
52. Kamlah, M., Tsakmakis, C.: Phenomenological modeling of the non-linear electromechanical coupling in ferroelectrics. Int. J. Solids Struct. **36**(5), 669–695 (1999)
53. Keip, M.A.: Modeling of electro-mechanically coupled materials on multiple scales. Ph.D. thesis, Universität Duisburg-Essen (2011)
54. Knuth, D.E.: The Art of Computer Programming: vol. 3. Addison-Wesley, Sorting and Searching (1998)
55. Krasnosel'skii, M.A., Pokrovskii, A.V.: Systems with Hysteresis. Springer, Berlin (1989)
56. Kurzhöfer, I.: Mehrskalen-Modellierung polykristalliner Ferroelektrika. Ph.D. thesis, Universität Duisburg-Essen (2007)

57. Landis, C.M.: Fully coupled, multi-axial, symmetric constitutive laws for polycrystalline ferroelectric ceramics. J. Mech. Phys. Solids **50**(1), 127–152 (2002)
58. Lenk, A., Ballas, R.G., Werthschützky, R., Pfeiefer, G.: Electromechanical Systems in Microtechnology and Mechatronics: Electrical, Mechanical and Acoustic Networks, their Interactions and Applications. Springer, Berlin (2010)
59. Lin, C.J., Yang, S.R.: Precise positioning of piezo-actuated stages using hysteresis-observer based control. Mechatronics **16**(7), 417–426 (2006)
60. Ma, Y., Mao, J.: On modeling and tracking control for a smart structure with stress-dependent hysteresis nonlinearity. Acta Autom. Sin. **36**(11), 1611–1619 (2010)
61. Macki, J.W., Nistri, P., Zecca, P.: Mathematical models for hysteresis. SIAM Rev. **35**(1), 94–123 (1993)
62. Mayergoyz, I.D.: Hysteresis models from the mathematical and control theory points of view. J. Appl. Phys. **57**(8), 3803–3805 (1985)
63. Mayergoyz, I.D.: Dynamic preisach models of hysteresis. IEEE Trans. Magn. **24**(6), 2925–2927 (1988)
64. Mayergoyz, I.D.: Mathematical Models of Hysteresis and their Applications. Elsevier, New York (2003)
65. Meggitt Sensing Systems: Product portfolio (2018). https://www.meggittsensingsystems.com
66. Mittal, S., Menq, C.H.: Hysteresis compensation in electromagnetic actuators through preisach model inversion. IEEE/ASME Trans. Mechatron. **5**(4), 394–409 (2000)
67. Mrad, R.B., Hu, H.: A model for voltage-to-displacement dynamics in piezoceramic actuators subject to dynamic-voltage excitations. IEEE/ASME Trans. Mechatron. **7**(4), 479–489 (2002)
68. Nierla, M., Sutor, A., Rupitsch, S.J., Kaltenbacher, M.: Stageless evaluation for a vector Preisach model based on rotational operators. COMPEL - Int. J. Comput. Math. Electr. Electron. Eng. **36**(5), 1501–1516 (2017)
69. Oppermann, K., Arminger, B.R., Zagar, B.G.: A contribution to the classical scalar preisach hysteresis model for magneto–elastic materials. In: Proceedings of IEEE/ASME International Conference on Mechatronics and Embedded Systems and Applications (MESA), pp. 180–185. IEEE (2010)
70. Padthe, A.K., Drincic, B., Oh, J.H., Rizos, D.D., Fassois, S.D., Bernstein, D.S.: Duhem modeling of friction-induced hysteresis. IEEE Control Syst. Mag. **28**(5), 90–107 (2008)
71. PI Ceramic GmbH: Product portfolio (2018). https://www.piceramic.com
72. Polomoff, N.A., Premnath, R.N., Bosse, J.L., Huey, B.D.: Ferroelectric domain switching dynamics with combined 20 nm and 10 ns resolution. J. Mater. Sci. **44**(19), 5189–5196 (2009)
73. Preisach, F.: über die magnetische Nachwirkung. Zeitschrift für Physik **94**(5–6), 277–302 (1935)
74. Quant, M., Elizalde, H., Flores, A., Ramírez, R., Orta, P., Song, G.: A comprehensive model for piezoceramic actuators: modelling, validation and application. Smart Mater. Struct. **18**(12), 1–16 (2009)
75. Rakotondrabe, M., Clévy, C., Lutz, P.: Complete open loop control of hysteretic, creeped, and oscillating piezoelectric cantilevers. IEEE Trans. Autom. Sci. Eng. **7**(3), 440–450 (2010)
76. Reiländer, U.: Das Großsignalverhalten piezoelektrischer Aktoren. Ph.D. thesis, Technische Universität München (2003)
77. Richter, H., Misawa, E.A., Lucca, D.A., Lu, H.: Modeling nonlinear behavior in a piezoelectric actuator. Precis. Eng. **25**(2), 128–137 (2001)
78. Rupitsch, S.J., Wolf, F., Sutor, A., Lerch, R.: Reliable modeling of piezoceramic materials utilized in sensors and actuators. Acta Mech. **223**, 1809–1821 (2012)
79. Sawyer, C.B., Tower, C.H.: Rochelle salt as a dielectric. Phys. Rev. **35**(3), 269–273 (1930)
80. Schrade, D.: Microstructural modeling of ferroelectric material behavior. Ph.D. thesis, Technische Universität Kaiserslautern (2011)
81. Schröder, J., Romanowski, H.: A simple coordinate invariant thermodynamic consistent model for nonlinear electro-mechanical coupled ferroelectrica. In: Proceedings of European Congress on Computational Methods in Applied Science and Engineering (ECCOMAS) (2004)

82. Schwaab, H., Grünbichler, H., Supancic, P., Kamlah, M.: Macroscopical non-linear material model for ferroelectric materials inside a hybrid finite element formulation. Int. J. Solids Struct. **49**(3–4), 457–469 (2012)
83. Sepliarsky, M., Asthagiri, A., Phillpot, S.R., Stachiotti, M.G., Migoni, R.L.: Atomic-level simulation of ferroelectricity in oxide materials. Curr. Opin. Solid State Mater. Sci. **9**(3), 107–113 (2005)
84. Smith, R.C., Hatch, A.G., Mukherjee, B., Liu, S.: A homogenized energy model for hysteresis in ferroelectric materials: General density formulation. J. Intell. Mater. Syst. Struct. **16**(9), 713–732 (2005)
85. Song, D., Li, C.J.: Modeling of piezo actuator's nonlinear and frequency dependent dynamics. Mechatronics **9**(4), 391–410 (1999)
86. Stancu, A., Ricinschi, D., Mitoseriu, L., Postolache, P., Okuyama, M.: First-order reversal curves diagrams for the characterization of ferroelectric switching. Appl. Phys. Lett. **83**(18), 3767–3769 (2003)
87. Stoleriu, L., Stancu, A., Mitoseriu, L., Piazza, D., Galassi, C.: Analysis of switching properties of porous ferroelectric ceramics by means of first-order reversal curve diagrams. Phys. Rev. B **74**(17), 174,107 (2006)
88. Sutor, A., Bi, S., Lerch, R.: Identification and verification of a Preisach-based vector model for ferromagnetic materials. Appl. Phys. A: Mater. Sci. Process. **118**(3), 939–944 (2014)
89. Sutor, A., Rupitsch, S.J., Lerch, R.: A Preisach-based hysteresis model for magnetic and ferroelectric hysteresis. Appl. Phys. A: Mater. Sci. Process. **100**, 425–430 (2010)
90. Szabó, Z.: Preisach functions leading to closed form permeability. Phys. B: Condens. Matter **372**(1), 61–67 (2006)
91. Takahashi, N., Miyabarn, S.I., Fojiwara, K.: Problems in practical finite element analysis using preisach hysteresis model. IEEE Trans. Magn. **35**, 1243–1246 (1999)
92. Tan, X., Baras, J.S.: Modeling and control of hysteresis in magnetostrictive actuators. Automatica **40**(9), 1469–1480 (2004)
93. Tellini, B., Bologna, M., Pelliccia, D.: A new analytic approach for dealing with hysteretic materials. IEEE Trans. Magn. **41**(1), 2–7 (2005)
94. Visintin, A.: Differential Models of Hysteresis. Springer, Berlin (1994)
95. Viswamurthy, S.R., Ganguli, R.: Modeling and compensation of piezoceramic actuator hysteresis for helicopter vibration control. Sens. Actuators A: Phys. **135**(2), 801–810 (2007)
96. Völker, B.: Phase-field modeling for ferroelectrics in a multi-scale approach. Ph.D. thesis, Karlsruher Institut für Technologie (KIT) (2010)
97. Völker, B., Marton, P., Elsässer, C., Kamlah, M.: Multiscale modeling for ferroelectric materials: A transition from the atomic level to phase-field modeling. Contin. Mech. Thermodyn. **23**(5), 435–451 (2011)
98. Wang, D.H., Zhu, W.: A phenomenological model for pre-stressed piezoelectric ceramic stack actuators. Smart Mater. Struct. **20**(3), 035,018 (2011)
99. Webster, J.G.: The Measurement, Instrumentation, and Sensors Handbook. CRC Press, Boca Raton (1999)
100. Wen, Y.K.: Method for random vibration of hysteretic systems. J. Eng. Mech. Div. **102**(2), 249–263 (1976)
101. Wolf, F.: Generalisiertes Preisach-Modell für die Simulation und Kompensation der Hysterese piezokeramischer Aktoren. Ph.D. thesis, Friedrich-Alexander-University Erlangen-Nuremberg (2014)
102. Wolf, F., Hirsch, H., Sutor, A., Rupitsch, S.J., Lerch, R.: Efficient compensation of nonlinear transfer characteristics for piezoceramic actuators. In: Proceedings of Joint IEEE International Symposium on Applications of Ferroelectric and Workshop on Piezoresponse Force Microscopy (ISAF-PFM), pp. 171–174 (2013)
103. Wolf, F., Sutor, A., Rupitsch, S.J., Lerch, R.: Modeling and measurement of hysteresis of ferroelectric actuators considering time-dependent behavior. Procedia Eng. **5**, 87–90 (2010)
104. Wolf, F., Sutor, A., Rupitsch, S.J., Lerch, R.: Modeling and measurement of creep- and rate-dependent hysteresis in ferroelectric actuators. Sens. Actuators A: Phys. **172**, 245–252 (2011)

105. Wolf, F., Sutor, A., Rupitsch, S.J., Lerch, R.: Modeling and measurement of influence of mechanical prestress on hysteresis of ferroelectric actuators. Procedia Eng. **25**, 1613–1616 (2011)
106. Wolf, F., Sutor, A., Rupitsch, S.J., Lerch, R.: A generalized Preisach approach for piezoceramic materials incorporating uniaxial compressive stress. Sens. Actuators A: Phys. **186**, 223–229 (2012)
107. Xu, Q., Li, Y.: Dahl model-based hysteresis compensation and precise positioning control of an xy parallel micromanipulator with piezoelectric actuation. J. Dyn. Syst. Meas. Control **132**(4), 1–12 (2010)
108. Yi, K.A., Veillette, R.J.: A charge controller for linear operation of a piezoelectric stack actuator. IEEE Trans. Control Syst. Technol. **13**(4), 517–526 (2005)
109. Zhou, D., Kamlah, M.: Room-temperature creep of soft PZT under static electrical and compressive stress loading. Acta Mater. **54**(5), 1389–1396 (2006)
110. Zhou, D., Kamlah, M., Munz, D.: Effects of uniaxial prestress on the ferroelectric hysteretic response of soft PZT. J. Eur. Ceram. Soc. **25**(4), 425–432 (2005)

Chapter 7
Piezoelectric Ultrasonic Transducers

An ultrasonic transducer is a device, which generates acoustic waves above audible frequencies (i.e., $f > 20\,\text{kHz}$) from electrical inputs and provides electrical outputs for incident ultrasonic waves. Such transducers are used in various applications like medical diagnostics, parking assistance systems as well as nondestructive testing. Mostly, ultrasonic transducers are based on piezoelectricity. They contain piezoelectric materials (e.g., piezoceramics) allowing efficient conversions of electrical quantities into acoustic waves and vice versa. The resulting devices are usually called *piezoelectric ultrasonic transducers.*

In principle, we distinguish between two fundamental operation modes of ultrasonic transducers containing a single piezoelectric element, namely the (i) pulse-echo mode and the (ii) pitch-catch mode (see Fig. 7.1) [4]. The *pulse-echo mode* is based on one transducer, which enables both emitting ultrasonic waves and receiving reflections from a target. Thereby, the transducer emits ultrasonic waves due to the electrical excitation u_I and, subsequently, converts reflected waves into the electrical output u_O. By contrast, the *pitch-catch mode* requires two ultrasonic transducers. The first transducer exclusively emits ultrasonic waves, while the second transducer receives reflected, refracted, or transmitted waves. That is the reason why the first and second transducers are commonly also referred to as transmitter and receiver, respectively.

There exist various piezoelectric ultrasonic transducers, whereby their internal structure strongly depends on the propagation medium of ultrasonic waves. For instance, one can exploit so-called interdigital transducers for generating waves that travel along the surface of a solid, i.e., for generating surface acoustic waves (SAW; cf. Fig. 9.1 on p. 408) also known as Rayleigh waves. Interdigital transducers consist of interlocking comb-shaped arrays of metallic electrodes on the surface of a piezoelectric substrate. Such devices, which are oftentimes termed surface acoustic wave sensors, can be applied to determine chemical conditions, temperature, and mechanical quantities (e.g., [4, 42, 58]).

© Springer-Verlag GmbH Germany, part of Springer Nature 2019   261
S. J. Rupitsch, *Piezoelectric Sensors and Actuators*, Topics in Mining, Metallurgy and Materials Engineering, https://doi.org/10.1007/978-3-662-57534-5_7

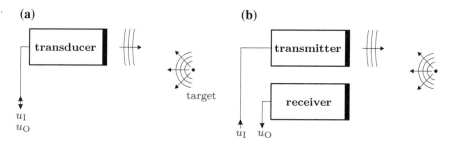

Fig. 7.1 Fundamental operation modes of piezoelectric ultrasonic transducers containing single piezoelectric element; **a** pulse-echo mode; **b** pitch-catch mode; transmitter and receiver correspond to transducer that exclusively generates and receives ultrasonic waves, respectively

This chapter addresses piezoelectric ultrasonic transducers that are specially designed to generate and receive sound waves in fluid media, i.e., air or water. Moreover, we will also discuss ultrasonic transducers for medical diagnostics. Section 7.1 deals with a semi-analytical approach to calculate sound fields and electrical transducer outputs for common transducer shapes, e.g., piston-type transducers. In doing so, the complex structure of a piezoelectric ultrasonic transducer is reduced to an active surface, which can generate and receive sound pressure waves. The semi-analytical approach will be used in Sect. 7.2 to determine sound fields as well as directional characteristics of common transducers. Section 7.3 details the axial and lateral spatial resolution of spherically focused transducers operating in pulse-echo mode. In Sect. 7.4, we will study the general structure of piezoelectric ultrasonic transducers. This includes single-element transducers, transducer arrays as well as piezoelectric composite transducers. Afterward, a simple one-dimensional modeling approach is shown that allows analytical description of basic physical relationships for piezoelectric transducers under consideration of their internal structure. Section 7.6 contains selected examples for piezoelectric ultrasonic transducers. Finally, a brief introduction to the fundamental imaging modes of ultrasonic imaging will be given which is an important application of piezoelectric ultrasonic transducers.

7.1 Calculation of Sound Fields and Electrical Transducer Outputs

The cost-effective development and optimization of ultrasonic transducers demand the prediction of the generated sound fields. Moreover, it will be very helpful to predict electrical transducer outputs if an ultrasonic transducer is utilized for receiving sound pressure waves. For both tasks, one can apply finite element (FE) simulations because such simulations allow the consideration of the whole configuration including the piezoelectric material (see Chap. 4). Although coupled FE simulations yield the desired quantities (e.g., generated sound pressure) directly from electrical

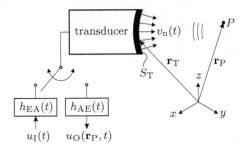

Fig. 7.2 Active surface S_T of transducer parameterized by $\mathbf{r}_T$ oscillates with surface normal velocity $v_n(t)$; ideal point-like target P at $\mathbf{r}_P$ [63]; electrical transducer input $u_I(t)$ and output $u_O(\mathbf{r}_P, t)$; electroacoustical impulse response $h_{EA}(t)$; acousto-electrical impulse response $h_{AE}(t)$

transducer inputs, the required spatial and temporal discretizations are oftentimes accompanied by a remarkable computational effort. Especially in case of large configurations, which additionally call for a three-dimensional model, this problem gains in importance. Therefore, let us present here a semi-analytical approach that enables time-efficient calculations of generated sound fields and electrical transducer outputs. Before the so-called *spatial impulse response* of a transducer is introduced in Sect. 7.1.2, we will discuss the underlying approach on the basis of sound diffraction at an ideal point-like target, which is located in a fluid propagation medium. Sections 7.1.3 and 7.1.4 deal with piecewise analytical solutions of the spatial impulse response for piston-type transducers and spherically focused transducers, respectively.

7.1.1 Diffraction at Point-Like Target

Let us consider an ultrasonic transducer being located in a rigid baffle [41, 59, 63]. The active surface S_T of the transducer[1] is parameterized by the vector $\mathbf{r}_T$ (see Fig. 7.2). It generates in and receives acoustic waves from a nonviscous as well as lossless fluid medium that features the sound velocity (speed of sound) c_0 and the equilibrium density ϱ_0. The fluid propagation medium contains an ideal point-like target at position $\mathbf{r}_P$. This point-like target can be interpreted as a rigid sphere, which is small compared to the wavelength λ of generated acoustic waves and exhibits the reflection coefficient $r_p = 1$ for incident acoustic waves. In a first step, the transducer is excited by the time-dependent electrical input $u_I(t)$. We suppose that $u_I(t)$ causes exclusively deformations of the active transducer surface in direction of its normal vectors. By means of the electroacoustical impulse response $h_{EA}(t)$, the surface normal velocity $v_n(t)$ can be expressed as (time t; temporal convolution $*$)

---

[1]Following technical optics, the active transducer surface S_T is also referred to as aperture.

$$v_n(t) = h_{EA}(t) * u_I(t) \ . \tag{7.1}$$

Note that this equation will only hold if the normal velocity over the active surface is assumed to be uniform; i.e., each element has the same instantaneous velocity. To generalize (7.1), it is possible to insert a location-dependent weighting function $w_E(\mathbf{r}_T)$ for generating sound waves, which leads to the surface normal velocity

$$V_n(\mathbf{r}_T, t) = w_E(\mathbf{r}_T) \cdot v_n(t) \tag{7.2}$$

depending on both position on S_T and time. When $w_E(\mathbf{r}_T)$ is real-valued and fulfills $w_E(\mathbf{r}_T) \geq 0$, the whole active transducer surface will move in phase.

According to the Huygens–Fresnel principle, we may model S_T as combination of point sources that emit spherical waves in the half space (e.g., [20, 26]). The resulting Rayleigh surface integral for the acoustic velocity potential $\Psi_T(\mathbf{r}_P, t)$ at position $\mathbf{r}_P$ takes the form

$$\Psi_T(\mathbf{r}_P, t) = \int_{S_T} \frac{V_n\left(\mathbf{r}_T, t - \frac{|\mathbf{r}_P - \mathbf{r}_T|}{c_0}\right)}{2\pi |\mathbf{r}_P - \mathbf{r}_T|} \, dS_T(\mathbf{r}_T) \ . \tag{7.3}$$

The sound pressure $p_{I\sim}(\mathbf{r}_P, t)$ at this position directly follows from (cf. (2.123), p. 34)

$$p_{I\sim}(\mathbf{r}_P, t) = \varrho_0 \frac{\partial \Psi_T(\mathbf{r}_P, t)}{\partial t} \ . \tag{7.4}$$

Using the properties of the Dirac delta distribution $\delta(\cdot)$ as well as inserting (7.2) and (7.3) into (7.4) then yields

$$p_{I\sim}(\mathbf{r}_P, t) = \varrho_0 v_n(t) * \frac{\partial}{\partial t} \left[\int_{S_T} w_E(\mathbf{r}_T) \frac{\delta\left(t - \frac{|\mathbf{r}_P - \mathbf{r}_T|}{c_0}\right)}{2\pi |\mathbf{r}_P - \mathbf{r}_T|} \, dS_T(\mathbf{r}_T) \right] \ . \tag{7.5}$$

As mentioned above, we consider a rigid sphere at $\mathbf{r}_P$ serving as ideal point-like target, which perfectly reflects the incident sound pressure waves. Of course, one can treat such target as a point source that emits spherical waves. The acoustic velocity potential $\Psi_P(\mathbf{r}, t)$ originating from this point source becomes

$$\Psi_P(\mathbf{r}, t) \approx \frac{s_P v_P\left(\mathbf{r}_P, t - \frac{|\mathbf{r} - \mathbf{r}_P|}{c_0}\right)}{4\pi |\mathbf{r} - \mathbf{r}_P|} \quad \text{with} \quad v_P(\mathbf{r}_P, t) \approx \frac{p_{I\sim}(\mathbf{r}_P, t)}{\varrho_0 c_0} \tag{7.6}$$

at the position $\mathbf{r}$. The expressions s_P and v_P stand for the surface and the velocity of the sphere's surface, respectively. Strictly speaking, the relations (7.6) solely represent approximations because the sphere surface is not excited cophasally from the incident sound pressure wave. However, when the diameter of the sphere is much smaller than λ, the phase deviations will be negligible. Just like in (7.4), the sound pressure $p_{R\sim}(\mathbf{r}, t)$ of the reflected sound wave at $\mathbf{r}$ results from the time derivative of (7.6). By additionally using $\delta(\cdot)$, $p_{R\sim}(\mathbf{r}, t)$ reads as

$$
p_{R\sim}(\mathbf{r}, t) = \frac{s_P}{2c_0} \frac{\partial}{\partial t} \left[p_{I\sim}(\mathbf{r}_P, t) * \frac{\delta\left(t - \frac{|\mathbf{r} - \mathbf{r}_P|}{c_0}\right)}{2\pi |\mathbf{r} - \mathbf{r}_P|} \right] .
$$
(7.7)

The considered ultrasonic transducer serves as transmitter and receiver of acoustic waves. Similar to emitting sound waves, each point of the active transducer surface S_T is assumed to be able to receive pressure waves from the entire half space. To compute the resulting electrical output $u_O(\mathbf{r}_P, t)$ of the transducer due to the point-like target, one has to evaluate the sound pressure at S_T. The averaged sound pressure $\overline{p}_{T\sim}(\mathbf{r}_P, t)$ along S_T is given by (surface area $|S_T|$)

$$
\overline{p}_{T\sim}(\mathbf{r}_P, t) = \frac{1}{|S_T|} \int_{S_T} w_R(\mathbf{r}_T)\, p_{R\sim}(\mathbf{r}_T, t)\, dS_T(\mathbf{r}_T) .
$$
(7.8)

Here, $w_R(\mathbf{r}_T)$ denotes a location-dependent weighting function for receiving sound waves. If (7.5) and (7.7) are inserted into (7.8), one will obtain

$$
\overline{p}_{T\sim}(\mathbf{r}_P, t) = \frac{s_P \varrho_0 v_n(t)}{2c_0 |S_T|} * \frac{\partial}{\partial t} \left[\int_{S_T} w_E(\mathbf{r}_T) \frac{\delta\left(t - \frac{|\mathbf{r}_P - \mathbf{r}_T|}{c_0}\right)}{2\pi |\mathbf{r}_P - \mathbf{r}_T|} dS_T(\mathbf{r}_T) \right]
$$
$$
* \frac{\partial}{\partial t} \left[\int_{S_T} w_R(\mathbf{r}_T) \frac{\delta\left(t - \frac{|\mathbf{r}_T - \mathbf{r}_P|}{c_0}\right)}{2\pi |\mathbf{r}_T - \mathbf{r}_P|} dS_T(\mathbf{r}_T) \right] .
$$
(7.9)

The acousto-electrical impulse response $h_{AE}(t)$ finally leads to the electrical output of the ultrasonic transducer

$$
u_O(\mathbf{r}_P, t) = h_{AE}(t) * \overline{p}_{T\sim}(\mathbf{r}_P, t) .
$$
(7.10)

7.1.2   Spatial Impulse Response (SIR)

In order to achieve a compact formulation of the previous equations, let us introduce the spatial impulse response (SIR) of the ultrasonic transducer. Without limiting the generality, we suppose that the location-dependent weighting functions for generating and receiving sound waves are identical, i.e., $w_{ER}(\mathbf{r}_T) = w_E(\mathbf{r}_T) = w_R(\mathbf{r}_T)$ along S_T [41, 59, 63]. Therewith, the spatial impulse response $h_{SIR}(\mathbf{r}, t)$ of the ultrasonic transducer for the generated velocity potential $\Psi_T(\mathbf{r}, t)$ at position $\mathbf{r}$ becomes (cf. (7.3))

$$h_{\text{SIR}}(\mathbf{r}, t) = \int_{S_{\text{T}}} w_{\text{ER}}(\mathbf{r}_{\text{T}}) \frac{\delta\left(t - \frac{|\mathbf{r} - \mathbf{r}_{\text{T}}|}{c_0}\right)}{2\pi\,|\mathbf{r} - \mathbf{r}_{\text{T}}|} \mathrm{d}S_{\text{T}}(\mathbf{r}_{\text{T}}) \ . \qquad (7.11)$$

It can be clearly seen that the scalar quantity $h_{\text{SIR}}(\mathbf{r}, t)$ with unit ms^{-1} depends on both space and time, which justifies the naming. The temporal convolution of the SIR and the time-dependent normal velocity $v_{\text{n}}(t)$ of the active transducer surface S_{T} yields $\Psi_{\text{T}}(\mathbf{r}_{\text{P}}, t)$ at the position $\mathbf{r}_{\text{P}}$ of the point-like target, i.e.,

$$\Psi_{\text{T}}(\mathbf{r}_{\text{P}}, t) = v_{\text{n}}(t) * h_{\text{SIR}}(\mathbf{r}_{\text{P}}, t) \ . \qquad (7.12)$$

Owing to the mathematical relation (7.4), $h_{\text{SIR}}(\mathbf{r}_{\text{P}}, t)$ can also be utilized for calculating the incident sound pressure $p_{\text{I}\sim}(\mathbf{r}_{\text{P}}, t)$

$$p_{\text{I}\sim}(\mathbf{r}_{\text{P}}, t) = \varrho_0 v_{\text{n}}(t) * \frac{\partial h_{\text{SIR}}(\mathbf{r}_{\text{P}}, t)}{\partial t} \qquad (7.13)$$

at this position. Moreover, it is possible to directly exploit $h_{\text{SIR}}(\mathbf{r}_{\text{P}}, t)$ in (7.9), which yields

$$\overline{p}_{\text{T}\sim}(\mathbf{r}_{\text{P}}, t) = \frac{s_{\text{P}}\varrho_0 v_{\text{n}}(t)}{2c_0\,|S_{\text{T}}|} * \frac{\partial h_{\text{SIR}}(\mathbf{r}_{\text{P}}, t)}{\partial t} * \frac{\partial h_{\text{SIR}}(\mathbf{r}_{\text{P}}, t)}{\partial t} \qquad (7.14)$$

for the averaged sound pressure along S_{T} due to sound reflections at the ideal point-like target. By inserting (7.1) and (7.14) into (7.10), we end up with

$$u_{\text{O}}(\mathbf{r}_{\text{P}}, t) = \frac{s_{\text{P}}\varrho_0}{2c_0\,|S_{\text{T}}|} h_{\text{EA}}(t) * h_{\text{AE}}(t)$$
$$* \frac{\partial h_{\text{SIR}}(\mathbf{r}_{\text{P}}, t)}{\partial t} * \frac{\partial h_{\text{SIR}}(\mathbf{r}_{\text{P}}, t)}{\partial t} * u_{\text{I}}(t) \ . \qquad (7.15)$$

The SIR of an ultrasonic transducer allows, thus, the calculation of the generated sound pressure $p_{\text{I}\sim}(\mathbf{r}, t)$ at an arbitrary point in a homogeneous propagation medium. Furthermore, we achieve a compact description of the resulting electrical transducer output $u_{\text{O}}(\mathbf{r}_{\text{P}}, t)$ for an ideal point-like target that is located within the propagation medium. In both cases, one can start from the electrical excitation signal $u_{\text{I}}(t)$ of the transducer since according to (7.1), the normal velocity $v_{\text{n}}(t)$ of S_{T} is directly linked to $u_{\text{I}}(t)$. The underlying computations of $p_{\text{I}\sim}(\mathbf{r}, t)$ as well as $u_{\text{O}}(\mathbf{r}_{\text{P}}, t)$ are highly efficient, especially when there exist analytical solutions for h_{SIR}. Such solutions can be found in the literature for some shapes of the active transducer surface S_{T}, e.g., piston-type transducers [26, 74], spherically focused transducers [1, 53], and for transducers with rectangular-shaped active surfaces [68]. Besides, Jensen and Svendsen [33] suggest dividing the active surface into small pieces in order to apply the SIR for arbitrarily shaped ultrasonic transducers. In Sects. 7.1.3 and 7.1.4, we will address analytical formulations for the spatial impulse responses of piston-type and spherically focused transducers, respectively.

As a matter of fact, the SIR of an ultrasonic transducer is not restricted to a single point-like target but also enables compact descriptions of the electrical transducer output for a finite-sized solid structure that is immersed in a homogeneous fluid propagation medium. In doing so, one should divide the structure's surface into a sufficient amount of single elements [34, 61, 63]. By evaluating $h_{SIR}(\mathbf{r}, t)$ for these elements separately (cf. (7.15)) and superimposing the individual results, we are able to compute $u_O(t)$ due to the surface part, which is facing the transducer. Strictly speaking, this procedure will make only sense if we are exclusively interested in transducer outputs originating from the structure's surface. Reflections from inhomogeneities (e.g., flaws) and mode conversions within the solid structure cannot be taken into account because the SIR in its classical form supposes a homogeneous propagation medium. Note that this assumption is already violated at an interface of homogeneous fluid and homogeneous solid. When we want to study reflections as well as mode conversions within the solid structure, alternative approaches like the FE method (see Chap. 4) or hybrid simulations such as the so-called SIRFEM [48, 62] will be indispensable.

7.1.3   SIR of Piston-Type Transducer

Many ultrasonic transducers in practical applications (e.g., parking sensors) have a circular active surface S_T. Since this planar aperture often oscillates like a piston, let us detail the analytical solution of the SIR for so-called piston-type transducers. Stepanishen [74] and Harris [26] deduced a piecewise continuous solution of the SIR for piston-type transducers, which are assumed to oscillate and receive uniformly, i.e., $w_{ER}(\mathbf{r}_T) = 1$ along S_T. Consequently, one can treat the piston-type transducer as a rotationally symmetric configuration. Moreover, each point P in the sound propagation medium is completely described by means of the two coordinates ρ and z (see Fig. 7.3). While ρ refers to the off-axis distance with respect to the center of S_T, z stands for the on-axis distance of the point P.

Fig. 7.3 Configuration and geometrical variables to compute piecewise continuous spatial impulse response $h_{SIR}(\rho, z, t)$ for piston-type transducer at point P (coordinates ρ and z); radius R_T of active transducer area S_T

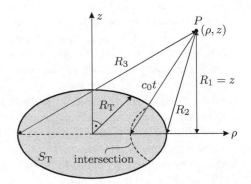

The piecewise continuous solution of the SIR $h_{SIR}(\rho, z, t)$ results from the intersection of a sphere surface (radius $c_0 t$; origin P) with S_T. For a circular active surface of radius R_T, $h_{SIR}(\rho, z, t)$ in P reads as (sound velocity c_0)

$$
h_{SIR}(\rho, z, t) = \begin{cases} 0 & \text{for } t \leq t_1 \\ c_0 & \text{for } t_1 < t \leq t_2 \\ \dfrac{c_0}{\pi} \arccos\left[\dfrac{(c_0 t)^2 - z^2 + \rho^2 - R_T^2}{2\rho\sqrt{(c_0 t)^2 - z^2}} \right] & \text{for } t_2 < t \leq t_3 \\ 0 & \text{for } t_3 < t \end{cases} \tag{7.16}
$$

if $\rho \leq R_T$ and otherwise (i.e., $\rho > R_T$)

$$
h_{SIR}(\rho, z, t) = \begin{cases} 0 & \text{for } t \leq t_2 \\ \dfrac{c_0}{\pi} \arccos\left[\dfrac{(c_0 t)^2 - z^2 + \rho^2 - R_T^2}{2\rho\sqrt{(c_0 t)^2 - z^2}} \right] & \text{for } t_2 < t \leq t_3 \\ 0 & \text{for } t_3 < t \end{cases} . \tag{7.17}
$$

The instants of time t_i are given by (see Fig. 7.3)

$$
t_1 = \frac{R_1}{c_0} = \frac{z}{c_0}
$$

$$
t_2 = \frac{R_2}{c_0} = \frac{\sqrt{z^2 + (\rho - R_T)^2}}{c_0}
$$

$$
t_3 = \frac{R_3}{c_0} = \frac{\sqrt{z^2 + (\rho + R_T)^2}}{c_0} .
$$

Figure 7.4 depicts normalized spatial impulse responses $h_{SIR}(\rho, z, t)/c_0$ for a piston-type transducer, which generates sound pressure waves in a fluid propagation medium. In Fig. 7.4a and b, the observer point P is located at the on-axis distance $z = R_T$ and $z = 6R_T$, respectively. The off-axis distances ρ of P were selected from the values 0, $2R_T/3$, and $4R_T/3$. The curves clearly demonstrate the piecewise continuous solution of $h_{SIR}(\rho, z, t)$ with respect to time. If $h_{SIR}(\rho, z, t) \neq 0$ is fulfilled, $h_{SIR}(\rho, z, t) = c_0$ at on-axis points, and thus, the SIR will be constant. At points being located off-axis, $h_{SIR}(\rho, z, t)$ also takes values according to the partial solution containing the arccosine function.

The analytical solution of $h_{SIR}(\rho, z, t)$ in (7.16) is, however, only valid for uniformly oscillating and receiving piston-type transducers. Harris [27] extended this solution with a view to considering an arbitrary location-dependent weighting function $w_{ER}(\mathbf{r}_T)$ along S_T. The resulting generalized formulation will provide again a piecewise continuous solution of the SIR when $w_{ER}(\mathbf{r}_T)$ is rotationally symmetric with respect to the center of S_T. Moreover, $w_{ER}(\mathbf{r}_T)$ has to be defined by a specific mathematical function, e.g., a Gaussian function. Alternatively, one may divide S_T

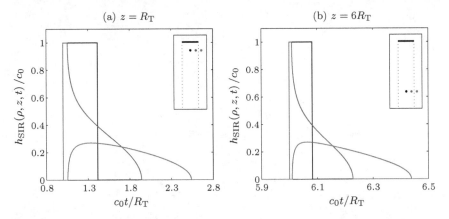

Fig. 7.4 Normalized spatial impulse responses $h_{SIR}(\rho, z, t)/c_0$ for piston-type transducer of radius R_T at selected points P; **a** P at on-axis distance $z = R_T$; **b** P at on-axis distance $z = 6R_T$; off-axis distance $\rho \in \{0, 2R_T/3, 4R_T/3\}$; insets assign colors of $h_{SIR}(\rho, z, t)$ and explain positions of P with respect to S_T

of the piston-type transducer into concentric annuli that possess individual weights, respectively [1, 63]. The SIR $h_{SIR,a}(\rho, z, t)$ for a single annulus with inner radius R_i and outer radius R_o follows from

$$h_{SIR,a}(\rho, z, t) = h_{SIR}(\rho, z, t)|_{R_T=R_o} - h_{SIR}(\rho, z, t)|_{R_T=R_i} . \tag{7.18}$$

In doing so, it is possible to spatially discretize the weighting function $w_{ER}(\mathbf{r}_T)$ along S_T.

7.1.4 SIR of Spherically Focused Transducer

Besides piston-type ultrasonic transducers, spherically focused transducers are frequently utilized in practical applications because such transducers concentrate the sound energy in a small region that is named focal volume. This fact is especially desirable for imaging systems, which are based on acoustic waves (e.g., acoustic microscopes [43, 60, 82]). Spherically focused transducers feature a concave active surface S_T, which can be interpreted as a section of a sphere surface (see Fig. 7.5). The section is parameterized by its radial size R_T and its radius F_T of curvature representing the geometrical focus of the configuration. That is the reason why F_T is commonly termed as geometrical focal length of a spherically focused transducer. The depth H_T of S_T is given by

$$H_T = F_T - \sqrt{F_T^2 - R_T^2} . \tag{7.19}$$

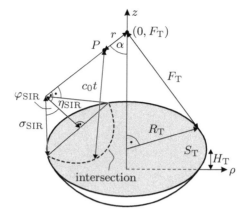

Fig. 7.5 Configuration and geometrical variables to compute piecewise continuous spatial impulse response $h_{SIR}(\rho, z, t)$ for spherically focused transducer at point P (coordinates ρ and z) [53]; radial size R_T of active transducer area S_T; geometrical focal length F_T

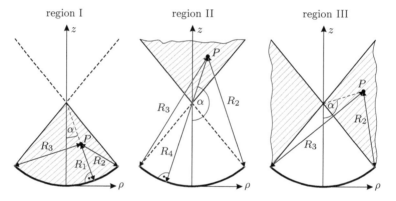

Fig. 7.6 Geometrical variables for calculating $h_{SIR}(\rho, z, t)$; point P (except geometrical focus) within propagation medium lies in one of three regions; hatched area indicates region

Penttinen and Luukkala [53] and Arditi et al. [1] deduced a piecewise continuous solution of the SIR for gently curved spherically focused transducers. Let us detail their solution hereinafter. Just as for piston-type transducers, it is supposed that each point of S_T oscillates and receives uniformly, i.e., $w_{ER}(\mathbf{r}_T) = 1$. Therefore, S_T of the spherically focused transducer is rotationally symmetric. Two coordinates (i.e., off-axis distance ρ as well as on-axis distance z) are sufficient again to completely describe the position of a point P in the sound propagation medium. Contrary to piston-type transducers, we are, however, not able to determine piecewise continuous solutions of the SIR $h_{SIR}(\rho, z, t)$ for $\rho \leq R_T$ and $\rho > R_T$, respectively. With the exception of the geometrical focus at $(\rho, z) = (0, F_T)$, one has to divide the propagation medium into three regions (see Fig. 7.6). Nevertheless, $h_{SIR}(\rho, z, t)$ always results from the intersection of a sphere surface (radius $c_0 t$; origin P) with S_T and

Table 7.1 Piecewise continuous solution for angle $\varphi_{SIR}(t)$ with respect to region and time t

$\varphi_{SIR}(t)$	Region I	Region II	Region III
$= 0$	$t \leq t_1$	$t \leq t_2$	$t \leq t_2$
$= \pi$	$t_1 < t \leq t_2$	$t_3 \leq t < t_4$	—
$= \arccos\left[\dfrac{\eta_{SIR}(t)}{\sigma_{SIR}(t)}\right]$	$t_2 < t < t_3$	$t_2 < t < t_3$	$t_2 < t < t_3$
$= 0$	$t_3 \leq t$	$t_4 \leq t$	$t_3 \leq t$

takes the form

$$h_{SIR}(\rho, z, t) = \frac{c_0 F_T}{\pi \sqrt{\rho^2 + (F_T - z)^2}} \varphi_{SIR}(t) \ . \tag{7.20}$$

Table 7.1 contains for each region the piecewise continuous solution for the angle $\varphi_{SIR}(t)$ between adjacent $\eta_{SIR}(t)$ and hypotenuse $\sigma_{SIR}(t)$ of the right-angled triangle, which is shown in Fig. 7.5. These quantities and the distance r become

$$\eta_{SIR}(t) = F_T\left[\frac{F_T - H_T}{F_T \sin\alpha} - \frac{F_T^2 + r^2 - (c_0 t)^2}{2 F_T r \tan\alpha}\right] \tag{7.21}$$

$$\sigma_{SIR}(t) = F_T \sqrt{1 - \left[\frac{F_T^2 + r^2 - (c_0 t)^2}{2 F_T r}\right]^2} \tag{7.22}$$

$$r = \sqrt{\rho^2 + (F_T - z)^2} \tag{7.23}$$

with the angle α (see Fig. 7.5)

$$\alpha = \begin{cases} 0 & \text{for } t \leq t_1 \\ \arctan\left(\dfrac{\rho}{F_T - z}\right) & \text{for } z < F_T \\ \pi - \arctan\left(\dfrac{\rho}{z - F_T}\right) & \text{for } z > F_T \\ \dfrac{\pi}{2} & \text{else} \end{cases} \ .$$

The required instants of time t_i in Table 7.1 follow from (see Fig. 7.6)

$$t_1 = \frac{R_1}{c_0} = \frac{F_T - \sqrt{\rho^2 + (F_T - z)^2}}{c_0}$$

$$t_1 = \frac{R_2}{c_0} = \frac{\sqrt{(R_T - \rho)^2 + (z - H_T)^2}}{c_0}$$

$$t_3 = \frac{R_3}{c_0} = \frac{\sqrt{(R_T + \rho)^2 + (z - H_T)^2}}{c_0}$$

$$t_4 = \frac{R_4}{c_0} = \frac{F_T + \sqrt{\rho^2 + (F_T - z)^2}}{c_0}.$$

When the coordinates ρ and z of P satisfy the condition

$$\frac{\rho}{|F_T - z|} < \frac{R_T}{\sqrt{F_T^2 - R_T^2}}, \tag{7.24}$$

P will lie either in region I if $z < F_T$ or in region II if $z > F_T$ is fulfilled. Otherwise, P is located in region III. At the geometrical focus of S_T, the SIR is defined as

$$h_{SIR}(0, F_T, t) = H_T \delta\left(t - \frac{F_T}{c_0}\right). \tag{7.25}$$

Figure 7.7 displays normalized spatial impulse responses $h_{SIR}(\rho, z, t)/c_0$ for a spherically focused transducer featuring radial size $R_T = 5\,\text{mm}$ and geometrical focal length $F_T = 20\,\text{mm}$. Due to the chosen normalization, the curve progression solely depends on the ratio R_T/F_T, which takes here the value $1/4$. In Fig. 7.7a and b, the observer point P in the fluid propagation medium is located at the on-axis distance $z = F_T/3$ and $z = 7F_T/3$, respectively. The off-axis distances ρ of P were selected from the values 0, $R_T/2$, and R_T. Basically, the SIR $h_{SIR}(\rho, z, t)$ of the spherically focused transducer behaves similar to that of the piston-type transducer (cf. Fig. 7.4). When $h_{SIR}(\rho, z, t) \neq 0$, the focusing property of S_T will cause, however, varying values of $h_{SIR}(\rho, z, t)$ at points P being located on-axis, i.e., $\rho = 0$.

Guided by [27], Verhoef et al. [79] extended the piecewise analytical solution of $h_{SIR}(\rho, z, t)$ for spherically focused transducers to take into account an arbitrary location-dependent weighting $w_{ER}(\mathbf{r}_T)$ along S_T. They describe the rotationally symmetric progression of $w_{ER}(\mathbf{r}_T)$ through a polynomial being defined by even-numbered exponents. To become more flexible, one can also divide S_T into concentric annuli that possesses individual weights (cf. (7.18)).

7.2 Sound Fields and Directional Characteristics

The generated sound field and directional characteristic represent decisive quantities of ultrasonic transducers concerning practical applications like medical diagnostics. Below, we will concentrate on calculated quantities for piston-type (see Sect. 7.2.1) as well as spherically focused transducers (see Sect. 7.2.2) since it is possible to demonstrate essential facts through those specific transducer shapes. This includes distributions of sound pressure and acoustic intensity in space as well as the resulting directivity pattern of ultrasonic transducers.

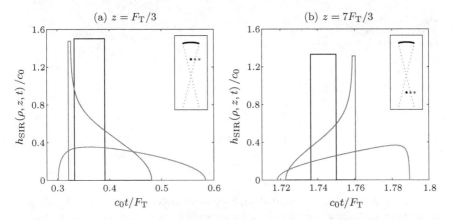

Fig. 7.7 Normalized spatial impulse responses $h_{SIR}(\rho, z, t)/c_0$ for spherically focused transducer of radial size $R_T = 5\,\text{mm}$ and geometric focal length $F_T = 20\,\text{mm}$ at selected points P; **a** P at on-axis distance $z = F_T/3$; **b** P at on-axis distance $z = 7F_T/3$; off-axis distance $\rho \in \{0, R_T/2, R_T\}$; insets assign colors of $h_{SIR}(\rho, z, t)$ and explain positions of P with respect to S_T

For the following calculations, let us assume that the whole active surface S_T of the transducer oscillates sinusoidally at frequency f, uniformly and in phase, i.e., $w_E(\mathbf{r}_T) = 1$ along S_T. The resulting normal velocity $v_n(t)$ can then be written as

$$v_n(t) = \hat{v}_n \cos(\omega t) = \hat{v}_n \Re\{e^{j\omega t}\} \tag{7.26}$$

with $\hat{v}_n$ and $\omega = 2\pi f$ denoting velocity amplitude and angular frequency, respectively. By exploiting the spatial impulse response $h_{SIR}(\mathbf{r}, t)$ of the considered transducer, we can compute the generated sound pressure $p_\sim(\mathbf{r}, t)$ at position $\mathbf{r}$ from (7.13), which leads to

$$
\begin{aligned}
p_\sim(\mathbf{r}, t) &= \varrho_0 \hat{v}_n \Re\left\{ e^{j\omega t} * \frac{\partial h_{SIR}(\mathbf{r}, t)}{\partial t} \right\} \\
&= \omega \varrho_0 \hat{v}_n \Re\left\{ j e^{j\omega t} * h_{SIR}(\mathbf{r}, t) \right\} .
\end{aligned}
\tag{7.27}
$$

The complex representation $\underline{p}_\sim(\mathbf{r}, t)$ of the generated sound pressure takes the form

$$\underline{p}_\sim(\mathbf{r}, t) = j\omega \varrho_0 \hat{v}_n e^{j\omega t} * h_{SIR}(\mathbf{r}, t) . \tag{7.28}$$

Owing to the arccosine functions in the piecewise continuous solutions for the SIR (e.g., (7.16)), there does, however, not exist an analytical solution to (7.27) at each position $\mathbf{r}$. Nevertheless, the smooth progression of h_{SIR} allows a simple numerical evaluation of the temporal convolution by discretizing the relevant time interval, e.g., $[t_2, t_3]$ in (7.16).

Fig. 7.8 Directivity pattern
$\hat{p}_\sim(\theta)$ of ultrasonic
transducer results from sound
pressure amplitudes along
circumference of circle with
radius R_{dir}; rotationally
symmetric config-
uration (e.g., piston-type
transducer): $\rho = R_{\mathrm{dir}} \sin \theta$
and $z = R_{\mathrm{dir}} \cos \theta$

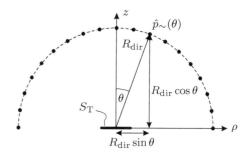

The computed sound pressure distribution and directivity pattern of a transducer result from variations of **r**. In case of rotationally symmetric piston-type and spherically focused transducers, this means that one has to vary both off-axis distance ρ and on-axis distance z in $h_{\mathrm{SIR}}(\rho, z, t)$ accordingly. Usually, the sound pressure distribution relates to the sound pressure amplitudes $\hat{p}_\sim = |\underline{p}_\sim|$ in a prescribed plane, e.g., $\hat{p}_\sim(x, z)$ in the xz-plane. In contrast, the directivity pattern of an ultrasonic transducer is typically defined as the sound pressure amplitudes $\hat{p}_\sim(\theta)$ along the circumference of a circle, whose center is located in the middle of S_{T}. The position onto this circle with radius R_{dir} is parameterized by the angle θ (see Fig. 7.8).

Besides calculations that are based on the SIR, we will additionally discuss common approximations for on-axis sound fields and off-axis sound fields of piston-type as well as spherically focused transducers.

7.2.1 Piston-Type Transducer

Figure 7.9 depicts normalized sound pressure distributions $\hat{p}_\sim(x, z)$ for piston-type transducers. The panels differ in the product of wave number $k = \omega/c_0 = 2\pi/\lambda_{\mathrm{aco}}$ and radius R_{T} of the active transducer surface S_{T}, i.e., kR_{T}. Due to the chosen normalization of x- and z-axis, the sound pressure distributions exclusively depend on this product. Small values for kR_{T} (e.g., 1 and 2) indicate that the wavelength λ_{aco} of the emitted sound waves is large compared to the geometric dimensions of S_{T}. In such cases, the piston-type transducer emits almost spherical sound waves and, thus, behaves like a point source (cf. Fig. 7.10). Larger values for kR_{T} (e.g., 10 and 20) cause, however, focused sound fields, even though we are considering a planar piston-type transducer. Moreover, there arise several local minima as well as maxima, especially close to S_{T}. Both the focusing behavior and the formation of such local extrema are an immediate consequence of the model assumption that each point of S_{T} emits spherical waves, which get superimposed in the sound propagation medium. Depending on the position (x, z), those superimpositions can be accompanied by destructive and constructive interferences that become visible in sound pressure fields through local minima and maxima, respectively. The greater kR_{T}, the larger R_{T} will

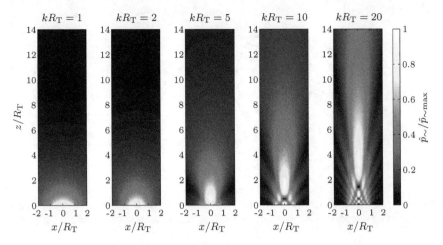

Fig. 7.9 Normalized sound pressure distribution $\hat{p}_\sim(x, z)$ in xz-plane of piston-type transducer for various products kR_T; normalization with respect to maximum amplitude $\hat{p}_{\sim\text{max}}$ of directivity pattern

be compared to λ_{aco}, and therefore, the number of destructive as well as constructive interferences increases because the amount of wave trains per unit length ($\widehat{=}k$) also increases.

In Fig. 7.10, one can see normalized directivity patterns $\hat{p}_\sim(\theta)$ of a piston-type transducer, which were determined along the circumference of a circle in the transducer's far field, i.e., $R_{\text{dir}} \gg R_T$. We can draw the same conclusions as for the sound pressure distributions in Fig. 7.9. While the piston-type transducer will act as point source when the product kR_T takes small values, the directivity is strongly pronounced for large values of this product. Additionally, it becomes apparent that there will arise certain side lobes if $kR_T \gg 1$, which also exist far away from the transducer. A greater value of kR_T entails more side lobes in the directivity pattern.

On-Axis Sound Field

Now, let us take a closer look at the generated sound pressure amplitudes $\hat{p}_\sim(z)$ on the symmetry axis (i.e., z-axis) of a piston-type transducer. For such so-called on-axis sound field, one can find simple approximations in the literature (e.g., [35, 40]), which will be deduced in the following. The starting point here is the acoustic velocity potential $\underline{\Psi}_T(z, t)$ in complex representation

$$\underline{\Psi}(z, t) = \int_{S_T} e^{j\omega t} \underbrace{\frac{\hat{v}_n e^{-jkr}}{2\pi r}}_{G(r)} \, dS_T \tag{7.29}$$

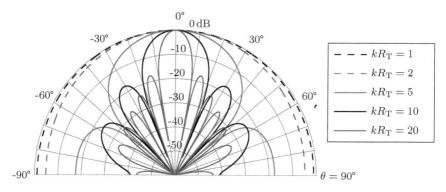

Fig. 7.10 Normalized directivity pattern $20 \cdot \log_{10}\left(\hat{p}_{\sim}(\theta) / \hat{p}_{\sim\mathrm{max}}\right)$ in far field of piston-type transducer for various products $k\,R_{\mathrm{T}}$; normalization with respect to maximum amplitude $\hat{p}_{\sim\mathrm{max}}$

with Green's function $G(r)$. The expression $r = \sqrt{\rho^2 + z^2}$ stands for the distance between an element $\mathrm{d}S_{\mathrm{T}}$ of the active transducer surface S_{T} and the considered on-axis point P (see Fig. 7.11). Since $\mathrm{e}^{\mathrm{j}\omega t}$ accounts for the time dependence, this term will often be omitted in the literature when the physical quantity (e.g., sound pressure) features sinusoidal progression. The reduced version of (7.29) reads then as

$$\underline{\Psi}(z) = \frac{1}{2\pi} \int_{S_{\mathrm{T}}} \frac{\hat{v}_{\mathrm{n}} \mathrm{e}^{-\mathrm{j}kr}}{r} \mathrm{d}S_{\mathrm{T}} \tag{7.30}$$

and will be absolutely sufficient if we are only interested in the amplitude $\hat{\Psi}(z) = \left|\underline{\Psi}(z)\right|$. For a piston-type transducer that behaves rotationally symmetric, (7.30) simplifies to

$$\begin{aligned}
\underline{\Psi}(z) &= \frac{\hat{v}_{\mathrm{n}}}{2\pi} \int_0^{2\pi} \mathrm{d}\varphi \int_0^{R_{\mathrm{T}}} \frac{\mathrm{e}^{-\mathrm{j}r}}{r} \rho \, \mathrm{d}\rho \\
&= \frac{\hat{v}_{\mathrm{n}}}{\mathrm{j}k}\left(\mathrm{e}^{-\mathrm{j}kz} - \mathrm{e}^{-\mathrm{j}k\sqrt{R_{\mathrm{T}}^2 + z^2}}\right) .
\end{aligned} \tag{7.31}$$

It is possible to interpret the expression inside the bracket as follows: While the first term describes a plane wave, which propagates perpendicular to S_{T}, the second one stems from waves arising at the edge of S_{T}. Both waves are superimposed resulting in various local minima and maxima along the z-axis, i.e., destructive and constructive interference, respectively. By means of the mathematical identity

$$\mathrm{e}^{\mathrm{j}\alpha} - \mathrm{e}^{\mathrm{j}\beta} = 2\mathrm{j} \sin\left(\frac{\alpha - \beta}{2}\right) \mathrm{e}^{\mathrm{j}(\alpha+\beta)/2} ,$$

(7.31) takes the form

Fig. 7.11 Configuration and geometrical variables to compute the on-axis sound field (i.e., along z-axis) of piston-type transducer with active surface S_T

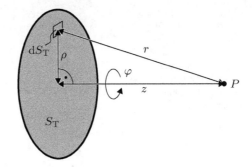

$$\underline{\Psi}(z) = \frac{2\hat{v}_n}{k} \sin\left[\frac{k}{2}\left(\sqrt{R_T^2 + z^2} - z\right)\right] e^{-\frac{jk}{2}\left(\sqrt{R_T^2 + z^2} + z\right)}. \tag{7.32}$$

The relation $\Psi(z, t) = \Re\{\underline{\Psi}(z)\, e^{j\omega t}\}$ leads finally to the time-dependent sound pressure values $p_\sim(z, t)$

$$p_\sim(z, t) = \varrho_0 \frac{\partial \Psi(z, t)}{\partial t} \tag{7.33}$$

$$= \frac{2\varrho_0 \omega \hat{v}_n}{k} \sin\left[\frac{k}{2}\left(\sqrt{R_T^2 + z^2} - z\right)\right] \Re\left\{ je^{j\omega t} e^{-\frac{jk}{2}\left(\sqrt{R_T^2 + z^2} + z\right)} \right\}$$

and with $k = \omega/c_0$ to the sound pressure amplitude $\hat{p}_\sim(z)$

$$\hat{p}_\sim(z) = 2\varrho_0 c_0 \hat{v}_n \left| \sin\left[\frac{k}{2}\left(\sqrt{R_T^2 + z^2} - z\right)\right]\right| \tag{7.34}$$

along the z-axis. Note that the spatial impulse response $h_{SIR}(\rho, z, t)$ of the piston-type transducer will, of course, yield the same results for $p_\sim(z, t)$ and $\hat{p}_\sim(z)$ [59].

Figure 7.12 depicts normalized values for the sound pressure amplitudes $\hat{p}_\sim(z)$ along the z-axis in case of different values kR_T (cf. Fig. 7.9). According to (7.34), the curves will exhibit local minima at z_{min} if

$$\frac{k}{2}\left(\sqrt{R_T^2 + z_{min}^2} - z_{min}\right) = n\pi \quad \forall\, n \in \mathbb{N}_+ \tag{7.35}$$

and local maxima at z_{max} if

$$\frac{k}{2}\left(\sqrt{R_T^2 + z_{max}^2} - z_{max}\right) = \frac{2n - 1}{2}\pi \quad \forall\, n \in \mathbb{N}_+ \tag{7.36}$$

is fulfilled. Consequently, z_{min} and z_{max} are given by

Fig. 7.12 Normalized sound pressure amplitudes $\hat{p}_\sim(z)$ along z-axis (i.e., on-axis) of piston-type transducer for various products kR_T; exact and approximated curves from (7.34) and (7.40), respectively

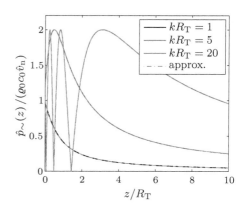

$$z_{min} = \frac{1}{4kn\pi}\left[(kR_T)^2 - (2n\pi)^2\right] \tag{7.37}$$

$$z_{max} = \frac{1}{2k(2n-1)\pi}\left[(kR_T)^2 - (2n-1)^2\pi^2\right] . \tag{7.38}$$

On this account, there will exist at least one local minimum along the z-axis when $kR_T > 2\pi$ and at least one local maximum when $kR_T > \pi$ is satisfied.

From (7.38), we can deduce with $k = 2\pi/\lambda_{aco}$ that the last maximum (i.e., $n = 1$) of $\hat{p}_\sim(z)$ along the z-axis arises at

$$N_{near} = \frac{1}{2k\pi}\left[(kR_T)^2 - \pi^2\right] = \frac{(2R_T)^2 - \lambda_{aco}^2}{4\lambda_{aco}} \approx \frac{R_T^2}{\lambda_{aco}} , \tag{7.39}$$

which is usually known as *near-field length* or natural focal length of a piston-type transducer. The region $z < N_{near}$ in the propagation medium is called near field or, alternatively, Fresnel zone. Beyond the near field, there occur neither local minima nor local maxima, and thus, the acoustic quantities (e.g., sound pressure) decrease monotonically. If $z \gg R_T$ additionally holds, we can exploit the Taylor approximation $\sqrt{1+x} = 1 + \frac{x}{2}$ to estimate the sound pressure amplitude (see (7.34)) on the z-axis by

$$\hat{p}_\sim(z) \approx 2\varrho_0 c_0 \hat{v}_n \left|\sin\left(\frac{kR_T^2}{4z}\right)\right| \tag{7.40}$$

and the averaged acoustic intensity[2] $\overline{I}_{aco}(z) = \left\|\overline{\mathbf{I}}_{aco}(z)\right\|_2$ by

$$\overline{I}_{aco}(z) \approx 4\overline{I}_{aco}(0)\left[\sin\left(\frac{kR_T^2}{4z}\right)\right]^2 . \tag{7.41}$$

---

[2]Averaging refers to one sine period in the time domain (see Sect. 2.3.1).

Fig. 7.13 Configuration and geometrical variables to compute the off-axis sound field in transducer's far field; active surface S_T of piston-type transducer

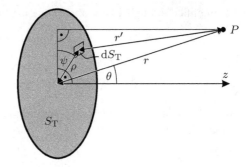

Here, $\overline{I}_{aco}(0) \approx Z_{aco}\hat{v}_n^2/2$ stands for the averaged acoustic intensity at S_T, whereas $Z_{aco} = \varrho_0 c_0$ is the acoustic impedance of the propagation medium. Figure 7.12 clearly illustrates that these approximations seem to be reasonable for the entire on-axis sound field. Both approximations can be simplified further to

$$\hat{p}_\sim(z) \approx \varrho_0 c_0 \hat{v}_n \frac{k R_T^2}{2z} \quad \text{and} \quad \overline{I}_{aco}(z) \approx \overline{I}_{aco}(0)\left(\frac{k R_T^2}{2z}\right)^2 \qquad (7.42)$$

when $z \gg N_{near}$ is also fulfilled which relates to the far field of the piston-type transducer. As a result, the sound pressure amplitude is inversely proportional to the on-axis distance z, while the averaged acoustic intensity decreases quadratically with increasing z.

Off-Axis Sound Field in Far Field

Contrary to on-axis sound fields, we cannot deduce an analytical solution for off-axis sound fields of piston-type transducers due to the complicated sound pressure distribution, which is especially present in the near field (cf. Fig. 7.9). However, it is possible to conduct a reliable approximation for the off-axis sound field in the far field (i.e., $z \gg N_{near}$) of the transducer. The approximation follows from the acoustic velocity potential $\underline{\Psi}(r, \theta)$ (reduced version; cf. (7.30))

$$\underline{\Psi}(r, \theta) = \frac{1}{2\pi}\int_{S_T} \frac{\hat{v}_n e^{-jkr'}}{r'} dS_T = \frac{1}{2\pi}\int_{\rho=0}^{R_T}\int_{\psi=0}^{2\pi} \frac{\hat{v}_n e^{-jkr'}}{r'}\rho\, d\psi\, d\rho \qquad (7.43)$$

with the geometric distance (see Fig. 7.13)

$$r'^2 = r^2 + \rho^2 - 2r\rho\sin\theta\cos\psi \ . \qquad (7.44)$$

In case of large distances r from the active transducer surface S_T (i.e., $r \gg R_T$ and $r \gg \rho$), which is satisfied in the far field of a piston-type transducer, (7.44) can be simplified through Taylor approximation to

$$r' = r\sqrt{1 + \left(\frac{\rho}{r}\right)^2 - \frac{2\rho}{r} \sin\theta \cos\psi}$$
$$\approx r - \rho \sin\theta \cos\psi .$$ (7.45)

By replacing $1/r'$ with $1/r$ and inserting the approximation for r' into (7.43), one obtains the relation

$$\underline{\Psi}(r,\theta) \approx \frac{\hat{v}_n e^{-jkr}}{2\pi r} \int\limits_{\rho=0}^{R_T} \int\limits_{\psi=0}^{2\pi} e^{jk\rho \sin\theta \cos\psi} \rho \, d\psi \, d\rho$$ (7.46)

that is integrable in closed form. After some mathematical treatment [24], we end up with

$$\underline{\Psi}(r,\theta) \approx \frac{\hat{v}_n R_T^2 e^{-jkr}}{r} \frac{J_1(kR_T \sin\theta)}{kR_T \sin\theta}$$ (7.47)

for the reduced version of the acoustic velocity potential. The time derivative of (7.47) finally yields the time-dependent sound pressure values $p_\sim(r,\theta,t)$

$$p_\sim(r,\theta,t) \approx \frac{\omega \varrho_0 \hat{v}_n R_T^2}{r} \frac{J_1(kR_T \sin\theta)}{kR_T \sin\theta} \Re\{je^{j\omega t} e^{-jkr}\}$$ (7.48)

and the sound pressure amplitude $\hat{p}_\sim(r,\theta)$

$$\hat{p}_\sim(r,\theta) \approx \frac{\omega \varrho_0 \hat{v}_n R_T^2}{r} \left| \frac{J_1(kR_T \sin\theta)}{kR_T \sin\theta} \right|$$ (7.49)
$$= \frac{k\varrho_0 c_0 \hat{v}_n R_T^2}{r} \left| \frac{J_1(kR_T \sin\theta)}{kR_T \sin\theta} \right| .$$

The approximation of the averaged acoustic intensity $\overline{I}_{aco}(r,\theta)$ becomes

$$\overline{I}_{aco}(r,\theta) \approx \overline{I}_{aco}(0) \left(\frac{kR_T^2}{r}\right)^2 \left[\frac{J_1(kR_T \sin\theta)}{kR_T \sin\theta}\right]^2$$ (7.50)

with $\overline{I}_{aco}(0) \approx Z_{aco} \hat{v}_n^2/2$. For $\theta = 0°$ (i.e., on-axis $r = z$), these results coincide with (7.42). The additional expression $J_1(kR_T \sin\theta)/(kR_T \sin\theta)$ containing the Bessel function $J_1(\cdot)$ of the first kind and order 1 represents a directionality factor that causes side lobes in the directivity patterns of a piston-type transducer (cf. Fig. 7.10). For small values of kR_T, the directionality factor is negligible, and therefore, the transducer behaves in the far field like a point source.

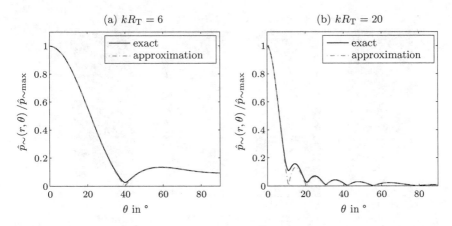

Fig. 7.14 Normalized sound pressure amplitudes $\hat{p}_\sim(r, \theta)$ in far field of piston-type transducer at $r = 10R_T$; **a** product $kR_T = 6$; **b** product $kR_T = 20$; exact and approximated curves from (7.28) and (7.49), respectively

Figure 7.14 displays exact and approximated sound pressure amplitude $\hat{p}_\sim(r, \theta)$ for two piston-type transducers differing in the product kR_T. The distance r between the center of S_T and the observer point equals $10R_T$, which is beyond the near-field length N_{near}. As already mentioned, there does not exist an analytical solution for $\hat{p}_\sim(r, \theta)$, but if (7.28) is numerically evaluated by fine time discretization (e.g., 100 time steps), the obtained results can be considered as exact solution. The comparison of the curves clearly reveals that (7.49) yields precise estimations of the sound pressure amplitudes. For greater values of r, the approximation converges to the exact solution since the applied simplifications (e.g., replacing $1/r'$ with $1/r$) cause less errors. Apart from the excellent match of approximation and exact solution in the far field, Fig. 7.14 demonstrates that the maximum of the first side lobe emerging at

$$\theta = \arcsin\left(\frac{5.14}{kR_T}\right) \tag{7.51}$$

always exhibits the relative value $\hat{p}_\sim(r, \theta) / \hat{p}_{\sim max} = 0.1323$, which corresponds to $-17.6\,\mathrm{dB}$. This fact will be discussed in more detail for spherically focused transducers.

7.2.2 Spherically Focused Transducer

In Fig. 7.15, one can see normalized sound pressure distributions $\hat{p}_\sim(x, z)$ for spherically focused transducers, which differ in the product kR_T. The top panels refer to the ratio $F_T/R_T = 4$ of geometrical focal length F_T and radial size R_T of the active

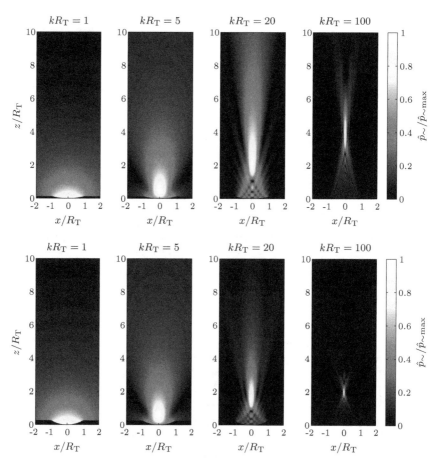

Fig. 7.15 Normalized sound pressure distribution $\hat{p}_\sim(x, z)$ in xz-plane of spherically focused transducer for various products kR_T; top panels for ratio $F_T/R_T = 4$; bottom panels for ratio $F_T/R_T = 2$; normalization with respect to maximum amplitude $\hat{p}_{\sim\text{max}}$

transducer area S_T, whereas the results for $F_T/R_T = 2$ are depicted in the bottom panels. Due to the chosen normalization, the sound pressure amplitudes exclusively depend on the product kR_T and this ratio. Just like piston-type transducers (cf. Fig. 7.9), spherically focused transducers will emit spherical sound waves if kR_T takes small values. For larger values of kR_T (i.e., 20 and 100), focusing becomes effective, and therefore, the sound energy is concentrated within a small area, the so-called focal volume. The greater the product kR_T and the smaller the ratio F_T/R_T, the more pronounced focusing will be. Furthermore, there arise several local minima and maxima between S_T and the focal volume.

According to the directivity pattern in Fig. 7.16, a spherically focused transducer does not offer significant focusing properties in the far field, even when the product kR_T takes large values. This behavior is completely different from that of a

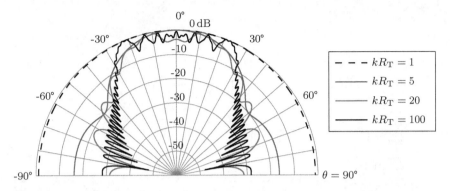

Fig. 7.16 Normalized directivity pattern $20 \cdot \log_{10}\left(\hat{p}_{\sim}(\theta) / \hat{p}_{\sim\mathrm{max}}\right)$ in far field of spherically focused transducer for various products $k R_{\mathrm{T}}$; ratio $F_{\mathrm{T}}/R_{\mathrm{T}} = 2$; normalization with respect to maximum amplitude $\hat{p}_{\sim\mathrm{max}}$ of directivity pattern

piston-type transducer and originates from the focusing property of a spherically focused transducer. Its focal volume can be interpreted as source of sound waves. Especially if $k R_{\mathrm{T}}$ is large and $F_{\mathrm{T}}/R_{\mathrm{T}}$ is small, the focal volume will become small and, consequently, approaches a point source, which emits almost spherical sound waves. Hence, it is not surprising that a spherically focused transducer appears in the far field as source of unfocused sound waves.

Before we will detail generated on-axis sound fields as well as sound fields in the geometrical focal plane, let us introduce two dimensionless quantities, which are inspired by technical optics (e.g., [8, 28]) and can also be found in the context of spherically focused ultrasonic transducers. The first quantity, the so-called *Fresnel parameter* S_{F}, is defined as

$$S_{\mathrm{F}} = \frac{F_{\mathrm{T}} \lambda_{\mathrm{aco}}}{R_{\mathrm{T}}^2} = \frac{2\pi F_{\mathrm{T}}}{k R_{\mathrm{T}}^2} \tag{7.52}$$

and rates the focusing behavior of a transducer. In accordance with the definition, a small value of S_{F} indicates a strong focus. For instance, the upper left and lower right panels in Fig. 7.15 refer to $S_{\mathrm{F}} = 25.1$ and $S_{\mathrm{F}} = 0.1$, respectively. The second quantity, the so-called f-number $f^{\#} = F_{\mathrm{T}}/(2R_{\mathrm{T}})$, solely depends on the geometric dimensions of the active transducer surface S_{T}. Therefore, this quantity does not take into account the emitted sound waves, i.e., their wavelength λ_{aco} or frequency.

On-Axis Sound Field

By exploiting the spatial impulse response $h_{\mathrm{SIR}}(0, z, t)$, one can derive an analytical relation for the time-dependent sound pressure values $p_{\sim}(z, t)$ on the symmetry axis (i.e., z-axis) of a spherically focused transducer [59]. In complex representation, (7.28) leads to

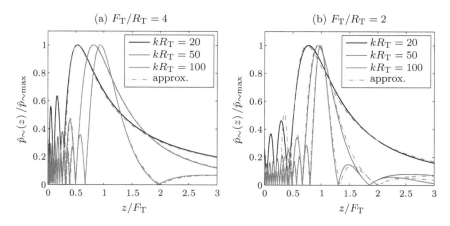

Fig. 7.17 Normalized sound pressure amplitudes $\hat{p}_\sim(z)$ along z-axis (i.e., on-axis) of spherically focused transducer for various products kR_T; **a** ratio $F_T/R_T = 4$; **b** ratio $F_T/R_T = 2$; exact and approximated curves from (7.53) and (7.57), respectively; approximation shown for $z > 0.3F_T$

$$\underline{p}_\sim(z, t) = \begin{cases} \dfrac{\varrho_0 c_0 F_T \hat{v}_n}{F_T - z}\left[e^{j\omega(t-\tau_1)} - e^{j\omega(t-\tau_2)}\right] & \text{for } z \neq F_T \\[2mm] \varrho_0 \hat{v}_n H_T \omega e^{j\omega(t-F_T/c_0)} & \text{for } z = F_T \end{cases} \tag{7.53}$$

with the depth H_T of S_T (see (7.19)), $\tau_1 = z/c_0$ and

$$\tau_2 = \frac{\sqrt{R_T^2 + z^2 - 2zH_T + H_T^2}}{c_0}. \tag{7.54}$$

Similar to piston-type transducers (cf. (7.31)), we can draw the following conclusion: The first expression $e^{j\omega(t-\tau_1)}$ inside the bracket indicates a plane wave and the second one $e^{j\omega(t-\tau_2)}$ a wave arising at the edge of S_T. Both waves are superimposed resulting in various local minima and local maxima along the z-axis.

Figure 7.17 illustrates normalized sound pressure amplitudes $\hat{p}_\sim(z) = |\underline{p}_\sim(z, t)|$ for two ratios F_T/R_T and various products kR_T. As expected from Fig. 7.15, the number of local minima and maxima along the z-axis increases remarkably with rising values kR_T. Even though total constructive interference only exists at the geometrical focus F_T of the spherically focused transducer, the maximum $\hat{p}_{max}$ of the sound pressure amplitudes always occurs between S_T and F_T. For increasing kR_T, this true focus moves toward F_T because of the smaller region for constructive interference close to the geometrical focus.

The curve progressions in Fig. 7.17 also clearly reveal that an increasing ratio F_T/R_T and product kR_T is accompanied by a smaller extension of the main lobe. The larger kR_T for a fixed ratio F_T/R_T, the smaller the Fresnel parameter S_F will be, and consequently, the focusing behavior is more pronounced. Thus, the emitted sound energy is concentrated in a smaller area, whereby $\hat{p}_{max}$ has to rise because of

Fig. 7.18 Normalized maximum $\hat{p}_{\sim\text{max}}$ of sound pressure amplitude along z-axis of spherically focused transducer with respect to product kR_T and ratio F_T/R_T

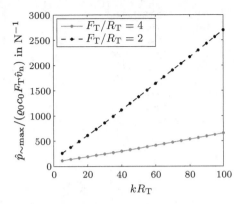

focusing. This fact is confirmed by Fig. 7.18, which shows the normalized maximum of sound pressure amplitude along the z-axis of spherically focused transducers. Already starting from small values of kR_T, $\hat{p}_{\text{max}}$ grows linearly with kR_T. It is not surprising that the increase of $\hat{p}_{\text{max}}$ will be stronger for a decreasing ratio F_T/R_T, i.e., for a lower f-number $f^{\#}$.

Just as for piston-type transducers, one can find approximations for the on-axis sound field of spherically focused transducers (e.g., [35]). The starting point for those approximations is the transformation of the uniform surface normal velocity $v_n(t)$ along the active transducer surface S_T to a planar surface being flush with the surrounding rigid baffle (see Fig. 7.19). Therefore, the axial distance between the origin of S_T and the planar surface corresponds to the depth H_T. By supposing acoustic beams, which propagate perpendicular to S_T, the velocity amplitudes $\hat{v}_z(\theta)$ in z-direction at this surface can be estimated through

$$\hat{v}_z(\theta) \approx \hat{v}_n \frac{F_T}{F_T - R_1} \cos\theta . \tag{7.55}$$

Since the so-called aperture angle θ_0 of a spherically focused transducer yields the trigonometrical relation $F_T - H_T = F_T \cos\theta_0$ and $F_T - R_1 = (F_T - H_T)/\cos\theta$ has to be satisfied, (7.55) becomes

$$\hat{v}_z(\theta) \approx \hat{v}_n \frac{\cos^2\theta}{\cos\theta_0} . \tag{7.56}$$

The expression $\hat{v}_z(\theta)$ can be assumed to serve as nonuniform surface normal velocity of a planar piston-type transducer, which is located at $z = H_T$ and features the radius R_T. It is, therefore, possible to conduct similar steps to approximate sound quantities for spherically focused transducers as for piston-type transducers. Owing to this fact, let us detail solely essential results. If $z^2 \gg R_T^2$ is fulfilled, the curve progression of the normalized sound pressure amplitude $\hat{p}_{\sim}(z)$ along the z-axis can be estimated as

Fig. 7.19 Configuration and geometrical variables to estimate on-axis sound field (i.e., along z-axis) and sound field at geometrical focal plane (i.e., $z = F_T$) of spherically focused transducer with active surface S_T [35]

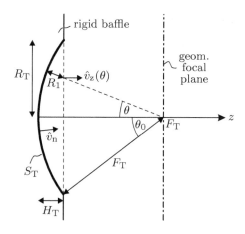

$$\frac{\hat{p}_\sim(z)}{\hat{p}_{\sim \text{max}}} \approx \frac{F_T}{z} \left| \text{sinc} \left[\frac{1}{2 S_F} \left(\frac{F_T}{z} \right) - 1 \right] \right| , \tag{7.57}$$

whereby $\text{sinc}(x) = \sin(\pi x)/(\pi x)$ stands for the sinc function. As Fig. 7.17 points out, this approximation coincides very well with the exact solution, especially close to the maximum of $\hat{p}_\sim(z)$. When $F_T \ll R_T^2/\lambda_{\text{aco}}$ (i.e., $S_F \ll 1$) holds in addition, (7.57) will simplify to

$$\frac{\hat{p}_\sim(z)}{\hat{p}_{\sim \text{max}}} \approx \left| \text{sinc} \left(\frac{z - F_T}{2 S_F F_T} \right) \right| . \tag{7.58}$$

The approximation of the averaged acoustic intensity $\overline{I}_{\text{aco}}(z = F_T)$ at the geometrical focus F_T of the spherically focused transducers reads as

$$\overline{I}_{\text{aco}}(F_T) \approx \overline{I}_{\text{aco}}(0) \left[\frac{\pi R_T^2}{F_T \lambda_{\text{aco}}} \right]^2 = \overline{I}_{\text{aco}}(0) \left[\frac{\pi}{S_F} \right]^2 \tag{7.59}$$

with the averaged acoustic intensity $\overline{I}_{\text{aco}}(0)$ at S_T. Hence, the acoustic intensity is much higher at F_T than at S_T for strongly focusing transducers, i.e., $S_F \ll 1$.

Now, let us derive the so-called depth of focus $d_z(-3\,\text{dB})$ representing the geometric distance between the two points along the z-axis at which the sound pressure amplitude $\hat{p}_\sim(z)$ takes the value $\hat{p}_{\sim \text{max}}/\sqrt{2}$. In order to deduce a simple mathematical relation for $d_z(-3\,\text{dB})$, one should use (7.58). This approximation leads to

$$d_z(-3\,\text{dB}) \approx 1.772 \cdot S_F F_T = 7.089 \cdot \lambda_{\text{aco}} \left(f^\# \right)^2 \tag{7.60}$$

for the depth of focus of a spherically focused transducer, which satisfies the condition $S_F \ll 1$. Accordingly, $d_z(-3\,\text{dB})$ will be rather short if the transducer is focusing sound waves strongly.

Sound Field in Geometrical Focal Plane

In contrast to on-axis sound fields, there does not exist an analytical relation for sound fields in the geometrical focal plane (i.e., $z = F_T$; see Fig. 7.19) of spherically focusing transducers. As it is the case for the far field of piston-type transducers, we can, however, numerically evaluate (7.28). By a sufficiently fine time discretization, this procedure provides again an exact solution for the time-dependent sound pressure values $p_\sim(\rho, t) = p_\sim(\rho, F_T, t)$. Apart from the numerical evaluation, it is possible to estimate such sound fields through the velocity amplitudes $\hat{v}_z(\theta)$ (cf. (7.55)) of a planar surface being flush with the surrounding rigid baffle. After conducting several simplifications that were already shown for piston-type transducers, one ends up with the approximations

$$\hat{p}_\sim(\rho) \approx \hat{p}_{\sim\max} \left| \frac{J_1(\nu)}{\nu} \right| \tag{7.61}$$

$$\overline{I}_{\text{aco}}(\rho, F_T) \approx \overline{I}_{\text{aco}}(0) \left[\frac{\pi R_T^2}{F_T \lambda_{\text{aco}}} \right]^2 \left[\frac{J_1(\nu)}{\nu} \right]^2 \tag{7.62}$$

$$\nu = \frac{\rho k R_T}{F_T} = \frac{2\pi \rho R_T}{\lambda_{\text{aco}} F_T} \tag{7.63}$$

for the sound pressure amplitude $\hat{p}_\sim(\rho)$ and averaged acoustic intensity $\overline{I}_{\text{aco}}(\rho, F_T)$ in the geometrical focal plane, respectively. The expression $\hat{p}_{\sim\max} = \hat{p}_\sim(0, F_T)$ is the maximum sound pressure amplitude in the geometrical focal plane, which always arises at $\rho = 0$ and $\overline{I}_{\text{aco}}(0)$ stands for the averaged acoustic intensity at S_T. The comparison of these approximations with (7.49) and (7.50) makes clear that the sound quantities exhibit identical curve progressions in the far field of piston-type transducers and in the geometrical focal plane of spherically focused transducers.

Figure 7.20 depicts exact as well as approximated sound pressure amplitudes $\hat{p}_\sim(\rho)$ in the geometrical focal plane of two spherically focused transducers, which differ in the ratio F_T/R_T. The deviations between both curve progressions are extremely small in a wide range of products $k R_T$. It can, therefore, be stated that the approximations (7.61) and (7.62) lead to reliable results. On this account, let us utilize the approximations to deduce simple mathematical formulas for important quantities, e.g., the lateral beam width $d_\rho(-3\,\text{dB})$.

As Fig. 7.20 demonstrates, the radial extension of the main lobe in the geometrical focal plane decreases for increasing products $k R_T$ as well as decreasing ratios F_T/R_T. From (7.61) and (7.62), it follows that the zeros and local maxima of $\hat{p}_\sim(\rho)$ and $\overline{I}_{\text{aco}}(\rho, F_T)$ occur at the radial positions

$$\rho_{\text{zero}} = \{3.83; 7.02; 10.17; 13.32; \ldots\} \cdot \frac{F_T}{k R_T}$$

$$\rho_{\max} = \{5.14; 8.42; 11.62; 14.80; \ldots\} \cdot \frac{F_T}{k R_T} \,.$$

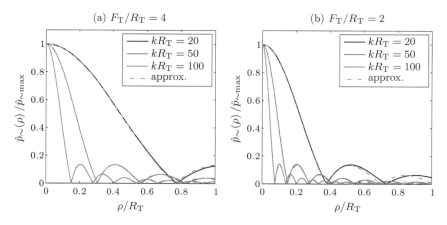

Fig. 7.20 Normalized sound pressure amplitudes $\hat{p}_\sim(\rho)$ in geometrical focal plane of spherically focused transducer for various products kR_T; **a** ratio $F_T/R_T = 4$; **b** ratio $F_T/R_T = 2$; exact and approximated curves from (7.28) and (7.61), respectively

Related to the global maximum $\hat{p}_{\sim\text{max}}$ in the geometrical focal plane, the first four local maxima of $\hat{p}_\sim(\rho)$ take the values

$$\frac{\hat{p}_\sim(\rho_{\text{max}})}{\hat{p}_{\sim\text{max}}} = \{0.1323;\ 0.0645;\ 0.040;\ 0.028\}$$

$$\cong \{-17.6\,\text{dB};\ -23.8\,\text{dB};\ -28.0\,\text{dB};\ -31.1\,\text{dB}\}$$

and, thus, depend neither on kR_T nor on F_T/R_T. This fact naturally applies to all subsequent local maxima in the geometrical focal plane of a spherically focused transducer.

Finally, let us define the so-called lateral beam width (beam diameter) $d_\rho(-3\,\text{dB})$ denoting the geometric distance between the two points in the geometrical focal plane at which the sound pressure amplitude $\hat{p}_\sim(\rho)$ equals $\hat{p}_{\sim\text{max}}/\sqrt{2}$. By exploiting the approximation (7.61), this distance will result from

$$d_\rho(-3\,\text{dB}) \approx 0.515 \cdot S_F R_T = 1.029 \cdot \lambda_{\text{aco}} f^{\#} \tag{7.64}$$

when $S_F \ll 1$ holds. Just as the depth of focus $d_z(-3\,\text{dB})$, the lateral beam width $d_\rho(-3\,\text{dB})$ will be rather small if the transducer is focusing sound waves strongly. In any case, $d_z(-3\,\text{dB})$ is much greater than $d_\rho(-3\,\text{dB})$. The depth of focus and lateral beam width are linked through (see (7.60) and (7.64))

$$d_z(-3\,\text{dB}) \approx 3.5 \frac{F_T}{R_T} d_\rho(-3\,\text{dB}) \ . \tag{7.65}$$

Both distances can be interpreted as doubled semi-principal axes of a prolate ellipsoid representing the focal volume of the spherically focused transducer (see

Fig. 7.21 Prolate ellipsoid representing focal volume of spherically focused transducer; depth of focus $d_z(-3\,\mathrm{dB})$ and lateral beam width $d_\rho(-3\,\mathrm{dB})$ denote doubled semi-principal axes; local coordinate system $x_F y_F z_F$ with origin at geometrical focus F_T

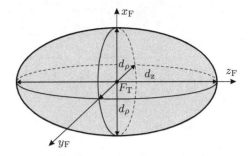

Fig. 7.21). Within this prolate ellipsoid, the sound pressure amplitudes always exceed $\hat{p}_{\sim\mathrm{max}}/\sqrt{2}$.

7.3   Spatial Resolution in Pulse-Echo Mode

In the previous section, we concentrated on generated sound fields of piston-type and spherically focused ultrasonic transducers, which are excited by a pure sinusoidal electrical input $u_I(t)$. Concerning practical applications, ultrasonic transducers are, however, commonly operated in pulse-echo mode. Thereby, the transducer serves as transmitter and receiver of sound waves (cf. Fig. 7.1). The time-of-flight of the reflected sound waves (i.e., echo) and their intensity deliver information about the sound reflector like its geometric distance from the transducer or the geometrical structure of the reflector. Not surprisingly, the achievable spatial resolution constitutes a decisive quantity for these so-called pulse-echo measurements. The spatial resolution of an imaging system measures basically the closest geometric distance of two point-like targets, which still allows separating them in the recorded image (e.g., [9, 35, 43]).

Below, we will discuss the influence of the transducer excitation on both the generated sound fields and the resulting electrical outputs from the theoretical point of view. Afterward, Sects. 7.3.2 and 7.3.3 detail axial as well as lateral resolutions of piston-type and spherically focused transducers being operated in pulse-echo mode.

7.3.1   Transducer Excitation and Resulting Output

Owing to the fact that it is impossible to emit sound waves and analyze reflected sound waves simultaneously through a single transducer, we have to use several transducers or alternative electrical inputs instead of a pure sinusoidal excitation. For instance, short pulses as well as a finite number of sine periods represent such alternative electrical excitation signals for a single transducer. In contrast to a pure sinusoidal

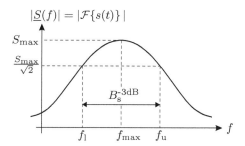

Fig. 7.22 Spectral magnitude (amplitude response; frequency response) $\left|\underline{S}(f)\right|$ of time signal $s(t)$; maximum spectral magnitude $S_{max} = \left|\underline{S}(f_{max})\right|$; lower cutoff frequency f_l and upper cutoff frequency f_u; bandwidth $B_s^{-3\,dB} = f_u - f_l$; center frequency f_c does not necessarily coincide with frequency f_{max}

signal consisting of only one frequency, those alternative excitation signals feature a certain bandwidth $B_s^{-3\,dB}$ in the frequency domain because of their limited signal duration T_s. The bandwidth of a band-limited time signal $s(t)$ corresponds prevalently to the frequency range between lower and upper cutoff frequency, in which the signal's spectral magnitude[3] $\left|\underline{S}(f)\right| = |\mathcal{F}\{s(t)\}|$ stays above $S_{max}/\sqrt{2}$ ($\hat{=} - 3\,dB$), whereby S_{max} stands for the maximum signal's spectral magnitude (see Fig. 7.22). Regarding signal energy, $B_s^{-3\,dB}$ also indicates the frequency range, in which the signal's spectral energy is more than half of its maximum. Generally speaking, a signal of short duration T_s (e.g., short pulse) offers a large bandwidth. Against that, the bandwidth equals zero for a pure sine wave since it is of infinite duration.

The spatial resolution of an ultrasonic transducer that operates in pulse-echo mode relates to the resulting electrical output $u_O(t)$ for a given surface normal velocity $v_n(t)$ of the active transducer area S_T. According to (7.15), the transducer output due to sound reflections at an ideal point-like target at position $\mathbf{r}$ depends on the transducer's spatial impulse response $h_{SIR}(\mathbf{r}, t)$. The essential part of the underlying formula reads as

$$u_O(\mathbf{r}, t) \propto v_n(t) * \frac{\partial h_{SIR}(\mathbf{r}, t)}{\partial t} * \frac{\partial h_{SIR}(\mathbf{r}, t)}{\partial t} \tag{7.66}$$

and is, therefore, very similar to those for the generated sound pressure at $\mathbf{r}$ (see (7.13)). The main difference lies in the additional temporal convolution with $\partial h_{SIR}(\mathbf{r}, t) / \partial t$. Obviously, this expression is the only one in (7.13) as well as (7.15) relating to the position $\mathbf{r}$. The spatial impulse response $h_{SIR}(\mathbf{r}, t)$ can, thus, be interpreted as spatial filter, which is applied once for the generated sound pressure and twice for the electrical transducer output in pulse-echo mode, respectively [59]. The amplitude $\hat{u}_O$ of the transducer output will be, consequently, reduced to half of its maximum $\hat{u}_{O;max}$ ($\hat{=} - 6\,dB$) at points, where the sound pressure amplitude $\hat{p}_\sim$ takes the value $\hat{p}_{\sim max}/\sqrt{2}$. Strictly speaking, this holds solely for pure sinusoidal

---

[3]The spectral magnitude results from the Fourier transform (operator $\mathcal{F}\{\cdot\}$).

transducer excitation, which does not make sense in case of pulse-echo mode. If we utilize the center frequency f_c of $v_n(t)$ for estimating characteristic quantities (e.g., dimensions of focal volume), it will be, nevertheless, possible to exploit the same approximations for the transducer outputs as for the generated sound fields. The center frequency of a band-limited signal is defined as geometric mean

$$f_c = \sqrt{f_l \cdot f_u} \tag{7.67}$$

of its lower cutoff frequency f_l and upper cutoff frequency f_u (see Fig. 7.22). When the bandwidth $B_s^{-3\,\mathrm{dB}} = f_u - f_l$ is small compared to the center frequency, one can use the arithmetic mean $f_c \approx (f_l + f_u)/2$ instead.

Finally, let us address an important point concerning the center frequency f_c of the transducer's surface normal velocity $v_n(t)$. Even though f_c yields reliable approximations for characteristic transducer quantities, f_c is not sufficient to deduce unambiguous sound pressure distributions of ultrasonic transducers. This is attributed to the fact that both a pure sinusoidal and a pulse-shaped signal may feature identical values for f_c. Contrary to a pure sinusoidal signal for $v_n(t)$, a short pulse will, however, only cause slight destructive and constructive interferences in the sound pressure distribution because of its limited duration T_s. Hence, we always have to consider the time behavior of $v_n(t)$ or its spectral composition with a view to precisely predicting generated sound pressure distributions as well as electrical transducer outputs.

7.3.2 Axial Resolution

As initially mentioned, the spatial resolution indicates the closest geometric distance of two point-like targets, which still enables separation of those targets in the recorded data. In the context of piston-type and spherically focused ultrasonic transducers being operated in pulse-echo mode, axial resolution refers to the closest on-axis distance of two ideal point-like targets along the z-axis. Without any signal processing, the axial resolution d_{ax} for such transducers can be defined as [57]

$$d_{ax} = \frac{c_0 \tau_p(-20\,\mathrm{dB})}{2} \tag{7.68}$$

if a short pulse serves as electrical excitation $u_I(t)$. The expression $\tau_p(-20\,\mathrm{dB})$ stands for the time difference between the two instants of time, at which the envelope of the electrical transducer output $u_O(t)$ for a single point-like target takes the value $u_{ev;max}/10$ ($\hat{=} - 20\,\mathrm{dB}$). Here, $u_{ev;max}$ represents the maximum of the envelope. A very efficient way to determine the envelope $u_{ev}(t)$ of $u_O(t)$ is based on the so-called *Hilbert transform* [49, 51]. By means of the *Hilbert operator* $\mathcal{H}\{\cdot\}$ that provides the corresponding imaginary part of a real-valued part, the envelope of the electrical transducer output is calculated from

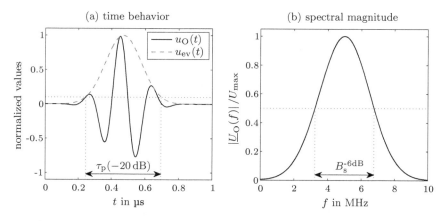

Fig. 7.23 **a** Time behavior of electrical transducer output $u_O(t)$ and its envelope $u_{ev}(t)$ for single point-like target; normalization with respect to maximum $u_{ev;max}$ of envelope; **b** normalized spectral magnitude $|\underline{U}_O(f)| = |\mathcal{F}\{u_O(t)\}|$ of surface normal velocity; characteristic data from curves: $\tau_p(-20\,dB) = 474\,ns$, $f_l = 3.24\,MHz$, $f_u = 6.77\,MHz$, $B_s^{-6\,dB} = f_u - f_l = 3.53\,MHz$, $f_{max} = 5.00\,MHz$, $f_c = 4.68\,MHz$

$$u_{ev}(t) = |u_O(t) + j\mathcal{H}\{u_O(t)\}| \; . \tag{7.69}$$

Let us continue with a short example. Figure 7.23a and b show a typical time behavior of $u_O(t)$ and its spectral magnitude $|\underline{U}_O(f)|$ for pulse-like transducer excitation. From the resulting envelope $u_{ev}(t)$, we can deduce the characteristic time difference $\tau_p(-20\,dB) = 474\,ns$. Assuming the sound velocity $c_0 = 1500\,ms^{-1}$ of the propagation medium, (7.68) leads then to the axial resolution $d_{ax} = 355\,\mu m$ of the ultrasonic transducer in pulse-echo mode.

It seems only natural that we have to reduce the signal duration T_s of $u_O(t)$ for a single point-like target and, therefore, $\tau_p(-20\,dB)$ when the axial resolution of a transducer should be improved. However, owing to the available bandwidth of transducer, electrical excitation as well as read-out electronics, such reduction is possible only to a limited extent. If an ultrasonic transducer is forced to oscillate with frequencies far away from its center frequency, the generated sound pressure waves and electrical outputs will be rather small. As a result, one has to manage output signals exhibiting a low signal-to-noise ratio.

7.3.3 Lateral Resolution

Besides the axial resolution d_{ax}, the lateral resolution d_{lat} represents a decisive quantity concerning pulse-echo measurements in ultrasound-based imaging systems (e.g., [9, 43]). Since such imaging systems typically utilize focused transducer devices, let us show the lateral resolution in the focal plane, i.e., in the plane where the best lateral resolution can be achieved. For spherically focused transducers featuring a low f-number $f^{\#}$ as well as a low Fresnel parameter (i.e., $S_F \ll 1$), the true focus nearly coincides with the geometric focus at $(\rho, z) = (0, F_T)$. The resulting electrical output signal $u_\Sigma(x, t)$ due to two ideal point-like targets being located in the geometrical focal plane at $(x, y, z) = (\pm d_{\text{lat}}/2, 0, F_T)$ reads as

$$u_\Sigma(x, t) = u_O(x + d_{\text{lat}}/2, F_T, t) + u_O(x - d_{\text{lat}}/2, F_T, t) \qquad (7.70)$$

with the lateral distance d_{lat} between both targets. If we suppose an excitation pulse yielding also a pulse-like electrical output $u_O(x \pm d_{\text{lat}}/2, F_T, t)$ for a single point-like target, the magnitude $\hat{u}_\Sigma(x)$ of the sum signal can be approximated by

$$\hat{u}_\Sigma(x) = \underbrace{\hat{u}_O(x + d_{\text{lat}}/2, F_T)}_{\hat{u}_{O;1}} + \underbrace{\hat{u}_O(x - d_{\text{lat}}/2, F_T)}_{\hat{u}_{O;2}} \qquad (7.71)$$

and, therefore, by the sum of the signal magnitudes originating from the individual targets. Without limiting the generality, it is, moreover, possible to describe the pulse-like time signals in the frequency domain. Because we are hereinafter interested solely in normalized magnitudes, let us substitute the broadband signals $u_O(x \pm d_{\text{lat}}/2, F_T, t)$ by their center frequency f_c. The magnitude for a single point-like target results from the approximations (7.61) and (7.62) through the relation $\hat{u}_O \propto \hat{p}_\sim^2 \propto \bar{I}_{\text{aco}}$, which is, strictly speaking, only satisfied for pure sinusoidal signals. According to these simplifications, the normalized curve progression $\hat{u}_\Sigma(x)$ for a spherically focused transducer becomes

$$\hat{u}_\Sigma(x) \propto \left| \frac{J_1(\nu_1)}{\nu_1} \right|^2 + \left| \frac{J_1(\nu_2)}{\nu_2} \right|^2 \qquad (7.72)$$

with

$$\nu_1 = \frac{2\pi(x + d_{\text{lat}}/2) R_T}{\lambda_{\text{aco}} F_T} \quad , \quad \nu_2 = \frac{2\pi(x - d_{\text{lat}}/2) R_T}{\lambda_{\text{aco}} F_T} \quad \text{and} \quad \lambda_{\text{aco}} = \frac{c_0}{f_c} \,.$$

Just as the axial resolution, the lateral resolution indicates the closest geometric distance of two ideal point-like targets, which still enables their separation. One can find two different definitions for the lateral resolution in the literature, namely the (i) *Rayleigh two-point definition* and the (ii) *Sparrow two-point definition* [28, 35]. The Rayleigh two-point definition states that it will be possible to separate two point-like targets when the maximum response of one target arises at the position, where the response of the other target equals zero (cf. Fig. 7.24a). In the geometrical focal plane of a spherically focused transducer, the Rayleigh two-point definition leads to the lateral resolution

$$d_{\text{lat}}(\text{Rayleigh}) = 1.22 \cdot \lambda_{\text{aco}} f^{\#} \,, \qquad (7.73)$$

which follows from the first zero $\rho_{zero} = 3.83 \cdot F_T/(k R_T)$ of $J_1(\cdot)$.

If the lateral distance d_{lat} between both ideal point-like targets is reduced further, the local minimum in the sum signal $\hat{u}_\Sigma(x)$ will disappear at $x = 0$. The Sparrow two-point definition refers to the lateral distance at which the second-order derivative of $\hat{u}_\Sigma(x)$ with respect to x equals zero for the first time. Therefore, $\hat{u}_\Sigma(x)$ offers a broad maximum (cf. Fig. 7.24b). By numerically evaluating (7.72), we obtain

$$d_{lat}(\text{Sparrow}) = 0.95 \cdot \lambda_{aco} f^\# \tag{7.74}$$

for the lateral resolution in the geometrical focal plane according to the Sparrow two-point definition. As the comparison of (7.73) and (7.74) reveals, the Sparrow two-point definition always yields smaller values than the Rayleigh two-point definition. Instead of applying one of those definitions, the lateral resolution d_{lat} of a spherically focused transducer is oftentimes approximated by the lateral beam width $d_\rho(-3\,\text{dB})$ in the focal plane (see (7.64)) [52, 57]. We can state that the lateral beam width lies between both two-point definitions, i.e., $d_{lat}(\text{Sparrow}) < d_\rho(-3\,\text{dB}) < d_{lat}(\text{Rayleigh})$. In any case, the lateral resolution of a spherically focused transducer will be improved when the lateral dimension of its focal volume decreases. This implies that one has to reduce the f-number $f^\#$ of the transducer and/or to increase the center frequency f_c of the generated pulse-like sound wave.

Finally, we will compute the lateral resolution of a spherically focused transducer, which has the ratio $F_T/R_T = 2$ of geometrical focal length F_T and radial size R_T, i.e., $f^\# = 1$. As discussed above, the simple relations for the Rayleigh as well as Sparrow two-point definition are based on the center frequency f_c. When the electrical outputs of a transducer are utilized for evaluating its center frequency, the lower and upper cutoff frequency should be defined at $\left|\underline{U}_O(f)\right|/U_{max} = 0.5$ instead of $\left|\underline{U}_O(f)\right|/U_{max} = 1/\sqrt{2}$. This is justified by the fact that the transducer characteristic is included twice in pulse-echo mode, i.e., once for transmitting and receiving, respectively. For the signal shown in Fig. 7.23, the maximum spectral magnitude U_{max} arises at $f_{max} = 5.00\,\text{MHz}$, while the center frequency f_c equals $4.68\,\text{MHz}$. Figure 7.24a and b depict the resulting normalized output magnitudes $\hat{u}_O$ due to the individual point-like target as well as the sum signal $\hat{u}_\Sigma$ for both two-point definitions. At the center of the two targets (i.e., $x = 0$), the Rayleigh two-point definition yields a local minimum of $\hat{u}_\Sigma$ that is reduced by 26.5 % in comparison with the global maximum. The Sparrow two-point definition leads to a broad global maximum without a local minimum at $x = 0$. Assuming again the sound velocity $c_0 = 1500\,\text{ms}^{-1}$ of the propagation medium (i.e., $\lambda_{aco} = c_0/f_c = 321\,\mu\text{m}$), the lateral resolutions amount to $d_{lat}(\text{Rayleigh}) = 391\,\mu\text{m}$ and $d_{lat}(\text{Sparrow}) = 305\,\mu\text{m}$, respectively.

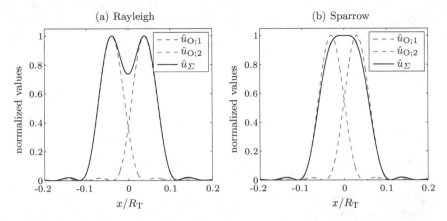

Fig. 7.24 Normalized magnitudes of transducer outputs for ideal point-like targets being located in focal plane of spherically focused transducer ($R_T = 5$ mm; $F_T = 10$ mm) according to **a** Rayleigh two-point definition and **b** Sparrow two-point definition; $\hat{u}_O(x)$ due to individual point-like targets at $\pm d_{lat}/2$ and resulting sum signal (see (7.71))

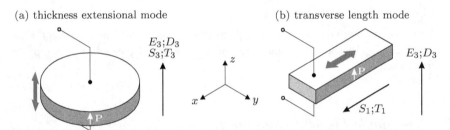

Fig. 7.25 Thin piezoelectric element operating in **a** thickness extensional mode and **b** transverse length mode; bottom and top surface completely covered with electrodes; red arrow shows direction of mechanical vibrations; **P** refers to electrical polarization, e.g., for piezoceramic materials

7.4   General Structure

Ultrasonic transducers that are based on piezoelectric materials usually exploit either the thickness extensional mode or the transverse length mode (see Fig. 7.25). If the piezoelectric material is thin compared to its lateral dimensions, we can simplify the material law for linear piezoelectricity remarkably because various quantities become negligible. For the thickness extensional mode, this means in the assumed coordinate system that it is possible to omit the electrical and mechanical quantities in x- and y-direction, e.g., E_1 and S_2. By contrast, one can omit the electrical quantities in x- and y-direction as well as the mechanical quantities in y- and z-direction for the transverse length mode. Thus, the relevant constitutive equations in d-form read as (cf. (3.30), p. 51)

$$D_3 = \varepsilon_{33}^T E_3 + d_{33} T_3 \qquad (7.75)$$

$$S_3 = d_{33} E_3 + s_{33}^E T_3 \qquad (7.76)$$

for the thickness extensional mode and

$$D_3 = \varepsilon_{33}^T E_3 + d_{31} T_1 \qquad (7.77)$$

$$S_1 = d_{31} E_3 + s_{11}^E T_1 \qquad (7.78)$$

for the transverse length mode, respectively. The listed variables indicate the following physical quantities (see Sect. 3.3):

- Electric field intensity E_i.
- Electric flux density D_i.
- Mechanical stress T_p.
- Mechanical strain S_p.
- Electric permittivity ε_{ii}^T for constant mechanical stress.
- Elastic compliance constant s_{pq}^E for constant electric field intensity.
- Piezoelectric strain constant d_{ip}.

Hereinafter, we will discuss the general structure and setup of single-element transducers as well as the idea behind transducer arrays and piezoelectric composite transducers.

7.4.1 Single-Element Transducers

A single-element transducer contains only one piezoelectric element, which is used to generate and receive sound pressure waves. Let us explain the setup of single-element-based ultrasonic transducers that exploit either the thickness extensional mode or transverse length mode of piezoelectric materials.

Thickness Extensional Mode

The thickness extensional mode of piezoelectric materials is applied for many ultrasonic transducers, especially if a transducer should be used in liquid media such as water (e.g., [35, 65]). It makes sense to operate the transducer at frequencies nearby the resonance of the thickness extensional mode since the resulting amplitudes of both the surface normal velocity $\hat{v}_n$ and the generated sound pressure $\hat{p}_\sim$ (cf., e.g., (7.34)) will be comparatively large for such frequencies. Not surprisingly, the electrical output signal of the transducer will be large too when the incident sound pressure waves are of similar frequency. According to the fundamentals of continuum mechanics (see Sect. 2.2), the resonance frequency f_r of the thickness extensional mode of a thin piezoelectric element (e.g., disk) is given by

$$f_r = \frac{c_P}{\lambda_P} = \frac{c_P}{2t_S} . \tag{7.79}$$

The variables c_P, λ_P, and t_S stand for the wave propagation velocity within the piezo-electric material, the corresponding wavelength, and the material thickness, respectively. In other words, the resonance of mechanical vibrations within the piezoelectric element will arise when t_S equals $\lambda_P/2$.

Besides, one has to consider the fact that common piezoelectric materials (e.g., piezoceramics) and typical sound propagation media exhibit great differences in the acoustic impedance Z_{aco}. For instance, Z_{aco} of water and common piezoceramic materials take approximately the values $1.5 \cdot 10^6$ and $30 \cdot 10^6 \, \mathrm{Nsm^{-3}}$, respectively. Because the majority of incident waves will be, therefore, reflected at the interface of piezoelectric element/wave propagation medium (see Sect. 2.3.4), an ultrasonic transducer consisting exclusively of the piezoelectric element cannot be applied for efficient radiation and reception of sound pressure waves. That is the reason why we need additional components being placed between piezoelectric element and wave propagation media. These components are usually referred to as *matching layers* (cf. Fig. 7.26) since they should enable matching of different acoustic properties [3, 40]. Mainly, the matching layers have to satisfy two conditions. The first condition specifies the layer thickness, whereas the second condition refers to its acoustic impedance. Just as for transmission lines in electrical engineering, the layers should be of thickness $t_M = \lambda_M/4 = c_M/(4f)$, whereby λ_M and c_M denote the wavelength and wave propagation velocity of longitudinal mechanical waves within a layer.[4] There can be found a limited set of design criteria in the literature for the second condition, i.e., the acoustic impedance of the matching layers [15, 35]. If a single matching layer is utilized, its optimal acoustic impedance $Z_{aco;M}$ will compute as

$$Z_{aco;M} = \sqrt{Z_{aco;P} \, Z_{aco;W}} \tag{7.80}$$

with the acoustic impedances $Z_{aco;P}$ and $Z_{aco;W}$ of the piezoelectric material and the sound propagation medium. In case of two matching layers, the optimal acoustic impedances results in

$$Z_{aco;M1} = Z_{aco;P}^{3/4} \, Z_{aco;W}^{1/4} \tag{7.81}$$

$$Z_{aco;M2} = Z_{aco;P}^{1/4} \, Z_{aco;W}^{3/4} . \tag{7.82}$$

While $Z_{aco;M1}$ indicates the acoustic impedance of the matching layer close to the piezoelectric element, $Z_{aco;M2}$ is the one for the matching layer being placed close to the sound propagation medium. Both design criteria follow directly from the maximum power transfer theorem in electrical engineering. Alternatively, Desilets et al. [15] suggested the acoustic impedance

---

[4] In accordance with electrical engineering, layers of thickness $\lambda_M/4$ are oftentimes termed $\lambda/4$ transformer.

$$Z_{\text{aco};M} = Z_{\text{aco};P}{}^{1/3} Z_{\text{aco};W}{}^{2/3} \tag{7.83}$$

for a single matching layer and

$$Z_{\text{aco};M1} = Z_{\text{aco};P}{}^{4/7} Z_{\text{aco};W}{}^{3/7} \tag{7.84}$$

$$Z_{\text{aco};M2} = Z_{\text{aco};P}{}^{1/7} Z_{\text{aco};W}{}^{6/7} \tag{7.85}$$

when two matching layers are utilized for the ultrasonic transducer. Even though the alternative design criteria for the matching layer(s) do not lead to maximum sound pressures and electrical transducer outputs, those criteria offer advantages for generation and reception of broadband (i.e., short) ultrasonic pulses.

Matching layers at the front of the piezoelectric element permit an efficient energy transfer between piezoelectric material and sound propagation medium. However, the back of the piezoelectric element also plays an important role regarding the performance of the ultrasonic transducer. Owing to the location, components at the back of the piezoelectric element are usually called *backing* (cf. Fig. 7.26) [3, 65]. In practical situations, we can distinguish two limits for the acoustic impedance $Z_{\text{aco};B}$ of this backing, namely (i) $Z_{\text{aco};B} \ll Z_{\text{aco};P}$ and (ii) $Z_{\text{aco};B} \approx Z_{\text{aco};P}$. The first limit will arise if there is not any solid material at the back of the piezoelectric element; i.e., the element is terminated by air. Consequently, the incident waves get almost fully reflected at the interface piezoelectric element/air, which leads to additional waves propagating to the front of the piezoelectric element. The additional waves increase the generated sound pressure values as well as electric transducer outputs but reduce the effective bandwidth of the ultrasonic transducer.

Because the acoustic impedances of both piezoelectric element and backing approximately coincide for the second limit (i.e., $Z_{\text{aco};B} \approx Z_{\text{aco};P}$), the incident waves are almost completely transmitted through the interface piezoelectric element/backing. Hence, waves will be hardly reflected at this interface. If the backing material provides, moreover, a great attenuation along its thickness t_B for propagating mechanical waves, the intensity of waves being reflected at the rear end of the backing will be negligible when reaching again the interface backing/piezoelectric element [47]. This means that there does not occur a further pulse for pulse-like transducer excitation. The effective bandwidth of the ultrasonic transducer is, therefore, rather large, but the generated sound pressure values and electric transducer outputs will be lower than without backing. On these grounds, one has to design the backing with respect to the practical application of the ultrasonic transducer. While air should be used for high sound pressure values and transducer outputs, the increased bandwidth in case of a matched backing yields a larger transducer bandwidth, which can be advantageous for ultrasonic imaging systems [11, 35].

Figure 7.26 illustrates the typical structure of an ultrasonic transducer that is based on the thickness extensional mode of a piezoelectric material. The piezoelectric material (e.g., piezoceramic) is covered with electrodes on front and back. Following the above explanations, we need an appropriate matching layer at the front and, depending on the application, a backing material at the back of the piezoelectric element.

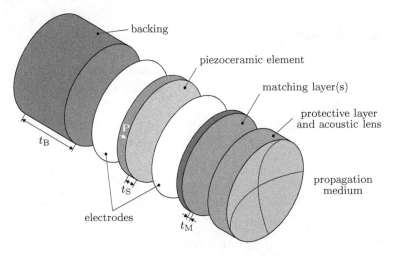

Fig. 7.26 Typical structure of piezoelectric ultrasonic transducer consisting of piezoceramic element covered with electrodes, matching layer(s), backing and protective layer incorporating acoustic lens

In addition to these components, ultrasonic transducers are often equipped with a further layer that should protect the piezoelectric element against the sound propagation medium and can be, moreover, exploited to achieve a focused ultrasonic transducer [11, 40]. Both the curvature radius of such a protective layer serving as acoustic lens and the acoustic properties of the involved media (i.e., layer material and sound propagation medium) affect the resulting focal length. As a matter of course, ultrasonic transducers will only perform well if the additional components (i.e., matching layer, backing material, and protective layer) are carefully joined with the piezoelectric element during the manufacturing process.

Transverse Length Mode

A piezoelectric element will feature relatively high displacements as well as velocities when the element is operated close to its resonance frequency f_r of mechanical vibrations. As already explained, this means for the thickness extensional mode that the piezoelectric element should be of thickness $t_S \approx \lambda_P/2$. Because small ultrasonic frequencies imply large wavelengths λ_P, we will need rather thick elements. Especially when the application calls for a large active area, thick elements would be rather expensive due to the required quantity of piezoelectric material.

Instead of the thickness extensional mode, one can use the transverse length mode of a piezoelectric element for generation and reception of sound pressure waves. To explain the fundamental idea behind the transverse length mode, let us consider a thin circular membrane that is mechanically clamped at its edge. The first eigenfrequency f_{r1} (corresponding to the first resonance frequency) of bending waves within such a membrane computes as [55]

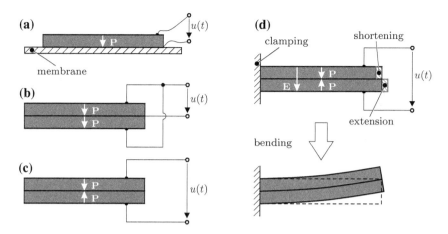

Fig. 7.27 **a** Piezoelectric unimorph transducer; **b** piezoelectric parallel bimorph transducer; **c** piezoelectric serial bimorph transducer; **d** functional principle of piezoelectric bimorph transducers; electrical excitation $u(t)$; polarization **P** of piezoceramic material

$$f_{r1} = \frac{10.216}{2\pi R_{mem}^2} \sqrt{\frac{E_M\, t_{mem}^2}{12(1 - \nu_P^2)\varrho_0}}. \tag{7.86}$$

The expressions R_{mem}, t_{mem}, ϱ_0, E_M, μ_P denote the membrane radius and thickness, its density, Young's modulus, and Poisson's ratio, respectively. For an aluminum membrane[5] of radius $R_{mem} = 5$ mm and thickness $t_{mem} = 0.5$ mm, the first resonance frequency f_{r1} takes the value 51 kHz, which represents a typical ultrasonic frequency in airborne ultrasound. It is possible to excite bending waves within a membrane through an appropriate piezoelectric element (e.g., thin disk) that is attached to the membrane. The mechanical deformations of the piezoelectric element due to its transverse length mode induce bending torques, which lead to deflections of a clamped membrane. By a time-varying electrical excitation of the piezoelectric element, we can, therefore, produce periodic membrane oscillations yielding a sound field in the wave propagation medium. Owing to their simple structure and little amount of piezoelectric material, these so-called *piezoelectric unimorph transducers* (see Fig. 7.27a) are often utilized in practical applications, especially for airborne ultrasound [18, 40]. Piezoelectric unimorph transducers are not restricted to sound field generation but can, moreover, serve as receiver of sound pressure waves; i.e., such a transducer is applicable for pulse-echo mode.

Figure 7.27b and c display alternative structures for ultrasonic transducers, which also exploit the transverse length mode of piezoelectric materials. Because both structures consist of two piezoelectric elements that have to be mechanically connected,

---

[5]Material data of aluminum for (7.86): $\varrho_0 = 2700$ kg m^{-3}, $E_M = 67.6$ kN mm^{-2}, and $\nu_P = 0.36$ (cf. Table 2.5 on p. 26).

these transducers are commonly referred to as *piezoelectric bimorph transducers* [40, 67]. In contrast to piezoelectric unimorph transducers, bimorph transducers do not contain an additional membrane. Guided from the direction of polarization and electrical wiring in case of a piezoceramic material, the first transducer type (see Fig. 7.27b) is also called *parallel bimorph* and the second one (see Fig. 7.27c) *serial bimorph*. The electric polarizations **P** of both piezoceramic materials are aligned in the same direction for parallel bimorphs, whereas **P** points in opposite directions for serial bimorphs. Figure 7.27d demonstrates the functional principle of a piezo-electric serial bimorph transducer that gets electrically excited by the voltage $u(t)$. When an electric voltage is applied to the bimorph, one piezoelectric element will expand, while the other one will contract in lateral direction. Due to the extension and shortening of the single elements, the piezoelectric bimorph undergoes a bending, which results in a certain deflection. Just as it is the case for piezoelectric unimorph transducers, we can use this functional principle for ultrasonic transducers operating in pulse-echo mode.

Finally, it should be stressed that flexural transducers such as piezoelectric uni-morph and bimorph transducers are not limited to the ultrasonic range. By means of a suitable transducer design (i.e., in particular the geometric dimensions), they can also be employed within the audible range, i.e., for frequencies $f < 20\,\mathrm{kHz}$ [23].

7.4.2 Transducer Arrays

By combining several single-element transducers, one obtains a transducer array. When it is possible to electrically excite and read out these single-element transduc-ers separately, the resulting array will be commonly called phased array[6] (e.g., [29, 77]). Figure 7.28 displays fundamental operation modes of a phased array consisting of seven elements, which are arranged in a straight line. Apart from the pulse-echo and pitch-catch mode (cf. Fig. 7.1), we can basically distinguish between four fun-damental operation modes, namely (i) synchronous beam, (ii) beam steering, (iii) focused beam, and (iv) steering and focusing. The mode *synchronous beam* relates to simultaneous excitation and readout of all array elements or a subgroup, while the other operation modes demand appropriate time delays Δt_i. These time delays directly follow from the sound velocity c_0 in the propagation medium. For instance, the time shift of the excitation signals between two neighboring elements has to be constant for the mode *beam steering* in order to generate a tilted wave front. In contrast, the operation modes *focused beam* as well as *steering and focusing* require different time shifts between two neighboring elements. We have to consider the indi-vidual time delays during array excitation and readout if the operation modes (ii)–(iv) should be applied for sound field generation and evaluation, respectively.

Besides the time delays Δt_i for the array elements, one can additionally vary the amplitudes of electrical excitation as well as the contribution to the summed receive

---

[6]The name *phased array* originates from the fact that time delays correspond to phase shifts.

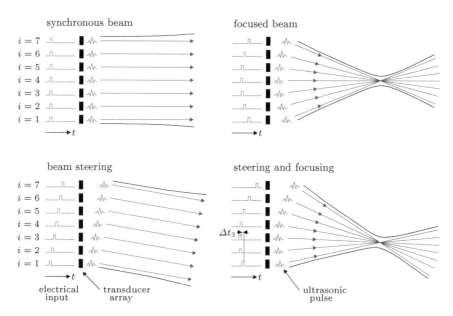

Fig. 7.28 Fundamental operation modes of phased arrays (transducer arrays) for generating ultrasonic pulses; linear array consisting of seven single elements, i.e., $i = 1, \ldots, 7$; time delay Δt_i for array element i

signal individually. This so-called apodization allows modifying both the transmit and the receive behavior of phased arrays [12, 77]. In doing so, we are able, for example, to generate a Gaussian ultrasonic beam by means of linear phased arrays. The combination of individual time delays and apodization during sound emission is also termed beam shaping, whereas it is frequently referred to as beam forming during sound reception.

Phased arrays are especially helpful for ultrasonic imaging systems, which operate in pulse-echo mode, e.g., as in case of medical diagnostics. Contrary to focused single-element transducers, one does not have to mechanically move such arrays across the entire object area that should be imaged but can apply the fundamental operation modes (e.g., steering and focusing) instead. Thereby, the examination time is reduced remarkably.

There exist various possibilities for the arrangement and geometrical shape of the single-element transducers within ultrasonic phased arrays. Figure 7.29 shows the three most well-known types of arrays, which are based on the thickness extensional mode of single-element transducers: (i) annular arrays, (ii) linear arrays, and (iii) two-dimensional arrays [29, 40]. Below, we will discuss these types of ultrasonic phased arrays.

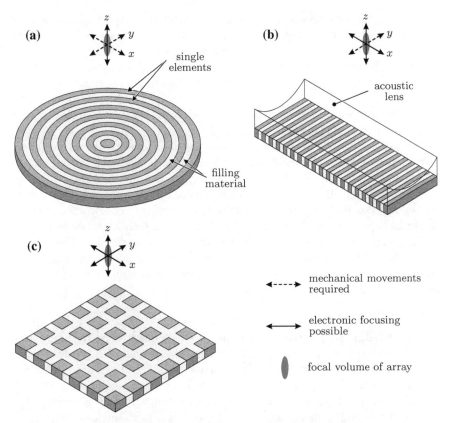

Fig. 7.29 **a** Annular array; **b** linear array (linear phased array) equipped with acoustic lens exhibiting fixed focus in y-direction; **c** two-dimensional phased array; arrays consist of single-element transducers (operating in thickness extensional mode) with appropriate filling between; backing and matching omitted; dashed parts of Cartesian coordinate system at focal volume require mechanical movements of array; solid parts allow electronic focusing; electrodes at bottom and top surface, matching layer(s) as well as backing omitted

Annular Arrays

An ultrasonic annular array consists of several rings that are concentrically arranged and mostly made of piezoceramic materials (see Fig. 7.29a). With the aid of annular arrays, one can imitate piston-type and spherically focused transducers operating in pulse-echo mode [4, 65]. If a piston-type transducer should be imitated, the individual rings have to be excited as well as read out simultaneously. However, a spherically focused annular array demands varying time delays Δt_i for the individual rings (cf. Fig. 7.28), which have to be applied for both pulse generation and reception. The inner rings of the annular array get electrically excited as well as read out later than the outer ones. In doing so, the different time-of-flights through the acoustic lens of a spherically focused single-element transducer are emulated. The time delays Δt_i

along with the sound velocity c_0 determine the depth (i.e., z-position) of the focal volume. The greater Δt_i, the smaller the z-position of the focal volume will be, and thus, the axial distance from the annular array decreases. While the time delays clearly specify the position of the focal volume for sound generation, it is possible to vary them independently for the reception mode. This follows from the fact that the evaluation of reflected ultrasonic waves can be performed offline for different Δt_i. As a result, we are able to dynamically focus at different z-positions in the reception mode for a single emitted ultrasonic pulse [29].

Due to the possibility of electronic focusing in z-direction and without any further signal processing (e.g., synthetic aperture focusing technique), the investigation of an object by means of an annular array takes less time than with a spherically focused transducer. Nevertheless, we still have to mechanically move the array at least in one further direction (here x- and y-direction) during imaging because annular arrays do not allow steering. Such mechanically movements usually implicate long examination times, which constitutes a problem for practical applications like medical diagnostics. That is the reason why ultrasonic annular arrays are nowadays only employed for special applications such as simultaneous determination of thickness and sound velocities of layered structures [38].

Linear Arrays

Figure 7.29b depicts the typical setup of an ultrasonic linear array. To some extent, linear arrays comprise up to a few hundred single-element transducers made of piezo-ceramic materials, which are arranged along a straight line [29]. As already discussed, such linear phased arrays will enable various fundamental operation modes (cf. Fig. 7.28) when we can excite and read out each single element separately. The fundamental operation modes are often not executed with all single elements at the same time but only with a subgroup. This means, for example, that the first ten elements of the linear array (i.e., $i = 1, \ldots, 10$) are used to generate a focused beam and to evaluate the resulting reflections. Afterward, the same procedure is conducted with the ten array elements starting from the second element, i.e., $i = 2, \ldots, 11$. By means of these parallel scans (see Fig. 7.30a), one can investigate a rather large cross section of an object by the ultrasonic linear array without any mechanical movements and, thus, in a short time. Parallel scans require, however, a sufficient acoustic coupling to the object along the whole linear array, which is almost impossible, e.g., if a large body area should be examined in medical diagnostics. Owing to this fact, the single elements are frequently arranged along a curved line yielding a convex sonic head, also called curved phased array (see Fig. 7.30b) [40, 77]. Compared to a conventional ultrasonic linear array, a curved phased array needs a rather small coupling area and is, therefore, particularly suited for examining the abdominal cavity in medical diagnostics.

Alternatively to conventional linear arrays and curved phased arrays, we can use special linear arrays, so-called sector phased arrays (see Fig. 7.30c), that demand only a small coupling area but consist of up to a few hundred single-element transducers [29]. Because parallel scans do not make sense for such short linear arrays, ultrasonic imaging has to be based on the fundamental operation mode *steering and*

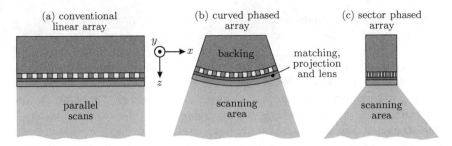

Fig. 7.30 **a** Parallel scans of conventional linear array and scanning areas of **b** curved phased array as well as **c** sector phased array

focusing. That is the reason why the resulting B-mode images (see Sect. 7.7) take the form of a circle segment and are commonly termed sector scans. Steering and focusing will only be possible if each array element offers almost a spherical directivity pattern, which limits their lateral element size to $\lambda_{aco}/2$ (is satisfied for $kR_T \leq \pi$; cf. Fig. 7.10). From the practical point of view, array elements of small size imply a small sensitivity in pulse-echo images yielding a low signal-to-noise ratio. Furthermore, to avoid grating lobes in the directivity pattern of the whole phased array, which may cause ambiguities in the resulting images, the spacing between two neighboring array elements (i.e., from center to center) is also limited to $\lambda_{aco}/2$ [72]. When the spacing takes larger values, one will have to cope with a reduced scanning area [81]. In other words, the possible angles θ (cf. Fig. 7.8) for steering and focusing of the phased array will be much smaller than 90°. According to these basic requirements, the small array elements have to be densely arranged, which may lead to disturbing electrical as well as mechanical crosstalks between neighboring elements. Even though sector phased arrays pose great challenges for the manufacturing process, the advantages such as small coupling area and short examination times outweigh both the production expenditure and the mandatory production accuracy. Hence, it is not surprising that sector phased arrays are oftentimes used, especially in medical diagnostics.

While linear phased array enables electronic focusing in x-direction, focusing in y-direction is usually achieved by an acoustic lens exhibiting a fixed focal length (see Fig. 7.29b). Pulse-echo images may, therefore, appear blurred since mechanical focusing is just ideal in a certain depth, i.e., z-position. Moreover, three-dimensional investigations of objects still demand mechanical movements of the linear array in one direction, here the y-direction.

Two-Dimensional Phased Arrays

For several practical applications of ultrasonic imaging such as real-time imaging in medical diagnostics, it is desirable to have the possibility of three-dimensional (3-D) investigations without any mechanical movements of the array. Annular arrays and linear phased arrays can, however, not be used for this task because one has to move them mechanically. In principle, two-dimensional (2-D) phased arrays (see

Fig. 7.29c) enable electronic focusing and steering in all spatial directions [40, 77]. Such arrays should, therefore, be applicable for 3-D ultrasound-based investigations of objects.

As a matter of course, 2-D phased arrays pose similar challenges for the manufacturing process as sector phased arrays. The great difference lies in the fact that we have to handle the challenges in two directions in space for 2-D phased arrays. Furthermore, the number of single-element transducers will increase drastically if the 2-D phased array should exhibit a reasonable number of elements in both directions, i.e., x- and y-direction. For instance, 128 elements in each direction of the array result altogether in $128 \times 128 = 16384$ single elements that should operate independently regarding the fundamental operation modes of phased arrays. A large number of single elements also goes hand in hand with considerable challenges for signal processing and conditioning. In addition, we need independent control and read-out electronics for the array elements, which can be managed through multiplexers and demultiplexers being located directly within the transducer head. On these grounds, there is much research and development ongoing in the field of 2-D phased arrays for ultrasonic imaging during the last years. Currently, piezoelectric micromachined ultrasonic transducers (pMUTs) represent an interesting and promising approach for 2-D phased arrays that exploit piezoelectricity [14, 45, 80].

7.4.3 Piezoelectric Composite Transducers

The piezoelectrically active part of so-called piezoelectric composite transducers is not a piezoelectric element in the classical sense (e.g., a piezoceramic disk) but contains small or thin piezoelectric elements that are embedded in a passive material matrix such as epoxy. Piezoelectric composite transducers are frequently utilized in practical applications as an alternative to common transducers consisting of bulk piezoelectric ceramics (e.g., [25, 70]). The reason for this lies in the following advantages of piezoelectric composite materials over conventional piezoceramics [65]:

- Altogether, the acoustic impedance Z_{aco} of piezoelectric composites is smaller than of piezoceramics since the passive material matrix offers comparably low values for Z_{aco}. This fact facilitates acoustic matching of the piezoelectrically active part to the wave propagation medium, e.g., water or human body.
- To some extent, ultrasonic transducers based on piezoelectric composite materials provide higher efficiency of electromechanical coupling, which is rather important for emitting as well as receiving of ultrasonic waves.
- The passive material can reduce unwanted coupling of mechanical vibration modes within piezoelectric composites. For example, the radial modes of an appropriately designed piezoelectric composite disk hardly influence its thickness extensional mode.

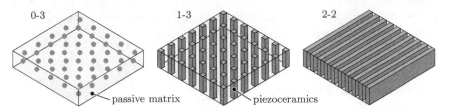

Fig. 7.31 Typical structures of piezoelectric composite materials featuring connectivity patterns 0-3, 1-3, and 2-2; composites consist of passive matrix material and piezoceramic material serving as active part; electrodes at top and bottom surface as well as partially passive materials omitted for better illustration

- Due to the soft passive material matrix, the resulting piezoelectric composites are mechanically flexible and it is, moreover, much easier to fabricate curved ultrasonic transducers.

In general, there exist various configurations for piezoelectric composite materials. With a view to uniquely defining the different configurations, Newnham et al. [46] introduced the so-called *connectivity pattern* for piezoelectric composites. According to this definition, we distinguish between ten connectivity patterns of piezoelectric composites that contain two different materials, i.e., a piezoelectric material serving as active material and a passive material. Each material can be continuous (self-connected) in zero, one, two, or three directions in space within the composite. When a material is continuous in one direction, the material can only move along this direction without affecting the other material. Hence, the number of directions in space referring to continuity of a material also corresponds to the degrees of freedom for mechanical movements.

It is common to specify connectivity patterns of piezoelectric composites in the form i-j with $\{i, j\} = \{1, 2, 3\}$, which yields ten possibilities, namely 0-0, 0-1, 0-2, 0-3, 1-1, 1-3, 2-1, 2-2, 2-3 as well as 3-3 [46, 66]. The first digit (i.e., i) indicates the degrees of freedom in space of the piezoelectric material, whereas the second digit (i.e., j) denotes the degrees of freedom of the passive material. Figure 7.31 displays typical structures of the three connectivity patterns 0-3, 1-3, and 2-2 that are frequently used for piezoelectric composite materials in ultrasonic transducers. Especially, 1-3 composites being based on piezoceramic fibers represent an outstanding candidate for ultrasonic transducers because of the cost-efficient manufacturing process [21, 65, 73]. The resulting behavior of such piezoelectric composite (usually termed *1-3 fiber composite*) mainly depends on three points: (i) material properties of fibers and passive material, (ii) geometric dimensions and arrangement (evenly or unevenly distributed), and (iii) volume fraction of the piezoceramic fibers within the composite.

As the comparison of Figs. 7.29 and 7.31 reveals, the general structure of transducer arrays and piezoelectric composite transducers is very similar, e.g., 2-D phased array and 1-3 composite. The main difference lies, however, in the electrical contacting of the individual piezoelectric elements. While the piezoelectric elements within composite materials can be contacted in common, transducer arrays require separate contacting and, therefore, electrically isolated electrodes.

7.5 Analytical Modeling

With a view to predicting the behavior of piezoelectric ultrasonic transducers, one can exploit finite element (FE) simulations as shown in Chap. 4. Thereby, it is of utmost importance to couple different physical fields (e.g., mechanical and acoustic fields) at the surface of the piezoelectric material as well as within the material. In the following, we will apply a simplified one-dimensional modeling approach in the frequency domain instead, which allows analytical description of basic physical relationships for piezoelectric ultrasonic transducers. Contrary to FE simulations, this one-dimensional modeling does not provide, however, spatially distributed quantities in all three directions but is restricted to one direction in space.

Let us regard a thin piezoelectric disk being polarized in z-direction (i.e., thickness direction) and completely covered with electrodes at its front as well as back (see Fig. 7.32a). The disk with base area A_S and thickness t_S is assumed to vibrate exclusively in the direction of polarization, which means that we can neglect dependencies of physical quantities in the other directions in space, i.e., in x- and y-direction. At the front and back, the disk is loaded with the mechanical forces F_F and F_B in z-direction, respectively. The velocities of the disk at front and back are termed v_F and v_B. In the frequency domain, the complex representation of these boundary conditions reads as

$$\underline{F}_B = A_S\underline{T}_3(-t_S/2) \; ; \qquad\qquad \underline{F}_F = -A_S\underline{T}_3(t_S/2) \qquad (7.87)$$
$$\underline{v}_B = \underline{v}_3(-t_S/2) \; ; \qquad\qquad \underline{v}_F = -\underline{v}_3(t_S/2) \; ,$$

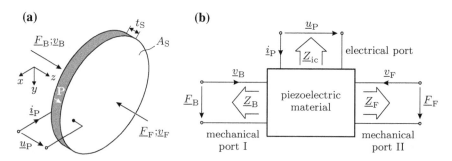

Fig. 7.32 **a** Regarded thin piezoelectric disk (thickness t_S; polarized in z-direction) with electrodes at its front and back; **b** representation of piezoelectric material as three-port network

whereby $\underline{T}_3(z)$ represents the mechanical stress within the disk and $\underline{v}_3(z)$ the velocity, both quantities pointing in positive z-direction. From the physical basics about continuum mechanics (see Sect. 2.2), we can deduce

$$\frac{\partial \underline{T}_3}{\partial z} = j\omega \varrho_P \underline{v}_3 \quad \text{and} \quad \frac{\partial \underline{v}_3}{\partial z} = j\omega \underline{S}_3 \tag{7.88}$$

with the angular frequency $\omega = 2\pi f$, the material density ϱ_P, and the mechanical strain $\underline{S}_3(z)$ within the piezoelectric element. The physical basics about electromagnetics in Sect. 2.1 lead to

$$\underline{i}_P = j\omega A_S \underline{D}_3 \quad \text{and} \quad \underline{u}_P = \int_{-t_S/2}^{t_S/2} \underline{E}_3 dz \tag{7.89}$$

for the complex representations of the electric current $\underline{i}_P$ flowing through and the electric voltage $\underline{u}_P$ across the piezoelectric element. According to the constitutive equations for linear piezoelectricity (see (3.21), p. 50), the relevant part of the h-form for vibrations in thickness direction becomes

$$\underline{T}_3 = c_{33}^D \underline{S}_3 - h_{33} \underline{D}_3 \tag{7.90}$$

with the elastic stiffness constant c_{33}^D for constant electric flux density D. The expression h_{33} denotes a so-called piezoelectric h constant, which is also termed transmitting constants. Because piezoelectric materials do not contain free electric charges, the electric flux density $\underline{D}_3$ stays constant with respect to the z-position, i.e., $\partial D_3/\partial z = 0$. This fact and (7.88) and (7.90) yield the differential equation [35]

$$\frac{\partial^2 \underline{v}_3}{\partial z^2} + \frac{\omega^2 \varrho_P}{c_{33}^D} \underline{v}_3 = 0 \tag{7.91}$$

that can be solved by the ansatz (wave number $k = \omega\sqrt{\varrho_P/c_{33}^D}$)

$$\underline{v}_3(z) = \underbrace{\underline{v}_3^+ \cdot e^{-jkz}}_{\text{forward}} + \underbrace{\underline{v}_3^- \cdot e^{jkz}}_{\text{backward}} \tag{7.92}$$

consisting of a forward and a backward wave. Now, it is possible to combine all these equations with the boundary conditions (7.87). After some mathematical treatment, we arrive at the matrix system[7] [3, 39]

---

[7] Hyperbolic cosecant $\operatorname{csch}(x) \equiv 1/\sinh(x)$.

$$
\begin{bmatrix} \underline{F}_B \\[2ex] \underline{F}_F \\[2ex] \underline{u}_P \end{bmatrix} = \begin{bmatrix} Z_P\coth(kt_S) & Z_P\mathrm{csch}(kt_S) & \dfrac{h_{33}}{j\omega} \\[2ex] Z_P\mathrm{csch}(kt_S) & Z_P\coth(kt_S) & \dfrac{h_{33}}{j\omega} \\[2ex] \dfrac{h_{33}}{j\omega} & \dfrac{h_{33}}{j\omega} & \dfrac{1}{j\omega C_0} \end{bmatrix} \begin{bmatrix} \underline{v}_B \\[2ex] \underline{v}_F \\[2ex] \underline{i}_P \end{bmatrix} . \tag{7.93}
$$

Here, $Z_P = Z_{\mathrm{aco;P}} A_S$ (unit Nsm^{-1}) indicates the radiation impedance[8] with the acoustic impedance $Z_{\mathrm{aco;P}} = \sqrt{\varrho_P c_{33}^D}$ of the piezoelectric material. Its clamped capacitance C_0 (electric permittivity ε_{33}^S for constant mechanical strain S) is given by

$$
C_0 = \frac{\varepsilon_{33}^S A_S}{t_S} . \tag{7.94}
$$

The upper left 2×2 submatrix of (7.93) describes the transmission of mechanical waves within the piezoelectric disk that propagate with the phase velocity $c_P = \sqrt{c_{33}^D / \varrho_P}$. By contrast, the terms containing the h constant h_{33} rate the electromechanical coupling.

In the current form, the matrix system does not consider losses within the piezoelectric material. To approximate such losses, one has to replace the radiation impedance Z_P, the wave number k as well as C_0 by

$$
\begin{aligned}
\underline{Z}_P &= \sqrt{\varrho_P c_{33}^D}\left(1 + \frac{j}{2Q_P}\right) A_S \\[1ex]
\underline{k} &= \omega \sqrt{\frac{\varrho_P}{c_{33}^D}}\left(\frac{1}{2Q_P} + j\right) \\[1ex]
C_0 &= \frac{\varepsilon_{33}^S A_S (1 - \tan\delta_d)}{t_S}
\end{aligned} \tag{7.95}
$$

with the mechanical quality factor Q_P and the loss factor $\tan\delta_d$ (cf. (5.56), p. 167) of the piezoelectric material [39]. For piezoelectric materials like piezoceramics, it is advisable to exploit the fundamental connection

$$
\tan\delta_d = \frac{1}{Q_P} = \alpha_d \tag{7.96}
$$

between these factors and the damping coefficient α_d. We are able to identify the material-specific quantity α_d through the inverse method (see Sect. 5.3). The remaining expressions $\underline{Z}_P$ and $\underline{k}$ in (7.95) stand for complex-valued versions of Z_P and k, respectively.

---

[8]The radiation impedance is also known as mechanical impedance.

From the system point of view, (7.93) can be interpreted as three-port network consisting of two mechanical (acoustic) ports and one electrical port. The mechanical ports relate to mechanical forces and velocities, while the electric port links electric voltage and electric current (see Fig. 7.32b). Since forces correspond to voltages and velocities to currents, the chosen analogy of mechanical systems and electrical networks is the so-called *force–voltage analogy* that is also named *impedance analogy* [40].

Below, we will discuss two well-known electrical networks, which exactly reflect the matrix system (7.93). Section 7.5.2 explains the calculation procedure to predict decisive information about piezoelectric ultrasonic transducers in advance, e.g., time response of generated sound pressure. Finally, exemplary computation results will be shown.

7.5.1 Equivalent Electrical Circuits

Through the mentioned force–voltage analogy, one can define equivalent electrical circuits for piezoelectric ultrasonic transducers. These equivalent circuits aim to exactly emulate the transducer behavior at the mechanical and electrical ports by means of lumped elements of electrical networks. Let us briefly discuss two equivalent circuits, which are widely used for simulations, design as well as optimization of piezoelectric ultrasonic transducers operating in thickness extensional mode, namely (i) Mason's and (ii) KLM equivalent circuits.

Mason's Equivalent Circuit

Figure 7.33 illustrates Mason's equivalent circuit for piezoelectric elements [6, 44]. It consists of a T-network that represents the propagation of mechanical waves within the piezoelectric material. Moreover, the equivalent circuit contains an ideal transformer, which exhibits the constant transmission ratio $N_P = h_{33}C_0$. Together with the negative capacitance $-C_0$, this ideal transformer accounts for the coupling of the mechanical ports and the electrical port.

KLM Equivalent Circuit

Krimholtz, Leedom, and Matthaei [37] proposed the so-called KLM equivalent circuit (see Fig. 7.34) in 1970. The wave propagation of mechanical waves within the piezoelectric element is modeled by a mechanical transmission line of length t_S. Again, the electrical and mechanical ports get coupled through a transformer that is connected to the center of this transmission line. In contrast to Mason's equivalent circuit, the transmission ratio $\underline{\Phi}_P(\omega)$ of the transformer being defined as

$$\underline{\Phi}_P(\omega) = \frac{\omega Z_P}{2h_{33}} \csc\left(\frac{kt_S}{2}\right) \tag{7.97}$$

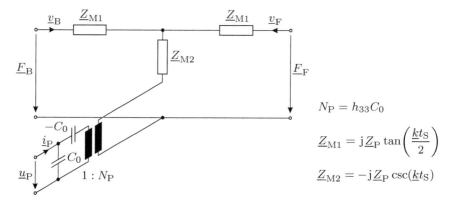

$$N_P = h_{33} C_0$$

$$\underline{Z}_{M1} = j \underline{Z}_P \tan\left(\frac{\underline{k} t_S}{2}\right)$$

$$\underline{Z}_{M2} = -j \underline{Z}_P \csc(\underline{k} t_S)$$

Fig. 7.33 Mason's equivalent circuit for piezoelectric element operating in thickness extensional mode

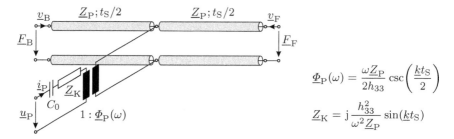

$$\underline{\Phi}_P(\omega) = \frac{\omega \underline{Z}_P}{2 h_{33}} \csc\left(\frac{\underline{k} t_S}{2}\right)$$

$$\underline{Z}_K = j \frac{h_{33}^2}{\omega^2 \underline{Z}_P} \sin(\underline{k} t_S)$$

Fig. 7.34 KLM equivalent circuit for piezoelectric element operating in thickness extensional mode

is not constant but depends on the considered frequency.

Both equivalent circuits entirely correspond to the matrix system (7.93). However, their different structure facilitates understanding and describing specific behavior of piezoelectric ultrasonic transducers [3, 39]. Since the electrical port in Mason's equivalent circuit is equipped with the clamped capacitance C_0 in parallel, it enables a better representation of the electrical behavior. On the other hand, the KLM equivalent circuit simplifies consideration of additional components (e.g., matching layers) at the front and back of the piezoelectric element because this circuit version is based on mechanical transmission lines.

7.5.2 Calculation Procedure

Apart from geometric dimensions and properties of the piezoelectric material, the analytical modeling of an ultrasonic transducer demands the radiation impedances $\underline{Z}_B$ and $\underline{Z}_F$ at both mechanical ports (see Fig. 7.32b). For the presumed arrow directions of forces $\underline{F}_{\mathrm{B;F}}$ and velocities $\underline{v}_{\mathrm{B;F}}$, these radiation impedances are given by

$$\underline{Z}_{\mathrm{B}} = -\frac{\underline{F}_{\mathrm{B}}}{\underline{v}_{\mathrm{B}}} \quad \text{as well as} \quad \underline{Z}_{\mathrm{F}} = -\frac{\underline{F}_{\mathrm{F}}}{\underline{v}_{\mathrm{F}}} . \tag{7.98}$$

We can also take an additional electrical circuit at the electrical port into account if the electrical input impedance $\underline{Z}_{\mathrm{ic}}$ of the circuit is known.

The major interest in many applications of piezoelectric ultrasonic transducers lies in the electrical impedance as well as the transmission behavior from the electrical port to a mechanical port and vice versa. In order to derive such quantities, let us start from the matrix system (7.93), which reads as

$$\begin{bmatrix} \underline{F}_{\mathrm{B}} \\ \underline{F}_{\mathrm{F}} \\ \underline{u}_{\mathrm{P}} \end{bmatrix} = \begin{bmatrix} \underline{z}_{11} & \underline{z}_{12} & \underline{z}_{13} \\ \underline{z}_{21} & \underline{z}_{22} & \underline{z}_{23} \\ \underline{z}_{31} & \underline{z}_{32} & \underline{z}_{33} \end{bmatrix} \begin{bmatrix} \underline{v}_{\mathrm{B}} \\ \underline{v}_{\mathrm{F}} \\ \underline{i}_{\mathrm{P}} \end{bmatrix} = \underline{\mathbf{Z}} \begin{bmatrix} \underline{v}_{\mathrm{B}} \\ \underline{v}_{\mathrm{F}} \\ \underline{i}_{\mathrm{P}} \end{bmatrix} \tag{7.99}$$

in compact form with the 3×3 impedance matrix $\underline{\mathbf{Z}}$ (components $\underline{z}_{ij}$) of the piezo-electric element. Without limiting the generality, let us, furthermore, assume that the element's front (i.e., mechanical port II) is used for emitting and receiving sound pressure waves. By introducing the radiation impedance $\underline{Z}_{\mathrm{B}} = -\underline{F}_{\mathrm{B}}/\underline{v}_{\mathrm{B}}$ (see (7.98)) at the element's back, one can reduce the three-port network to a two-port network exclusively consisting of mechanical port I and the electrical port. The matrix system for this two-port network results in

$$\begin{bmatrix} \underline{F}_{\mathrm{F}} \\ \underline{u}_{\mathrm{P}} \end{bmatrix} = \begin{bmatrix} \underline{z}_{22} - \dfrac{\underline{z}_{12}\underline{z}_{21}}{\underline{Z}_{\mathrm{B}} + \underline{z}_{11}} & \underline{z}_{23} - \dfrac{\underline{z}_{13}\underline{z}_{21}}{\underline{Z}_{\mathrm{B}} + \underline{z}_{11}} \\ \underline{z}_{32} - \dfrac{\underline{z}_{12}\underline{z}_{31}}{\underline{Z}_{\mathrm{B}} + \underline{z}_{11}} & \underline{z}_{33} - \dfrac{\underline{z}_{13}\underline{z}_{31}}{\underline{Z}_{\mathrm{B}} + \underline{z}_{11}} \end{bmatrix} \begin{bmatrix} \underline{v}_{\mathrm{F}} \\ \underline{i}_{\mathrm{P}} \end{bmatrix} \tag{7.100}$$

or in compact form with the 2×2 impedance matrix $\tilde{\underline{\mathbf{Z}}}$ (components $\tilde{\underline{z}}_{ij}$)

$$\begin{bmatrix} \underline{F}_{\mathrm{F}} \\ \underline{u}_{\mathrm{P}} \end{bmatrix} = \begin{bmatrix} \tilde{\underline{z}}_{11} & \tilde{\underline{z}}_{12} \\ \tilde{\underline{z}}_{21} & \tilde{\underline{z}}_{22} \end{bmatrix} \begin{bmatrix} \underline{v}_{\mathrm{F}} \\ \underline{i}_{\mathrm{P}} \end{bmatrix} = \tilde{\underline{\mathbf{Z}}} \begin{bmatrix} \underline{v}_{\mathrm{F}} \\ \underline{i}_{\mathrm{P}} \end{bmatrix} . \tag{7.101}$$

For determining electrical impedance and transmission behavior of a piezoelectric element, it is recommended to convert $\tilde{\underline{\mathbf{Z}}}$ into the 2×2 inverse chain matrix $\underline{\mathbf{B}}$ (components $\underline{b}_{ij}$). The matrix system with mechanical and electrical quantities serving as input and output, respectively, becomes then [7]

$$\begin{bmatrix} \underline{u}_{\mathrm{P}} \\ \underline{i}_{\mathrm{P}} \end{bmatrix} = \frac{1}{\tilde{\underline{z}}_{12}} \begin{bmatrix} \tilde{\underline{z}}_{22} & \det \tilde{\underline{\mathbf{Z}}} \\ 1 & \tilde{\underline{z}}_{11} \end{bmatrix} \begin{bmatrix} \underline{F}_{\mathrm{F}} \\ -\underline{v}_{\mathrm{F}} \end{bmatrix}$$

$$= \begin{bmatrix} \underline{b}_{11} & \underline{b}_{12} \\ \underline{b}_{21} & \underline{b}_{22} \end{bmatrix} \begin{bmatrix} \underline{F}_F \\ -\underline{v}_F \end{bmatrix} = \mathbf{B} \begin{bmatrix} \underline{F}_F \\ -\underline{v}_F \end{bmatrix} . \tag{7.102}$$

For the further computation steps, one has to additionally know the radiation impedance at the front. By inserting $\underline{Z}_F = -\underline{F}_F/\underline{v}_F$ from (7.98), the matrix system (7.102) takes the form

$$\begin{bmatrix} \underline{u}_P \\ \underline{i}_P \end{bmatrix} = \begin{bmatrix} \underline{b}_{11} & \underline{b}_{12} \\ \underline{b}_{21} & \underline{b}_{22} \end{bmatrix} \begin{bmatrix} -\underline{Z}_F \\ -1 \end{bmatrix} \underline{v}_F = \begin{bmatrix} -\underline{Z}_F\underline{b}_{11} - \underline{b}_{12} \\ -\underline{Z}_F\underline{b}_{21} - \underline{b}_{22} \end{bmatrix} \underline{v}_F . \tag{7.103}$$

This matrix system leads directly to the electrical impedance $\underline{Z}_{el}$ of the piezoelectric element. Dividing the first by the second line yields [35]

$$\underline{Z}_{el} = \frac{\underline{u}_P}{\underline{i}_P} \tag{7.104}$$

$$= \frac{1}{j\omega C_0} \left\{ 1 + \frac{k_t^2}{k t_S} \frac{j[\underline{Z}_B + \underline{Z}_F]\underline{Z}_P \sin(\underline{k}t_S) - 2\underline{Z}_P^2[1 - \cos(\underline{k}t_S)]}{[\underline{Z}_P^2 + \underline{Z}_B\underline{Z}_F] \sin(\underline{k}t_S) - j[\underline{Z}_B + \underline{Z}_F]\underline{Z}_P \cos(\underline{k}t_S)} \right\} .$$

Here, losses within the piezoelectric material have been considered. The expression k_t stands for the electromechanical coupling factor in thickness direction of a thin piezoelectric material like a disk (cf. (5.16), p. 133). When both mechanical ports are short-circuited (i.e., $\underline{F}_B = \underline{F}_F = 0$), the radiation impedances $\underline{Z}_B$ and $\underline{Z}_F$ will equal zero. As a result, (7.104) simplifies to

$$\underline{Z}_{el} = \frac{1}{j\omega C_0} \left[1 - k_t^2 \frac{\tan(\underline{k}t_S/2)}{\underline{k}t_S/2} \right] . \tag{7.105}$$

The term $1/(j\omega C_0)$ in (7.104) and (7.105) relates to the capacitive behavior of the piezoelectric element, while the second term in the brackets originates from electromechanical couplings within piezoelectric materials. Because the radiation impedance of air is much smaller than of common piezoelectric materials, (7.105) represents the electrical impedance of a piezoelectric element operating in air. From the practical point of view, this means that neither front nor back of the piezoelectric element are loaded with a material.

To evaluate the transmission behavior of a piezoelectric element for emitting and receiving sound pressure waves, the inverse chain matrix $\mathbf{B}$ from (7.102) should be converted to the hybrid matrix $\mathbf{H}$ (components $\underline{h}_{ij}$). In doing so, we arrive at [7]

$$\begin{bmatrix} \underline{F}_F \\ \underline{i}_P \end{bmatrix} = \frac{1}{\underline{b}_{11}} \begin{bmatrix} \underline{b}_{12} & 1 \\ -\det \mathbf{B} & \underline{b}_{21} \end{bmatrix} \begin{bmatrix} \underline{v}_F \\ \underline{u}_P \end{bmatrix} = \mathbf{H} \begin{bmatrix} \underline{v}_F \\ \underline{u}_P \end{bmatrix} . \tag{7.106}$$

By inserting $\underline{v}_F = -\underline{F}_F/\underline{Z}_F$ in the first line and utilizing the connection $\underline{F}_F = \underline{p}_\sim A_S$ between mechanical force and sound pressure $\underline{p}_\sim$, it is possible to compute the aimed transmission behavior. The transfer function $\underline{M}_e$ of a piezoelectric transducer for emitting sound pressure waves reads as

$$\underline{M}_e = \frac{\underline{p}_\sim}{\underline{u}_P} = \frac{\underline{h}_{12}}{A_S\left(1 + \underline{h}_{11}/\underline{Z}_F\right)} . \tag{7.107}$$

The transfer function for receiving sound pressure waves can be determined in a similar manner [39].

As stated in Sect. 7.4, piezoelectric elements of ultrasonic transducers operating in thickness extensional mode are usually located between a backing material and appropriate matching as well as protective layers. We can directly incorporate the backing of thickness t_B in the analytical modeling by simply replacing $\underline{Z}_B$ with $\underline{Z}_{aco;B} A_S$. This step is permitted since wave reflections arising at the rear end of the backing are negligible because of the great attenuation along the backing material. However, the wave propagation within matching and protective layers calls for special treatment. For the sake of simplicity, let us regard an ultrasonic transducer that is only equipped with a single matching layer of thickness t_M between piezoelectric element and sound propagation medium. To take into account this matching layer during analytical modeling, one should describe the underlying wave propagation by a transmission line for mechanical waves. From the upper left 2×2 submatrix of (7.93), it is possible to deduce the matrix system (chain matrix $\underline{\mathbf{A}}_M$) [39]

$$\begin{bmatrix} \underline{F}_F \\ -\underline{v}_F \end{bmatrix} = \begin{bmatrix} \cosh(\underline{k}_M t_M) & \underline{Z}_M \sinh(\underline{k}_M t_M) \\ \frac{1}{\underline{Z}_M}\sinh(\underline{k}_M t_M) & \cosh(\underline{k}_M t_M) \end{bmatrix} \begin{bmatrix} \underline{F}_W \\ \underline{v}_W \end{bmatrix} = \underline{\mathbf{A}}_M \begin{bmatrix} \underline{F}_W \\ \underline{v}_W \end{bmatrix} \tag{7.108}$$

for the lossy transfer behavior of a transmission line, which represents a two-port network (see Fig. 7.35). The expressions $\underline{Z}_M = \underline{Z}_{aco;M} A_S$ and $\underline{k}_M$ stand for the radiation impedance of the matching layer and the complex-valued wave number within this layer. While the left port $[\underline{F}_F, -\underline{v}_F]$ of the transmission line is connected to the front of the piezoelectric element, the right port $[\underline{F}_W, \underline{v}_W]$ is followed by the sound propagation medium featuring the radiation impedance $\underline{Z}_W = \underline{Z}_{aco;W} A_S$. Inserting $\underline{Z}_W = -\underline{F}_W/\underline{v}_W$ and dividing the first line of (7.108) by the second line leads to

$$\underline{Z}_F = -\frac{\underline{F}_F}{\underline{v}_F} = \underline{Z}_M \frac{\tanh(\underline{k}_M t_M) - \frac{\underline{Z}_W}{\underline{Z}_M}}{1 - \frac{\underline{Z}_W}{\underline{Z}_M}\tanh(\underline{k}_M t_M)} \tag{7.109}$$

for the resulting radiation impedance $\underline{Z}_F$ at the element's front. If $\underline{Z}_F$ in (7.104) is substituted by this relation, one can calculate the electrical impedance $\underline{Z}_{el}$ of a piezoelectric element that is equipped with backing as well as matching layer under consideration of the sound propagation medium. For determining the transfer

Fig. 7.35 Equivalent circuit
for matching layer followed
by propagation medium;
matching layer modeled as
transmission line

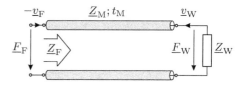

function $\underline{M}_e$ for such arrangement, we have to perform the matrix multiplication

$$\begin{bmatrix} \underline{u}_P \\ \underline{i}_P \end{bmatrix} = \mathbf{B}\,\underbrace{\mathbf{A}_M}_{\mathbf{B}_M} \begin{bmatrix} \underline{F}_W \\ \underline{v}_W \end{bmatrix} \tag{7.110}$$

in a first step. The matrices $\mathbf{B}$ and $\mathbf{A}_M$ stem from (7.102) and (7.108), respectively. After converting the matrix product $\mathbf{B}_M$ into the hybrid matrix $\mathbf{H}_M$ according to (7.106), $\underline{M}_e$ results again from (7.107) by replacing $\underline{Z}_F$ with the radiation impedance $\underline{Z}_W$ of the sound propagation medium. In this way, the calculated transfer function of a piezoelectric ultrasonic transducer for emitting sound pressure waves directly relates to the sound propagation medium.

7.5.3 Exemplary Results

At the end of the section, let us take a look at exemplary computation results of the analytical modeling approach for piezoelectric ultrasonic transducers. The regarded disk-shaped piezoelectric element (radius $R_T = 15\,\mathrm{mm}$; thickness $t_S = 2\,\mathrm{mm}$) is made of the piezoceramic material PIC255 and operates exclusively in thickness extensional mode. Table 5.3 on p. 160 contains the material parameters in e-form and d-form, which were identified through the inverse method. The parameter conversion from (3.33, p. 52) leads to the required piezoelectric h constant $h_{33} = 2.36 \cdot 10^9\,\mathrm{Vm}^{-1}$ and elastic stiffness constant $c_{33}^D = 1.59 \cdot 10^{11}\,\mathrm{Nm}^{-2}$. Furthermore, the analytical modeling demands material density $\varrho_P = 7.8 \cdot 10^3\,\mathrm{kg\,m}^{-3}$ and damping coefficient $\alpha_d = 0.0129$ of the piezoelectric material. Inserting these parameters in (7.95) yields

$$\underline{Z}_P = (2.49 \cdot 10^4 + \mathrm{j}\,1.60 \cdot 10^2)\,\mathrm{Nsm}^{-1}$$
$$\underline{k} = (1.43 \cdot 10^{-6} + \mathrm{j}\,2.22 \cdot 10^{-4})\,\mathrm{m}^{-1}$$
$$C_0 = 2.38\,\mathrm{nF}$$

for the radiation impedance $\underline{Z}_P$, the complex-valued wave number $\underline{k}$, and the clamped capacitance C_0 of the piezoelectric disk. The calculated resonance frequency f_r for the thickness extensional mode equals $\approx 1\,\mathrm{MHz}$.

Table 7.2 Decisive parameters for analytical modeling of regarded piezoelectric ultrasonic transducer; entries "−" not required for calculation procedure

	Damping coefficient α_i	Radiation impedance $\Re\{\underline{Z}_i\}$	$\Im\{\underline{Z}_i\}$	Propagation velocity c_i	Thickness t_i
		10^3 Nsm^{-1}		ms^{-1}	mm
Backing	$1/Q_B = 0.3$	24.9	3.7	−	−
Matching	$1/Q_M = 0.1$	5.1	0.3	2000	0.44
Water	−	1.1	0	−	−

Apart from the piezoelectric element, the simulated ultrasonic transducer consists of a backing material and a single matching layer. The wave propagation medium water was assumed to be lossless, i.e., $\Im\{\underline{Z}_W\} = 0$. Table 7.2 lists the decisive parameters of the individual components. Note that the real parts of the radiation impedance of matching layer as well as backing were chosen according to the real part of $\underline{Z}_P$. For optimal matching and backing, this means $\Re\{\underline{Z}_M\} = \sqrt{\Re\{\underline{Z}_P\} Z_W}$ (cf. (7.80)) and $\Re\{\underline{Z}_B\} = \Re\{\underline{Z}_P\}$. The thickness t_M of the matching layer results from the condition $c_M/(4 f_r)$, whereby c_M stands for the assumed wave propagation velocity within the matching layer (see Sect. 7.4.1).

The following simulation results refer to four different configurations of the piezoelectric ultrasonic transducer:

1. Without backing material (i.e., solely air) and without matching layer.
2. Without backing material (i.e., solely air) and with matching layer.
3. With backing material and without matching layer.
4. With backing material and with matching layer.

For each configuration, water serves as sound propagation medium. Figure 7.36a and b display magnitude $|\underline{Z}_{el}(f)|$ and phase $\arg\{\underline{Z}_{el}(f)\}$ of the frequency-resolved electrical impedance, which was simulated through the analytical modeling approach. One can clearly see that configuration 1 exhibits strongly pronounced resonance–antiresonance pairs in $\underline{Z}_{el}(f)$ for the fundamental vibration mode as well as its overtones. Contrary to that, the resonance–antiresonance pairs get remarkably attenuated for the remaining configurations. This behavior is a consequence of the additional material layers being attached to the piezoelectric disk. Owing to reduced wave reflections at the disk's back and front in case of appropriate backing and matching, the vibration behavior of the piezoelectric disk changes which gets also visible in the electrical impedance.

Figure 7.36c depicts normalized magnitudes $|\underline{M}_e(f)|$ with respect to frequency f of the simulated transfer function for emitting sound pressure waves into water. As expected from the considerations in Sect. 7.4.1, configuration 1 features the highest maximum but smallest bandwidth of $|\underline{M}_e(f)|$. By attaching the backing material to the piezoelectric disk (i.e., configuration 3), we are able to substantially improve

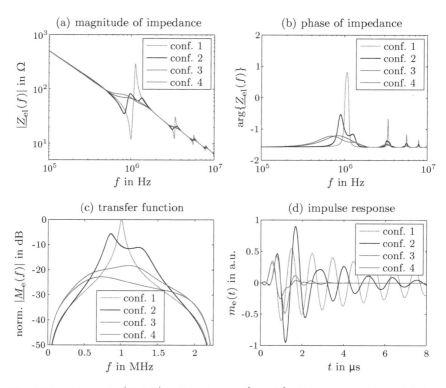

Fig. 7.36 **a** Magnitude $\left|\underline{Z}_{el}(f)\right|$ and **b** phase $\arg\left\{\underline{Z}_{el}(f)\right\}$ of frequency-resolved electrical impedances; **c** normalized transfer function $\left|\underline{M}_e(f)\right|$ (magnitude) for emitting sound pressure waves; **d** resulting impulse response $m_e(t)$; calculations for different configurations of piezoelectric ultrasonic transducer as mentioned in text

the $-3\,\mathrm{dB}$ bandwidth $B_s^{-3\,\mathrm{dB}}$ from 49 to 767 kHz. Even though $B_s^{-3\,\mathrm{dB}}$ slightly decreases to 762 kHz in case of the matching layer at disk's front (i.e., configuration 4), the maximum of $\left|\underline{M}_e(f)\right|$ increases by 4.5 dB, which constitutes an advantage concerning efficient piezoelectric ultrasonic transducer.

The observed behavior in $\left|\underline{M}_e(f)\right|$ is also confirmed by the impulse response $m_e(t)$ (see Fig. 7.36d) that follows from the inverse Fourier transform of $\underline{M}_e(f)$, i.e., $m_e(t) = \mathcal{F}^{-1}\left\{\underline{M}_e(f)\right\}$. Since $f \in [0, 2f_r]$ covers the working area of conventional piezoelectric ultrasonic transducers, the inverse Fourier transform was restricted to this frequency range. Compared to configuration 1, the matching layer in configuration 2 leads to higher magnitudes of $m_e(t)$. The impulse responses for configuration 1 and 2 show, however, remarkable post-pulse oscillations, while the resulting curves demonstrate the pulse-like characteristic for configuration 3 and 4. In accordance with the transfer functions $\left|\underline{M}_e(f)\right|$ from Fig. 7.36c, the magnitude of $m_e(t)$ is larger for configuration 4 than for configuration 3. To sum up, it can, thus, be stated that the presented simulation results definitely prove the importance of both backing material and matching layers for piezoelectric ultrasonic transducers.

7.6 Examples for Piezoelectric Ultrasonic Transducers

Hereafter, let us show selected piezoelectric ultrasonic transducers as well as their internal structure and exemplary measurement results. We concentrate on ultrasonic transducers for fluid wave propagation media and the human body. Section 7.6.1 deals with ultrasonic transducers (e.g., parking sensors) for airborne ultrasound. In Sects. 7.6.2 and 7.6.3, ultrasonic transducers for underwater use and medical diagnostics will be detailed, respectively.

7.6.1 Airborne Ultrasound

Airborne ultrasound is utilized in various technical and industrial applications, e.g., liquid level meters. Below, we will study a few examples of piezoelectric ultrasonic transducers for such applications. This includes conventional air-coupled piezoelectric transducers for emitting and receiving ultrasonic waves in air, parking sensors in motor vehicles as well as broadband ultrasonic transducers that are based on the EMFi material.

Conventional Piezoelectric Transducers

Figure 7.37a depicts a conventional air-coupled piezoelectric transducer that allows emitting ultrasonic waves in air with center frequencies of $f_c \approx 40\,\text{kHz}$. This low-cost ultrasonic transmitter is placed within an aluminum case and exploits the transverse length mode of a piezoceramic disk.[9] Because this disk is glued on a metallic membrane featuring a larger diameter (see Fig. 7.37b and c), the resulting transducer represents a piezoelectric unimorph (cf. Fig. 7.27a). In order to generate high sound pressure levels up to $100\,\text{dB}$, the air-coupled ultrasonic transmitter is additionally equipped with a horn facilitating sound radiation in air [4, 40]. The stainless steel mesh at the front should protect horn, membrane as well as piezoceramic disk against damage and dirt. Section 8.5.3 contains selected experimental results for the generated sound pressure amplitudes $\hat{p}_\sim$ of a conventional piezoelectric transmitter for airborne ultrasound.

Together with a corresponding air-coupled ultrasonic receiver, which is identically constructed but differs in the electrical input impedance, such transmitter can be applied for burglar alarm systems, liquid level meters, and anticollision devices. In former times, the transmitter–receiver combination was also utilized in remote controls for television sets.

---

[9]The transverse length mode of a disk corresponds to the radial mode (cf. Fig. 5.3 on p. 131).

(a)   **(b)**   **(c)**

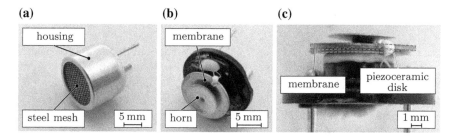

Fig. 7.37 a Conventional piezoelectric ultrasonic transmitter Sanwa SCS-401T for generating airborne ultrasound; **b, c** images without housing show piezoelectric unimorph consisting of piezoceramic disk and metallic membrane

Parking Sensors

Parking assistance systems in motor vehicles are usually based on airborne ultrasound. By evaluating the simple relation $z = c_0 t /2$ of geometric distance z, sound velocity c_0, and time-of-flight t of sound waves for pulse-echo mode, the driver can be informed about the current geometric distance between car and obstacles. Furthermore, such parking assistance systems will enable automatic parking if a car is equipped with several parking sensors. The utilized parking sensors are commonly located in the back and front bumper of a car (see Fig. 7.38a). Figure 7.38b shows a typical parking sensor of cylindrical shape that exploits again the transverse length mode (i.e., radial mode) of a single piezoceramic disk for emitting and receiving ultrasonic waves. As the cross-sectional views in Fig. 7.38c and d demonstrate, the disk is glued to the bottom of the cylindrical pot, which serves as membrane. Depending on both the desired directivity pattern and the operating range of these piezoelectric unimorph transducers (cf. Fig. 7.27a), the center frequencies of the ultrasonic waves currently can reach values up to 68 kHz. Such center frequencies lead to a wavelength of $\lambda_{\mathrm{aco}} \approx 5\,$mm in air.

EMFi Material

As already mentioned in Sect. 3.6.3, ferroelectrets like the so-called electromechanical film (EMFi) material offer a rather high piezoelectric strain constant d_{33} and piezoelectric voltage constant g_{33}. Due to the cellular structure, ferroelectret materials are, furthermore, mechanically flexible and feature a low mechanical stiffness as well as material density, which leads to a small acoustic impedance Z_{aco}. According to the considerations in Sects. 7.4 and 7.5, small values of Z_{aco} can be a major advantage regarding acoustic matching of piezoelectric materials with wave propagation media such as air. Ultrasonic transducers based on ferroelectrets should, therefore, provide a large bandwidth (see, e.g., [5, 17]).

Figure 7.39a shows displacement amplitudes $\hat{u}_z$ in thickness direction of an EMFi foil (thickness $70\,\mu$m; $d_{33} \approx 200\,$pC N^{-1}) with respect to both excitation voltage and

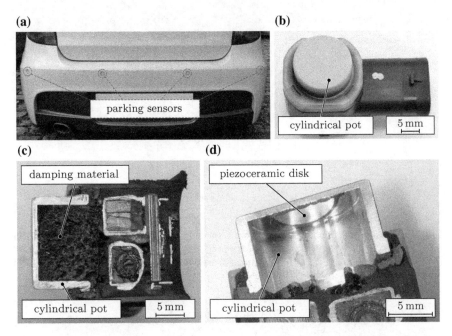

Fig. 7.38 **a** Parking sensors in back bumper of car; **b** typical ultrasound-based parking sensor from automotive supplier Valeo [78]; **c** cross-sectional view of parking sensor from **b**; **d** cross-sectional view without damping material; piezoceramic disk is glued to bottom of cylindrical pot

excitation frequency f. The displacements were acquired with the laser scanning vibrometer Polytec PSV-300 [56] and represent the arithmetic mean over the surface of the EMFi foil. For a given excitation voltage, the displacements stay nearly constant in the frequency range from 20 to 200 kHz and are in phase. Above the resonance frequency $f_r \approx 300$ kHz for the thickness extensional mode, there occur, however, remarkable differences in the amplitudes as well as phase angles, which can be ascribed to the inhomogeneous structure of ferroelectrets [64, 75]. The behavior in the lower frequency band should allow both emitting and receiving of short ultrasonic pulses in air. To prove this statement, let us take a look at the generated sound pressure level L_p of a disk-shaped EMFi foil with radius $R_T = 10$ mm. Figure 7.39b depicts the measured L_p at an axial distance of $R_{dir} = 0.5$ m (i.e., far field) with respect to excitation frequency f. Thereby, an excitation voltage of 640 V_{PP} was chosen. The sound pressure values were recorded in an anechoic room with a 1/8-inch condenser microphone from the company Brüel & Kjær [10]. As the measurement curve clearly reveals, one can generate high and constant sound pressure levels over a wide frequency range with a single foil. The resulting normalized directivity pattern $\hat{p}_{\sim}(\theta)$ for three different excitation frequencies is shown in Fig. 7.40. In accordance with the findings from Sect. 7.2.1, the directivity of the disk-shaped transducer will be pronounced stronger if f increases (cf. Fig. 7.10). The main lobes of the measured

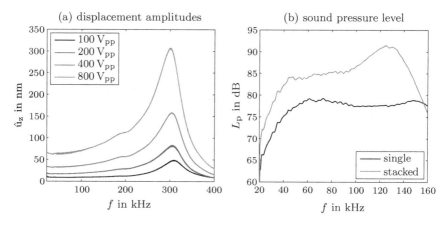

Fig. 7.39 a Measured amplitudes of averaged displacements $\hat{u}_z$ in thickness direction of EMFi foil with respect to excitation voltage and excitation frequency f; **b** measured sound pressure level L_p at axial distance $R_{dir} = 0.5\,\text{m}$ of single and two stacked disk-shaped EMFi foils with radius $R_T = 10\,\text{mm}$; excitation voltage $640\,\text{V}_{PP}$

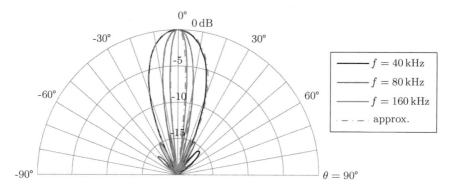

Fig. 7.40 Measured and approximated directivity pattern $20 \cdot \log_{10}\left(\hat{p}_\sim(\theta) / \hat{p}_{\sim\text{max}}\right)$ at distance $R_{dir} = 0.5\,\text{m}$ of EMFi foil with radius $R_T = 10\,\text{mm}$ for three different excitation frequencies f; approximations from (7.49); normalization with respect to maximum amplitude $\hat{p}_{\sim\text{max}}$

directivity patterns coincide very well with the corresponding approximations from (7.49).

It is also possible to exploit the large bandwidth of ferroelectret materials for receiving ultrasonic waves [64, 75]. By utilizing a preamplifier circuit, EMFi materials provide sensitivities of a few $\text{mV}\,\text{Pa}^{-1}$ in a wide frequency range, e.g., from 20 kHz up to 200 kHz. An appropriate preamplifier circuit consists of an impedance matching circuit followed by voltage amplification.

Even though the material parameters d_{33} and g_{33} are rather high, the small mechanical stiffness of ferroelectrets usually yields low electromechanical coupling factors, e.g., $k_{33} \approx 0.1$ for EMFi materials. This fact constitutes problems, especially when

Fig. 7.41 Artificial bat head
containing three EMFi-based
ultrasonic transducers, i.e.,
one transmitter of radius
$R_T = 7.5$ mm as well as two
receivers of radius
$R_T = 5.0$ mm; pinnas and
receivers can be rotated by
means of electric motors [64]

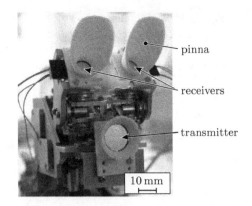

such a material should be used as ultrasonic emitter because high excitation voltages
are required. However, by stacking single EMFi foils and connecting them electrically in parallel, we can reduce the excitation voltage of a transmitter and increase
the receiver sensitivity [30, 64]. As Fig. 7.39b demonstrates, the generated sound
pressure level L_p will rise remarkably if two stacked EMFi foils are used instead of
a single foil. Although one would expect that L_p increases by 6 dB for two stacked
EMFi foils, there arise differences of more than 20 dB, which is a consequence of
the reduced resonance frequency $f_r < 150$ kHz for the two stacked EMFi foils. The
reduction directly follows from the greater mass of stacked foils compared to a single
foil. Since the resulting displacements u_z are relatively high nearby f_r, the generated
sound pressure level will also take high values. Hence, the difference in L_p between
stacked EMFi foils and a single foil can exceed 6 dB in the considered frequency
range.

The provided broadband characteristic of EMFi materials makes them particularly suited for generation and reception of airborne ultrasound. That was the reason
why EMFi foils were exploited to build up an artificial bat head in the course of the
project Chiroptera Inspired Robotic Cephaloid (CIRCE), which aimed at functional
reproduction of the biosonar system found in bats [75, 76]. Figure 7.41 illustrates
the realized setup of the artificial bat head mainly consisting of one ultrasonic transmitter ($R_T = 7.5$ mm) as well as two ultrasonic receivers ($R_T = 5.0$ mm). Each of
those transducers was fabricated of a single EMFi foil. For concentrating reflected
sound waves, the artificial bat head is additionally equipped with two pinnas, which
can be rotated by electric motors.

Recently, Álvarez-Arenas [2] suggested a promising design for an air-coupled
ultrasonic transducer that combines the advantages of both piezoceramic and ferroelectret materials. From the inside out, the transducer is composed of three main
components: (i) piezoelectric disk made of 1-3 composite, (ii) appropriate matching
layers, and (iii) ferroelectret foil. Generally speaking, this special transducer design
enables several different operation modes, which are listed below:

- The conventional pulse-echo mode is executed with either the piezoelectric composite disk or the ferroelectret foil; i.e., one of these components serves as both transmitter and receiver. When the piezoelectric composite is used as active element, the ferroelectret foil will represent a further passive matching layer of the ultrasonic transducer.
- The pitch-catch mode can be accomplished in two different ways, i.e., either the piezoelectric composite disk or the ferroelectret foil serves as transmitter, while the other component acts as receiver.
- Both the piezoelectric composite disk and ferroelectret foil are operated simultaneously as transmitter and receiver. Therewith, it is possible to increase the generated sound pressure and the bandwidth of the air-coupled ultrasonic transducer.

7.6.2   Underwater Ultrasound

Nondestructive testing, material characterization and acoustic microscopy are often based on underwater ultrasound because water allows contactless and low-loss transmission of ultrasonic waves between transducer and specimen. Hereinafter, we will show various examples for piezoelectric ultrasonic transducers (so-called *immersion transducers*) that are specifically designed for underwater use. This includes conventional transducer structures as well as special transducers being equipped with a delay line. Moreover, measurement results for the pulse-echo characteristic of a spherically focused immersion transducer are presented.

Conventional Immersion Transducers

Figure 7.42 illustrates seven different immersion transducers that are commercially available and contain a single piezoelectric element operating in thickness extensional mode. These transducers do not only differ in geometric dimension as well as provided center frequency for emitting and receiving ultrasonic waves but also in the focusing behavior and internal structure. Transducer 1–5 represents conventional immersion transducers. While transducer 1 and 5 offer a spherically focused characteristic, transducer 2 and 4 are piston-type transducers due to the planar active surface. In contrast, transducer 3 features a cylindrically focused behavior; i.e., this immersion transducer generates nonsymmetric sound pressure fields (cf. Fig. 8.21 on p. 377). Table 7.3 lists the focusing behavior, the diameter $2R_T$ of the active surface, the actual focal length $F_{T,act}$ as well as the center frequency f_c for all immersion transducers from Fig. 7.42.

The internal structure of a typical piston-type transducer (e.g., transducer 2 from Fig. 7.42) is displayed in Fig. 7.43a. Besides the piezoceramic disk and a waterproof housing, the transducer is equipped with a backing material and a single matching layer at the disk's front. According to Sect. 7.4.1, the backing material should increase the bandwidth of both emitted and receivable ultrasonic waves. The thin matching layer allows impedance matching of the piezoceramic material and the

Fig. 7.42 Different commercially available immersion transducers based on piezoelectric single elements; manufacturer Olympus [50] and Krautkramer [22]; transducer 6 and 7 are equipped with fused silica delay line; Table 7.3 lists focusing behavior and contains characteristic parameters of these ultrasonic transducers

Table 7.3 Focusing behavior and decisive parameters of piezoelectric ultrasonic transducers from Fig. 7.42; diameter $2R_T$ of active surface; actual focal length $F_{T,act}$; center frequency f_c for emitting and receiving ultrasonic waves

Transducer number	Focusing behavior	$2R_T$ mm	$F_{T,act}$ mm	f_c MHz
1	Spherically focused	38.1	88.9	2.25
2	Piston-type	12.7	–	2.25
3	Cylindrically focused	12.7	25.4	2.25
4	Piston-type	9.5	–	5
5	Spherically focused	12.7	88.9	10
6	Spherically focused	6.4	19.1	20
7	Spherically focused	6.4	12.7	50

sound propagation medium water. On the other hand, the matching material also serves as protective layer for the piezoceramic disk that is covered with electrodes.

Immersion Transducers with Delay Lines

As stated above, the seven immersion transducers in Fig. 7.42 also differ in their internal structure. This particularly refers to transducer 6 and 7, which are equipped with a fused silica delay line between piezoelectric disk and sound propagation medium (see Fig. 7.43b). One reason for the application of fused silica delay lines lies in inspections of specimens being placed close to the active transducer surface. If a conventional immersion transducer (i.e., without delay line) is employed for such

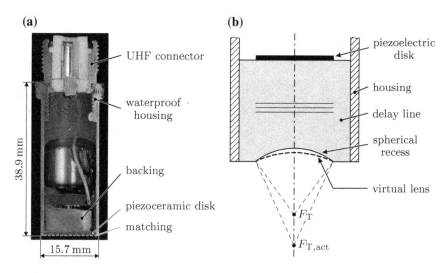

Fig. 7.43 **a** Cross-sectional view of internal structure of typical piston-type immersion transducer (cf. transducer 2 from Fig. 7.42); **b** spherically focused immersion transducer equipped with delay line between piezoelectric disk and sound propagation medium; virtual lens emulates piezoelectric disk and delay line; geometrical focus F_T of spherical recess; actual focal length $F_{T,act}$ representing radius of virtual lens

inspections, it might happen that information-bearing sound reflections arrive at the transducer although the emitted pulse has not decayed. Consequently, we are not able to recognize those reflections in a reliable way. By means of an appropriate delay line, this problem does not arise because the sound waves have to propagate twice through the delay line in pulse-echo mode. As a further advantage, a delay line of sufficient length ensures that the actual wave propagation medium (here water) and, therefore, the investigated specimen are located in the far field of the piezoelectric disk representing a simple piston-type transducer. Contrary to the near field, the far field of a piston-type transducer does not contain local minima and maxima (see Sect. 7.2.1), which can be a great benefit for ultrasonic imaging systems. However, one has to keep in mind that there always occur reflections of incident sound waves at the interface delay line/water. Such reflections significantly reduce the electrical transducer outputs in pulse-echo mode and, thus, lower the available signal-to-noise ratio.

When the fused silica delay line features a spherical recess (radius F_T; see Fig. 7.43b) at the front end, the emitted and received ultrasonic waves will be focused; i.e., the immersion transducer shows spherically focusing behavior. The great difference in the sound velocities of delay line and water causes substantial deviations between geometrical focal length F_T and actual focal length $F_{T,act}$. The deviations have two reasons, namely (i) different axial distances between the piezoelectric disk and points along the surface of the spherical recess as well as (ii) refraction of incident sound pressure waves at the interface delay line/water. We can consider both reasons

Fig. 7.44 Procedure to acquire 2-D pulse-echo characteristic of immersion transducer by means of wire target; wire diameter D_{wire}; position (x, z) of wire with respect to active transducer surface; wire orientated in parallel to y-axis

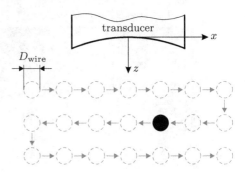

for semi-analytical calculation approaches (see, e.g., Sect. 7.1) of sound fields and electrical transducer outputs by introducing a virtual lens of radius $F_{T,act}$ for the ultrasonic transducer [54, 59]. This virtual lens corresponds then to the combination of piezoelectric disk and delay line (see Fig. 7.43b). Note that deviations of F_T and $F_{T,act}$ will also emerge for conventional focused immersion transducers without fused silica delay line whenever the curvature of both piezoelectric element and active transducer surface differ.

Pulse-Echo Characteristic

Since 3-D sound field measurements of immersion transducers require expensive equipment (e.g., hydrophones, see Chap. 8) and are, furthermore, hardly feasible in case of high center frequencies, it makes sense to acquire the pulse-echo characteristic. In doing so, we measure electrical transducer outputs $u_O(t)$ for a defined target instead of sound pressure values $p_\sim$. The pulse-echo characteristic results then from collecting the maximum $u_{ev;max}$ of the output's envelope at different distances between transducer and specimen. When an immersion transducer is exploited for an ultrasonic imaging system, this procedure will provide the advantage that one obtains information about the imaging properties, which also includes the utilized read-out electronics. Principally, the pulse-echo characteristic of an immersion transducer can be acquired by means of various specimens, e.g., spheres, plate targets as well as wire targets [57, 59]. To achieve quantitative data for a spherically focused transducer in a short time, it is recommended to use a wire target. By altering both on-axis and off-axis distances between transducer and wire (see Fig. 7.44), we can create a 2-D image for the pulse-echo characteristic. This data will also allow us to draw conclusions regarding generated sound fields if the wire diameter D_{wire} is much smaller than the wavelength λ_{aco} in the surrounding wave propagation medium.

Figure 7.45 displays the acquired pulse-echo characteristic $u_{ev;max}(x, z)$ of the spherically focused immersion transducer Olympus V311-SU [50], which was operated in a water tank by the pulser/receiver Olympus PR5900 [50]. A tungsten wire with diameter $D_{wire} = 50\,\mu m$ served as specimen for this measurement. The pulse-echo characteristic in Fig. 7.45a clearly indicates the focal volume of the spherically focused transducer. Moreover, it is noticeable that the transducer slightly squints

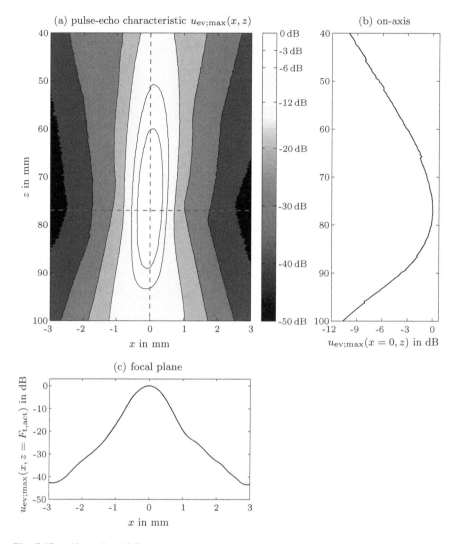

Fig. 7.45 **a** Normalized 2-D pulse-echo characteristic $u_{\mathrm{ev;max}}(x, z)$ for wire target at (x, z); **b** pulse-echo characteristic on-axis, i.e., $u_{\mathrm{ev;max}}(x = 0, z)$; **c** pulse-echo characteristic in focal plane, i.e., $u_{\mathrm{ev;max}}(x, z = F_{\mathrm{t,act}})$; maximum of $u_{\mathrm{ev;max}}(x, z)$ at $x = 0$; spherically focused immersion transducer Olympus V311-SU

which stems from production-related asymmetries. The maximum of $u_{\mathrm{ev;max}}(x, z)$ representing the true focus arises at the axial distance $F_{\mathrm{T,act}} = 77.0\,\mathrm{mm}$ from the active transducer surface.

Alternatively to a wire target, one can measure the pulse-echo signal $u_{\mathrm{O}}(t)$ of the spherically focused transducer for a plane plate target that is placed in the focal plane, i.e., at the axial distance of the true focus. Figure 7.46a and b illustrate

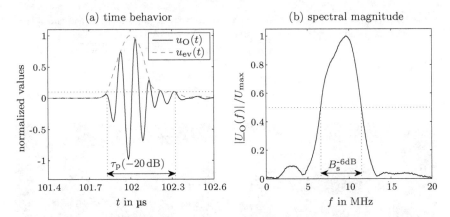

Fig. 7.46 a Time behavior of electrical transducer output $u_O(t)$ and its envelope $u_{ev}(t)$ for plane plate target being placed in focal plane; normalization with respect to maximum $u_{ev;max}$ of envelope; **b** normalized spectral magnitude $|\underline{U}_O(f)| = |\mathcal{F}\{u_O(t)\}|$; characteristic data from curves: $\tau_p(-20\,dB) = 495\,ns$, $f_l = 6.60\,MHz$, $f_u = 11.51\,MHz$, $B_s^{-6dB} = f_u - f_l = 4.91\,MHz$, $f_{max} = 9.66\,MHz$, $f_c = 8.72\,MHz$; spherically focused immersion transducer Olympus V311-SU

the acquired time behavior of $u_O(t)$ and its spectral magnitude $|\underline{U}_O(f)| = |\mathcal{F}\{u_O(t)\}|$, respectively. From the curves, we are able to deduce quantities like time difference $\tau_p(-20\,dB) = 495\,ns$ and center frequency $f_c = 8.72\,MHz$. Together with $F_{T,act}$, the transducer radius $R_T = 6.5\,mm$, and the sound velocity $c_0 = 1483\,ms^{-1}$ of water, these quantities yield the wavelength $\lambda_{aco} = 0.17\,mm$ as well as the dimensionless Fresnel parameter $S_F = 0.31$ and the dimensionless f-number $f^\# = 5.9$ of the immersion transducer. Furthermore, one can estimate the achievable spatial resolution of the spherically focused transducer in axial and lateral direction through (7.68), (7.73), and (7.74). The axial resolution takes the value $d_{ax} = 0.37\,mm$, whereas the lateral resolution equals $d_{lat}(Rayleigh) = 1.23\,mm$ and $d_{lat}(Sparrow) = 0.96\,mm$ according to the Rayleigh and Sparrow two-point definition, respectively.

By evaluating the pulse-echo characteristic on-axis (see Fig. 7.45b) and in the focal plane (see Fig. 7.45c), we can deduce quantities like depth of focus d_z and lateral beam width d_ρ. As explained in Sect. 7.2.2, these geometric distances refer to reductions of sound pressure values by 3 dB. Owing to the fact that the pulse-echo characteristic $u_{ev;max}(x, z)$ contains the transducer behavior twice, a reduction by 6 dB has to be considered instead. Therefore, $d_z(-3\,dB)$ and $d_\rho(-3\,dB)$ in the generated sound field correspond to $d_z(-6\,dB)$ and $d_\rho(-6\,dB)$ in $u_{ev;max}(x, z)$, respectively. Table 7.4 contains d_z and d_ρ resulting from the pulse-echo characteristic as well as those, which are derived from the approximations (7.60) and (7.64). As the comparison of the entries clearly points out, measured values and approximations coincide very well.

Table 7.4 Results of measurements and approximations for depth of focus d_z and lateral beam width d_ρ of spherically focused immersion transducer Olympus V311-SU

	Depth of focus d_z in mm	Lateral beam width d_ρ in mm
Measurement	42.1	1.00
Approximation	42.3	1.04

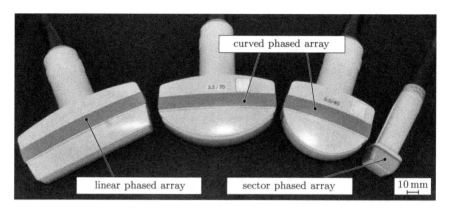

Fig. 7.47 Different ultrasonic linear arrays based on piezoceramic materials for medical diagnostics; curved phased arrays differ in curvature radius

7.6.3 Medical Diagnostics

Figure 7.47 shows four conventional piezoelectric transducer arrays for medical diagnostics. Such transducer arrays are utilized for checking internal organs, pregnancy examinations as well as detecting kidney stones or gallstones (e.g., [12, 29, 77]). The linear phased array and both curved phased arrays contain more than a hundred single-element transducers made of piezoceramics, which are either partially operated in parallel or can be excited and read out separately. By contrast, the sector phased array consists only of 64 single-element transducers that have to be operated separately for 2-D examinations (cf. Sect. 7.4.2). While the linear and both curved phased arrays offer the center frequency $f_c = 3.5$ MHz for emitting and receiving ultrasonic waves, the sector phased array is limited to $f_c = 2.25$ MHz. The curvature radius of the left and right curved phased array equals 70 mm and 40 mm, respectively.

In Fig. 7.48, one can see a cross-sectional view of the curved phased array featuring the curvature radius 70 mm. Just as typical single-element transducers (see Sect. 7.6.2), the array elements operating in thickness extensional mode are equipped with a backing material to increase the transducer bandwidth. A thin matching layer at the element's front enables again matching of acoustic impedances, and a protective layer serves as acoustic lens for emitted as well as received ultrasonic waves. Since the sound velocity in the protective layer is smaller than in the human body, the

Fig. 7.48 Cross-sectional view of curved phased array with curvature radius 70 mm from Fig. 7.47

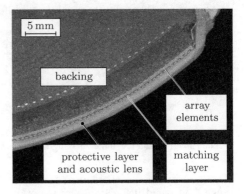

acoustic lens exhibits a concave shape. The efficient acoustic coupling of protective layer and human body demands an additional liquid coupling gel to avoid disturbing air between them.

7.7 Ultrasonic Imaging

Ultrasonic imaging represents a very important application area of piezoelectric ultrasonic transducers. It is often conducted in noninvasive medical diagnostics and in nondestructive testing. For such applications, the ultrasonic transducers are commonly operated in the pulse-echo mode. When there exist inhomogeneities of the acoustic impedance Z_{aco} along the propagation path, the incident ultrasonic waves will become reflected (cf. Sect. 2.3.4). The time-of-flight of these reflections and their magnitudes can be exploited to deduce certain information about the internal structure of the investigated area. For example, we will be able to localize inhomogeneities such as flaws inside of solids if the sound velocity is known.

Apart from pregnancy examinations, ultrasonic imaging in medical diagnostics[10] can be found in anesthesiology, cardiology, gastroenterology, neurology, and urology [16, 40, 77]. Compared to other imaging techniques (e.g., radiography and magnetic resonance imaging), ultrasonic imaging is very cheap regarding equipment costs and examination costs. Since the utilized acoustic intensities in medical diagnostics stay below $100\,\mathrm{mW\,cm^{-2}}$, ultrasonic imaging does not pose a threat to the examined patient. The center frequencies of the used piezoelectric ultrasonic transducers (e.g., phased arrays; cf. Fig. 7.47) typically range from 1 MHz up to 15 MHz.

As mentioned above, a further important application area of ultrasonic imaging is nondestructive testing[11] (NDT). If nondestructive testing is based on ultrasonic waves, it will also be called *ultrasonic testing*. The applications of ultrasonic testing

---

[10]Ultrasonic imaging in medical diagnostics is also known as medical ultrasound, diagnostic sonography, and ultrasonography.

[11]Nondestructive testing is also called nondestructive evaluation (NDE).

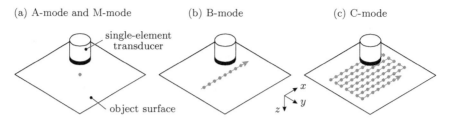

Fig. 7.49 Scan lines and scan positions on object surface for **a** A-mode and M-mode imaging, **b** B-mode imaging, and **c** C-mode imaging with single-element ultrasonic transducer

range from weld inspection over material characterization to detection and localization of small flaws [19, 36]. Ultrasonic testing makes it possible to investigate areas below the object surface. Just as in case of ultrasound-based medical diagnostics, the major advantage of ultrasonic testing lies in the low examination costs. Ultrasonic testing devices are often portable and enable highly automated operation. The typical range of the transducer's center frequencies is comparable to that in medical diagnostics. However, acoustic microscopy as a special application of NDT requires piezoelectric ultrasonic transducers with center frequencies much higher than 10 MHz [9, 43]. For instance, some acoustic microscopes operate with 1 GHz, which leads to the wavelength $\lambda \approx 1.5\,\mu\text{m}$ in water.

The acoustic coupling of ultrasonic transducer and investigated object is decisive for ultrasonic imaging and depends on the application. While acoustic microscopy commonly exploits water for coupling, a special gel serves as coupling medium in noninvasive medical diagnostics (cf. Sect. 7.6.3). Air pockets between transducer and investigated object may distort the resulting image or even impede ultrasonic imaging.

Below, let us briefly discuss the most important imaging modes that are applied in ultrasonic imaging, namely A-mode and M-mode, B-mode as well as C-mode imaging. Figure 7.49 illustrates typical transducer movements, which will be required for the four imaging modes if a single-element transducer is used to investigate a specimen. Especially in medical diagnostics, there exist further imaging modes like continuous wave (CW) Doppler and pulsed wave (PW) Doppler imaging for blood flow measurements [32, 71].

7.7.1 A-Mode and M-Mode Imaging

The name A-mode imaging originates from the word *amplitude*. For this imaging mode, the ultrasonic transducer stays at the same position with respect to the surface of the investigated object (see Fig. 7.49a). A single A-mode line represents the envelope $u_{\text{ev}}(t)$ of the received echo signal $u_{\text{O}}(t)$ (i.e., electrical transducer output) due to reflected ultrasonic waves. The envelope can be calculated with the aid of the

Fig. 7.50 Amplified
transducer output $u_O(t)$ and
envelope $u_{ev}(t)$ representing
single A-mode line; surface
of plate reflector made of
acrylic glass located at
$z = 12.7\,mm$; spherically
focused immersion
transducer Olympus
VU390-SU/RM

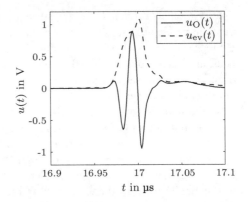

Hilbert transform (cf. (7.69)). Figure 7.50 depicts an amplified output $u_O(t)$ and the resulting envelope $u_{ev}(t)$ of the spherically focused immersion transducer Olympus VU390-SU/RM (see transducer number 7 in Fig. 7.42 [50]), which was operated in a water tank by the pulser/receiver Olympus PR5900 [50]. Thereby, the surface of a plate reflector made of acrylic glass was placed in the transducer's focal plane, i.e., at the axial distance $z = 12.7\,mm$ from the active transducer surface.

By arranging several A-mode lines from subsequent instants of time column by column, one obtains a M-mode image, whereby the letter M stands for *motion*. The current A-mode line is either brightness-coded or color-coded in the M-mode image. Mostly, high amplitude values are assigned to bright colors, while low values are visualized by dark colors. When there arise changes in the internal structure of the object, the A-mode lines and, consequently, the resulting M-mode image will be altered. Temporal changes in the A-mode lines were already used in the 1950s to detect heart motions [77].

7.7.2 B-Mode Imaging

B-mode imaging is the most commonly used mode in ultrasonic imaging. The name originates from the word *brightness*. In contrast to A-mode and M-mode imaging, the single-element transducer has to be mechanically moved along a line for B-mode imaging (see Fig. 7.49b). The B-mode image results then from arranging the A-mode lines at subsequent transducer positions column by column. Similar to M-mode images, the A-mode lines are either brightness-coded or color-coded in B-mode images. A B-mode image can be interpreted as cross-sectional view of the investigated object along the prescribed line, which is also called scan line. The image will provide information about the internal structure if the sound velocities are known.

The movement of a single-element transducer does not constitute a problem for NDT because the investigated specimen remains unchanged and the examination time is often not critical. However, since the movement is a time-consuming procedure, single-element transducers are not utilized in medical diagnostics for generating B-mode images. Nowadays, the required real-time capability in medical diagnostics is usually achieved with phased arrays [31, 77]. As mentioned in Sect. 7.4.2, transducer arrays allow steering and focusing, which lead to a reasonable spatial resolution in the whole B-mode image.

In the majority of cases, acoustic microscopes use spherically focused single-element transducers [9, 43, 82]. This is a consequence of the fact that the desired spatial resolution demands ultrasonic frequencies considerably higher than 10 MHz. Transducer arrays operating at such high frequencies still pose a great challenge for research and transducer manufacturer (cf. Sect. 7.4.2). Owing to the limitation to spherically focused single-element transducers, the spatial resolution of acoustic microscopes is only ideal in the focal zone. To improve the spatial resolution outside the focal zone, one can apply the so-called *synthetic aperture focusing technique* (SAFT) [52, 60, 69]. The underlying idea of this technique lies in coherently summing the transducer output signals at different transducer positions. From the theoretical point of view, the SAFT yields depth-independent spatial resolutions in B-mode images of homogeneous wave propagation media. The application of the SAFT to inhomogeneous specimens (e.g., layered structures) is subject to ongoing research at the Chair of Sensor Technology (Friedrich-Alexander-University Erlangen-Nuremberg) in the framework of the Collaborative Research Center TRR39 [13].

As a simple example, let us study a B-mode image of a thin wire reflector being aligned in parallel to y-axis at $x = 0$ mm and at the axial distance $z \approx 9.5$ mm from the active transducer surface. Since the used spherically focused immersion transducer Olympus VU390-SU/RM features the focal length $F_{T,act} = 12.7$ mm, the wire reflector is located in the negative defocus region. Figure 7.51a shows the amplified transducer outputs $u_O(t)$ and the resulting A-mode lines $u_{ev}(t)$ at two different lateral transducer positions. One can clearly see that both the temporal position and the height of the maximum in $u_O(t)$ and $u_{ev}(t)$ strongly depend on the lateral transducer position x. Apart from the main reflection, the A-mode lines contain signal components stemming from multiple reflections. The resulting B-mode image in Fig. 7.51b shows sickle-shaped reflections, which are characteristic for small reflectors being located outside the transducer's focal plane. The time axis from the A-mode line was rescaled by $z = c_0 t/2$ with the sound velocity c_0 in water.

7.7.3 C-Mode Imaging

The name C-mode imaging originates from the word *complex*. For recording a C-mode image, the single-element ultrasonic transducer has to be moved along the xy-plane over the surface of the investigated object (see Fig. 7.49c). The C-mode image results from evaluating the maximum of the A-mode line at each transducer position. Brightness coding or color coding finally leads to the C-mode image. When

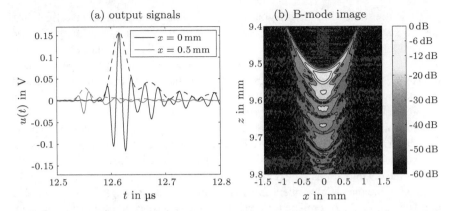

Fig. 7.51 a Amplified transducer output $u_O(t)$ and $u_{ev}(t)$ (representing A-mode line) for two different lateral distances x of transducer and wire reflector (copper wire with 45 μm in diameter); wire reflector located at $z \approx 9.5$ mm; **b** normalized B-mode image of wire reflector; spherically focused immersion transducer Olympus VU390-SU/RM

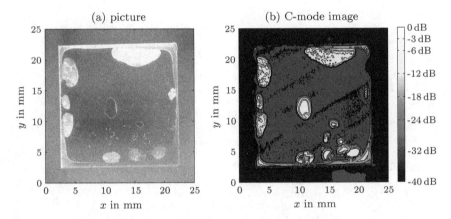

Fig. 7.52 a Picture of joined acrylic glass plates with geometric dimensions 20.0 mm × 20.0 mm × 4.75 mm; **b** normalized C-mode image of adhesive layer; bright areas mark delaminations; spherically focused immersion transducer Olympus VU390-SU/RM

the evaluation of the maximum is limited to a short time window, we will obtain a cross-sectional view of the investigated object in a certain depth z with respect to the active transducer surface. Thus, one can generate depth-dependent information about the object by altering the short time window. That is the reason why C-mode imaging is often used in acoustic microscopy. Just as in case of B-mode imaging, the spatial resolution in the C-mode images will only be ideal if the selected cross-sectional view is close to the transducer's focal plane. Again, the SAFT can be applied to enhance spatial resolution outside the focal plane.

To demonstrate C-mode imaging in acoustic microscopy, let us investigate the adhesive area of two optically transparent acrylic glass plates with the geometric dimensions 20.0 mm × 20.0 mm × 4.75 mm. The acrylic glass plates were joined with an optically transparent plastic glue. The thickness of the adhesive layer equals ≈ 0.25 mm. As the pictures in Fig. 7.52a indicates, there exist discontinuous areas of the adhesive layer, which are usually accompanied by trapped air. These areas can be interpreted as delamination of the joined plates.

Ultrasonic imaging was again conducted with the spherically focused immersion transducer Olympus VU390-SU/RM. For C-mode imaging, the considered time window was set to the depth of the adhesive layer. The C-mode image in Fig. 7.52b clearly reveals the delaminations. The reason for this lies in the acoustic impedance Z_{aco}. In contrast to the plastic glue, the acoustic impedance of air strongly differs from that of acrylic glass. Therefore, the incident ultrasonic waves become almost completely reflected at delaminations, whereas the reflections are rather small at areas of good plate joint. The large reflections appear as large peaks of the A-mode lines in the considered time window.

References

1. Arditi, M., Foster, F.S., Hunt, J.W.: Transient fields of concave annular arrays. Ultrason. Imaging **3**(1), 37–61 (1981)
2. Álvarez Arenas, T.E.G.: Air-coupled piezoelectric transducers with active polypropylene foam matching layers. Sensors (Switzerland) **13**(5), 5996–6013 (2013)
3. Arnau, A.: Piezoelectric Transducers and Applications, 2nd edn. Springer, Berlin (2008)
4. Asher, R.C.: Ultrasonic Sensors. Institute of Physics Publishing, Bristol (1997)
5. Bauer, S., Gerhard-Multhaupt, R., Sessler, G.M.: Ferroelectrets: soft electroactive foams for transducers. Phys. Today **57**(2), 37–43 (2004)
6. Berlincourt, D.A., Curran, D.R., Jaffe, H.: Piezoelectric and Piezomagnetic Materials and Their Function in Transducers. Academic Press, New York (1964)
7. Bernstein, H.: NF- und HF-Messtechnik. Springer, Wiesbaden (2015)
8. Born, M., Wolf, E., Bhatia, A.B., Clemmow, P.C., Gabor, D., Stokes, A.R., Taylor, A.M., Wayman, P.A., Wilcock, W.L.: Principles of Optics: Electromagnetic Theory of Propagation, Interference and Diffraction of Light. Cambridge University Press, Cambridge (2000)
9. Briggs, A.: Advances in Acoustic Microscopy, vol. 1. Springer, Berlin (1995)
10. Brüel & Kjær: Product portfolio (2018). http://www.bksv.com
11. Cannata, J.M., Ritter, T.A., Chen, W.H., Silverman, R.H., Shung, K.K.: Design of efficient, broadband single-element (20–80 MHz) ultrasonic transducers for medical imaging applications. IEEE Trans. Ultrason. Ferroelectr. Freq. Control **50**(11), 1548–1557 (2003)
12. Cobbold, R.S.C.: Foundations of Biomedical Ultrasound. Oxford University Press, Oxford (2007)
13. Collaborative Research Center TRR 39: Production Technologies for Lightmetal and Fiber Reinforced Composite based Components with Integrated Piezoceramic Sensors and Actuators (PT-PIESA) (2018). http://www.pt-piesa.tu-chemnitz.de
14. Dausch, D.E., Castellucci, J.B., Chou, D.R., Von Ramm, O.T.: Theory and operation of 2-D array piezoelectric micromachined ultrasound transducers. IEEE Trans. Ultrason. Ferroelectr. Freq. Control **55**(11), 2484–2492 (2008)
15. Desilets, C.S., Fraser, J.D., Kino, G.S.: The design of efficient broad-band piezoelectric transducers. IEEE Trans. Sonics Ultrason. **25**(3), 115–125 (1978)

16. Dössel, O.: Bildgebende Verfahren in der Medizin. Springer, Berlin (2000)
17. Ealo, J.L., Seco, F., Jimenez, A.R.: Broadband EMFi-based transducers for ultrasonic air applications. IEEE Trans. Ultrason. Ferroelectr. Freq. Control **55**(4), 919–929 (2008)
18. Eriksson, T.J.R., Ramadas, S.N., Dixon, S.M.: Experimental and simulation characterisation of flexural vibration modes in unimorph ultrasound transducers. Ultrasonics **65**, 242–248 (2016)
19. Fendt, K., Mooshofer, H., Rupitsch, S.J., Ermert, H.: Ultrasonic defect characterization in heavy rotor forgings by means of the synthetic aperture focusing technique and optimization methods. IEEE Trans. Ultrason. Ferroelectr. Freq. Control **63**(6), 874–885 (2016)
20. Fink, M., Cardoso, J.F.: Diffraction effects in pulse-echo measurement. IEEE Trans. Sonics Ultrason. **31**(4), 313–329 (1984)
21. Gebhardt, S., Schönecker, A., Steinhausen, R., Seifert, W., Beige, H.: Quasistatic and dynamic properties of 1–3 composites made by soft molding. J. Eur. Ceram. Soc. **23**(1), 153–159 (2003)
22. General Electrics (GE): Product portfolio (2018). https://www.gemeasurement.com
23. Germano, C.P.: Flexure mode piezoelectric transducers. IEEE Trans. Audio Electroacoust. **19**(1), 6–12 (1971)
24. Gradshteyn, I.S., Ryzhik, I.M.: Tables of Integrals, Series, and Products, 8th edn. Academic Press, Amsterdam (2014)
25. Gururaja, T.R., Cross, L.E., Newnham, R.E., Auld, B.A., Wang, Y.J., Schulze, W.A.: Piezoelectric composite materials for ultrasonic transducer applications. Part I: resonant modes of vibration of PZT rod-polymer composites. IEEE Trans. Sonics Ultrason. **32**(4), 481–498 (1985)
26. Harris, G.R.: Review of transient field theory for a baffled planar piston. J. Acoust. Soc. Am. **70**(1), 10–20 (1981)
27. Harris, G.R.: Transient field of a baffled planar piston having an arbitrary vibration amplitude distribution. J. Acoust. Soc. Am. **70**(1), 186–204 (1981)
28. Hecht, E.: Optics, 5th edn. Pearson (2016)
29. Heywang, W., Lubitz, K., Wersing, W.: Piezoelectricity: Evolution and Future of a Technology. Springer, Berlin (2008)
30. Hillenbrand, J., Sessler, G.M.: High-sensitivity piezoelectric microphones based on stacked cellular polymer films (l). J. Acoust. Soc. Am. **116**(6), 3267–3270 (2004)
31. Honskins, P., Martin, K., Thrush, A.: Diagnostic Ultrasound - Physics and Equipment, 2nd edn. Cambridge University Press, Cambridge (2010)
32. Jensen, J.A.: Estimation of Blood Velocities Using Ultrasound. Cambridge University Press, Cambridge (1996)
33. Jensen, J.A., Svendsen, N.B.: Calculation of pressure fields from arbitrarily shaped, apodized, and excited ultrasound transducers. IEEE Trans. Ultrason. Ferroelectr. Freq. Control **39**(2), 262–267 (1992)
34. Jespersen, S.K., Pedersen, P.C., Wilhjelm, J.E.: The diffraction response interpolation method. IEEE Trans. Ultrason. Ferroelectr. Freq. Control **45**(6), 1461–1475 (1998)
35. Kino, G.S.: Acoustic Waves, Devices, Imaging and Analog Signal Processing. Prentice Hall, New Jersey (1987)
36. Krautkrämer, J., Krautkrämer, H.: Werkstoffprüfung mit Ultraschall. Springer, Berlin (1986)
37. Krimholtz, R., Leedom, D.A., Matthaei, G.L.: New equivalent circuits for elementary piezoelectric transducers. Electron. Lett. **6**(12), 398–399 (1970)
38. Kümmritz, S., Wolf, M., Kühnicke, E.: Simultaneous determination of thicknesses and sound velocities of layered structures. Tech. Messen **82**(3), 127–134 (2015)
39. Lerch, R.: Simulation von Ultraschall-Wandlern. Acustica **57**(4–5), 205–217 (1985)
40. Lerch, R., Sessler, G.M., Wolf, D.: Technische Akustik: Grundlagen und Anwendungen. Springer, Berlin (2009)
41. Lhemery, A.: Impulse-response method to predict echo-responses from targets of complex geometry. Part I: theory. J. Acoust. Soc. Am. **90**(5), 2799–2807 (1991)
42. Länge, K., Rapp, B.E., Rapp, M.: Surface acoustic wave biosensors: a review. Anal. Bioanal. Chem. **391**(5), 1509–1519 (2008)
43. Maev, G.: Advances in Acoustic Microscopy and High Resolution Imaging. Wiley-VCH, Weinheim (2012)

44. Mason, W.P.: Electro-mechanical Transducers and Wave Filters, 3rd edn. D. van Nostrand, Princeton (1964)
45. Muralt, P., Ledermann, N., Paborowski, J., Barzegar, A., Gentil, S., Belgacem, B., Petitgrand, S., Bosseboeuf, A., Setter, N.: Piezoelectric micromachined ultrasonic transducers based on PZT thin films. IEEE Trans. Ultrason. Ferroelectr. Freq. Control **52**(12), 2276–2288 (2005)
46. Newnham, R.E., Skinner, D.P., Cross, L.E.: Connectivity and piezoelectric-pyroelectric composites. Mater. Res. Bull. **13**(5), 525–536 (1978)
47. Nguyen, N.T., Lethiecq, M., Karlsson, B., Patat, F.: Highly attenuative rubber modified epoxy for ultrasonic transducer backing applications. Ultrasonics **34**(6), 669–675 (1996)
48. Nierla, M., Rupitsch, S.J.: Hybrid seminumerical simulation scheme to predict transducer outputs of acoustic microscopes. IEEE Trans. Ultrason. Ferroelectr. Freq. Control **63**(2), 275–289 (2016)
49. Olver, F.W.J., Lozier, D.W., Boisvert, R.F., Clark, C.W.: NIST Handbook of Mathematical Functions. Cambridge University Press, Cambridge (2010)
50. Olympus Corporation: Product portfolio (2018). https://www.olympus-ims.com
51. Oppenheim, A.V., Schafer, R.W., Buck, J.R.: Discrete-Time Signal Processing. Prentice Hall, New Jersey (1999)
52. Passmann, C., Ermert, H.: A 100-MHz ultrasound imaging system for dermatologic and ophthalmologic diagnostics. IEEE Trans. Ultrason. Ferroelectr. Freq. Control **43**(4), 545–552 (1996)
53. Penttinen, A., Luukkala, M.: The impulse response and pressure nearfield of a curved ultrasonic radiator. J. Phys. D Appl. Phys. **9**(10), 1547–1557 (1976)
54. Penttinen, A., Luukkala, M.: Sound pressure near the focal area of an ultrasonic lens. J. Phys. D Appl. Phys. **9**(13), 1927–1936 (1976)
55. Pilkey, W.D.: Stress, Strain, and Structural Matrices. Wiley, New York (1994)
56. Polytec GmbH: Product portfolio (2018). http://www.polytec.com
57. Raum, K., O'Brien, W.D.: Pulse-echo field distribution measurement technique for high-frequency ultrasound sources. IEEE Trans. Ultrason. Ferroelectr. Freq. Control **44**(4), 810–815 (1997)
58. Reindl, L.M.: Theory and application of passive SAW radio transponders as sensors. IEEE Trans. Ultrason. Ferroelectr. Freq. Control **45**(5), 1281–1292 (1998)
59. Rupitsch, S.J.: Entwicklung eines hochauflösenden Ultraschall-Mikroskops für den Einsatz in der zerstörungsfreien Werkstoffprüfung. Ph.D. thesis, Johannes Kepler University Linz (2008)
60. Rupitsch, S.J., Zagar, B.G.: Acoustic microscopy technique to precisely locate layer delamination. IEEE Trans. Instrum. Meas. **56**(4), 1429–1434 (2007)
61. Rupitsch, S.J., Zagar, B.G.: A method to increase the spatial resolution of synthetically focussed ultrasound transducers. Tech. Messen **75**(4), 259–267 (2008)
62. Rupitsch, S.J., Nierla, M.: Efficient numerical simulation of transducer outputs for acoustic microscopes. In: Proceedings of IEEE Sensors, pp. 1656–1659 (2014)
63. Rupitsch, S.J., Kindermann, S., Zagar, B.G.: Estimation of the surface normal velocity of high frequency ultrasound transducers. IEEE Trans. Ultrason. Ferroelectr. Freq. Control **55**(1), 225–235 (2008)
64. Rupitsch, S.J., Lerch, R., Strobel, J., Streicher, A.: Ultrasound transducers based on ferroelectret materials. IEEE Trans. Dielectr. Electr. Insul. **18**(1), 69–80 (2011)
65. Safari, A., Akdogan, E.K.: Piezoelectric and Acoustic Materials for Transducer Applications. Springer, Berlin (2010)
66. Safari, A., Allahverdi, M., Akdogan, E.K.: Solid freeform fabrication of piezoelectric sensors and actuators. J. Mater. Sci. **41**(1), 177–198 (2006)
67. Sammoura, F., Kim, S.G.: Theoretical modeling and equivalent electric circuit of a bimorph piezoelectric micromachined ultrasonic transducer. IEEE Trans. Ultrason. Ferroelectr. Freq. Control **59**(5), 990–998 (2012)
68. San Emeterio, J.L., Ullate, L.G.: Diffraction impulse response of rectangular transducers. J. Acoust. Soc. Am. **92**(2), 651–662 (1992)

69. Scharrer, T., Schrapp, M., Rupitsch, S.J., Sutor, A., Lerch, R.: Ultrasonic imaging of complex specimens by processing multiple incident angles in full-angle synthetic aperture focusing technique. IEEE Trans. Ultrason. Ferroelectr. Freq. Control **61**(5), 830–839 (2014)
70. Smith, W.A.: Role of piezocomposites in ultrasonic transducers. Proc. Int. IEEE Ultrason. Symp. (IUS) **2**, 755–766 (1989)
71. Smythe, W.B.: Diagnostic Ultrasound - Imaging and Blood Flow Measurements. CRC Press, Boca Raton (2006)
72. Steinberg, B.D.: Principles of Aperture and Array System Design: Including Random and Adaptive Arrays. Wiley, New York (1976)
73. Steinhausen, R., Hauke, T., Seifert, W., Beige, H., Watzka, W., Seifert, S., Sporn, D., Starke, S., Schönecker, A.: Finescaled piezoelectric 1–3 composites: properties and modeling. J. Eur. Ceram. Soc. **19**(6–7), 1289–1293 (1999)
74. Stepanishen, P.R.: The time-dependent force and radiation impedance on a piston in a rigid infinte planar baffle. J. Acoust. Soc. Am. **49**(3), 841–849 (1971)
75. Streicher, A.: Luftultraschall-Sender-Empfänger-System für einen künstlichen Fledermauskopf. Ph.D. thesis, Friedrich-Alexander-University Erlangen-Nuremberg (2008)
76. Streicher, A., Müller, R., Peremans, H., Lerch, R.: Broadband ultrasonic transducer for a artificial bat head. Proc. Int. IEEE Ultrason. Symp. (IUS) **2**, 1364–1367 (2003)
77. Szabo, T.L.: Diagnostic Ultrasound Imaging: Inside Out, 2nd edn. Academic Press, Amsterdam (2014)
78. Valeo: Product portfolio (2018). http://www.valeo.com/en/
79. Verhoef, W.A., Cloostermans, M.J.T.M., Thijssen, J.M.: The impulse response of a focused source with an arbitrary axisymmetric surface velocity distribution. J. Acoust. Soc. Am. **75**(6), 1716–1721 (1984)
80. Wang, Z., Zhu, W., Miao, J., Zhu, H., Chao, C., Tan, O.K.: Micromachined thick film piezoelectric ultrasonic transducer array. Sens. Actuators A Phys. **130–131**, 485–490 (2006)
81. Wooh, S.C., Shi, Y.: Optimum beam steering of linear phased arrays. Wave Motion **29**(3), 245–265 (1999)
82. Wüst, M., Eisenhart, J., Rief, A., Rupitsch, S.J.: System for acoustic microscopy measurements of curved structures. Tech. Messen **84**(4), 251–262 (2017)

Chapter 8
Characterization of Sound Fields Generated by Ultrasonic Transducers

The metrological characterization of sound fields represents an important step in the design and optimization of ultrasonic transducers. In this chapter, we will concentrate on the so-called light refractive tomography (LRT), which is an optical-based measurement principle. It allows noninvasive, spatially as well as temporally resolved acquisition of both, sound fields in fluids and mechanical waves in optical transparent solids. Before the history and fundamentals (e.g., tomographic reconstruction) of LRT are studied in Sects. 8.2 and 8.3, we will discuss conventional measurement principles (e.g., hydrophones) for such measuring tasks. Section 8.4 addresses the application of LRT for investigating sound fields in water. For instance, the disturbed sound field due to a capsule hydrophone will be quantified. In Sect. 8.5, LRT results for airborne ultrasound are shown and verified through microphone measurements. Finally, LRT will be exploited to quantitatively acquire the propagation of mechanical waves in optically transparent solids, which is currently impossible by means of conventional measurement principles.

8.1 Conventional Measurement Principles

In this section, let us briefly describe conventional measurement principles for analyzing sound fields in fluids as well as mechanical waves in optically transparent solids. The measurements principles are categorized into five groups: (i) hydrophones, (ii) microphones, (iii) pellicle-based optical interferometry, (iv) Schlieren optical method, and (v) light diffraction tomography. At the end, the different measurement principles are compared regarding important requirements (e.g., spatially resolved results) in practical applications.

© Springer-Verlag GmbH Germany, part of Springer Nature 2019
S. J. Rupitsch, *Piezoelectric Sensors and Actuators*, Topics in Mining, Metallurgy and Materials Engineering, https://doi.org/10.1007/978-3-662-57534-5_8

8.1.1 Hydrophones

Sound fields in water and water-like liquids are frequently analyzed by means of so-called hydrophones [2, 26]. Since this measurement device has always to be immersed in the liquid, the incident sound waves will be reflected as well as diffracted at the hydrophone body. Therefore, the underlying measurement principle is invasive (see Sect. 8.4.4 and [21]). Hydrophones are usually based on piezoelectric materials such as thin piezoceramics or PVDF (polyvinylidene fluoride) foils. Piezoceramic materials provide higher coupling factors for converting mechanical into electrical energy, but due to its low acoustic impedance, PVDF is much better suited for water (see Sect. 3.6.3). Consequently, PVDF does not require a $\lambda_{aco}/4$ layer for matching acoustic impedances of piezoelectric material and water. As a result, PVDF hydrophones feature higher measurement bandwidths than those exploiting piezoceramics and, thus, PVDF hydrophones are more often used in practical applications.

In general, we can distinguish between three different types of piezoelectric hydrophones, namely (i) needle, (ii) capsule, and (iii) membrane hydrophones. As the name already suggests, needle hydrophones have the form of a needle. A piezoelectric material with a typical effective diameter of $\ll 1\,mm$ is directly located on the needle tip. Depending on the utilized piezoelectric material and without a preamplifier, these hydrophones offer currently nominal sensitivities ranging from 12 to $1200\,nV\,Pa^{-1}$ (i.e., -278 to $-238\,dB$ re $1\,V\,\mu Pa^{-1}$) and provide measurement bandwidths of $1-20\,MHz$ [30]. A larger effective diameter of the piezoelectric material yields a better hydrophone sensitivity but reduces the acceptance angle for incident sound pressure waves. For example, a needle hydrophone with the nominal sensitivity $1200\,nV\,Pa^{-1}$ exhibits only an acceptance angle of $15°$ at the sound frequency $5\,MHz$. Besides, a large effective diameter is crucial because sound pressure values are averaged over the hydrophone's active area. Note that this fact refers to all three types of piezoelectric hydrophones and may pose especially problems in the near field of ultrasonic transducers, where sound fields exhibit high spatial frequencies (cf. Fig. 8.10).

The second type of hydrophones, so-called capsule hydrophones, looks like a projectile and uses PVDF as piezoelectric material (see Fig. 8.1a). The designs as well as specifications of capsule and needle hydrophones are quite similar, but the sensitivity of capsule hydrophones does not depend so strongly on sound frequency. Although its special geometry allows a solid construction even for small PVDF diameters, there occur only minor reflections as well as diffractions of the incident sound waves at the hydrophone body.

Membrane hydrophones (see Fig. 8.1b), which represent the last hydrophone type, consist of an acoustically transparent PVDF membrane (thickness of a single PVDF foil $<30\,\mu m$; diameter $100\,mm$). Each side of the membrane is covered by electrodes in a manner that a small area at the membrane's center with a typical diameter $<1\,mm$ gets piezoelectric after poling [3, 50]. In doing so, this area can be used to convert incident sound pressures waves into corresponding electrical output signals. Compared to the other two types, membrane hydrophones provide a larger measurement bandwidth ranging from 1 up to $50\,MHz$ and more. For this

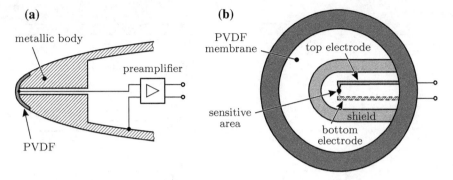

Fig. 8.1 Illustration of **a** capsule hydrophone and **b** membrane hydrophone [2]

reason, such hydrophones are perfectly suited for undistorted acquisition of pulse-shaped ultrasonic waves in water. Continuous sound waves may, however, lead to standing mechanical waves inside the thin membrane that distort sound field as well as electrical output signals [21]. Membrane hydrophones are also rather sensitive to the angle of incident sound pressure waves, which results in a small acceptance angle of typically <30°. Apart from piezoelectric hydrophones, fiber-optic hydrophones are sometimes utilized for characterizing sound fields in water since they enable sound pressure measurements far above 10 MPa [45].

8.1.2 Microphones

Devices for acquiring sound fields in air are called microphones. For precise as well as accurate measurements of sound pressure values and levels, one commonly utilizes electrostatic capacitor-based microphones such as condenser and electret microphones [26]. While condenser microphones require an external voltage supply U_{bias} for polarization, electret microphones exploit a permanently charged material. Depending on the condenser microphone, the values for U_{bias} lie in the range $20-200$ V. Figure 8.2a depicts the schematic structure of a condenser microphone. A moveable as well as mechanically prestressed circular membrane that oscillates with the incident sound waves serves as one plate of the capacitor. The membrane is commonly made of nickel or duraluminum of $\approx 10\,\mu$m thickness. For special measurement applications, a metallized polymer foil of less thickness is used instead. The air gap between membrane and backing electrode equals typically $30\,\mu$m. To increase the amplitude of membrane oscillations, the backing electrode oftentimes contains small holes, which increase the air volume within the microphone but barely alter its capacitance C_{mic}. For the air gap s_0 in equilibrium state (i.e., without sound pressure wave) and the active area A_{mic} of the microphone, C_{mic} becomes

$$C_{mic}(s_\sim) = \frac{\varepsilon_0 A_{mic}}{s_0 + s_\sim}, \tag{8.1}$$

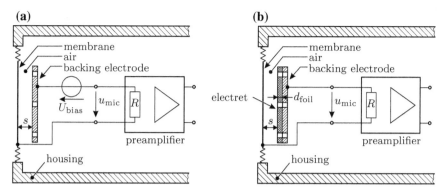

Fig. 8.2 Illustration of **a** condenser microphone and **b** electret microphone; air gap $s = s_0 + s_\sim$ [26]

whereby ε_0 stands for the electric permittivity of air and $s_\sim$ represents the membrane deflection[1] due to incident sound pressure waves. If the input resistance R of the preamplifier (see Fig. 8.2a) fulfills the condition $R \gg (2\pi f C_{mic})^{-1}$ (sound frequency f), the electric charges Q_0 on the capacitor plates will remain constant. Under additional assumption of small membrane deflections (i.e., $s_\sim \ll s_0$), the electric output voltage $u_{mic}(s_\sim)$ of the condenser microphones simplifies to

$$u_{mic}(s_\sim) = \frac{Q_0}{C_{mic}(s_\sim)} - U_{bias} = \frac{Q_0(s_0 + s_\sim)}{\varepsilon_0 A_{mic}} - \underbrace{\frac{Q_0 s_0}{\varepsilon_0 A_{mic}}}_{U_{bias}} \approx E_0 s_\sim \qquad (8.2)$$

with the (constant) electric field intensity $E_0 = Q_0(\varepsilon_0 A_{mic})^{-1}$ in the air gap. Therefore, $u_{mic}(s_\sim)$ depends on the incident sound pressure wave.

In 1962, Sessler and West [42] invented the so-called electret microphones. Contrary to condenser microphones, the electric field within the air gap results from a permanently charged dielectric foil (electret) that is located between circular membrane and backing electrode. The dielectric foil is a fluoropolymer (e.g., PVDF) of $d_{foil} = 6$ to $25\,\mu$m thickness, which is metallized at one side. By means of corona discharge, the fluoropolymer gets negatively charged at the other side. Figure 8.2b shows an embodiment of an electret microphone for the electret being located at the backing electrode. In the equilibrium state, the electric field intensity E_0 in the air gap of thickness s_0 computes as

$$E_0 = \frac{\sigma_{cor} d_{foil}}{\varepsilon_0 (d_{foil} + \varepsilon_r s_0)} . \qquad (8.3)$$

Here, σ_{cor} expresses the electric surface charge on the electret and ε_r its relative electric permittivity, respectively. Just as for condenser microphones, the electric

---

[1]For the sake of simplicity, the membrane deflection $s_\sim$ is assumed to be uniform over the active microphone surface.

Fig. 8.3 Illustration of pellicle-based optical interferometry [4]; ultrasonic transducer UT

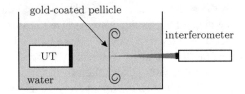

output signals of electret microphones due to incident sound pressure waves can be approximated by the simple relation $u_{mic}(s{\sim}) \approx E_0 s{\sim}$. Although such microphones do not need an external voltage supply, their performance (e.g., sensitivity) is comparable or even better than that of conventional condenser microphones. Under normal operating conditions, the permanently charged fluoropolymer almost entirely retains its electric surface charge [41]. The sensitivity of electret microphones to incident sound pressure waves is only reduced by less than 1 dB per year.

For measurement applications, condenser and electret microphones are commercially available in different sizes of the circular membrane. The membrane diameter commonly ranges from 1/8 inch to 1 inch. In fact, large membranes provide high microphone sensitivities for incident sound pressure waves, but they are not suitable for the acquisition of ultrasonic waves. Similar to hydrophones, this can be ascribed to the fact that the electric microphone output relates to the averaged deflection of the membrane. When the wavelength λ_{aco} of the sound waves is close to the membrane diameter, averaging over the membrane surface will lead to remarkable deviations between acquired and actually occurring sound pressure values. The membrane size affects, moreover, the acceptance angle for incident sound pressure waves. If the membrane size is large compared to λ_{aco}, the acceptance angle will be small. Those are the reasons why one has to carefully select the utilized microphone. While 1 inch condenser microphones offer currently sensitivities up to $100\,\mathrm{mV\,Pa^{-1}}$ and can be used in the frequency range $10\,\mathrm{Hz}-10\,\mathrm{kHz}$, $1/8$ inch versions provide only sensitivities of $1\,\mathrm{mV\,Pa^{-1}}$ [9]. Due to its small size, a $1/8$ inch condenser microphone is, however, suitable for reliable sound pressure measurements up to $140\,\mathrm{kHz}$.

8.1.3 Pellicle-Based Optical Interferometry

The pellicle-based optical interferometry exploits particle displacements that are caused by propagating sound pressure waves. In 1988, Bacon [4] suggested this measurement approach for primary calibration of hydrophones. He utilized a thin gold-coated pellicle of $3\,\mu m$ thickness, which was immersed in water. Owing to its low thickness, the pellicle is acoustically transparent but optically opaque. It should, therefore, be able to follow sound pressure waves passing through. If the pellicle movement is acquired by an appropriate device such as a Michelson interferometer (see Fig. 8.3), one can deduce sound pressure values with respect to time. In the presented implementation, the measurement uncertainty varies from 2.3 to 6.6% for the frequency range $0.5-15\,\mathrm{MHz}$ of the investigated sound field.

Koch and Molkenstruck [25] enhanced the experimental arrangement by mounting the pellicle on the water surface to extend the upper frequency limit to 70 MHz. Because of its high accuracy, this enhanced pellicle-based optical interferometry has become the standard for primary calibration of hydrophones in several countries, e.g., Germany [24]. Nevertheless, the measurement approach exhibits several limitations. Even though the thin pellicle is nearly nonperturbing, it has to be immersed in the sound propagation medium and, thus, the approach is, strictly speaking, invasive. Moreover, the output of the interferometer strongly depends on the incidence angle of the sound pressure waves on the pellicle. When the sound waves do not impinge perpendicular to the pellicle surface, the determined sound pressure amplitudes seem to be smaller than they actually are. On these grounds, pellicle-based optical interferometry should be only applied for primary calibration of hydrophones.

8.1.4 Schlieren Optical Methods

Schlieren optical methods exploit interactions between electromagnetic waves and acoustic waves in a sound field that is present in an optically transparent medium, i.e., fluid or solid [43, 54]. As illustrated in Fig. 8.4, a collimated electromagnetic wave (e.g., laser beam) propagates through the investigated sound field. Since sound pressure waves cause local variations of the density in the sound propagation medium, its optical refractive index also varies locally. This fact leads to a phase grating for the electromagnetic waves, which, therefore, get diffracted into different orders. In other words, the diffracted electromagnetic waves contain information about the sound field. After passing the sound propagation medium, the electromagnetic waves are focused by a lens. With a view to isolating the high-order diffractions of those waves, the zeroth diffraction order is removed by placing an optical stop as spatial filter at the focal plane of the lens. For instance, the necessary spatial filtering can be realized through a digital micromirror device, which allows applying different filters (e.g., knife-edge or low-pass filter) sequentially [49]. The remaining part (i.e., high-order diffractions) of the electromagnetic waves that contains sound field information is finally captured with an appropriate camera (Fig. 8.4).

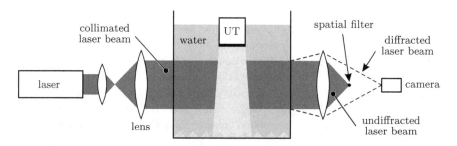

Fig. 8.4 Illustration of Schlieren optical method [43]; ultrasonic transducer UT

Compared to hydrophones in water and microphones in air, Schlieren optical methods are noninvasive and provide two-dimensional sound field information in real time. Each pixel in a Schlieren image is, however, proportional to the integration of the acoustic intensity along the path of the corresponding electromagnetic waves [39]. As a result, Schlieren optical methods do not yield spatially resolved sound pressure values even if one applies tomographic imaging in addition. Instead, sound power distributions and normalized sound pressure values are commonly reconstructed (e.g., [28, 53]). To sum up, it can be stated that Schlieren optical methods are excellently suited for visualizing sound fields in optically transparent media, but those methods currently do not deliver absolute values for the sound pressure.

8.1.5 Light Diffraction Tomography

In 1984, Reibold and Molkenstruck [37] presented the so-called light diffraction tomography, which also exploits interactions between electromagnetic waves and acoustic waves. Contrary to Schlieren optical methods, this noninvasive measurement approach provides absolute values for the sound pressure. Figure 8.5 depicts the experimental setup of light diffraction tomography that mainly differs in two points from setups of Schlieren optical methods: (i) The optical stop is replaced by a slit aperture and (ii) the camera is replaced by a combination of pinhole and photodiode. With the aid of an appropriately shaped slit aperture, the zeroth and first positive or negative diffraction orders are only allowed to pass through. Their spatial intensity and phase distributions are acquired by moving pinhole and photodiode in parallel to the xy-plane. By repeating these measurements for different projection angles (e.g., through rotating sound source), tomographic reconstruction leads to spatially resolved information about the sound field. Enhanced experimental setups enable sound pressure measurements up to sound frequencies of 10 MHz with uncertainties <10% [1]. However, light diffraction tomography does not deliver temporally resolved results and, therefore, can be utilized solely in case of continuous harmonic or standing sound pressure waves [36, 37].

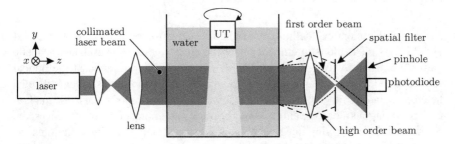

Fig. 8.5 Illustration of light diffraction tomography [1]; ultrasonic transducer UT

Table 8.1 Comparison of conventional measurement principles for analyzing sound fields in fluids and mechanical waves in solids

Property	Hydrophones microphones	Pellicle-based interferometry	Schlieren methods	Light diffraction
Noninvasive	No	No	Yes	Yes
Temporally resolved	Yes	Yes	Yes	No
Spatially resolved	Yes	Yes	No	Yes
Absolute values	Yes	Yes	No	Yes
Solid medium	No	No	Yes	Unknown
Omnidirectional	No	No	Yes	Yes
Accuracy	High	Very high	Unknown	High

8.1.6 Comparison

Finally, let us compare the mentioned measurement principles with regard to important requirements in practical applications (see Table 8.1). Actually, we desire a highly precise measurement approach that is noninvasive and provides temporally as well as spatially resolved absolute values of sound pressure. The utilized approach should not be limited to measurements in fluids but should also enable investigations of mechanical waves in solids. Moreover, since waves may propagate in different directions (e.g., reflections), the measurement approach is desired to be omnidirectional, i.e., equal sensitivity in all directions.

Even though each of those conventional measurement principles offers significant benefits, none of them fulfills all requirements in Table 8.1. Consequently, there is a great demand for alternative measurement approaches that can cope with the listed requirements. The remaining part of this chapter is dedicated to light refractive tomography, which represents such an approach.

8.2 History of Light Refractive Tomography

Just as Schlieren optical methods and light diffraction tomography (see Sects. 8.1.4 and 8.1.5), LRT exploits interactions between electromagnetic waves and acoustic waves, i.e., sound waves. Jia et al. [22] firstly measured variations of the optical refractive index in water and in air due to propagating sound waves. They utilized a heterodyne interferometer, whose output signal is directly proportional to the integral of these variations along the emitted laser beam. By assuming plane sound waves, the solution of the underlying integral equation is considerably simplified. In doing so, one can directly relate the interferometer output to sound pressure values. To get rid of this assumption, Matar et al. [27] additionally applied tomographic imaging, which enables spatially resolved reconstruction of the investigated sound field. Harvey and Gachagan [18] replaced the heterodyne interferometer with a commercial single-point laser Doppler vibrometer and, thus, reduced complexity of the experimental setup. Zipser and Franke [55] used a scanning vibrometer instead to lower measuring time. However, they exclusively concentrated on visualization of sound propagation in various practical applications but did not intend quantitative reconstruction of spatially resolved sound pressure. In summary, several researchers developed and utilized LRT for analyzing sound fields in water and in air. Nevertheless, one can barely find quantitative verifications of LRT results through conventional measurement principles such as hydrophones and microphones.

In 2006, Bahr started to research on LRT at the Chair of Sensor Technology (Friedrich-Alexander-University Erlangen-Nuremberg). Together with Lerch, he figured out that the filtered back projection algorithm provides the most reliable results for reconstructing spatially resolved sound fields [5]. They acquired sound fields of rotationally symmetric ultrasonic transducers operating in water and visualized mechanical waves in a PMMA block. Chen et al. (e.g., [12–14]) extended both the experimental setup and the reconstruction approach to investigate sound fields of arbitrarily shaped ultrasonic transducers as well as mechanical waves in optically transparent solids. In the following, fundamentals of LRT and important steps toward the extended version will be detailed.

8.3   Fundamentals of Light Refractive Tomography

In this section, the most important fundamentals of LRT will be given. We start with the underlying measurement principle and specify physical quantities (e.g., sound pressure) that can be determined in sound propagation media, i.e., fluids and solids. Section 8.3.2 details tomographic imaging, which allows spatially as well as temporally resolved reconstruction of the physical quantities through LRT measurements. In Sect. 8.3.3, the measurement procedure will be explained. Furthermore, the measurement setup will be presented that was realized at the Chair of Sensor Technology. Afterward, decisive parameters (e.g., number of projections) for LRT measurements are theoretically determined as well as optimized. Section 8.3.5 deals with sources for measurement deviations such as placement errors. Finally, we will discuss the range of sound frequencies, which can be acquired by means of LRT.

8.3.1   Measurement Principle

As stated above, LRT exploits interactions between electromagnetic waves and sound waves. Such interactions arise in each optically transparent medium (e.g., water) through which sound waves propagate [43]. A sound pressure wave causes changes of the density $\varrho(x, y, z, t)$ in the propagation medium that depend on both

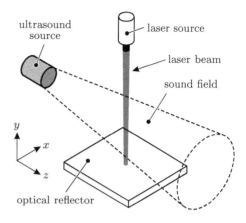

Fig. 8.6 Schematic representation of LRT principle; laser beam of laser sources (e.g., laser Doppler vibrometer) propagates in y-direction through sound field of ultrasound source; laser beam is reflected back to laser source by optical reflector

space (coordinates x, y and z) and time t. Owing to those changes, the optical refractive index $n(x, y, z, t)$ of the propagation medium also varies with respect to space and time. In a homogeneous medium, this fact leads to the deviation $\Delta n(x, y, z, t)$ of the optical refractive index from its value n_0 in the equilibrium state, i.e., without sound wave. Generally speaking, $n(x, y, z, t)$ will rise when the sound pressure $p_\sim(x, y, z, t)$ and, consequently, $\varrho(x, y, z, t)$ increase.

Let us consider a laser beam propagating in y-direction through the sound field of an ultrasound source (see Fig. 8.6). In case of LRT, the laser beam is usually reflected back to the laser source by means of an optical reflector (e.g., [5, 14]). Since $n(x, y, z, t)$ varies along the laser beam due to the sound waves, the optical path length changes. Under the assumption that electromagnetic waves propagate much faster than sound waves (i.e., wave propagation velocity $c_{em} \gg c_{aco}$), which is always fulfilled, the optical path difference ΔL along the laser beam becomes

$$\Delta L(x, z, t) = 2 \int \Delta n(x, y, z, t) \, dy \, . \tag{8.4}$$

Therefore, the optical reflector undergoes the virtual displacement $\Delta L(x, z, t)$ in y-direction although laser source and reflector exhibit a constant geometric distance. If $\Delta L(x, z, t)$ is acquired with an appropriate measurement device (e.g., laser Doppler vibrometer) and tomographic imaging (see Sect. 8.3.2) is applied in addition, one can reconstruct spatially as well as temporally resolved refractive index changes $\Delta n(x, y, z, t)$.

Sound Pressure in Fluids

The refractive index change $\Delta n(x, y, z, t)$ can be utilized to calculate the sound pressure $p_\sim(x, y, z, t)$ in optically transparent fluids, i.e., gases and liquids. According to the so-called *piezo-optic effect* [48], $\Delta n(x, y, z, t)$ is directly proportional to $p_\sim(x, y, z, t)$, which is demonstrated by the relation

$$\Delta n(x, y, z, t) = \left(\frac{\partial n}{\partial p}\right)_{S} \cdot p_{\sim}(x, y, z, t) \tag{8.5}$$

with the *piezo-optic coefficient* $(\partial n / \partial p)_S$ (index S for adiabatic conditions). For the sound pressure amplitudes $\hat{p}_{\sim}$ commonly occurring in acoustic wave propagation, this coefficient remains nearly constant [40, 51]. We are able to reconstruct the spatially as well as temporally resolved sound pressure $p_{\sim}(x, y, z, t)$ with the aid of LRT.

Dilatation in Solids

While solely longitudinal waves propagate in nonviscous fluids, there additionally exist mechanical transverse waves in solid media (see Sect. 2.2). That is why the full description of propagating mechanical waves in solids demands tensor quantities (e.g., mechanical strain **S**), which cannot be uniquely reconstructed from the scalar quantity refractive index change $\Delta n(x, y, z, t)$. Nevertheless, we are able to determine density changes $\Delta\varrho(x, y, z, t)$ in a homogenous optically transparent solid from $\Delta n(x, y, z, t)$. By assuming constant electric polarizability of atoms or molecules within the medium, Maxwell's equations yield the so-called *Lorentz–Lorenz equation* [7]

$$R_{LL} = \frac{n^2(x, y, z, t) - 1}{n^2(x, y, z, t) + 2} \cdot \frac{1}{\varrho(x, y, z, t)} = \text{const.} \tag{8.6}$$

with

$$\begin{cases} n(x, y, z, t) = n_0 + \Delta n(x, y, z, t) \\ \varrho(x, y, z, t) = \varrho_0 + \Delta\varrho(x, y, z, t) \end{cases}.$$

Here, n_0 and ϱ_0 stand for optical refractive index and density of the solid in the equilibrium state, respectively. The expression R_{LL} is the *Lorentz–Lorenz specific refraction* of the medium changing only by <1% even under extreme variations of temperature and pressure [6]. One may utilize $\Delta\varrho(x, y, z, t)$ of the solid to additionally calculate its relative volume change $\Delta V / V_0$, which is referred to as *dilatation* δ_{dil} [26]. Since propagation of mechanical waves in solids is commonly accompanied by extremely small relative volume changes (i.e., $V_0 \gg \Delta V$), δ_{dil} takes the form

$$\delta_{dil} = \frac{\Delta V}{V_0} \approx \frac{\Delta V}{V_0 + \Delta V} = -\frac{\Delta\varrho}{\varrho_0}. \tag{8.7}$$

Therefore, $\Delta n(x, y, z, t)$ leads to the spatially as well as temporally resolved dilatation $\delta_{dil}(x, y, z, t)$ in an optically transparent isotropic solid (e.g., [11, 12]). Note that only longitudinal waves alter the medium volume and, consequently, the dilatation, which is also shown in (normal strains S_{ii})

$$\delta_{dil} = S_{xx} + S_{yy} + S_{zz}. \tag{8.8}$$

For this reason, LRT is restricted to the acquisition of mechanical longitudinal waves in optically transparent solids.

8.3.2 Tomographic Imaging

To reconstruct spatially resolved refractive index changes $\Delta n(x, y, z, t)$ from the optical path differences $\Delta L(x, z, t)$, tomographic imaging has to be applied. The basis for tomographic imaging is the so-called *Radon transform* published by Radon in 1917 [35]. He suggested a mathematical formulation enabling reconstruction of a function from its projections. The *filtered back projection* (FBP) represents the best-known reconstruction approach in tomographic imaging and is oftentimes utilized in medical examinations as well as nondestructive testing (e.g., [10, 20, 23]). Below, the fundamentals of tomographic imaging are briefly outlined. This includes the *Fourier slice theorem* and the reconstruction procedure for parallel projections through FBP algorithms.

Fourier Slice Theorem

The Fourier slice theorem is a fundamental principle in tomographic imaging because it links object projections in the spatial domain and distributions in the spatial frequency domain. Figure 8.7 illustrates a graphical interpretation of this theorem. In order to explain the mathematical background, let us introduce the two-dimensional (2-D) object function $f(x, y)$ defined by the Cartesian coordinates xy and its 2-D Fourier transform $F(u, v)$ in the spatial frequency domain

$$F(u, v) = \int_{-\infty}^{+\infty} \int_{-\infty}^{+\infty} f(x, y)\, e^{-j2\pi(ux+vy)}\, dx\, dy \ . \tag{8.9}$$

Here, the arguments u and v stand for spatial frequencies, respectively. The projection[2] $o_\Theta(x)$ of $f(x, y)$ at x along the y-axis results in

$$o(x) = \int_{-\infty}^{+\infty} f(x, y)\, dy \ . \tag{8.10}$$

Without limiting the generality, we are able to transform the object function to the coordinate system (ξ, η), which should represent a rotated version of (x, y) with the rotation angle Θ; i.e., the underlying coordinate transform reads as

$$\begin{bmatrix} \xi \\ \eta \end{bmatrix} = \begin{bmatrix} \cos\Theta & \sin\Theta \\ -\sin\Theta & \cos\Theta \end{bmatrix} \begin{bmatrix} x \\ y \end{bmatrix} \ . \tag{8.11}$$

---

[2] Such projections are also named Radon-transformed of $f(x, y)$.

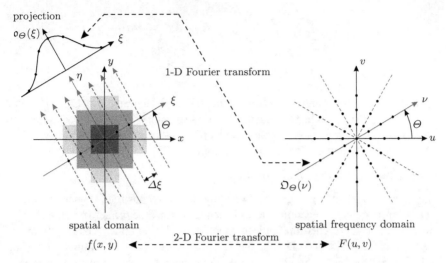

Fig. 8.7 Graphical representation of Fourier slice theorem for 2-D object function $f(x, y)$ and its 2-D Fourier transform $F(u, v)$; 1-D Fourier transform $\mathfrak{D}_\Theta(\nu)$ of projections $\mathfrak{o}_\Theta(\xi)$ in spatial domain is equal to distribution along corresponding radial lines in spatial frequency domain; step size $\Delta\xi$ in ξ-direction

For the rotated coordinate system, the projection $\mathfrak{o}_\Theta(\xi)$ at ξ along the η-axis becomes (see Fig. 8.7)

$$\mathfrak{o}_\Theta(\xi) = \int\limits_{-\infty}^{+\infty} f(\xi, \eta)\, d\eta \qquad (8.12)$$

with its one-dimensional (1-D) Fourier transform

$$\mathfrak{D}_\Theta(\nu) = \int\limits_{-\infty}^{+\infty} \mathfrak{o}_\Theta(\xi)\, e^{-j2\pi\nu\xi} d\xi$$

$$= \int\limits_{-\infty}^{+\infty} \left[\int\limits_{-\infty}^{+\infty} f(\xi, \eta)\, d\eta \right] e^{-j2\pi\nu\xi} d\xi$$

$$= \int\limits_{-\infty}^{+\infty} \int\limits_{-\infty}^{+\infty} f(\xi, \eta)\, e^{-j2\pi\nu\xi} d\eta d\xi \,. \qquad (8.13)$$

The expression ν denotes the spatial frequency and is a rotated version of u. Now, one can apply the coordinate transform (8.11) leading to the mathematical relation

$$\mathfrak{D}_\Theta(\nu) = \int\limits_{-\infty}^{+\infty}\int\limits_{-\infty}^{+\infty} f(x, y)\, e^{-j2\pi(x\nu\cos\Theta + y\nu\sin\Theta)}\mathrm{d}x\mathrm{d}y$$

$$= F(\nu\cos\Theta, \nu\sin\Theta) \ . \tag{8.14}$$

Because $u = \nu\cos\Theta$ and $v = \nu\sin\Theta$, the 1-D Fourier transform $\mathfrak{D}_\Theta(\nu)$ of the projection $\mathfrak{o}_\Theta(\xi)$ at angle Θ is equal to the linear intersection of the 2-D Fourier transform $F(u, v)$ at angle Θ, which results from the object function $f(x, y)$. This fact is commonly named Fourier slice theorem [10, 23].

Reconstruction for Parallel Projections

The general aim of tomographic imaging is to reconstruct object functions $f(x, y)$ from object projections. In LRT measurements, the refractive index change $\Delta n(x, y, z, t)$ represents the object function and the optical path difference $\Delta L(x, z, t)$ its projection. Due to the fact that $\Delta L(x, z, t)$ is acquired solely in y-direction (see Fig. 8.6), let us concentrate hereafter on parallel projections; i.e., $f(x, y)$ should be projected under different angles Θ in parallel yielding the projections $\mathfrak{o}_\Theta(\xi)$ (see (8.12)). According to the aforementioned Fourier slice theorem, one can determine the object information $F(u, v)$ in the whole spatial frequency domain if such projections are available for a sufficient number of projection angles Θ. By performing the 2-D inverse Fourier transform

$$f(x, y) = \int\limits_{-\infty}^{+\infty}\int\limits_{-\infty}^{+\infty} F(u, v)\, e^{j2\pi(ux+vy)}\mathrm{d}u\mathrm{d}v \ , \tag{8.15}$$

we finally obtain the desired quantity $f(x, y)$. However, regarding data acquisition in practical applications (e.g., LRT measurements), the object information in the spatial frequency domain is given rather in polar coordinates (ν, Θ) than in spatial frequencies (u, v). It makes, therefore, sense to rewrite $F(u, v)$ as $F(\nu, \Theta) \equiv F(\nu\cos\Theta, \nu\sin\Theta)$ through conducting the substitutions

$$u = \nu\cos\Theta \quad \text{and} \quad v = \nu\sin\Theta \ . \tag{8.16}$$

In doing so, (8.15) takes the form

$$f(x, y) = \int\limits_{0}^{2\pi}\int\limits_{0}^{+\infty} F(\nu, \Theta)\, e^{j2\pi\nu(x\cos\Theta + y\sin\Theta)}\nu\mathrm{d}\nu\mathrm{d}\Theta$$

$$= \int\limits_{0}^{\pi}\int\limits_{-\infty}^{+\infty} F(\nu, \Theta)\, e^{j2\pi\nu(x\cos\Theta + y\sin\Theta)}|\nu|\,\mathrm{d}\nu\mathrm{d}\Theta \ . \tag{8.17}$$

Since $F(\nu, \Theta)$ represents the linear intersection of $F(u, v)$ at angle Θ, $F(\nu, \Theta)$ can be replaced by the 1-D Fourier transform $\mathfrak{D}_\Theta(\nu)$ of the projection $\mathfrak{o}_\Theta(\xi)$, which leads to

$$f(x, y) = \int_0^\pi \mathfrak{T}_\Theta(x \cos \Theta + y \sin \Theta) \, d\Theta \qquad (8.18)$$

with

$$\mathfrak{T}_\Theta(r) = \int_{-\infty}^{+\infty} \mathfrak{D}_\Theta(\nu) \, |\nu| \, e^{j2\pi\nu r} d\nu . \qquad (8.19)$$

As a result, we will be able to reconstruct $f(x, y)$ if $\mathfrak{o}_\Theta(\xi)$ and, consequently, $\mathfrak{D}_\Theta(\nu)$ are known. This forms the basis for FBP algorithms [10, 23].

In practical applications of tomographic imaging, spatial and temporal sampling rates are actually limited due to data acquisition and measuring time. One has to cope with a limited number of spatial sampling points in both ξ-direction and Θ-direction for FBP meaning that discrete formulations of (8.18) and (8.19) are indispensable. To explain the discrete formulations, we consider N_{proj} projections of $f(x, y)$ exhibiting the angular increment $\Delta\Theta = \pi/N_{proj}$. Each projection is assumed to comprise N_{ray} sampling points in ξ-direction with equidistant step size $\Delta\xi$ (see Fig. 8.7). Under these assumptions, (8.18) gets modified to

$$f(x, y) = \frac{\pi}{N_{proj}} \sum_{i=1}^{N_{proj}} \mathfrak{T}_{\Theta_i}(x \cos \Theta_i + y \sin \Theta_i) \qquad (8.20)$$

with

$$\Theta_i = \frac{(i-1)\pi}{N_{proj}} \quad \forall \, i = 1, \ldots, N_{proj}$$

and (8.19) becomes

$$\mathfrak{T}_{\Theta_i}(m\Delta\xi) = \Delta\xi \cdot \text{IDFT}\Big\{ \text{DFT}\{\mathfrak{o}_{\Theta_i}(m\Delta\xi)\} \cdot \text{DFT}\{h_{ck}(m\Delta\xi)\} \cdot \text{window} \Big\}$$

$$\forall \, m = -\frac{N_{ray}}{2}, \ldots, \frac{N_{ray}}{2} - 1 . \qquad (8.21)$$

Here, $m\Delta\xi$ denotes the mth sampling point in ξ-direction and Θ_i is the angle of the ith projection. The operators $\text{DFT}\{\cdot\}$ and $\text{IDFT}\{\cdot\}$ in (8.21) stand for the 1-D discrete Fourier transform and 1-D inverse discrete Fourier transform, respectively. Instead of $|\nu|$ in (8.19), we use the 1-D discrete Fourier transform of an appropriate convolution kernel $h_{ck}(m\Delta\xi)$ that should serve as an additional filter to suppress noise in measurement data. For the implemented LRT setup, the so-called *Ram-Lak kernel* turned out to be a good choice because it is rather simple and provides excellent reconstruction results (e.g., [5, 11]). The Ram-Lak kernel is mathematically defined as

$$h_{ck}(m\Delta\xi) = \begin{cases} (2\Delta\xi)^{-2} & \text{for} \quad m = 0 \\ 0 & \text{for} \quad m \text{ is even} \\ -(m\pi\Delta\xi)^{-2} & \text{for} \quad m \text{ is odd} \end{cases} . \qquad (8.22)$$

Besides, (8.21) contains a window function, which is not necessarily required for reconstruction purpose but can significantly improve the imaging quality.

8.3.3 Measurement Procedure and Realized Setup

According to the previous subsections, one has to project the investigated sound field under different angles Θ to reconstruct spatially and temporally resolved field quantities by means of LRT. Basically, there exist two possibilities for this task, namely (i) simultaneous rotation of laser source as well as optical reflector and (ii) rotation of ultrasound source.[3] The first possibility is oftentimes applied for tomographic imaging principles in medical examinations and nondestructive testing (e.g., X-ray computed tomography [10]). While the investigated object retains its position and orientation, the measuring components rotate around the object. However, in case of LRT, the simultaneous rotation of laser source and optical reflector constitutes various problems. For example, it imposes high technical demands to precisely rotate both devices around a water tank that is necessary for acquiring sound fields in water. Besides, an additional optical path difference along the laser beam may occur during rotation due to optical refraction at the interfaces of different media (e.g., water tank and water) when the laser beam does not impinge orthogonal to those interfaces. The second possibility (i.e., rotation of ultrasound source) should, thus, be preferred for practical implementation of LRT.

Figure 8.8 illustrates an appropriate measurement procedure for LRT in order to obtain the spatially and temporally resolved refractive index change $\Delta n(x, y, z, t)$ in a single xy-plane. Note that the procedure will only work if the ultrasound source is periodically excited with the same signal. Taking into account the reconstruction of $\Delta n(x, y, z, t)$, the entire measurement process consists of three main steps:

1. The laser Doppler vibrometer (LDV), which emits a laser beam in y-direction, is moved in parallel to the x-axis with step size $\Delta\xi$. Hence, the z-distance between ultrasound source and LDV remains constant. The optical reflector being aligned

---

[3]LRT is applicable for ultrasonic waves and audible sound waves. Without limiting the generality, we will concentrate on sound fields generated by ultrasound sources.

Fig. 8.8 Measurement procedure for LRT to obtain refractive index change $\Delta n(x, y, z, t)$ in xy-plane (cf. Fig. 8.6); **1:** scanning of laser Doppler vibrometer (LDV) along x-direction with step size $\Delta\xi$; **2:** rotation of ultrasound source around its axis (i.e., parallel to z-axis) by angle increment $\Delta\Theta$

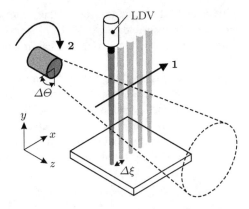

in parallel to the xz-plane reflects the laser beam back to the LDV. At each LDV position, the optical path difference $\Delta L(x, z, t)$ arising from the sound field is acquired and transferred to an evaluation unit.

2. In step 2, the ultrasound source is rotated around an axis parallel to the z-axis by the angle increment $\Delta\Theta$. Afterward, step 1 is conducted again, i.e., LDV signals are acquired and transferred at each LDV position. The sequence of step 1 and step 2 is repeated until the angular range $[0, 180° - \Delta\Theta]$ is completely covered. At this point, it should be mentioned that further angular steps (e.g., 180° and 180° + $\Delta\Theta$) do not provide extra information because $\Delta L(x, z, t)$ is independent of the direction in which the laser beam passes the sound field.

3. The stored LDV signals represent projections of the sound field under different angles. By combining these signals through FBP, one can finally reconstruct the refractive index change $\Delta n(x, y, z, t)$ in the investigated xy-plane.

When three-dimensional (3-D) information of the sound field is desired, several xy-planes will be required which means that the z-distance between ultrasound source and LDV needs to be altered. As a matter of course, step 1 to step 3 have to be performed for each xy-plane. In doing so, we are able to analyze sound fields of nearly arbitrarily shaped ultrasound sources in three spatial dimensions with respect to time.

In Fig. 8.9, one can see the experimental arrangement of LRT that was established at the Chair of Sensor Technology [11]. A differential LDV (Polytec OFV 512 [32]) containing two sensor heads serves as instrument to measure the optical path difference $\Delta L(x, z, t)$ along the laser beam. One of its fiber-optic sensor heads is mounted on a linear positioning system comprising three translation axes (Physik Instrumente M-531.DG [31]), which enable precise movements in xyz-direction. The other sensor head is mirrored and, therefore, does not contribute to the measurement procedure. To optimally reflect the laser beam back to the LDV, the optical reflector (glass plate coated with chrome) is placed onto an adjustable base. The analog output signal of the LDV is, depending on the applied decoder, either directly proportional to $\Delta L(x, z, t)$ or to the resulting velocity. With a view to rotating the investigated ultrasonic transducer that represents the ultrasound source, a rotation unit (Physik Instrumente M-037.DG [31]) is connected to a gear via a timing belt. The gear directly rotates a cylindrical mount in which the ultrasonic transducer is fixed.

A single substep of the LRT setup in Fig. 8.9 including LDV movement, waiting time as well as data acquisition and data transfer takes approximately 0.7 s. Let us assume that one xy-plane requires 5000 substeps. Therewith, the entire measurement procedure (step 1 and step 2) takes approximately one hour. In contrast, the reconstruction of the spatially as well as temporally resolved refractive index change through the FBP algorithm in step 3 takes only a few minutes on a commercial PC.

8.3.4 Decisive Parameters for LRT Measurements

Below, the decisive parameters (e.g., number of projections) for LRT measurements are determined from the theoretical point of view. Subsequently, we will optimize these parameters with regard to a short measuring time as well as reasonable reconstruction results. Finally, an appropriate window function is given which helps to filter images during the reconstruction stage.

Theoretical Determination of Measurement Parameters

For reliable investigations of sound fields by means of LRT, we have to fulfill the Nyquist sampling theorem in time domain and spatial domain [11, 14]. The sampling rate in both domains needs, thus, to be more than twice as fast as the highest frequency components of the signals. In the time domain, this can be simply guaranteed with conventional digital storage oscilloscopes (e.g., Tektronix TDS 3054 [46]). However, since most time in LRT measurements is spent for positioning tasks, let us take a closer look at the spatial domain. This domain is defined by the scanning area of the LDV as well as the number of sampling points N_{ray} along a single projection and the number of projections N_{proj}.

In order to theoretically determine the scanning area, N_{ray} and N_{proj}, we consider the sound field of a piston-type ultrasonic transducer in water (sound velocity $c_{aco} = 1480\,\text{m s}^{-1}$). The active circular area of the transducer featuring the radius $R_T = 6.35\,\text{mm}$ is assumed to oscillate uniformly at a frequency of $f = 1\,\text{MHz}$. The axisymmetric sound pressure field $p_{\sim}(x, z, t)$ was calculated through FE simulations, whereby absorbing boundary conditions suppressed unwanted reflections at the boundaries of the computational domain (see Sect. 4.4). Due to the fact that measurements are always affected by noise, white Gaussian noise was added to the simulated sound pressure field so that the signal-to-noise ratio (SNR) amounts 30 dB. Such SNR value can be easily reached in practical experiments by performing signal averaging. Figure 8.10a shows the resulting sound pressure distribution $p_{\sim}(x, z)$ normalized to its maximum $|p_{\sim}(x, z)|_{max}$ and at an arbitrary instant of time. As can be clearly seen, the sound field in the computational domain concentrates within a small area, whose geometric distance from the rotation axis (i.e., z-axis) corresponds approximately

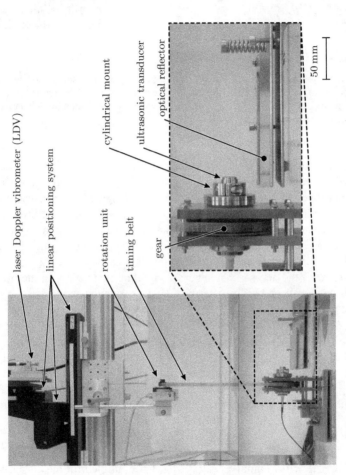

laser Doppler vibrometer (LDV)

linear positioning system

rotation unit

timing belt

cylindrical mount

ultrasonic transducer

optical reflector

gear

50 mm

Fig. 8.9 Realized experimental setup of LRT containing linear positioning system to move differential LDV and rotation unit to rotate ultrasonic transducer, which is fixed in cylindrical mount [11]

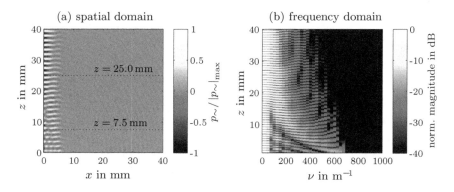

Fig. 8.10 **a** Snapshot of simulated sound pressure field $p_\sim(x, z)$ in spatial domain of piston-type transducer (radius $R_T = 6.35$ mm; excitation frequency $f = 1$ MHz) located at $z = 0$ mm; normalized to maximum $|p_\sim(x, z)|_{max}$; **b** resulting normalized distribution of each horizontal line from **a** in spatial frequency domain with spatial frequency ν

to R_T. It seems only natural that in regions of low sound pressure amplitudes, the optical path difference $\Delta L(x, z, t)$ caused by the propagating acoustic wave is also small. Hence, one can restrict the acquisition of LDV signals to areas, where remarkable sound energy arises. For the considered piston-type ultrasonic transducer, let us take into account $\Delta L(x, z, t)$ up to a distance of 20 mm from the z-axis, which is $\approx 3R_T$. Consequently, the LDV has to be moved for each projection along a line of 40 mm in x-direction representing the scanning area.

In a next step, the necessary number of sampling points N_{ray} in the previously identified scanning area will be determined. The excitation frequency $f = 1$ MHz leads to the wavelength $\lambda_{aco} = c_{aco}/f = 1.48$ mm and, thus, to the spatial frequency $\nu_{max} = \lambda_{aco}^{-1} = 676$ m^{-1}, which represents the highest spatial frequency that is possible. This can also be observed in Fig. 8.10b depicting the actual distribution of spatial frequencies ν in x-direction for the simulated sound field. Thereby, each horizontal line denotes the 1-D Fourier transform of the corresponding horizontal line in Fig. 8.10a. According to Nyquist sampling theorem for the spatial domain, the minimum spatial sampling rate ν_{samp} in radial direction becomes

$$\nu_{samp} > 2\nu_{max} = 2\lambda_{aco}^{-1} = 1351 \text{ m}^{-1} \tag{8.23}$$

meaning that $\Delta\xi < 0.74$ mm $(\hat{=}1/2\nu_{max})$ has to be fulfilled for the distance of two neighboring LDV positions. As a result, the scanning area of 40 mm requires at least $N_{ray} = 55$ sampling points for each projection.

A further decisive parameter in LRT measurements is the number of projections N_{proj}. Just as for the radial direction, the sound field of ultrasonic transducers may also exhibit in tangential direction spatial frequencies up to ν_{max}. The sampling rate in tangential direction should therefore be equal to the sampling rate in radial direction. Although this constitutes the worst-case scenario that does commonly not occur, we will determine N_{proj} for such case. Without limiting the generality, the projections are assumed to be evenly distributed with the angular increment $\Delta\Theta$, which yields [11, 23]

$$\nu_{max}\Delta\Theta = \nu_{max}\frac{\pi}{N_{proj}} . \tag{8.24}$$

Because the maximum spatial frequency ν_{max} also influences the number of sampling points N_{ray} in radial direction, one can deduce from $\Delta\Theta \approx 2/N_{ray}$ the relation

$$N_{proj} \approx \frac{\pi}{2}N_{ray} , \tag{8.25}$$

which links the sampling points in both directions. For the considered sound field of the piston-type transducer, this results in $N_{proj} = 86$ projections.

Optimization of Measurement Parameters

As the previous theoretical determination suggests, a sampling interval of $\Delta\xi = 0.7\,\text{mm}$ in radial direction should be sufficient to reconstruct the aimed values in a xy-plane unambiguously. However, in real measurements, $\Delta\xi$ has to be much smaller than the Nyquist rate. This can be attributed to the following facts:

- Each measurement is contaminated with noise. In order to avoid aliasing effects, $\Delta\xi$ has, therefore, to be remarkably reduced.
- Within the framework of tomographic reconstruction, an appropriate window function (see (8.21)) is commonly applied for filtering images. Such filters are difficult to be implemented if $\Delta\xi$ is close to the Nyquist rate.
- Sampling intervals, which should be theoretically sufficient, result in visually rough images. Smoothing can be achieved by means of various interpolation algorithms but may, on the other hand, hide important information in the reconstructions.

For these reasons, the sampling interval $\Delta\xi$ has to be significantly reduced. A value of $\Delta\xi = 0.2\,\text{mm}$ turned out to be excellent choice in case of the considered sound field since it leads to a good compromise between measuring time and accuracy of measurements [11]. Consequently, the number of sampling points in radial directions increases to $N_{ray} = 201$, which almost quadruples the entire measuring time compared to $N_{ray} = 55$.

Now, let us apply the specified scanning area (i.e., 40 mm) as well as the determined values $N_{ray} = 201$ and $N_{proj} = 86$ to emulate LRT measurements for the modeled ultrasonic transducer. In doing so, the optical path difference $\Delta L(x, z, t)$ from the sound pressure field in Fig. 8.10a is calculated through (8.4) and (8.5). Figure 8.11a and b show original normalized sound pressure curves with respect to x as well as reconstructed ones for the xy-planes at the axial distances $z = 7.5\,\text{mm}$ and $z = 25.0\,\text{mm}$, respectively. As the results illustrate, the reconstructed values coincide very well with the original sound pressure curves. The normalized relative

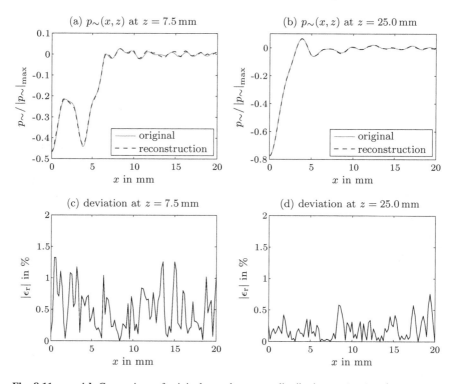

Fig. 8.11 **a** and **b** Comparison of original sound pressure distribution $p_\sim(x, z)$ and reconstruction with respect to x-position at axial distances $z = 7.5\,\mathrm{mm}$ and $z = 25.0\,\mathrm{mm}$, respectively (see Fig. 8.10a); normalized to maximum $|p_\sim(x, z)|_{\mathrm{max}}$; **c** and **d** normalized relative deviations $|\epsilon_{\mathrm{r}}|$ between reconstruction results and originals

deviation $|\epsilon_{\mathrm{r}}|$ (magnitude) of the reconstruction is always smaller than 1.5% in both xy-planes (see Fig. 8.11c and d).

Even though the selected scanning area and sampling parameters lead to convincing reconstruction results, it is desired to reduce measuring time in LRT measurements. Actually, this will constitute a crucial point especially when many xy-planes of the sound field should be investigated. While a decrease of the scanning area is not recommended, one is able to reduce N_{ray} as well as N_{proj}. Let us start with the idea behind reducing N_{ray}. Owing to the high concentration of ultrasonic energy nearby the rotation axis of the ultrasonic transducer (see Fig. 8.10a), nonequidistant sampling seems to be a proper way for decreasing N_{ray} [11, 15]. Regions of high energy demand fine sampling, but we can reduce the sampling rate outside such regions, which means skipping of sampling points. The skipped sampling points have to be filled up through suitable interpolation approaches like cubic spline interpolation [47]. For the considered sound field, the sampling intervals $\Delta\xi = 0.2\,\mathrm{mm}$ for $x \leq 6.4\,\mathrm{mm}$ (i.e., $\approx R_{\mathrm{T}}$) and $\Delta\xi = 0.8\,\mathrm{mm}$ beyond in the region $6.4 < x \leq 20.0\,\mathrm{mm}$ are a good choice. Therewith, the number of sampling points N_{ray} in radial direction decreases from 201 to 99 and, thus, measuring time will be halved.

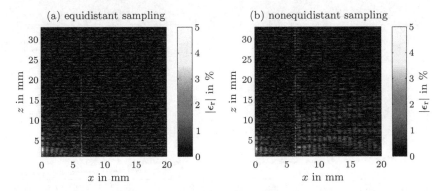

Fig. 8.12 Normalized relative deviations $|\epsilon_r|$ between reconstruction results and original sound pressure field (see Fig. 8.10a) for **a** equidistant sampling and **b** nonequidistant sampling in x-direction

To compare equidistant sampling and nonequidistant sampling, the simulated sound field was completely reconstructed in the region $x \times z = [0, 20\,\mathrm{mm}] \times [0, 33\,\mathrm{mm}]$ by individually computing sound pressure curves in each xy-plane. Figure 8.12a and b depict the normalized relative deviations $|\epsilon_r|$ (magnitude) between reconstruction results and original sound pressure field. Although $|\epsilon_r|$ slightly increases in the peripheral (i.e., $x > 6.4\,\mathrm{mm}$) for nonequidistant sampling, the maximum relative deviation for both sampling methods stays below 5%. On this account, nonequidistant sampling is a great opportunity to reduce measuring time in LRT measurements, particularly if the main emphasis lies on central areas of sound fields.

Besides the parameter N_{ray}, we may also reduce the number of projections N_{proj} in LRT measurements. To optimize N_{proj}, let us take a closer at the influence of that number on reconstruction results. For this task, LRT measurements were emulated again on the basis of simulated sound field (see Fig. 8.10a). The tomographic reconstruction was conducted with different amounts of projections N_{proj} ranging from 5 to 200 projections in steps of 5. Figure 8.13a and b display maximum relative deviations and mean relative deviations as function of N_{proj} for the entire reconstruction result, i.e., in the region $x \times z = [0, 20\,\mathrm{mm}] \times [0, 33\,\mathrm{mm}]$. Both deviations drop quickly at the beginning for increasing number of projections, while they remain almost constant for $N_{\mathrm{proj}} \geq 45$. Hence, it is reasonable to choose 50 projections in LRT measurements instead of $N_{\mathrm{proj}} = 86$. Apart from that, if one is exclusively interested in the central area of the sound field (here $x < 6.4\,\mathrm{mm}$), a much lower number of projections might be sufficient. This follows from the sampling point density, which is always higher close to the xy-plane's center than in its periphery [10]. For the considered sound field, $N_{\mathrm{proj}} = 15$ already yields reconstruction results, whose maximum and mean relative deviations are in the range of 6% and 1% (see Fig. 8.13a and b), respectively. In other words, when rough information is desired only in LRT measurements, we can utilize a rather small number of projections and, therefore, get rid of long measuring times.

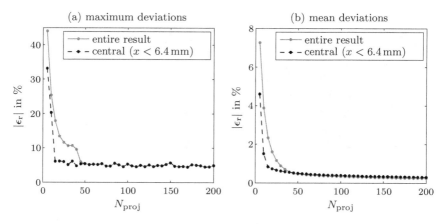

Fig. 8.13 a and **b** Maximum and mean of normalized relative deviations $|\epsilon_r|$ (magnitude) between reconstruction results and originals with respect to number of projections N_{proj} for entire sound pressure field (see Fig. 8.10a) and within central area (i.e., $x < 6.4\,\text{mm}$)

Window Function for Tomographic Imaging

One can suppress noise in tomographic imaging by means of an appropriate window function (see (8.21)). For example, the Hann window is oftentimes used in X-ray computed tomography because the spatial frequency components of the target spread in a large spatial frequency range [23]. However, this window may either damp wanted signals or insufficiently removes noise in LRT measurements. That is why we apply the so-called Turkey window, which combines rectangular and Hann window: Signal components with spatial frequencies $< \nu_{max}$ $(\hat{=}\lambda_{aco}^{-1})$ are not altered due to the rectangular window, while high-frequency components can be strongly damped through the Hann window [11, 14]. In doing so, we protect signals featuring a reasonable SNR and remove high-frequency noise without raising unwanted ringing effects in the spatial domain. Numerical studies revealed the transition band $[\nu_{max}, 3\nu_{max}]$ connecting passband and stopband as a proper choice for the Turkey window. Indeed, such transition band requires oversampling in the spatial domain; i.e., $\nu_{samp} > 6\nu_{max}$ has to be fulfilled. For the considered sound field, the minimum sampling rate results in $\nu_{samp} > 4054\,\text{m}^{-1}$, which yields the sampling interval $\Delta\xi < 0.25\,\text{mm}$. Note that the Turkey window was already applied in the previous reconstruction procedures (see Figs. 8.11, 8.12 and 8.13).

8.3.5 Sources for Measurement Deviations

For reconstructing spatially as well as temporally resolved quantities by means of LRT, projections from different angles have to be combined. All projections contribute to the final results. Owing to this fact, the result quality actually depends on the measurement accuracy over the whole cross section. Minor imperfections in the realized LRT setup accumulate and may cause substantial measurement deviations followed by completely distorted images. That is why we need to take care about potential sources for such measurement deviations. Here, two different types of sources are discussed in detail: (i) placement errors arising from misalignments of LRT components and (ii) optical errors originating from the nonideal laser beam of the LDV.

Placement Errors

The reliable reconstruction in LRT measurements demands specific knowledge of the sampling positions along a single projection and the projections angles. Therefore, one should utilize precise linear positioning systems as well as rotation units. Such components do not, however, ensure highly accurate reconstruction results because their geometric alignment is a further decisive point in LRT measurements. In order to study effects of misalignments, let us take a look at the geometric orientation of a LDV scanning plane relative to the cylindrical mount, which contains the ultrasound source. The Cartesian coordinate system xyz belongs to the scanning plane, whereas the front surface of the cylindrical mount represents the origin of the Cartesian coordinate system $x_c y_c z_c$ (see Fig. 8.14). If all components of the LRT setup are perfectly aligned, the z-axes (i.e., symmetry axes) of both coordinate systems will coincide. In practical setups, there always arise deviations that can be understood as geometric uncertainties and cause systematic errors in measurements. For the sake of simplicity, the laser beam of the LDV is here assumed to propagate in y-direction, which can be easily achieved in the realized setup. We then have to consider only three parameters defining relevant deviations of both coordinate systems in LRT measurements. Guided by nautical terms, the three parameters are named (i) sway distance, (ii) yaw angle, and (iii) pitch angle. Under the assumption that the coordinate system xyz of the scanning plane is spatially fixed, they can be interpreted as follows (see Fig. 8.14):

- The sway distance Δx_c stands for the horizontal distance between supposed and actual symmetry axes of the cylindrical mount.
- The yaw angle Φ_c indicates the angle around y_c by which the front surface of the cylindrical mount is rotated away from the xy-plane.
- The pitch angle Θ_c indicates the angle around x_c by which the front surface of the cylindrical mount is rotated away from the xy-plane.

To rate the impacts of these parameters on LRT measurements, the simulated sound pressure field of Fig. 8.10a was used. Depending on the parameter and its value, the original sound field has to be slightly shifted and rotated [14]. On the basis of the modified sound pressure field, LDV signals were emulated in the scanning planes leading to distorted projections, which served as input for the FBP algorithm. The reconstruction results can subsequently be compared to the original sound field. Table 8.2 contains maxima of the normalized relative deviations $|\epsilon_r|$ (magnitude) between reconstruction results and original sound pressure curves at the axial distances $z = 3.0\,\text{mm}$, $z = 7.5\,\text{mm}$ as well as $z = 25.0\,\text{mm}$. The parameters Δx_c, Φ_c,

Fig. 8.14 Illustration of sway distance Δx_c, yaw angle Φ_c, and pitch angle Θ_c representing geometric uncertainties in LRT setups [14]; Cartesian coordinate system $x_c y_c z_c$ of cylindrical mount; Cartesian coordinate system xyz of scanning plane

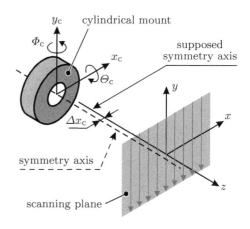

Table 8.2 Maxima of normalized relative deviations $|\epsilon_r|$ (magnitude) between reconstruction results and original sound pressure field (see Fig. 8.10a) at three xy-planes with different axial distances from transducer surface; Δx_c, Φ_c as well as Θ_c represent geometric uncertainties in LRT setups

Parameter	Value	Max. of $|\epsilon_r|$ in % $z = 3.0$ mm	Max. of $|\epsilon_r|$ in % $z = 7.5$ mm	Max. of $|\epsilon_r|$ in % $z = 25.0$ mm
Sway distance Δx_c	0.02 mm	0.9	0.7	0.2
	0.10 mm	3.4	3.1	1.0
	0.50 mm	18.9	19.5	5.4
Yaw angle Φ_c	0.02°	0.5	0.4	0.2
	0.10°	1.1	1.1	1.4
	0.50°	7.4	5.3	6.6
Pitch angle Θ_c	0.40°	0.9	0.9	0.7
	0.80°	4.0	3.6	2.9
	1.60°	10.6	7.5	8.4

and Θ_c were varied separately. As the table entries demonstrate, small sway distances x_c induce large deviations between reconstruction results and original sound pressure curves, especially in the near field of the transducer (i.e., $z = 3.0$ mm). This is due to the geometric shift of the projections against each other, which have to be combined in the reconstruction stage and, therefore, yield blurred images. Although the maxima of $|\epsilon_r|$ exhibit for the chosen yaw angles Φ_c and pitch angles Θ_c the same value range as for Δx_c, both parameters are actually not that critical for tomographic reconstruction. Instead of shifting projections against each other, Φ_c and Θ_c exclusively tilt the cylindrical mount. Consequently, a slightly different cross section of the sound field is projected and reconstructed in LRT measurements. Nevertheless, depending on the sound field, this may also cause high values for $|\epsilon_r|$ because the projected cross section does not coincide with the supposed one.

The previous investigations demonstrated that the LRT components need to be precisely adjusted. For the dedicated LRT setup (see Fig. 8.9), the alignment procedure was based on the intensity of the reflected LDV laser beam [11]. When the emitted laser beam is partially blocked, the intensity of the reflected beam will be decreased. In this way, one can detect edges of the cylindrical mount by moving the LDV in the xz-plane. Through corrections of component alignment and intensity measurements, it is possible to guarantee a sway distance $|\Delta x_c| < 0.01$ mm, a yaw angle $|\Phi_c| < 0.02°$, and a pitch angle $|\Theta_c| < 0.40°$. The remaining uncertainties in geometric component alignment induce only small deviations between reconstruction results and original sound pressure curves for the considered sound field (cf. Table 8.2).

Optical Errors

Besides placement errors, the nonzero spot size of the laser beam is a crucial point in LRT measurements. The helium–neon laser of the utilized LDV (Polytec OFV 512 [32]) emits a laser beam, whose beam profile is very close to an ideal Gaussian beam [44]. In order to describe beam properties (e.g., divergence) of the LDV, we can, therefore, apply fundamental relations that are valid for ideal Gaussian beams. For such a Gaussian beam, the spot size $w_{em}(\zeta)$ at which the beam intensity has decreased to $1/e^2$ times its value at the center becomes (see Fig. 8.15) [34, 48]

$$w_{em}(\zeta) = w_0 \sqrt{1 + \left(\frac{\zeta}{\zeta_0}\right)^2}. \tag{8.26}$$

The expression ζ stands for the axial position along the laser beam, and ζ_0 is the so-called Rayleigh range that computes as

$$\zeta_0 = \frac{w_0^2 \pi}{\lambda_{em}}. \tag{8.27}$$

The minimum spot size w_0 of the laser beam, which is named beam waist, appears at $\zeta = 0$. While a helium–neon laser emits in air laser beams at a wavelength of $\lambda_{em} = 632.8$ nm, the wavelength changes to 475.8 nm in water since its optical refractive index is $n_0 = 1.33$.

Actually, $w_{em}(\zeta)$ should be in LRT measurements as small as possible throughout the entire sound field of the ultrasound source. A low divergence of the laser beam is, however, accompanied by a large value of w_0 (cf. (8.26)). Hence, one has to find a compromise between beam waist extension and divergence so that the spot size $w_{em}(\zeta)$ of the laser beam stays sufficiently small in the investigated sound field. For the realized LRT setup, a small beam waist extension of the LDV turned out to be a good choice. As the laser beam propagates through the sound field twice, it makes sense to position the beam waist directly onto the surface of the optical reflector. The ultrasound source should be nearby the reflector in LRT measurements to achieve a tight laser beam in the sound field. However, when the reflector is located too close to the ultrasound source, the sound field will be strongly disturbed by the reflector. It is for this reason rather important to consider the investigated sound field for selecting an appropriate distance between ultrasound source and optical reflector.

Fig. 8.15 Increase of spot size $w_{\mathrm{em}}(\zeta)$ of laser beam along axial distance ζ from beam waist w_0 at $\zeta = 0$; Rayleigh range ζ_0

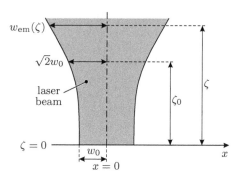

Now, let us evaluate optical errors arising in case of the dedicated LRT setup if we measure the sound field of Fig. 8.10a. In doing so, the spot size $w_{\mathrm{em}}(\zeta)$ of the laser beam and its normalized transverse energy distribution are required. Because the sound pressure almost completely vanishes at $x = 30\,\mathrm{mm}$, this distance represents a good choice for the y-spacing of ultrasound source and optical reflector. Related to the symmetry axis of cylindrical mount and, consequently, to the sound field distribution, the distance needs to be considered once more, which yields the range $\zeta = [0, 60\,\mathrm{mm}]$ for relevant axial positions along the laser beam. Taking into account the beam waist $w_0 = 94\,\mu\mathrm{m}$ of the emitted laser beam as well as its wavelength $\lambda_{\mathrm{em}} = 475.8\,\mathrm{nm}$ in water, $w_{\mathrm{em}}(\zeta)$ exhibits the extrema (see (8.26))

$$w_{\mathrm{em}}(\zeta) = \begin{cases} 94\,\mu\mathrm{m} & \text{at} \quad \zeta = 0\,\mathrm{mm} \\ 135\,\mu\mathrm{m} & \text{at} \quad \zeta = 60\,\mathrm{mm} \end{cases}. \tag{8.28}$$

The normalized transverse distribution $E_{\mathrm{n}}(x)$ of the electric field magnitude for a Gaussian laser beam is given by [19]

$$E_{\mathrm{n}}(x) = \mathrm{e}^{-x^2 \left(2\sigma_{\mathrm{em}}^2\right)^{-1}}. \tag{8.29}$$

As $w_{\mathrm{em}}(\zeta)$ specifies the distance at which the beam intensity ($\propto E_{\mathrm{n}}^2$) has decreased to $1/\mathrm{e}^2$ times its maximum, a single unknown remains in (8.29) that can be determined according to

$$E_{\mathrm{n}}^2(w_{\mathrm{em}}) = \mathrm{e}^{-2w_{\mathrm{em}}^2 \left(2\sigma_{\mathrm{em}}^2\right)^{-1}} \equiv E_{\mathrm{n}}^2(0) \cdot \mathrm{e}^{-2} = \mathrm{e}^{-2}$$
$$\Rightarrow \quad \sigma_{\mathrm{em}} = \frac{w_{\mathrm{em}}}{\sqrt{2}}. \tag{8.30}$$

Thus, it is possible to calculate $E_{\mathrm{n}}(x)$ for both spot sizes (8.28) of the laser beam and the normalized spatial Fourier transform $A_{\mathrm{n}}(\nu) \propto \mathcal{F}\{E_{\mathrm{n}}(x)\}$ (1-D; spatial frequency

Fig. 8.16 Magnitude of
Fourier transform $A_n(\nu)$ in
spatial frequency domain;
best case $w_{em} = 94\,\mu m$ and
worst case $w_{em} = 135\,\mu m$
refer to LDV beam profiles
for realized LRT setup;
hydrophone ($R_{HY} =$
0.2 mm) is assumed to be
uniformly sensitive across its
active surface

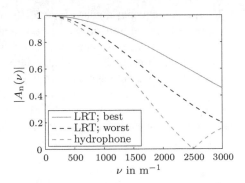

ν), respectively. The beam waist represents the best case, while $w_{em}(\zeta)$ at $\zeta = 60\,mm$
is the worst case. The resulting relations for the beam profiles read as[4]

$$\left.\begin{aligned} E_{n,\text{best}}(x) &= e^{-x^2 \cdot 1.13 \cdot 10^8\,m^{-2}} \\ E_{n,\text{worst}}(x) &= e^{-x^2 \cdot 5.50 \cdot 10^7\,m^{-2}} \end{aligned}\right\} \quad \xrightarrow{\mathcal{F}} \quad \left\{\begin{aligned} A_{n,\text{best}}(\nu) &= e^{-\nu^2 \cdot 8.72 \cdot 10^{-8}\,m^2} \\ A_{n,\text{worst}}(\nu) &= e^{-\nu^2 \cdot 1.79 \cdot 10^{-7}\,m^2} \end{aligned}\right. .$$

In the spatial frequency domain, LRT measurements can be understood as multiplica-
tion of ideal projections $\mathfrak{D}_\Theta(\nu)$ (see (8.13)) with the beam profile $A_n(\nu)$. It becomes
clear from Fig. 8.16 depicting $A_n(\nu)$ for the best and worst cases that the nonzero
spot size of the laser beam corresponds to a spatial low-pass filter. The larger the spot
size $w_{em}(\zeta)$, the lower will be the spatial cutoff frequency ν_{lp} of the low-pass filter
and, therefore, ideal projections will be more strongly affected. For the considered
sound field and realized setup, the maximum spatial frequency $\nu_{max} = 676\,m^{-1}$ is
attenuated by $\approx 8\%$ (worst case) in LRT measurements.

At this point, it should be mentioned that one is concerned with similar effects
in microphone and hydrophone measurements. The capsule hydrophone Onda
HGL-400, which is used for comparative measurements in Sect. 8.4, features the
radius $R_{HY} = 0.2\,mm$. Under the assumption of uniform sensitivity across the
hydrophone's active surface, the normalized sensitivity $a_{HY}(x)$ in the spatial domain
takes the form

$$a_{HY}(x) = \begin{cases} 1 & \text{for } |x| \leq R_{HY} \\ 0 & \text{elsewhere} \end{cases} \tag{8.31}$$

yielding the normalized Fourier transform [8]

$$A_n(\nu) = \frac{\sin(2\pi\nu R_{HY})}{2\pi\nu R_{HY}} = \text{sinc}(2\pi\nu R_{HY}) = \text{sinc}\left(\nu \cdot 1.26 \cdot 10^{-3}\,m\right) \tag{8.32}$$

---

[4]Function $f(x) = e^{-\alpha x^2}$; Fourier transform $F(\nu) = \sqrt{\pi/\alpha} \cdot e^{-(\pi\nu)^2/\alpha}$ [8].

in the spatial frequency domain. Figure 8.16 reveals that high-frequency components are more attenuated in hydrophone measurements due to averaging over the active surface than in LRT measurements, even for the worst case of the laser beam profile. In the far field of ultrasound sources, spatial frequencies in radial direction are, however, much lower than in propagation direction, i.e., axial direction. That is the reason why we are able to characterize also high-frequency sound fields through hydrophones, e.g., up to 20 MHz (i.e., $\nu_{max} = 1.35 \cdot 10^4 \, m^{-1}$) with the capsule hydrophone Onda HGL-400. Despite this fact, measurement data provided by typical hydrophones in transducer near fields is only useful to a limited extent.

8.3.6 Measurable Sound Frequency Range

Similar to conventional measurement principles (e.g., hydrophones) for sound field analysis, LRT has certain limits concerning measurable sound frequencies. The maximum measurable sound frequency f_{max} is mainly determined by the laser beam profile, whereas the spatial extension of the optical reflector defines the minimum measurable sound frequency f_{min}. Below, a theoretical analysis for both frequency limits is conducted.

Maximum Measurable Sound Frequency

As discussed in Sect. 8.3.5, the nonzero spatial extension of the LDV laser beam leads to a low-pass filter in LRT measurements. When its spatial cutoff frequency ν_{lp} is below decisive frequency components of the investigated sound field, those components will be attenuated and the measured sound pressure amplitude seems then to be decreased. On account of this fact, the spot size $w_{em}(\zeta)$ of the laser beam throughout the sound field determines the maximum measurable frequency f_{max} of sound. In the case that an attenuation of 3 dB is acceptable in LRT measurements, f_{max} directly relates to ν_{lp}. For the dedicated LRT setup and an assumed radial field extension of 30 mm, the largest laser spot size $w_{em}(\zeta = 60 \, mm) = 135 \, \mu m$ (worst case in Fig. 8.16) yields

$$f_{max} = c_{aco} \cdot \nu_{lp} = 1480 \, m \, s^{-1} \cdot 1390 \, m^{-1} = 2.1 \, MHz \qquad (8.33)$$

representing the maximum measurable frequency in water. To increase f_{max}, one may think about reducing $w_{em}(\zeta)$ throughout the sound field by raising the frequency of the laser light, which means reducing its wavelength λ_{em}. Major changes of λ_{em} are, however, hardly possible since commercial laser Doppler vibrometers (e.g., from the company Polytec GmbH [32]) usually work at fixed wavelengths within the visible range. Nevertheless, if an ultrasound source emits a strongly focused field, we may reliably analyze higher frequencies of sound with LRT. In such situation, the z-spacing between ultrasound source and optical reflector can be reduced. Owing to the small radial extension of the sound field, the beam waist w_0 can be decreased by increasing beam divergence. Anyway, LRT should not be used for investigations of sound fields in water with frequencies $f_{max} > 5 \, MHz$.

Fig. 8.17 Geometric
quantities for piston-type
transducer (radius R_T) to
compute sound pressure
amplitude $\hat{p}_\sim(r, \theta)$ at
arbitrary position given by
distance r and angle θ;
$r \sin \theta = z \tan \theta$

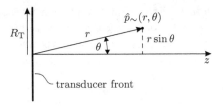

Minimum Measurable Sound Frequency

In tomographic imaging, it is of utmost importance to acquire projections of the
whole object under test. The object has, thus, to exhibit spatial extensions, which
can be completely covered by the projections. Due to the fact that there do not exist
clear boundaries for sound fields, an alternative criterion is needed for determining
the scanning area (i.e., in x-direction) in LRT measurements. The directivity pattern
of sound fields contains a dominant main lobe and several side lobes. With a view
to achieving reliable LRT results, it must be ensured that at least the main lobe is
entirely inside the scanning area [11]. Consequently, we are able to estimate for an
ultrasound source the minimum scanning area, which should be exceeded in LRT
measurements.

Let us take a look at the directivity pattern of a piston-type ultrasonic transducer
with the radius R_T. In case of a harmonically excitation, the sound pressure amplitude $\hat{p}_\sim(r, \theta)$ at an arbitrary position in the far field is proportional to (wave number k;
cf. (7.49, p. 280))

$$\hat{p}_\sim(r, \theta) \propto \frac{J_1(k R_T \sin \theta)}{k R_T \sin \theta} . \tag{8.34}$$

Here, r stands for the geometric distance from transducer center and Θ is the angle
between connecting line and z-axis (see Fig. 8.17). $J_1(\cdot)$ represents the Bessel function of the first kind and order 1. According to (8.34), $\hat{p}_\sim(r, \theta)$ will be zero if

$$\frac{J_1(k R_T \sin \theta)}{k R_T \sin \theta} = 0 , \tag{8.35}$$

which is fulfilled for the first time at $k R_T \sin \theta = 3.83$ rad. Because this zero point
defines the main lobe, one can calculate its diameter D_{main} at a given z-position in
the far field through

$$D_{main}(z) = 2z \tan \varphi = 2z \tan \left[\arcsin \left(\frac{3.83 \, \text{rad}}{k R_T} \right) \right] . \tag{8.36}$$

The scanning area in LRT measurements should cover at least the spatial extension
of the main lobe, i.e., D_{main}. Note that this fact does not only refer to the linear
positioning system but also to the optical reflector, which features an edge length
of $l_{opt} = 100$ mm in the realized LRT setup. The condition $D_{main} \leq l_{opt}$ and (8.36)

lead to the minimum wave number k_{min}, which can be just measured by means of LRT. For instance, at an axial distance $z = 50\,\mathrm{mm}$ of a harmonically excited ultrasonic transducer (piston-type; radius $R_T = 6.35\,\mathrm{mm}$) that operates in water, one obtains $k_{min} = 853\,\mathrm{rad\,m^{-1}}$. The minimum sound frequency f_{min} then results from

$$f_{min} = \frac{k_{min} c_{aco}}{2\pi} = \frac{853\,\mathrm{rad\,m^{-1}} \cdot 1480\,\mathrm{m\,s^{-1}}}{2\pi} = 201\,\mathrm{kHz}\,. \qquad (8.37)$$

When lower sound frequencies should be acquired, the reflector position would need to be changed during measurements or, alternatively, a larger optical reflector has to be used.

Additionally, there exists another frequency limitation that exclusively refers to the ultrasound source. The $\arcsin(\cdot)$ function in (8.36) demands an argument fulfilling $k R_T \geq 3.83\,\mathrm{rad}$. However, if this requirement is not fulfilled, the ultrasound source will not generate a sound field with pronounced side lobes. Instead, a wide main lobe will be emitted whose spatial extension is, strictly speaking, not limited. It is, therefore, complicated to analyze such sound fields through LRT. For the considered piston-type ultrasonic transducer operating in water, the critical wave number becomes $k = 603\,\mathrm{rad\,m^{-1}}$ yielding the minimum sound frequency $f_{min} = 142\,\mathrm{kHz}$ (see (8.37)) that can be measured in a reliable way.

8.4 Sound Fields in Water

In Sect. 8.3, we have studied the fundamentals of LRT including measurement principle, tomographic reconstruction, realized experimental setup as well as choosing decisive measurement parameters. Now, these fundamentals are applied to actual LRT measurements of sound fields in water, which is the most common propagation medium for ultrasonic waves. The piezo-optic coefficient $(\partial n/\partial p)_S$ is assumed to be (e.g., [11, 15])

$$\left(\frac{\partial n}{\partial p}\right)_S = 1.473 \cdot 10^{-10}\,\mathrm{Pa^{-1}} \qquad (8.38)$$

for the wavelength $\lambda_{em} = 475.8\,\mathrm{nm}$ of electromagnetic waves (i.e., laser beam of LDV) in water and the water temperature $20\,^\circ\mathrm{C}$.

Firstly, the sound pressure field of a piston-type ultrasonic transducer is analyzed in selected cross sections (see Sect. 8.4.1). The reconstructed sound pressure amplitudes will be compared to results of conventional hydrophone measurements. Afterward, we perform the same comparison for a cylindrically focused ultrasonic transducer. In Sect. 8.4.3, acceleration of the time-consuming measurement procedure is discussed. Finally, the disturbance of the sound field due to an immersed hydrophone will be quantified by means of LRT, which is hardly possible with other measurement approaches.

Fig. 8.18 Averaged output (mean of 16 signals) of LDV at position $(x, z) = (0, 32.8\,\text{mm})$ representing single point of projection $o_\Theta(\xi)$ for tomographic reconstruction; resulting SNR $\approx 34\,\text{dB}$; piston-type ultrasonic transducer Olympus V306-SU; excitation frequency $f = 1\,\text{MHz}$

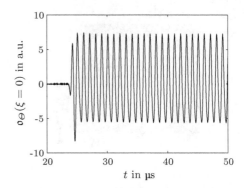

8.4.1   Piston-Type Ultrasonic Transducer

At the beginning, let us investigate the sound field of a piston-type ultrasonic transducer (Olympus V306-SU [29]) that was immersed in water. The ultrasonic transducer with a radius of $R_T = 6.35\,\text{mm}$ was excited by a sinusoidal burst signal of 40 cycles at $f = 1\,\text{MHz}$ and fixed in the cylindrical mount of the realized LRT setup (see Fig. 8.9). For the chosen transducer excitation, the near-field length equals $N_{\text{near}} = 27\,\text{mm}$. Several cross sections (i.e., in parallel to the xy-plane) of the arising sound field were acquired at different axial distances z, whereby $z = 0\,\text{mm}$ relates to the transducer front [14]. Figure 8.18 depicts the averaged output signal of the differential LDV at $(x, z) = (0, 32.8\,\text{mm})$, which represents data at a single point of a projection $o_\Theta(\xi)$ with respect to time t. By means of averaging 16 output signals directly within the utilized digital storage oscilloscopes Tektronix TDS 3054, the SNR exceeds 30 dB [11].

In Sect. 8.3.4, we determined decisive parameters for LRT measurements from the theoretical point of view. Thereby, a piston-type transducer was considered featuring the same radius R_T and excitation frequency as the investigated one. Here, exactly those parameters were applied to acquire sound pressure fields $p_\sim(x, y, t)$ in the cross sections $z = 32.8\,\text{mm}$ (far field) as well as $z = 8.4\,\text{mm}$ (near field). This refers to the scanning area but also to the number of projections N_{proj} and nonequidistant sampling between two neighboring LDV positions. Table 8.3 summarizes the most important parameters for the conducted LRT measurements. After recording the LDV output signals at the selected scanning positions, tomographic reconstruction was performed for each instant of time individually.

Figure 8.19a and c show the reconstructed sound pressure amplitudes $\hat{p}_\sim(x, y)$ in the cross sections[5] at $z = 32.8\,\text{mm}$ and $z = 8.4\,\text{mm}$, respectively. As expected from piston-type ultrasonic transducers, the sound pressure amplitudes are rotationally symmetric distributed in a cross section. Compared to the near field at $z = 8.4\,\text{mm}$,

---

[5]Geometric dimension of the plotted cross sections (in parallel to the xy-plane): $x \times y = [-10\,\text{mm}, 10\,\text{mm}] \times [-10\,\text{mm}, 10\,\text{mm}]$.

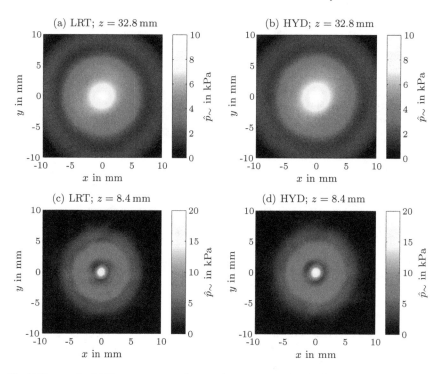

Fig. 8.19 a and **c** LRT measurements for sound pressure amplitudes $\hat{p}_\sim(x, y)$ in cross sections at $z = 32.8$ mm and $z = 8.4$ mm, respectively; **b** and **d** corresponding hydrophone (HYD) measurements; piston-type ultrasonic transducer Olympus V306-SU; excitation frequency $f = 1$ MHz

Table 8.3 Decisive parameters for LRT measurement of sound field arising from piston-type ultrasonic transducer

x-direction	$\Delta\xi = 0.2$ mm for $x \in [-6.4, 6.4]$ mm
	$\Delta\xi = 0.8$ mm for $x \in [-20.0, -6.4) \cup (6.4, 20.0]$ mm
z-direction	32.8 and 8.4 mm
Rotary direction	$\Delta\Theta = 1.8°$ per step
Temporal	100 MHz ($\Delta t = 10$ ns between sampling points)
Measurement	2 h per cross section
Reconstruction	4 min per cross section (5000 instants of time)

$\hat{p}_\sim(x, y)$ decreases in the far field at $z = 32.8$ mm, which coincides with the theory of sound propagation.

To verify the LRT results, hydrophone measurements were additionally carried out for both cross sections. Guided by the spatial resolution in LRT measurements, the utilized capsule hydrophone (Onda HGL-0400 [30]) was moved with the step size 0.2 mm in x- and y-direction. Because of the fine spatial resolution and a waiting time of 2 s that is required for mechanical vibrations to settle down after each hydrophone movement, such measurement takes approximately 12 h. Figure 8.19b

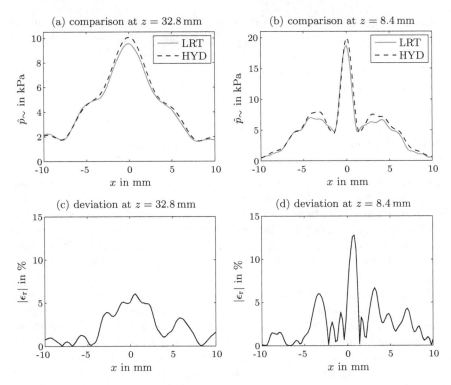

Fig. 8.20 **a** and **b** Comparison of LRT and hydrophone (HYD) measurements for sound pressure amplitudes $\hat{p}_{\sim}(x, y)$ along x-axis in cross sections at $z = 32.8$ mm and $z = 8.4$ mm, respectively; **c** and **d** relative deviations $|\epsilon_r|$ (magnitude) along x-axis between LRT and hydrophone measurements normalized to maximum of hydrophone output; piston-type ultrasonic transducer Olympus V306-SU

and d display the obtained results $\hat{p}_{\sim}(x, y)$ from hydrophone measurements. As the comparison clearly reveals, the results coincide very well with the corresponding LRT measurements. This also becomes obvious if we take a look at the acquired sound pressure amplitudes along the x-axis (see Fig. 8.20a and b) and the normalized relative deviations ϵ_r (see Fig. 8.20c and d) between the different measurement approaches. While $|\epsilon_r|$ is always smaller than 5.4% in the far field (i.e., $z = 32.8$ mm), there occur, however, relative deviations up to 12.3% in the near field (i.e., $z = 8.4$ mm). The pronounced deviations between LRT and hydrophone measurements in the near field can be mainly ascribed to three points. Firstly, the hydrophone is calibrated for ultrasound measurements in the far field, where the spatial frequencies ν in x- and y-direction are much smaller than in the near field (cf. Fig. 8.10b). As a second point, one has to keep in mind that a perfect alignment of the system components does not avoid spatial deviations of the coordinate systems for LRT and hydrophone measurements. Therefore, we always compare slightly different cross sections, which can be especially crucial in the near field of an ultrasonic transducer since sound fields are subject to strong fluctuations there. Lastly, hydrophone measurements are invasive and, thus, affect the analyzed sound field due to reflections of propagating sound waves, which is again critical in the near field. Nevertheless, the comparison of hydrophone and LRT measurements definitely proves that LRT is a reliable approach for investigating rotationally symmetric sound fields in water.

8.4.2 Cylindrically Focused Ultrasonic Transducer

LRT measurements usually concentrate on the inspection of rotationally symmetric sound fields (e.g., [5]). The dedicated LRT setup enables, however, the investigation of sound fields arising from nearly arbitrarily shaped sound sources. With a view to demonstrating this fact, a cylindrically focused ultrasonic transducer (Olympus V306-SU-CF1.00N [29]) exhibiting the focal length 25.4 mm was utilized to generate nonaxisymmetric sound fields in water [14]. Again, the transducer was excited by a sinusoidal burst of 40 cycles at $f = 1$ MHz. In contrast to the previous investigations, fairly conservative measurement parameters were chosen here. The radial scanning area was extended from 40 to 60 mm (i.e., $x \in [-30, 30]$ mm), and nonequidistant sampling between two neighboring LDV positions was abandoned. Consequently, the measuring time for a single cross section increases from 2 h to 6 h. Table 8.4 summarizes the most important parameters for the conducted LRT measurements.

Figure 8.21a depicts the reconstructed sound pressure amplitudes $\hat{p}_\sim(x, y)$ in the cross section at $z = 25.4$ mm representing the focal plane of the investigated cylindrically focused transducer. While the sound field is strongly focused in one direction, there hardly occurs focusing perpendicular to it, which is typical for such transducer shape.

To verify the LRT results, hydrophone measurements were carried out in the same cross section (see Fig. 8.21b), whereby the hydrophone was moved with the step size 0.2 mm in x- and y-direction. Both images show a fairly good agreement and have many fine details in common. This fact is also demonstrated in the sound pressure amplitudes along the x-axis (see Fig. 8.22a) and y-axis (see Fig. 8.22b), respectively. As displayed in Fig. 8.22c and d, the normalized relative deviations $|\epsilon_r|$ (magnitude) between LRT and hydrophone measurements along these axes are always smaller

Table 8.4 Decisive parameters for LRT measurement of sound field arising from cylindrically focused ultrasonic transducer

x-direction	$\Delta\xi = 0.2$ mm for $x \in [-30, 30]$ mm
z-direction	25.4 mm
Rotary direction	$\Delta\Theta = 1.8°$ per step
Temporal	100 MHz ($\Delta t = 10$ ns between sampling points)
Measurement	6 h per cross section
Reconstruction	6 min per cross section (5000 instants of time)

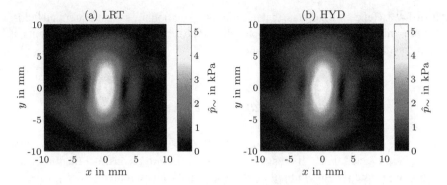

Fig. 8.21 a LRT measurement for sound pressure amplitudes $\hat{p}_\sim(x, y)$ in cross section at $z = 25.4$ mm; **b** corresponding hydrophone (HYD) measurement; cylindrically focused ultrasonic transducer Olympus V306-SU-CF1.00N; excitation frequency $f = 1$ MHz

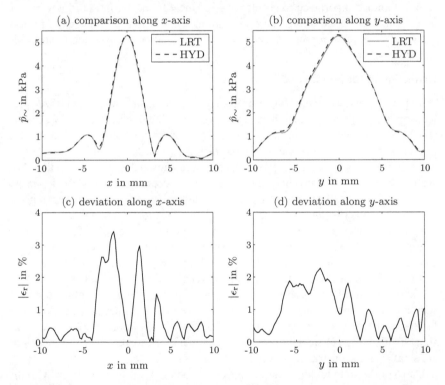

Fig. 8.22 a and **b** Comparison of LRT and hydrophone (HYD) measurements for sound pressure amplitudes $\hat{p}_\sim(x, y)$ along x-axis and along y-axis in cross section at $z = 25.4$ mm, respectively; **c** and **d** relative deviations $|\epsilon_r|$ (magnitude) between LRT and hydrophone measurements normalized to maximum of hydrophone output; cylindrically focused ultrasonic transducer Olympus V306-SU-CF1.00N

than 3.4%. Hence, we can state that in addition to rotationally symmetric sound fields, the presented LRT measurement approach is applicable for precise measurements of nonaxisymmetric sound fields in water.

8.4.3 Acceleration of Measurement Process

So far, the sound fields of piston-type and cylindrically focused ultrasonic transducers were projected under 100 angles per cross section to reconstruct field information (e.g., sound pressure amplitudes) through LRT measurements. The procedure comprising measurement and reconstruction takes a few hours for a single cross section. When sound propagation should be investigated with respect to space and time, we would need sound information in numerous cross sections [14]. It is, therefore, of utmost importance to accelerate LRT measurements. In the following, we will discuss two possibilities for acceleration: (i) reducing the number of projections and (ii) assumption-based reconstruction for rotationally symmetric ultrasonic transducers.

Reduction of the Number of Projections

A proper way to lower measuring time in LRT is reducing the number of utilized projections N_{proj}. The theoretical investigations in Sect. 8.3.4 indicated that $N_{proj} = 50$ should be sufficient to obtain reliable reconstruction results. Let us verify this value by means of LRT measurements for the piston-type ultrasonic transducer Olympus V306-SU. As an example, the sound pressure amplitudes $\hat{p}_\sim(x, y)$ in the cross section at $z = 32.8$ mm are considered. Figure 8.19a depicts the reconstruction result for $N_{proj} = 100$ acquired projections. Here, this value is reduced by picking out every Mth projection and ignoring the remaining ones for tomographic reconstruction. M was selected to be [2, 4, 5, 10, 20, 50] yielding $N_{proj} \in [50, 25, 20, 10, 5, 2]$ projections. In doing so, the reconstruction results for reduced numbers of projections can be compared to the full reconstruction, which follows from $N_{pro} = 100$.

Figure 8.23a and b illustrate maximum relative deviations and mean relative deviations from full reconstruction as functions of N_{proj}. Both deviations decrease with rising amount of projections for the considered cross section, i.e., $x \times y = [-10$ mm, 10 mm$] \times [-10$ mm, 10 mm$]$. The same holds in the central area of the sound field, which is here defined as $|x, y| < 6.4$ mm. However, since the deviations are smaller in the central area, reconstruction results in the periphery (i.e., $|x, y| \geq 6.4$ mm) of the sound field suffer more from reducing N_{proj}. In accordance with the theoretical investigations, $N_{proj} = 50$ leads in any case to precise LRT results and, therefore, constitutes a good compromise between measuring time and measurement accuracy. But, if solely information from the central area is desired, we will be able to additionally reduce the number of projections, e.g., $N_{proj} = 20$ or even less.

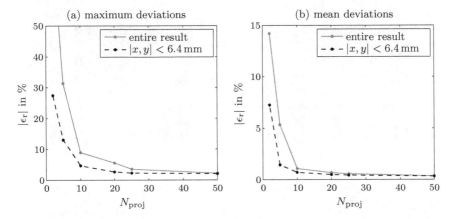

Fig. 8.23 a and **b** Maximum and mean of normalized relative deviations $|\epsilon_r|$ (magnitude) between reconstruction results and full reconstruction ($N_{\mathrm{proj}} = 100$; see Fig. 8.19a) versus number of projections N_{proj} for entire sound pressure field (i.e., $|x, y| \leq 10\,\mathrm{mm}$) and within central area (i.e., $|x, y| < 6.4\,\mathrm{mm}$), respectively

Axisymmetric Assumption

The realized LRT setup combined with tomographic reconstruction allows spatially and temporally resolved acquisition of sound fields generated by nearly arbitrarily shaped ultrasonic transducers. This is achieved by projecting the investigated sound field under a sufficient number of angles through a LDV. Here, let us study reconstruction results in cases where projections under different angles are not available; i.e., there exists only one projection $o_\Theta(\xi)$ of the sound field. For instance, such situation will be present when the sound source cannot be rotated or measuring time of LRT should be minimal. As a result, one has to reconstruct the entire sound field from a single projection, which means that an axisymmetric field has to be assumed. Such assumption will only make sense, however, if the sound source is of rotationally symmetric shape, e.g., piston-type or spherically focused.

As a first example, the sound field generated by the piston-type ultrasonic transducer Olympus V306-SU is considered. Again, we take a look at the sound pressure amplitudes $\hat{p}_\sim(x, y)$ in the cross sections $z = 32.8\,\mathrm{mm}$ (far field) and $z = 8.4\,\mathrm{mm}$ (near field), which are shown in Fig. 8.19a and c for full reconstruction, i.e., $N_{\mathrm{proj}} = 100$. To emulate the so-called assumption-based reconstruction that uses only one acquired projection, a single projection was picked out and replicated hundred times yielding $N_{\mathrm{proj}} = 100$ projections. These identical projections served then as input for tomographic reconstruction. Figure 8.24a and b display the results of the assumption-based reconstruction at $z = 32.8\,\mathrm{mm}$ and $z = 8.4\,\mathrm{mm}$, respectively. By comparing assumption-based and full reconstructions (see Fig. 8.19a and c), it becomes apparent that the spatial distributions of $\hat{p}_\sim(x, y)$ as well as their absolute values agree very well, especially in the far field. This can be also seen from the sound pressure amplitudes along the x-axis (see Fig. 8.24c and d) and the resulting

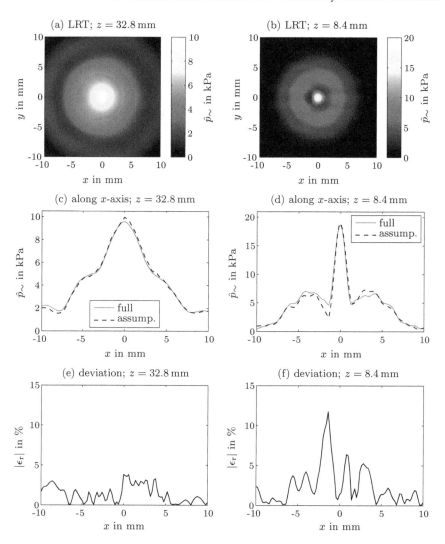

Fig. 8.24 **a** and **b** LRT results for sound pressure amplitudes $\hat{p}_\sim(x, y)$ in cross sections at $z = 32.8$ mm and $z = 8.4$ mm for assumption-based reconstruction, respectively; **c** and **d** $\hat{p}_\sim(x, y)$ along x-axis in cross sections at $z = 32.8$ mm and $z = 8.4$ mm for full and assumption-based reconstruction; (e) and (f) normalized relative deviations $|\epsilon_r|$ (magnitude) between both reconstruction approaches; piston-type ultrasonic transducer Olympus V306-SU; excitation frequency $f = 1$ MHz

normalized relative deviations ϵ_r (see Fig. 8.24e and f) between the different reconstruction approaches. Hence, the question arises why one should acquire projections under different angles, which is indeed a time-consuming procedure.

In order to answer the aforementioned question, let us investigate the sound field generated by the piston-type ultrasonic transducer Krautkramer Benchmark ISS 3.5 [17]. The transducer was excited again by a sinusoidal burst of 40 cycles at

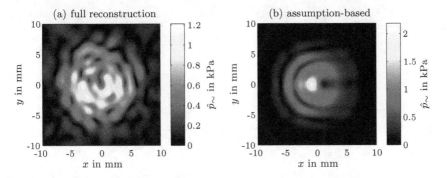

Fig. 8.25 LRT measurement for sound pressure amplitudes $\hat{p}_\sim(x, y)$ in cross section at $z = 6.5$ mm for **a** full reconstruction and **b** assumption-based reconstruction; piston-type ultrasonic transducer Krautkramer Benchmark ISS 3.5; excitation frequency $f = 1$ MHz

$f = 1$ MHz. Figure 8.25a depicts the resulting sound pressure amplitudes $\hat{p}_\sim(x, y)$ at the cross section $z = 6.5$ mm (near field) for full reconstruction; i.e., $N_{\text{proj}} = 100$ independent projections were acquired under 100 angles. Although the ultrasonic transducer features a rotationally symmetric shape, the spatial distribution of $\hat{p}_\sim(x, y)$ is by no means axisymmetric, which can be attributed to partial damages of the transducer's matching layer. In a next step, assumption-based reconstruction was performed for the same cross section by picking out a single projection and replicating it so that $N_{\text{proj}} = 100$ identical projections are available. Not surprisingly, the assumption-based reconstruction (see Fig. 8.25b) completely differs from full reconstruction regarding spatial distribution as well as absolute values. It can, therefore, be concluded that even though ultrasonic transducers are of rotationally symmetric shape, assumption-based reconstruction may cause remarkable deviations in LRT measurements of sound fields.

According to the previous discussions, we desire a simple criterion answering the question whether assumption-based reconstruction provides reliable results in LRT measurements. Such a criterion follows from a single sound field projection $\mathfrak{o}_\Theta(\xi)$ along a line, which is acquired by the LDV [11]. If that projection is symmetrical with respect to the rotation axis (i.e., z-axis), assumption-based reconstruction may be permitted. Figure 8.26 contains single projection magnitudes along the x-axis for both piston-type transducers. In contrast to the Olympus transducer, the projection for the Krautkramer transducer is completely asymmetrical around the rotation axis, which gets also clear by comparing mirrored projections from the left side (i.e., $x \in [-20 \text{ mm}, 0]$) with the original ones from the right side. It, thus, seems only natural that we cannot apply assumption-based reconstruction in LRT measurements for the Krautkramer transducer. However, the simple criterion is necessary but not sufficient for achieving reliable results with assumption-based reconstruction because the comparison of a single projection may lead to a wrong conclusion. For this reason, full reconstruction should be preferred even for rotationally symmetric transducer shapes, especially when precise information of the investigated sound fields is demanded.

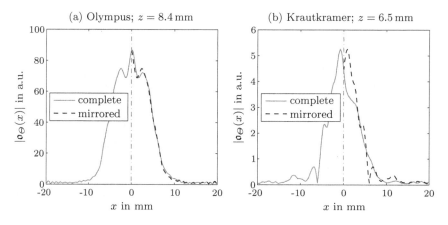

Fig. 8.26 Projections $|\varrho_\Theta(x)|$ (magnitude) of sound field along x-axis for **a** Olympus transducer V306-SU and **b** Krautkramer transducer Benchmark ISS 3.5; dashed lines represent mirrored version from left side, i.e., $x \in [-20\,\text{mm}, 0]$

8.4.4 Disturbed Sound Field due to Hydrophones

To verify LRT results in water, we compared them so far to hydrophone measurements, whereby the hydrophone had to be placed directly within the investigated sound field. Due to the different acoustic properties of hydrophone and the surrounding water, there occur, of course, reflections as well as diffractions of incident sound waves at the interface of water and hydrophone. Therefore, the sound field is influenced by the presence of the hydrophone, which especially constitutes a problem close to the transducer surface and medium boundaries. In the following, the disturbed sound field should be quantified. The applied measurement approach has to meet mainly three requirements:

- With a view to avoiding further disturbances of the sound field, the measurement has to be noninvasive and nonreactive.
- Spatially and temporally resolved as well as absolute measurement results should be provided.
- The measurement approach has to feature omnidirectional sensitivity because incident and reflected sound waves propagate in different directions.

In accordance with these requirements, LRT is an outstanding candidate for the quantitative analysis of sound reflections and diffractions at the hydrophone surface [11, 15].

Figure 8.27 shows the relevant part of the experimental setup comprising cylindrical mount, piston-type ultrasonic transducer Olympus V306-SU, optical reflector and capsule hydrophone Onda HGL-0400 as well as its preamplifier Onda AH-2010. The rotationally symmetric hydrophone was placed at the axial distance $z_H = 27.0\,\text{mm}$ in front of the transducer. Since LRT measurements demand sound field projections

Fig. 8.27 Relevant part of realized LRT setup to investigate disturbance of sound field at capsule hydrophone Onda HGL-0400 [15]; ultrasonic transducer UT

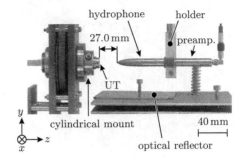

Table 8.5 Decisive parameters for LRT measurement of disturbed sound field between transducer front and hydrophone tip

x-direction	$\Delta \xi = 0.2$ mm for $x \in [-6.4, 6.4]$ mm
	$\Delta \xi = 0.8$ mm for $x \in [-20.0, -6.4) \cup (6.4, 20.0]$ mm
z-direction	$\Delta z = 0.2$ mm for $z \in [0.2, 26.8]$ mm
Rotary direction	$\Delta \Theta = 3.6°$ per step
Temporal	100 MHz ($\Delta t = 10$ ns between sampling points)
Reconstruction	134 cross sections at 5000 instants of time

under different angles and the realized setup allows only transducer rotation, the rotation axes of both cylindrical mount and hydrophone have to be aligned. The horizontal hydrophone alignments in the xz-plane were based again on the intensity of the reflected LDV laser beam (see Sect. 8.3.4). For vertical hydrophone alignments in the yz-plane, we analyzed pictures from a SLR camera (Nikon D80; 10.2 megapixels) that was located about 5 m away from the LRT setup [15]. In doing so, it is possible to ensure geometrical deviations of both rotation axes $<5\,\mu$m in x-direction and $<50\,\mu$m in y-direction, respectively.

The ultrasonic transducer was excited by a sinusoidal pulse of 8 cycles at $f = 1$ MHz. With the aid of LRT, the transient sound field was acquired in 134 cross sections (i.e., xy-planes), which were equidistantly distributed between transducer front and hydrophone tip (see Table 8.5 for measurement parameters). After measurement and tomographic reconstruction, the entire sound field at a selected instant of time follows from assembling sound pressure values of all cross sections for that instant. Beyond the hydrophone tip (i.e., $z \geq 27.0$ mm), LRT does not provide, however, absolute information about the sound field because the hydrophone blocks the LDV laser beam and this data is then missing for tomographic reconstruction. Nevertheless, each point of a projection $\mathfrak{o}_\Theta(\xi)$ is proportional to the integral of the sound pressure along the laser beam and, consequently, contains certain information about sound propagation. Instead of reconstructing spatially resolved sound pressure values for $z \geq 27.0$ mm, sound propagation was only visualized there through recording a single projection in every cross section [55].

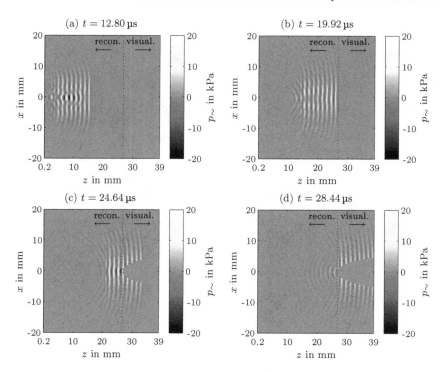

Fig. 8.28 Sound pressure fields in xz-plane (i.e., $y = 0$) at four instants of time after starting emitting sinusoidal burst (8 cycles at 1 MHz); reconstruction results for $z < 27.0$ mm; visualizations for $z \geq 27.0$ mm; **a** transducer finished emitting burst signal; **b** sound waves almost reached hydrophone; **c** constructive interference of incident and reflected sound waves; **d** all wave fronts passed hydrophone tip at $z_H = 27.0$ mm; piston-type ultrasonic transducer Olympus V306-SU

In Fig. 8.28, one can see the measured sound pressure fields in the xz-plane (i.e., $y = 0$) at four representative instants of time after starting pulse emission, namely 12.80, 19.92, 24.64, and 28.44 μs. Thereby, reconstructions between transducer front and hydrophone are combined with visualizations beyond hydrophone tip. To achieve meaningful images, visualizations were rescaled so that the absolute values of reconstructed sound pressure and visualized projections coincide nearby $z = 27.0$ mm. At the first two instants of time $t = 12.80$ and $t = 19.92$ μs (see Fig. 8.28a and b), the sound pressure waves have not reached the hydrophone tip, which means that there do not emerge disturbances of the sound field. The already existing constructive and destructive interference patterns stem from the beam characteristic of the piston-type ultrasonic transducer. In Fig. 8.28c, constructive interference of incident and reflected sound pressure waves arises since exactly four wave fronts passed the hydrophone tip. At the last instant of time $t = 28.44$ μs, all wave fronts have passed the hydrophone tip and the expected reflections propagate in negative z-direction toward to transducer. When the hydrophone is placed close to the transducer surfaces or medium boundaries, exactly such reflected waves can cause remarkably deviations in hydrophone measurements.

Fig. 8.29 Binarized version of sound pressure field at $t = 28.44\,\mu s$ (see Fig. 8.28d); positive and negative sound pressure values in black and white color; respectively

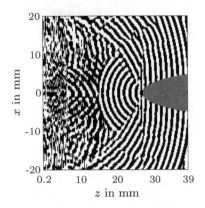

For better interpretation, Fig. 8.29 depicts a binarized version of Fig. 8.28d; i.e., negative and positive sound pressure values are shown in black-and-white color, respectively. It can be clearly observed that the reflected sound waves appear as eight concentric white circular regions in front of the hydrophone tip (i.e, $z < 27.0\,mm$). Besides, sound waves propagating in positive z-direction are still present for $z < 27.0\,mm$, which is a consequence of the transducer's beam characteristic. At this point, it should be mentioned again that the cross sections were individually reconstructed between transducer front and hydrophone tip. The spatial continuity and symmetry in the resulting images, thus, proves once more applicability of LRT even for challenging measurement tasks.

In addition to the spatial investigations of the disturbed sound field, let us now regard the reconstructed time-dependent values $p_\sim(t)$ at selected positions on the z-axis, i.e., $(x, y) = (0, 0)$. Figure 8.30a displays $p_\sim(t)$ at $z = 24.6\,mm$, which means 2.4 mm in front of the hydrophone tip. One can distinguish between three time periods within the sound pressure curve: (i) exclusively waves propagating in positive z-direction; (ii) interference between waves propagating in positive and negative z-direction; (iii) exclusively waves propagating in negative z-direction. In Fig. 8.30a, the time periods (i), (ii), and (iii) cover approximately the intervals $[18, 22]\,\mu s$, $[22, 27]\,\mu s$ and $[27, 31]\,\mu s$, respectively. Figure 8.30b shows $p_\sim(t)$ for the axial position $z = 26.6\,mm$, i.e., very close to the hydrophone tip. Contrary to $z = 24.6\,mm$ where constructive interference is present in the sound pressure curve, time period (ii) is dominating for $z = 26.6\,mm$ and refers to destructive interference. For these reasons, it can be stated that sound pressure amplitudes in the disturbed sound field vary greatly close to the hydrophone tip [15].

Finally, the hydrophone output is compared with the LRT result at the position of the hydrophone tip (see Fig. 8.30c and d). Note that for the LRT measurement, the hydrophone was removed from the water path and, consequently, there did not exist sources for disturbing the sound field anymore. The acquired sound pressure curves $p_\sim(t)$ of both measurement approaches coincide very well regarding time behavior and absolute values. Related to the hydrophone output, the relative deviation ϵ_r of the amplitudes becomes

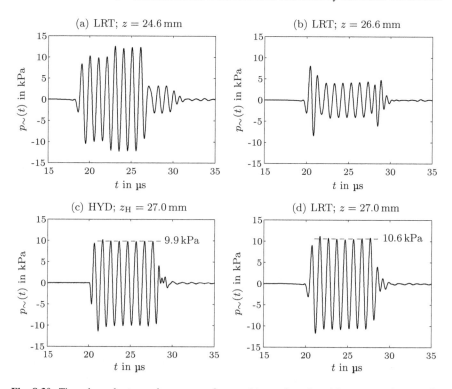

Fig. 8.30 Time-dependent sound pressure values $p_{\sim}(t)$ at selected positions on z-axis; **a** partly constructive interference of incident and reflected sound waves at $z = 24.6\,\mathrm{mm}$; **b** dominating destructive interference at $z = 26.6\,\mathrm{mm}$; **c** hydrophone output at $z_{\mathrm{H}} = 27.0\,\mathrm{mm}$; **d** LRT result at $z = 27.0\,\mathrm{mm}$ in the absence of hydrophone

$$\epsilon_{\mathrm{r}} = \frac{\hat{p}_{\mathrm{L}\sim} - \hat{p}_{\mathrm{H}\sim}}{\hat{p}_{\mathrm{H}\sim}} = \frac{10.6\,\mathrm{kPa} - 9.9\,\mathrm{kPa}}{9.9\,\mathrm{kPa}} = 7.1\% \ . \tag{8.39}$$

The expressions $\hat{p}_{\mathrm{H}\sim}$ and $\hat{p}_{\mathrm{L}\sim}$ stand for the determined values of hydrophone and LRT measurements, respectively. Taking into account the uncertainty of 10.4% in hydrophone measurements demonstrates once again the reliability of LRT results [33].

From the presented results in Sect. 8.4, it can be concluded that LRT is an excellent measurement approach providing spatially as well as temporally resolved data for sound fields in water. This also applies to situations where other measurement approaches fail, e.g., hydrophone for disturbed sound fields.

8.5 Sound Fields in Air

Air serves in various applications as propagation medium for ultrasonic waves. For example, airborne ultrasound is oftentimes employed in distance measurements such as parking sensors (see Sect. 7.6.1). Here, we will prove the applicability of LRT for quantitative measurements of airborne ultrasound. In Sect. 8.5.1, the piezo-optic coefficient $(\partial n/\partial p)_S$ in air is derived. Section 8.5.2 deals with the slightly modified LRT setup, which additionally contains foam to avoid disturbing reflections of ultrasound. Finally, we discuss reconstructed sound pressure amplitudes for an air-coupled ultrasonic transducer that operates at a frequency of $f = 40\,\mathrm{kHz}$. The results are verified by microphone measurements.

8.5.1 Piezo-optic Coefficient in Air

With a view to reconstructing spatially as well as temporally resolved sound pressure values for airborne ultrasound, LRT measurements require knowledge of the piezo-optic coefficient $(\partial n/\partial p)_S$ in air. Let us deduce this coefficient for an ideal gas, which constitutes a good approximation of air. The adiabatic state equation for an ideal gas is defined as (cf. (2.108, p. 32))

$$\frac{p_0 + \Delta p}{p_0} = \left(\frac{\varrho_0 + \Delta\varrho}{\varrho_0}\right)^{\kappa} \tag{8.40}$$

with the pressure p_0 and the density ϱ_0 in the equilibrium state, respectively. Δp and $\Delta\varrho$ stand for slight fluctuations around the equilibrium state, and κ is the adiabatic exponent. By utilizing the so-called *Gladstone–Dale relation* for gases [55]

$$K_{\mathrm{G}}\underbrace{(\varrho_0 + \Delta\varrho)}_{\varrho} = \underbrace{n_0 + \Delta n}_{n} - 1 \tag{8.41}$$

where K_{G} denotes the Gladstone–Dale constant, (8.40) becomes

$$\frac{\Delta p}{p_0} = \left(\frac{n_0 + \Delta n - 1}{n_0 - 1}\right)^{\kappa} - 1 = \left(1 + \frac{\Delta n}{n_0 - 1}\right)^{\kappa} - 1 . \tag{8.42}$$

Again, the expressions n_0 and Δn represent the optical refractive index of the gas in equilibrium state and its fluctuation, respectively. Owing to the fact that $\Delta n \ll n_0$ is usually fulfilled for airborne ultrasound, we can introduce the Taylor approximation $(1 + x)^k \approx 1 + kx$. Therewith, (8.42) simplifies to

$$\frac{\Delta p}{p_0} \approx 1 + \kappa\frac{\Delta n}{n_0 - 1} - 1 = \frac{\kappa\Delta n}{n_0 - 1} . \tag{8.43}$$

Rewriting this equation finally yields the piezo-optic coefficient of an ideal gas

$$\left(\frac{\partial n}{\partial p}\right)_S = \frac{\Delta n}{\Delta p} \approx \frac{n_0 - 1}{\kappa p_0} ,\tag{8.44}$$

which can be used to link refractive index changes Δn to sound pressure values $p_\sim$ ($\hat{=}\Delta p$) when the quantities n_0, p_0 as well as κ are known.

In LRT measurements of airborne ultrasound, we suppose the following conditions and parameters for air [11, 38]:

- wavelength of laser beam $\lambda_{em} = 632.8\,nm$
- air temperature $20\,°C$
- adiabatic exponent $\kappa = 1.4$ at $20\,°C$
- static air pressure $p_0 = 101.325\,kPa$
- relative humidity 40%
- carbon dioxide content 0.045%.

According to the empirical formula in [16], these values lead to the optical refractive index $n_0 = 1.000271$ in the equilibrium state and, consequently, to the piezo-optic coefficient

$$\left(\frac{\partial n}{\partial p}\right)_S \approx \frac{n_0 - 1}{\kappa p_0} = \frac{1.000271 - 1}{1.4 \cdot 101325\,Pa} = 1.91 \cdot 10^{-9}\,Pa^{-1}\tag{8.45}$$

in air. Actually, this coefficient is subject to certain fluctuations in practical situations. For instance, if the air temperature increases by $1\,°C$, n_0 will decrease by $1.1 \cdot 10^{-6}$ and $(\partial n/\partial p)_S$ will be reduced by 0.4% (e.g., [52]). It can be stated, however, that under normal ambient conditions, the piezo-optic coefficient in air exhibits a maximum uncertainty $<20\%$. Compared to water, whose piezo-optic coefficient amounts $\approx 10^{-10}\,Pa^{-1}$ (see (8.38)), the value in air is more than ten times larger.

8.5.2 Experimental Setup

The realized LRT setup for investigating airborne ultrasound is similar to the experimental arrangement that we utilized in water. This refers to the differential LDV, the linear positioning system, and rotation unit but also to the cylindrical mount as well as optical reflector (see Fig. 8.31). In order to avoid disturbing reflections of ultrasound, several components of the experimental setup have to be additionally surrounded and lined by a foam, which absorbs sound waves and, therefore, reduces echoes.

Fig. 8.31 Relevant part of realized LRT setup to investigate airborne ultrasound [38]; foam avoids disturbing reflections of sound waves; ultrasonic transducer UT

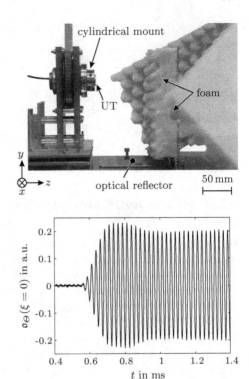

cylindrical mount

foam

UT

optical reflector   50 mm

Fig. 8.32 Averaged output (mean of 64 signals) of LDV at position $(x, z) = (0, 20\,\text{mm})$ representing single point of projection $o_\Theta(\xi)$ for tomographic reconstruction; transducer excitation started at 0.5 ms; resulting SNR $\approx$ 31 dB; piston-type ultrasonic transducer Sanwa SCS-401T; excitation frequency $f = 40\,\text{kHz}$

In fact, there arise significant differences between sound propagation in water and in air. On the one hand, attenuation of ultrasonic waves in air is much higher than in water (cf. Table 2.8 on p. 40). The acoustic impedance Z_{aco} of air is much lower than that of water on the other hand. While the acoustic impedance of water amounts $1.48 \cdot 10^6\,\text{N s m}^{-3}$, air exhibits a value of $413.5\,\text{N s m}^{-3}$ at $20\,^\circ\text{C}$. For these reasons, both the frequencies of airborne ultrasound and the resulting sound pressure values are usually comparatively small. In the present case, the air-coupled ultrasonic transducer Sanwa SCS-401T with a radius of $R_{\text{T}} = 6.5\,\text{mm}$ served as sound source. The piston-type transducer was fixed again in the cylindrical mount. The transducer was excited by a sinusoidal burst signal of 48 cycles at $f = 40\,\text{kHz}$, which is a typical frequency for airborne ultrasound in practical applications, e.g., parking sensors. To achieve a satisfactory SNR in LRT measurements, the excitation voltage was chosen to be $24\,\text{V}_{\text{pp}}$ (peak-to-peak). Figure 8.32 shows the averaged output signal (mean of 64 signals) of the differential LDV at $(x, z) = (0, 20\,\text{mm})$, which represents data at a single point of a projection $o_\Theta(\xi)$ with respect to time t. For such excitation voltage, the SNR exceeds 30 dB and, thus, is sufficient for LRT measurements. However, it should be noted that an appropriate microphone (e.g., 1/4 inch microphone from Brüel & Kjær) usually provides considerably higher SNR values [11, 38].

Table 8.6 Decisive parameters for LRT measurement of sound field in air arising from piston-type ultrasonic transducer

x-direction	$\Delta\xi = 1.0$ mm for $x \in [-45, 45]$ mm
z-direction	5, 20 and 90 mm
Rotary direction	$\Delta\Theta = 4.5°$ per step
Temporal	2.5 MHz ($\Delta t = 400$ ns between sampling points)
Measurement	50 min per cross section
Reconstruction	2 min per cross section (5000 instants of time)

8.5.3   Results for Piston-Type Ultrasonic Transducer

By means of LRT, we acquired sound pressure fields $p_\sim(x, y, t)$ of the air-coupled transducer in three different cross sections. The cross sections were located at the axial distances $z = 5$ mm, $z = 20$ mm and $z = 90$ mm from the transducer front. Compared to the LRT experiments in water, the scanning area in x-direction has been extended because wavelength λ_{aco} of the sound waves and, consequently, the main lobe's diameter D_{main} of the directivity pattern increases (see Sect. 8.3.6). Moreover, the higher value of λ_{aco} enables increasing the sampling interval $\Delta\xi$ between two neighboring LDV positions as well as reducing the number of projections N_{proj}, which is necessary for tomographic reconstruction. Table 8.6 summarizes the most important parameters for the conducted LRT measurements.

Figure 8.33a and c presents the reconstructed sound pressure amplitudes $\hat{p}_\sim(x, y)$ of the piston-type transducer in the cross sections[6] at $z = 5$ mm and $z = 20$ mm, respectively. The sound pressure amplitudes are rotationally symmetric distributed in both cross sections. For the chosen transducer excitation $f = 40$ kHz, the near-field length equals $N_{\mathrm{near}} = 2.8$ mm. Hence, the maximum of $\hat{p}_\sim(x, y)$ should be larger at $z = 5$ mm than at $z = 20$ mm, which is also confirmed in the LRT measurements.

With a view to verifying LRT results, we performed microphone measurements in addition [11, 38]. The utilized 1/4 inch condenser microphone (Brüel & Kjær; type 4939 [9]), whose output response to sound waves is almost constant up to 100 kHz, was moved with the step size 1.0 mm in x- and y-direction. Overall, this procedure takes 2.5 h and, thus, much more time than corresponding LRT measurements (see Table 8.6). In Fig. 8.33b and d, one can see the obtained amplitudes $\hat{p}_\sim(x, y)$ from microphone measurements. The comparison reveals that LRT and microphone results coincide very well in both cross sections. This is also demonstrated by the acquired sound pressure amplitudes along the x-axis (see Fig. 8.34a and b) and the normalized relative deviations ϵ_r (see Fig. 8.34c and d) between the different measurement approaches. The results exhibit a maximum difference of 11.7 and 8.9% at $z = 5$ mm and $z = 20$ mm, respectively. From there, LRT seems to be a suitable approach to

---

[6]Geometric dimension of the plotted cross sections (in parallel to the xy-plane): $x \times y = [-25$ mm, 25 mm$] \times [-25$ mm, 25 mm$]$.

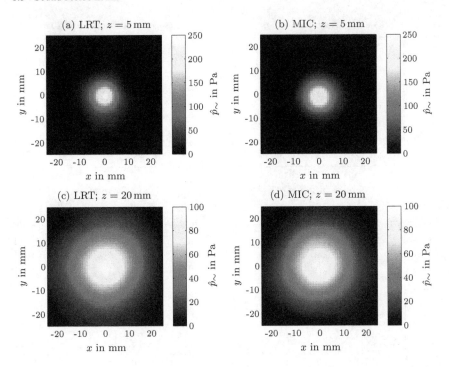

Fig. 8.33 **a** and **c** LRT measurements for sound pressure amplitudes $\hat{p}_\sim(x, y)$ in cross sections at $z = 5\,\text{mm}$ and $z = 20\,\text{mm}$, respectively; **b** and **d** corresponding microphone (MIC) measurements; piston-type ultrasonic transducer Sanwa SCS-401T; excitation frequency $f = 40\,\text{kHz}$

replace conventional microphone measurements, especially when we desire sound pressure information in the whole cross section.

Apart from the axial distances 5 and 20 mm, the sound field was investigated in the cross section at $z = 90\,\text{mm}$. Figure 8.35a and b display the achieved LRT and microphone results for the sound pressure amplitudes $\hat{p}_\sim(x, t)$. In contrast to the other cross sections, there emerge enormous deviations between the different measurement approaches, which gets also obvious in the acquired values along the x-axis (see Fig. 8.35c). The reason for the large deviations lies in the spatial main lobe extension of the sound field. At the axial distance $z = 90\,\text{mm}$, the main lobe has a diameter of $D_{\text{main}} = 244\,\text{mm}$ (see (8.36)), which, therefore, remarkably exceeds the geometric dimensions $l_{\text{opt}} = 100\,\text{mm}$ of the utilized optical reflector. Due to this fact, the main lobe cannot be completely covered by means of a single reflector position in LRT measurements. The projections $\mathfrak{o}_\Theta(\xi)$ for tomographic reconstruction may exhibit relatively large values at the boundaries of the scanning area (i.e., $x = \pm 45\,\text{mm}$), which is proven in Fig. 8.35d.

An insufficient scanning area has mainly two consequences for LRT results [11]. Firstly, nonzero projections at the scanning boundaries can be understood as spectral leakage in the spatial frequency domain. Such spectral leakage leads to spatial

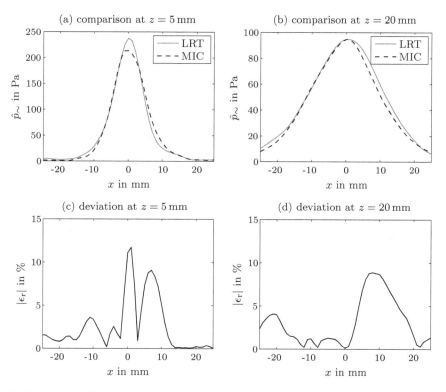

Fig. 8.34 **a** and **b** Comparison of LRT and microphone (MIC) measurements for sound pressure amplitudes $\hat{p}_{\sim}(x, y)$ along x-axis in cross sections at $z = 5$ and $z = 20$ mm, respectively; **c** and **d** relative deviations $|\epsilon_r|$ (magnitude) along x-axis between LRT and microphone measurements normalized to maximum of microphone output; piston-type ultrasonic transducer Sanwa SCS-401T

oscillations in the reconstructed images (see, e.g., Fig. 8.35a). As a second consequence, the energies of reconstructed and actual sound pressure field may differ in a cross section. Let us explain this circumstance through Fig. 8.36, which illustrates an insufficient scanning area in LRT measurements for a rotationally symmetric sound field. Within the scanning area, the entire sound information along the LDV beam contributes to the measurement. However, sound information is not available beyond this scanning area and, thus, sound energy and sound pressure values are set to zero there, regardless of their actual values. In other words, we are mixing up two situations differing in sound energy during tomographic reconstruction. It seems only natural that in case of an insufficient scanning area, LRT measurements are then accompanied by large deviations.

To sum up, LRT is also applicable for acquiring sound pressure values of airborne ultrasound. Nevertheless, this technique should be restricted to applications where sound pressure amplitudes allow a satisfactory SNR in the LDV outputs and the spatial extension of the sound field can be completely covered.

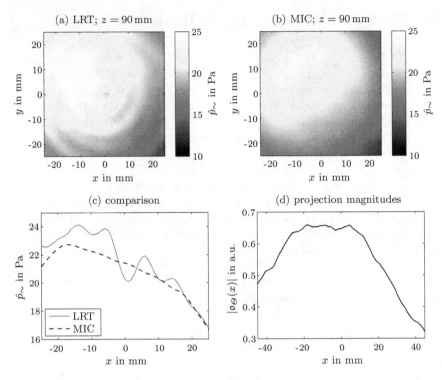

Fig. 8.35 a and **b** LRT and microphone (MIC) measurements for sound pressure amplitudes $\hat{p}_\sim(x, y)$ in cross section at $z = 90$ mm, respectively; **c** comparison of $\hat{p}_\sim(x, y)$ along x-axis; **d** magnitude $|o_\Theta(x)|$ of single projection; piston-type ultrasonic transducer Sanwa SCS-401T; excitation frequency $f = 40$ kHz

Fig. 8.36 Insufficient scanning area in LDV measurements for rotationally symmetric sound field [38]; extension of actual sound field remarkably exceeds reconstructed one

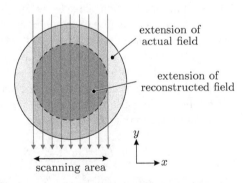

8.6 Mechanical Waves in Optically Transparent Solids

So far, LRT was exclusively applied to analyze sound propagation in fluids such as water and air. Apart from fluids, solid media are often involved in sound as well as ultrasound applications (e.g., nondestructive testing). At the boundary of a fluid and a solid, sound pressure waves propagating in the fluid get converted to mechanical waves, which propagate in the solid. Conventional measurement approaches (see Sect. 8.1) do not, however, allow quantitative investigations of mechanical waves within the solid. For instance, the Schlieren optical method can be utilized to visualize mechanical waves in optically transparent solids, but this method does not provide absolute information about physical quantities (e.g., mechanical stress) describing the waves. In this section, we will prove the applicability of LRT to determine such quantities in optically transparent solids with respect to both time and space.

As a starting point, the mechanical normal stress in an isotropic solid will be deduced from dilatation. Section 8.6.2 deals with the experimental setup that was used to excite mechanical waves in optically transparent solids. Finally, we present selected LRT results, which will be verified by numerical simulations and characteristic parameters (e.g., reflection coefficient).

8.6.1 Normal Stress in Isotropic Solids

As already discussed in Sect. 8.3.1, LRT enables acquisition of mechanical longitudinal waves in optically transparent solids. In doing so, we reconstruct the spatially as well as temporally resolved dilatation δ_{dil}, which corresponds to the sum of normal strains S_{ii} within the solid medium (cf. (8.8)). This quantity is, however, not common for describing propagation of mechanical waves. Besides, another physical quantity is desired instead of dilatation to compare sound fields in fluids with mechanical waves in solids. The mechanical stress tensor $[\mathbf{T}]$ seems to be such a quantity since it shares the same physical unit of measurement as the sound pressure $p_\sim$, i.e., $\mathrm{N\,m^{-2}} \cong \mathrm{Pa}$.

Let us deduce a normal component T_{ii} of the mechanical stress tensor from δ_{dil}. For the propagating mechanical waves, the normal stress T_{zz} in z-direction is assumed to be dominant in the following. According to Hooke's law (see Sect. 2.2.3), T_{zz} in an isotropic and homogeneous solid is given by

$$T_{zz} = \lambda_{\mathrm{L}}\left(S_{xx} + S_{yy} + S_{zz}\right) + 2\mu_{\mathrm{L}} S_{zz} \tag{8.46}$$

with the Lamé parameters λ_{L} and μ_{L}. Expressing these parameters with Young's modulus E_{M} as well as Poisson's ratio ν_{P} and inserting (8.8) leads to

$$
\begin{aligned}
T_{zz} &= \frac{E_{\mathrm{M}}}{(1+\nu_{\mathrm{P}})(1-2\nu_{\mathrm{P}})}\left[\nu_{\mathrm{P}} S_{xx} + \nu_{\mathrm{P}} S_{yy} + (1-\nu_{\mathrm{P}}) S_{zz}\right] \\
&= \frac{E_{\mathrm{M}}(1-\nu_{\mathrm{P}})}{(1+\nu_{\mathrm{P}})(1-2\nu_{\mathrm{P}})}\delta_{\mathrm{dil}} + \frac{E_{\mathrm{M}}(2\nu_{\mathrm{P}}-1)}{(1+\nu_{\mathrm{P}})(1-2\nu_{\mathrm{P}})}\left(S_{xx} + S_{yy}\right) .
\end{aligned} \tag{8.47}
$$

Actually, LRT measurements in optically transparent solids provide exclusively values for δ_{dil}. In other words, information about the normal strains S_{xx} and S_{yy} is not available [11, 12]. Owing to this fact, we approximate T_{zz} by neglecting the second term in (8.47), i.e.,

$$T_{zz} \approx k_m \delta_{\text{dil}} \quad \text{with} \quad k_m = \frac{E_M(1 - \nu_P)}{(1 + \nu_P)(1 - 2\nu_P)}. \tag{8.48}$$

The expression k_m stands for a material-dependent constant of the solid. Therewith, the relative deviation $|\epsilon_r|$ (magnitude) of the approximation becomes

$$\begin{aligned}
|\epsilon_r| &= \left| \frac{k_m \delta_{\text{dil}} - T_{zz}}{T_{zz}} \right| \\
&= \left| \frac{-(2\nu_P - 1)\left(S_{xx} + S_{yy}\right)}{(1 - \nu_P)\,\delta_{\text{dil}} + (2\nu_P - 1)\left(S_{xx} + S_{yy}\right)} \right| \\
&= \left| 1 + \frac{1 - \nu_P}{2\nu_P - 1}\left[1 + \frac{S_{zz}}{S_{xx} + S_{yy}}\right] \right|^{-1}
\end{aligned} \tag{8.49}$$

and, thus, strongly depends on Poisson's ratio of the solid. As (8.49) indicates, one can approximate a normal component of the stress tensor (e.g., T_{zz}) with $k_m \delta_{\text{dil}}$ very well when the normal strain in this direction dominates, e.g., $S_{zz} \gg S_{xx} + S_{yy}$. A sound pressure wave impinging perpendicular to an interface of fluid and solid causes such situation in the solid. By means of numerical simulations, this fact was proven in [12]. Summing up, it can be stated that LRT measurements should be applicable for reliably estimating normal stresses due to mechanical longitudinal waves, which propagate in an optically transparent solid.

8.6.2 Experimental Setup

The realized LRT setup for analyzing mechanical waves in solids is identical to the one, which was employed in water (cf. Fig. 8.9). As a sound source, we utilized either the piston-type ultrasonic transducer V306-SU or the cylindrically focused ultrasonic transducer V306-SU-CF1.00IN, both from the company Olympus Corporation [29] and optimized for generating sound fields in water. A PMMA (poly(methyl methacrylate)) block served as optically transparent solid in which the propagation of mechanical longitudinal waves should be investigated through LRT. This PMMA block with a geometric dimension of 160 mm × 60 mm × l_B (block length l_B in z-direction) was placed directly onto the optical reflector at the axial distance z_B from the transducer front (see Fig. 8.37). Both the PMMA block and the cylindrical mount containing the ultrasonic transducer were immersed in water.

Fig. 8.37 Relevant part of
realized LRT setup to
investigate mechanical waves
in optically transparent
PMMA block [12];
geometric block dimension
160 mm × 60 mm × l_B; axial
distance z_B of block surface
from transducer front;
ultrasonic transducer UT

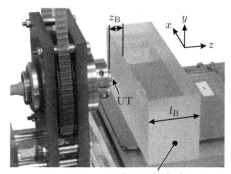

PMMA block

To reconstruct spatially as well as temporally resolved quantities (e.g., δ_{dil}), LRT measurements demand projections under different angles, which are acquired by the LDV. Note that this fact does not only refer to sound fields in fluids but also to mechanical waves in optically transparent solids. The dedicated experimental setup allows solely rotations of the ultrasonic transducer. It is, therefore, of utmost importance that such rotations do not alter sound fields as well as mechanical waves. Consequently, the PMMA block has to be aligned with respect to the cylindrical mount so that the block surface facing the transducer is in parallel to the scanning planes. Again, this alignment was conducted by evaluating the intensity of the reflected LDV beam (see Sect. 8.3.5). The PMMA block can then be treated as axisymmetric in LRT measurements.

8.6.3   Results for Different Ultrasonic Transducers

Two experiments will be presented in order to demonstrate applicability of LRT for quantitative measurements in the optically transparent solids. While one experiment exclusively deals with stress amplitudes in a cross section inside a PMMA block, the second experiment concentrates on transient field quantities for a wave propagating throughout water and PMMA [11, 12].

Mechanical Waves within a PMMA Block

In the first experiment, the mechanical longitudinal waves within a PMMA block (length $l_B = 45$ mm) were investigated through LRT. The block was placed at the axial distance $z_B = 11$ mm from the transducer front. The ultrasonic transducer (piston-type or cylindrically focused) was excited by a sinusoidal burst signal of 12 cycles at $f = 1$ MHz. At the interface water/PMMA, the generated sound pressure waves in water get reflected as well as converted to mechanical waves in the solid. Figure 8.38 depicts the averaged output signal (mean of 16 signals) of the differential LDV at $(x, z) = (0, 25.4$ mm$)$, which represents data at a single point of

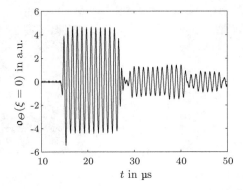

Fig. 8.38 Averaged LDV output (mean of 16 signals) within PMMA block at position $(x, z) =$ $(0, 25.4 \, \text{mm})$ representing single point of projection $\mathfrak{o}_\Theta(\xi)$ for tomographic reconstruction; incident mechanical waves as well as reflections at interface PMMA/water are present; resulting SNR $\approx$ 37 dB; piston-type ultrasonic transducer Olympus V306-SU; excitation frequency $f = 1 \, \text{MHz}$

a projection $\mathfrak{o}_\Theta(\xi)$ with respect to time t. Note that this position is within the PMMA block. In the acquired LDV signal, one can clearly see the incident mechanical waves as well as the reflections due to the interface of PMMA and water. Just as in water (cf. Fig. 8.18), the SNR of the LDV signal in the PMMA block exceeds 30 dB.

LRT was used to reconstruct amplitudes $\hat{T}_{zz}(x, y)$ of the mechanical stress in a cross section at the axial distance $z = 25.4 \, \text{mm}$, i.e., directly within the PMMA block. The most important parameters for the conducted LRT measurements can be found in Table 8.4. Figure 8.39a and b show the obtained results[7] for the piston-type and cylindrically focused transducer, respectively. The piston-type transducer produces an axisymmetric distribution of $\hat{T}_{zz}(x, y)$ in the PMMA block. In contrast, the cylindrically focused transducer causes a distribution that is slightly focused in one direction. However, compared to the distribution of sound pressure amplitudes $\hat{p}_\sim(x, y)$ in water at $z = 25.4 \, \text{mm}$ (cf. Fig. 8.21), focusing is less pronounced in the PMMA block. This can be ascribed to the fact that the cylindrically focused transducer is designed for operating in water and, thus, the focal length 25.4 mm also relates to water.

Up to now, conventional measurement approaches (e.g., Schlieren optical method) do not allow quantitative verification of the LRT results in PMMA. For this reason, let us compare the LRT results for the piston-type transducer with FE simulations [12]. In doing so, the ultrasonic transducer was modeled as a circular area of radius $R_T = 6.35 \, \text{mm}$ oscillating uniformly with the applied excitation signal in the experiments, i.e., 12 cycles at $f = 1 \, \text{MHz}$. The decisive material properties of the PMMA block were identified by measuring the wave propagation velocities of mechanical longitudinal and transverse waves in the block (see

---

[7]Geometric dimension of the plotted cross sections (in parallel to the xy-plane): $x \times y =$ $[-10, 10 \, \text{mm}] \times [-10, 10 \, \text{mm}]$.

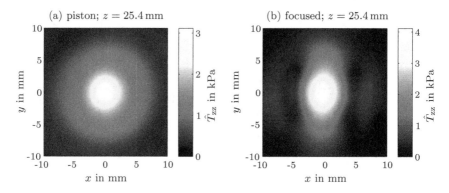

Fig. 8.39 LRT results for stress amplitudes $\hat{T}_{zz}$ within PMMA block in cross section at $z = 25.4$ mm; **a** piston-type ultrasonic transducer Olympus V306-SU; **b** cylindrically focused ultrasonic transducer Olympus V306-SU-CF1.00IN; excitation frequency $f = 1$ MHz

Sect. 5.1.2). These measurements lead to Young's modulus $E_m = 5.98$ GPa and Poisson's ratio $\nu_P = 0.33$.

Figure 8.40a compares measured (i.e., LRT results) and simulated amplitudes $\hat{T}_{zz}$ of the mechanical normal stress along the x-axis at $z = 25.4$ mm. Thereby, the simulation results have been scaled so that measured and simulated quantities feature the same energy within the PMMA block. The distribution of both quantities along the x-axis coincides very well, which is also demonstrated by their normalized relative deviation $|\epsilon_r|$ (see Fig. 8.40b). Besides the mechanical normal stress $\hat{T}_{zz}$, Fig. 8.40a contains the simulated dilatation $\hat{\delta}_{\text{dil;sim}}$ that was scaled with the material-dependent constant k_m. As discussed above, LRT measurements within optically transparent solids exploit the approximation of T_{ii} by this quantity (see (8.48)). Because the normalized relative deviation of $\hat{T}_{zz;\text{sim}}$ and $k_m \hat{\delta}_{\text{dil;sim}}$ is always smaller than 2% (see Fig. 8.40b), it can be stated that the applied approximation yields reliable LRT results for the normal stress, which is caused by mechanical longitudinal waves.

Fields Throughout Water and PMMA

In the second experiment, wave propagation phenomena throughout water and PMMA were analyzed by means of LRT. For this purpose, a PMMA block with the length $l_B = 22$ mm was placed onto the optical reflector at the axial distance $z_B = 24.4$ mm from the transducer front (see Fig. 8.37). To avoid overlaps of multiple reflections, the piston-type ultrasonic transducer Olympus V306-SU was excited by a sinusoidal burst signal consisting only of 8 cycles. Since this experiment required the acquisition of numerous cross sections in water as well as PMMA, we applied nonequidistant sampling between two neighboring LDV positions. Table 8.7 summarizes the most important parameters for the conducted LRT measurements.

The temporally resolved LRT results in the cross sections were assembled together yielding transient as well as spatially resolved information about wave propagation throughout water and PMMA. With a view to achieving visually continuous images,

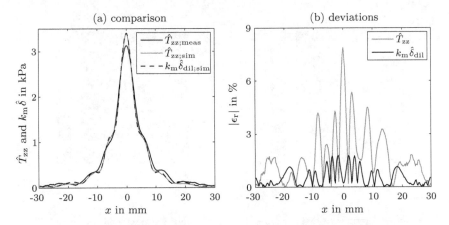

Fig. 8.40 **a** Comparison of measured stress amplitudes $\hat{T}_{zz;\text{meas}}$, simulated stress amplitudes $\hat{T}_{zz;\text{sim}}$ as well as scaled dilatations $k_m \hat{\delta}_{\text{dil};\text{sim}}$ within PMMA block along x-axis at $z = 25.4$ mm; **b** normalized relative deviations $|\epsilon_r|$ (magnitude) between measured and simulated stress amplitudes as well as between simulated stress amplitudes and scaled version of simulated dilatations; piston-type ultrasonic transducer Olympus V306-SU

Table 8.7 Decisive parameters for LRT measurement of sound pressure waves in water and mechanical waves in PMMA block

x-direction	$\Delta\xi = 0.4$ mm for $x \in [-8.0, 8.0]$ mm
	$\Delta\xi = 0.8$ mm for $x \in [-28.0, -8.0) \cup (8.0, 28.0]$ mm
z-direction	$\Delta z = 0.4$ mm for $z \in [9.4, 56.0]$ mm
Rotary direction	$\Delta\Theta = 3.6°$ per step
Temporal	100 MHz ($\Delta t = 10$ ns between sampling points)
Measurement	1 h per cross section
Reconstruction	4 min per cross section (5000 instants of time)

the spatial resolution in z-direction has been additionally doubled by cubic spline interpolation, i.e., from $\Delta z = 0.4$ mm to $\Delta z = 0.2$ mm. Figures 8.41a–c display the assembled data in the xz-plane (i.e., $y = 0$) for three different instants of time t after starting pulse emission, namely 18.2, 25.4, and 32.0 µs. Because sound pressure $p_\sim$ and mechanical stress T_{zz} have the same physical unit, they can be represented by a single color map in the images. Before discussing the LRT results, it should be pointed out that there is information missing in several cross sections ($z \in [24.4, 27.0] \cup [43.6, 46.2]$ mm) at the left and right sides of the PMMA block. The reason for this was the mechanical processing of the PMMA block. Cutting and milling alters the material density near the surfaces of the block and, consequently, the optical refractive index n changes permanently. Owing to these changes, the laser beam of the LDV gets deflected and is not reflected back to its sensor head anymore. Hence, projections are not available there and the field information cannot be reconstructed in LRT measurements.

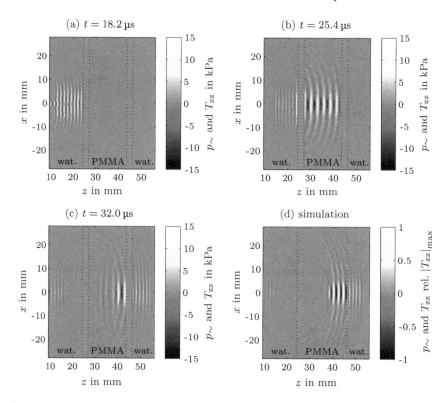

Fig. 8.41 Acquired sound pressure fields $p_\sim$ and mechanical fields T_{zz} in xz-plane (i.e., $y = 0$) throughout water and PMMA at three instants of time after starting emitting sinusoidal burst (8 cycles at 1 MHz); **a** sound pressure waves almost reached PMMA block; **b** mechanical waves reached right side of block; **c** mechanical waves got converted to sound pressure waves; no reconstructions at left and right block sides due to missing LDV signals; **d** normalized result of numerical simulation at $t = 32.0\,\mu s$; piston-type ultrasonic transducer Olympus V306-SU

At $t = 18.2\,\mu s$ (see Fig. 8.41a), the sound pressure waves in water have not reached the PMMA block and, therefore, neither sound reflections at the interface water/PMMA nor mechanical waves in the PMMA block arise. However, one can observe constructive and destructive interference patterns, which stem from the beam characteristic of the piston-type ultrasonic transducer (cf. Fig. 8.28). In Fig. 8.41b ($t = 25.4\,\mu s$), almost the whole wave front has passed the block surface facing the transducer. The reflected sound pressure waves propagate back to the transducer, i.e., in negative z-direction. Besides, the first wave front of mechanical waves resulting from converted sound pressure waves was about to reach the right side of the PMMA block. As the comparison of sound pressure waves before the block and longitudinal mechanical waves in PMMA reveals, the later ones exhibit a greater wavelength. This is also expected from the theoretical point of view since the wave propagation velocity in PMMA is higher than in water. Moreover, the energy of mechanical waves is distributed in a larger area in lateral direction, i.e., in x- and y-direction. The same behavior can be seen in Fig. 8.41c ($t = 32.0\,\mu s$), where most of the mechanical waves have reached the right block side and have been partially reflected as well as converted back to sound pressure waves again.

For qualitative comparison, a transient FE simulation was carried out in addition. Therewith, it is possible to predict the sound pressure field in water and the mechanical waves in PMMA. Figure 8.41d illustrates the normalized simulation result at the instant of time $t = 32.0\,\mu s$ (cf. Fig. 8.42c). Apart from the missing data at the border area of the PMMA block, simulations and LRT results share many fine details (e.g., waveforms), which demonstrates reliability of LRT measurements in optically transparent solids.

Now, let us regard the reconstructed time-dependent variations $p_\sim(t)$ as well as $T_{zz}(t)$ (see Fig. 8.42) at three selected positions on the z-axis, i.e., $(x, y) = (0, 0)$. The positions are located (i) in water at $z = 15.4\,mm$, i.e., between transducer and PMMA block, (ii) in the PMMA block at $z = 33.6\,mm$, and (iii) in water at $z = 46.8\,mm$, i.e., behind the PMMA block. In Fig. 8.42a, one can clearly observe the emitted sound pressure wave (group 1), which was firstly reflected by the left side of the PMMA block. Subsequently, these reflected waves got reflected by the transducer front again leading to a sound pressure wave (group 2) propagating in positive z-direction toward the PMMA block. Furthermore, several other wave groups are present that result from multiple reflections between medium boundaries. Both wave group 1 and wave group 2 also exist in the PMMA block (see Fig. 8.42b), i.e., in the time-dependent variation $T_{zz}(t)$. However, because the second position is further away from the transducer front than the first one, the dominant wave groups are shifted in time. An additional shift occurs at $z = 46.8\,mm$ (see Fig. 8.42c), which is located in water behind the PMMA block. Overall, the wave groups at the three positions are of similar shape with the exception that they appear upside-down at the third position. Note that this is a consequence of the continuity condition at a solid–fluid interface (see (4.104, p. 122)).

8.6.4 Verification of Experimental Results

Finally, the previous LRT results for the optically transparent PMMA block should be quantitatively verified. For this purpose, we determine the reflection coefficient at the interface water/PMMA as well as the amplitude ratio for dominating wave groups.

Reflection Coefficient

According to the LRT results in Fig. 8.42a, the reflected sound pressure waves feature the amplitude $\hat{p}_{LRT}(z = 15.4\,mm) = 3.84\,kPa$. Due to the fact that the PMMA block is located at $z_B = 24.4\,mm$, the reflected sound pressure waves propagate

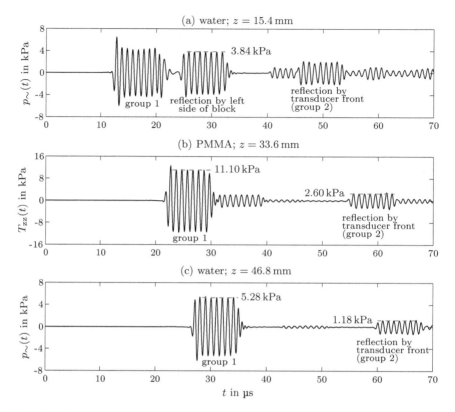

Fig. 8.42 Time-dependent sound pressure $p_\sim(t)$ and mechanical stress $T_{zz}(t)$ at selected positions on z-axis; **a** in water between transducer front and PMMA block at $z = 15.4\,\text{mm}$; **b** within PMMA block at $z = 33.6\,\text{mm}$; **c** in water behind PMMA block at $z = 46.8\,\text{mm}$; wave group 1 emitted by ultrasonic transducer; wave group 2 reflected by transducer front

altogether $33.4\,\text{mm}$ there[8] (see Fig. 8.43). In the absence of PMMA block, a hydrophone measurement yielded the sound pressure amplitude $\hat{p}_{\text{HYD}}(z = 33.4\,\text{mm}) = 10.78\,\text{kPa}$ at the axial distance $z = 33.4\,\text{mm}$ from the transducer front. We can utilize these two amplitudes to estimate the reflection coefficient r_p for an incident sound pressure wave at the interface water/PMMA, i.e.,

$$r_p = \frac{\hat{p}_{\text{LRT}}(z = 15.4\,\text{mm})}{\hat{p}_{\text{HYD}}(z = 33.4\,\text{mm})} = \frac{3.84\,\text{kPa}}{10.78\,\text{kPa}} = 0.356 \,. \tag{8.50}$$

From the theoretical point of view, the reflection coefficient r_p' in case of plane sound waves results in (see (2.139, p. 38))

---

[8] $2 \cdot 24.4 - 15.4\,\text{mm} = 33.4\,\text{mm}$

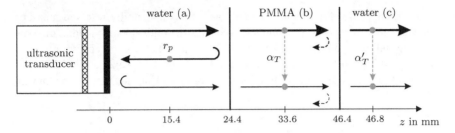

Fig. 8.43 Illustration of wave propagation throughout water and PMMA block; time-dependent curves of sound pressure $p_\sim(t)$ and mechanical stress $T_{zz}(t)$ are shown in Fig. 8.42a–c

$$r_p' = \frac{Z_{\text{PMMA}} - Z_{\text{water}}}{Z_{\text{PMMA}} + Z_{\text{water}}}$$
$$= \frac{3.26 \cdot 10^6\,\text{N s m}^{-3} - 1.48 \cdot 10^6\,\text{N s m}^{-3}}{3.26 \cdot 10^6\,\text{N s m}^{-3} + 1.48 \cdot 10^6\,\text{N s m}^{-3}} = 0.376 . \tag{8.51}$$

The expressions Z_{PMMA} and Z_{water} stand for the acoustic impedance of PMMA and water, respectively. Although two completely different approaches were applied to determine the reflection coefficient at the interface water/PMMA, both values coincide very well and exhibit a relative deviation of only -5.1% (rel. to r_p').

Amplitude Ratio

As the time-dependent curves in Fig. 8.42 indicate, there exist two dominant wave groups (group 1 and group 2), which originate from the emitted sinusoidal burst and the reflection by the transducer front. Let us evaluate the amplitude ratios of the wave groups in the PMMA block as well as in water behind the block (see Fig. 8.43). In the PMMA block at $z = 33.6\,\text{mm}$, the amplitude ratio α_T is defined by the amplitudes of the mechanical normal stress, i.e.,

$$\alpha_T = \frac{\hat{T}_{zz;1} - \hat{T}_{zz;2}}{\hat{T}_{zz;1}} = \frac{11.10\,\text{kPa} - 2.60\,\text{kPa}}{11.10\,\text{kPa}} = 0.766 \tag{8.52}$$

with $\hat{T}_{zz;1}$ and $\hat{T}_{zz;2}$ representing the stress amplitudes of incident and reflected waves, respectively. In water at $z = 46.8\,\text{mm}$, the amplitude ratio α_T' of the corresponding sound pressure waves results in

$$\alpha_T' = \frac{\hat{p}_{\sim;1} - \hat{p}_{\sim;2}}{\hat{p}_{\sim;1}} = \frac{5.28\,\text{kPa} - 1.18\,\text{kPa}}{5.28\,\text{kPa}} = 0.777 . \tag{8.53}$$

Since the relative deviation of the amplitude ratios is only -1.3% (rel. to α_T'), it can be stated once more that LRT measurements lead to reliable information about waves, which propagate throughout water and PMMA.

To sum up, LRT also enables spatially as well as temporally resolved measurements of mechanical waves in optically transparent solids. In such media, mechanical longitudinal waves locally alter the dilatation, which can be acquired through LRT. By additionally introducing a material-dependent approximation, LRT measurements provide absolute values for the mechanical normal stresses within the solid.

References

1. Almqvist, M., Holm, A., Jansson, T., Persson, H., Lindström, K.: High resolution light diffraction tomography: nearfield measurements of 10 MHz continuous wave ultrasound. Ultrasonics **37**(5), 343–353 (1999)
2. Asher, R.C.: Ultrasonic Sensors. Institute of Physics Publishing, Bristol (1997)
3. Bacon, D.R.: Characteristics of a PVDF membrane hydrophone for use in the range 1-100 MHz. IEEE Trans. Sonics Ultrason. **SU-29**(1), 18–25 (1982)
4. Bacon, D.R.: Primary calibration of ultrasonic hydrophone using optical interferometry. IEEE Trans. Ultrason. Ferroelectr. Freq. Control **35**(2), 152–161 (1988)
5. Bahr, L., Lerch, R.: Beam profile measurements using light refractive tomography. IEEE Trans. Ultrason. Ferroelectr. Freq. Control **55**(2), 405–414 (2008)
6. Bodmann, V.O.: Partielle spezifische Refraktionen von Polymethylmethacrylat und Polystyrol. I. Einfluss verschiedener Lösungsmittel. Die. Makromolekulare Chemie **122**(1), 196–209 (1969)
7. Born, M., Wolf, E., Bhatia, A.B., Clemmow, P.C., Gabor, D., Stokes, A.R., Taylor, A.M., Wayman, P.A., Wilcock, W.L.: Principles of Optics: Electromagnetic Theory of Propagation, Interference and Diffraction of Light. Cambridge University Press, Cambridge (2000)
8. Bronstein, I.N., Semendjajew, K.A., Musiol, G., Mühlig, H.: Handbook of Mathematics, 6h edn. Springer, Berlin (2015)
9. Brüel & Kjær: Product Portfolio (2018). Homepage: http://www.bksv.com
10. Buzug, T.M.: Computed Tomography, 6th edn. Springer, Berlin (2008)
11. Chen, L.: Light refractive tomography for noninvasive ultrasound measurements in various media. Ph.D. thesis, Friedrich-Alexander-University Erlangen-Nuremberg (2014)
12. Chen, L., Rupitsch, S.J., Grabinger, J., Lerch, R.: Quantitative reconstruction of ultrasound fields in optically transparent isotropic solids. IEEE Trans. Ultrason. Ferroelectr. Freq. Control **61**(4), 685–695 (2014)
13. Chen, L., Rupitsch, S.J., Lerch, R.: Application of light refractive tomography for reconstructing ultrasound fields in various media. Tech. Messen. **79**(10), 459–463 (2012)
14. Chen, L., Rupitsch, S.J., Lerch, R.: A reliability study of light refractive tomography utilized for noninvasive measurement of ultrasound pressure fields. IEEE Trans. Ultrason. Ferroelectr. Freq. Control **59**(5), 915–927 (2012)
15. Chen, L., Rupitsch, S.J., Lerch, R.: Quantitative reconstruction of a disturbed ultrasound pressure field in a conventional hydrophone measurement. IEEE Trans. Ultrason. Ferroelectr. Freq. Control **60**(6), 1199–1206 (2013)
16. Ciddor, P.E.: Refractive index of air: new equations for the visible and near infrared. Appl. Opt. **35**(9), 1566–1572 (1996)
17. General Electrics (GE): Product portfolio (2018). Homepage: https://www.gemeasurement.com
18. Harvey, G., Gachagan, A.: Noninvasive field measurement of low-frequency ultrasonic transducers operating in sealed vessels. IEEE Trans. Ultrason. Ferroelectr. Freq. Control **53**(10), 1749–1758 (2006)
19. Hecht, E.: Optics, 5th edn. Pearson, London (2016)

20. Herman, G.T.: Fundamentals of Computerized Tomography. Springer, Berlin (2009)
21. Huttunen, T., Kaipio, J.P., Hynynen, K.: Modeling of anomalies due to hydrophones in continuous-wave ultrasound fields. IEEE Trans. Ultrason. Ferroelectr. Freq. Control **50**(11), 1486–1500 (2003)
22. Jia, X., Quentin, G., Lassoued, M.: Optical heterodyne detection of pulsed ultrasonic pressures. IEEE Trans. Ultrason. Ferroelectr. Freq. Control **40**(1), 67–69 (1993)
23. Kak, A.C., Slaney, M.: Principles of Computerized Tomographic Imaging. Society of Industrial and Applied Mathematics (2001)
24. Koch, C.: Status report PTB. In: Consultative Committee for Acoustics, Ultrasound and Vibration (2008). CCAUV/08-09
25. Koch, C., Molkenstruck, W.: Primary calibration of hydrophones with extended frequency range 1 to 70 MHz using optical interferometry. IEEE Trans. Ultrason. Ferroelectr. Freq. Control **46**(5), 1303–1314 (1999)
26. Lerch, R., Sessler, G.M., Wolf, D.: Technische Akustik: Grundlagen und Anwendungen. Springer, Berlin (2009)
27. Matar, O.B., Pizarro, L., Certon, D., Remenieras, J.P., Patat, F.: Characterization of airborne transducers by optical tomography. Ultrasonics **38**(1), 787–793 (2000)
28. Neumann, T., Ermert, H.: A new designed schlieren system for the visualization of ultrasonic pulsed wave fields with high spatial and temporal resolution. In: Proceedings of International IEEE Ultrasonics Symposium (IUS), pp. 244–247 (2006)
29. Olympus Corporation: Product Portfolio (2018). Homepage: https://www.olympus-ims.com
30. ONDA Corporation: Product Portfolio of Hydrophones (2018). Homepage: http://www.ondacorp.com
31. Physik Instrumente (PI) GmbH & Co. KG: Product Portfolio (2018). Homepage: https://www.physikinstrumente.com/en/
32. Polytec GmbH: Product Portfolio (2018). Homepage: http://www.polytec.com
33. PTB: Calibration Certificate for HGL-0400 (sn:1375) with preamp AH-2010 (sn:1028) and DC Block AH-2010DCBNS (sn:0015). Physikalisch-Technische Bundesanstalt (PTB) (2012). Calibration mark: 1.62/16002 PTB 12
34. Quimby, R.S.: Photonics and Lasers. Wiley, New York (2006)
35. Radon, J.: Über die Bestimmung von Funktionen durch ihre Integralwerte längs gewisser Mannigfaltigkeiten. Berichte über die Verhandlungen der Königlich-Sächsischen Akademie der Wissenschaften zu Leipzig **69**, 262–277 (1917)
36. Reibold, R.: Light diffraction tomography applied to the investigation of ultrasonic fields. Part II: Standing waves. Acta Acustica united with Acustica **63**(4), 283–289 (1987)
37. Reibold, R., Molkenstruck, W.: Light diffraction tomography applied to the investigation of ultrasonic fields. Part I: Continuous waves. Acustica **56**(3), 180–192 (1984)
38. Rupitsch, S.J., Chen, L., Winter, P., Lerch, R.: Quantitative measurement of airborne ultrasound utilizing light refractive tomography. In: Proceedings of Sensors and Measuring Systems (ITG/GMA Symposium), pp. 1–5 (2014)
39. Schneider, B.: Quantitative analysis of pulsed ultrasonic beam patterns using a Schlieren system. IEEE Trans. Ultrason. Ferroelectr. Freq. Control **43**(6), 1181–1186 (1996)
40. Scruby, C.B., Drain, L.E.: Laser Ultrasonics. Adam Hilger (1990)
41. Sessler, G.M.: Electrets, 2nd edn. Springer, Berlin (1987)
42. Sessler, G.M., West, J.E.: Self-biased condenser microphone with high capacitance. J. Acoust. Soc. Am. **34**(11), 1787–1788 (1962)
43. Settles, G.S.: Schlieren and Shadowgraph Techniques: Visualizing Phenomena in Transparent Media. Experimental Fluid Mechanics. Springer, Berlin (2001)
44. Silfvast, W.T.: Laser Fundamentals, 2nd edn. Cambridge University Press, Cambridge (2004)
45. Staudenraus, J., Eisenmenger, W.: Fibre-optic probe hydrophone for ultrasonic and shock-wave measurements in water. Ultrasonics **31**(4), 267–273 (1993)
46. Tektronix, Inc.: Product Portfolio (2018). Homepage: https://www.tek.com
47. Thévenaz, P., Blu, T., Unser, M.: Interpolation revisited - medical images application. IEEE Trans. Med. Imaging **19**(7), 739–758 (2000)

48. Träger, F.: Handbook of Lasers and Optics. Springer, Berlin (2007)
49. Unverzagt, C., Olfert, S., Henning, B.: A new method of spatial filtering for schlieren visualization of ultrasound wave fields. Phys. Proc. **3**(1), 935–942 (2010)
50. Wilkens, V., Molkenstruck, W.: Broadband PVDF membrane hydrophone for comparisons of hydrophone calibration methods up to 140 MHz. IEEE Trans. Ultrason. Ferroelectr. Freq. Control **54**(9), 1784–1791 (2007)
51. Yadav, H.S., Murty, D.S., Verma, S.N., Sinha, K.H.C., Gupta, B.M., Chand, D.: Measurement of refractive index of water under high dynamic pressures. J. Appl. Phys. **44**(5), 2197–2200 (1973)
52. Zagar, B.G.: Laser interferometer displacement sensors. In: The Measurement, Instrumentation and Sensors Handbook, pp. 6–65–6–77. CRC Press, Boca Raton (2011)
53. Zakharin, B., Stricker, J.: Schlieren systems with coherent illumination for quantitative measurements. Appl. Opt. **43**(25), 4786–4795 (2004)
54. Zernike, F.: Phase contrast, a new method for the microscopic observation of transparent objects. Part II. Physica **9**(10), 974–986 (1942)
55. Zipser, L., Franke, H.: Laser-scanning vibrometry for ultrasonic transducer development. Sens. Actuators A Phys. **110**(1–3), 264–268 (2004)

Chapter 9
Measurement of Physical Quantities and Process Measurement Technology

Various physical and chemical quantities can be determined by piezoelectric sensors. That is the reason why piezoelectric sensors are frequently used in the process measurement technology. For example, they enable measuring

- force, torque, pressure, and acceleration (see Sect. 9.1),
- geometric distance and layer thickness (see Sect. 9.2),
- properties of liquids [1, 5],
- concentrations of substances in fluids [79],
- fluid flow (see Sect. 9.3),
- cavitation activity (see Sect. 9.4),
- temperature [44].

A large number of piezoelectric sensors exploit the impact of the aimed quantities on mechanical quantities, which influence the electrical sensor characteristic due to the direct piezoelectric effect. In case of a so-called quartz crystal microbalance (QCM or QMB), one evaluates the resonance frequency f_r of a specific vibration mode inside a disk-shaped quartz plate [96]. Since these sensors mostly operate in thickness shear mode (i.e., transverse shear mode; see Fig. 9.1a) of piezoelectricity, they are also referred to as thickness shear mode (TSM) resonators. When the quartz disk gets mechanically loaded by a mass, f_r will be shifted to lower values. The greater the mass, the larger the frequency shift Δf_r will be. This fact allows determining the thickness of a homogeneous material layer being located on the quartz disk. On the other hand, we can measure the material density when the layer thickness is known. If the quartz crystal is appropriately operated in a liquid, Δf_r will depend on the liquid density and viscosity [46, 66]. By equipping the quartz disk with a sensitive layer that changes its mass depending on the concentration of a substance in a fluid, a QCM also enables to measure substance concentration. Such sensors are often used for biological and chemical analyses, e.g., [79].

© Springer-Verlag GmbH Germany, part of Springer Nature 2019
S. J. Rupitsch, *Piezoelectric Sensors and Actuators*, Topics in Mining, Metallurgy and Materials Engineering, https://doi.org/10.1007/978-3-662-57534-5_9

Other piezoelectric sensor devices are based on the propagation of waves between transducers, i.e., between piezoelectric transmitters and piezoelectric receivers. From the time-of-flight of these waves, we will be able to calculate the sound velocity in the propagating medium when the geometric distance between transmitter and receiver is known. The ratio of emitted and received waves can be used to deduce further characteristic parameters like liquid density and viscosity [1]. Apart from a setup consisting of separated piezoelectric transducers, it is also possible to build up a compact sensor device, which includes both transmitter and receiver [107]. The setup of such a device commonly comprises either a thick plate of a piezoelectric material or a silicon substrate with an additional piezoelectric film, e.g., aluminum nitride (AlN) or zinc oxide (ZnO). The piezoelectric transmitters and receivers are formed by appropriate interdigital electrode structures on the piezoelectric material [60]. Depending on the type of waves that propagate inside the compact device, one can mainly differ between surface acoustic wave (SAW) sensors, love wave (LW) sensors, flexural plate wave (FPW) sensors as well as shear-horizontal acoustic plate mode (SH-APM) sensors [2]. Figure 9.1 depicts the principle setup of these piezoelectric

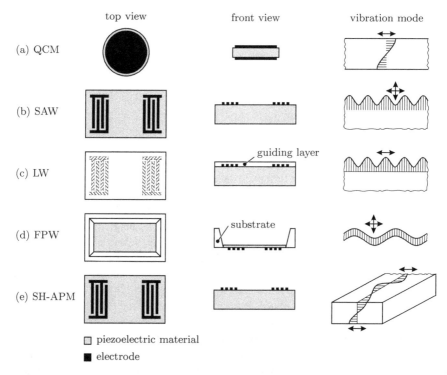

Fig. 9.1 Principle setup of **a** quartz crystal microbalance (QCM) sensors, **b** surface acoustic wave (SAW) sensors, **c** love wave (LW) sensors, **d** flexural plate wave (FPW) sensors, and **e** shear-horizontal acoustic plate mode (SH-APM) sensors; schematic representation of propagating waves in right panels; arrows indicate dominating directions of particle motion

sensors and a schematic representation of the propagating waves. The wave propagation between transmitter and receiver is not only influenced by the sensor materials but also by the surrounding medium. The time-of-flight (i.e., phase) and amplitude of the incident waves at the receiver provide, thus, characteristic information about the surrounding medium. If such a sensor is additionally equipped with a sensitive layer between transmitter and receiver, one can conduct biological and chemical analyses, too. Because all these sensors are based on wave propagation in solids, they are commonly termed bulk acoustic wave (BAW) sensors. According to the underlying operation principle and their geometric size, they are, moreover, referred to as microacoustic resonators.

In this chapter, we will concentrate on a few selected physical quantities that are important in process measurement technology. Section 9.1 deals with the typical designs of piezoelectric sensors for the mechanical quantities force, torque, pressure, and acceleration. Subsequently, an ultrasound-based method will be described which enables the simultaneous determination of plate thickness and speed of sound inside the plate. Section 9.3 treats the metrological registration of fluid flow by means of ultrasonic waves. At the end of the chapter, a piezoelectric device will be presented that can be used as cavitation sensor in ultrasonic cleaning.

9.1 Force, Torque, Pressure, and Acceleration

The mechanical quantities force, torque, pressure, and acceleration represent decisive process variables in a wide range of applications. Here, let us regard piezoelectric sensors for measuring these quantities. We will start with the underlying principle of such sensors. The basic designs as well as some selected special designs will be detailed in the subsequent subsections. Finally, methods for reading out piezoelectric sensors (e.g., charge amplifier) are described in Sect. 9.1.5.

9.1.1 Fundamentals

Generally speaking, a sensor contains both a transduction element for converting one form of energy into another and a sensing element. In piezoelectric sensors, the transduction element mostly corresponds to the sensing element, which is made of a material featuring piezoelectric properties, e.g., quartz. Such sensors are always based on the direct piezoelectric effect, i.e., conversion of mechanical inputs into electrical outputs. Because piezoelectric sensors do not require an external power supply to obtain an electrical output, they belong to the group of active sensors. The high rigidity of piezoelectric materials leads, moreover, to small disturbances of the measured quantity. It is, therefore, not surprising that piezoelectric sensors are often employed in practical applications for measuring physical quantities like mechanical forces.

As already explained in Sect. 3.4.3, there can arise four specific modes of piezo-electric coupling within a piezoelectric material, namely (i) the longitudinal mode, (ii) the transverse mode, (iii) the longitudinal shear mode, and (iv) the transverse shear mode. Depending on the actual implementation, piezoelectric sensors for measuring force, torque, pressure, or acceleration exploit one or more of those modes for converting mechanical quantities into electrical quantities. Note that torque, pressure as well as acceleration measurements can be traced back to force measurements since there always occur mechanical forces, which act on the piezoelectric material.

To illustrate the underlying principle of piezoelectric sensors for the mentioned quantities, let us take a closer look at two sensor elements in Fig. 9.2 that are based on the longitudinal mode and transverse mode, respectively. The first element is disk-shaped (diameter d_S; thickness t_S), while the second one is bar-shaped (length l_S; width w_S; thickness t_S). We suppose in both cases a piezoelectric material, whose axis of piezoelectricity (i.e., electric polarization) points in positive 3-direction. The bottom and top surfaces of the disk and bar are completely covered with electrodes. Furthermore, a charge amplifier circuit (see Sect. 9.1.5) is utilized to read out the resulting electric charges on the electrodes. In doing so, the piezoelectric elements become electrically short-circuited which leads to $\mathbf{E} = 0$, i.e., the electric field intensity inside the element vanishes. If the disk gets loaded uniformly at its top surface by a mechanical force F_z in negative z-direction, a mechanical stress T_3 will arise inside the disk. Under the assumption of force-freeness in the other directions (i.e., $T_1 = T_2 = T_4 = T_5 = T_6 = 0$), T_3 computes as (base area A_S of disk)

$$T_3 = -\frac{F_z}{A_S} = -\frac{4F_z}{d_S^2 \pi}. \tag{9.1}$$

This relation can now be inserted in the d-form of the material law for linear piezo-electricity (see Sect. 3.3). Since $\mathbf{E} = 0$ holds and force-freeness is supposed with the exception of T_3, we obtain from the piezoelectric strain constant d_{33}

$$D_3 = \underbrace{\varepsilon_{33}^T E_3}_{=0} + d_{33}T_3 = d_{33}T_3 = -\frac{4F_z d_{33}}{d_S^2 \pi} \tag{9.2}$$

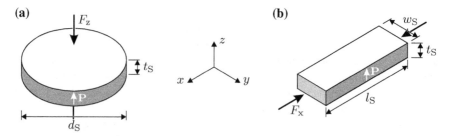

(a)   **(b)**

Fig. 9.2 **a** Disk-shaped piezoelectric element based on longitudinal mode; **b** bar-shaped piezoelec-tric element based on transverse mode; mechanical forces F_z and F_x; axis of piezoelectricity (i.e., vector $\mathbf{P}$) points in positive z-direction

for the electric flux density $\mathbf{D}$ in z-direction. Irrespective of the fact that d_{13} and d_{23} are zero for the most important crystal classes of piezoelectric materials, the components D_1 as well as D_2 do not generate electric charges on the electrodes of the considered disk. This stems from the orthogonal orientation of electrodes and these components. The resulting electric charge Q_S on the top electrode due to the mechanical force is then given by

$$Q_S = \int_{A_S} \mathbf{D} \cdot d\mathbf{A} = D_3 A_S = -F_z d_{33} \ . \tag{9.3}$$

If F_z represents the aimed quantity, the so-called *measurand*, and subsequent amplifier circuits are not taken into account, the ratio Q_S/F_z will denote the sensitivity of the piezoelectric sensor. In view of various practical applications, the sensitivity should be as large as possible. According to (9.3), the geometric dimensions of the sensor element do not, however, influence its sensitivity. Consequently, a piezoelectric disk with larger diameter and greater thickness would exhibit the same sensitivity. Its value is exclusively determined by the parameter d_{33} of the used piezoelectric material.

In a next step, we will perform the same analysis for the bar-shaped piezoelectric element, which gets loaded uniformly at the $w_S t_S$ surface by F_x in negative x-direction (see Fig. 9.2b). This force produces the mechanical stress T_1 inside the bar. The assumption of force-freeness in the other directions leads to $T_1 = -F_x/(w_S t_S)$ and for the electric flux density D_3 in z-direction to

$$D_3 = \underbrace{\varepsilon_{33}^T E_3}_{=0} + d_{31} T_1 = d_{31} T_1 = -\frac{F_x d_{31}}{w_S t_S} \tag{9.4}$$

with the piezoelectric strain constant d_{31}. Again, the remaining components of $\mathbf{D}$ (i.e., D_1 and D_2) are irrelevant. The resulting electric charge Q_S on the top electrode of the piezoelectric bar computes as

$$Q_S = D_3 l_S w_S = -\frac{F_x d_{31} l_S}{t_S} \tag{9.5}$$

and, therefore, d_{31} influences the sensitivity Q_S/F_x for the measurand F_x. In contrast to the longitudinal mode, where the force direction coincides with both the axis of piezoelectricity and the normal vector of the electrodes, the geometric dimensions of the bar also alter Q_S/F_x in case of the transverse mode. We are able to improve Q_S/F_x by increasing the ratio l_S/t_S, which means that a long and thin bar provides a high sensitivity.

These greatly simplified examples already demonstrate the significant role of material parameters and, thus, of the selected piezoelectric material for the sensitivity of piezoelectric sensors. However, in practical applications of such sensors, one does not only have to take care about the demanded sensor sensitivity but also about measuring range, cross-sensitivities as well as temperature dependence. That is the reason why there exist various designs of piezoelectric sensors. The subsections below will briefly explain popular implementations of piezoelectric sensors for measuring force, torque, pressure, or acceleration.

9.1.2   Force and Torque

Figure 9.3 depicts the internal structure of two simple piezoelectric force sensors of cylindrical shape for measuring forces in z-direction [33]. Both force sensors contain the main components top plate, cylindrical housing, piezoelectric elements, and electrical connector. The top plate and the cylindrical housing are made of electrically conductive materials like steel. Piezoelectric disks that exploit the longitudinal mode serve as transduction elements. By means of the top plate, the mechanical force F_z gets transmitted to the piezoelectric disks. The cylindrical housing is hermetically welded to the top plate and holds the piezoelectric disks under a certain mechanical preload. This preload ensures that the sensor components are properly fixed together and should eliminate gaps between contacting faces. In doing so, one can obtain a good sensor linearity as well as a very high rigidity, which is necessary for achieving high natural frequencies of the sensor.

The piezoelectric force sensors in Fig. 9.3 differ in the amount of piezoelectric elements. While the left setup consists of one disk, the right setup contains two disks. In both cases, the cylindrical housing exhibits the same electric potential as the top plate and acts either as ground electrode of the piezoelectric disks or is directly contacting one disk electrode. To avoid an electrical short-circuit across the disk, we need for the setup in Fig. 9.3a an additional insulating layer between top plate and the remaining electrode, i.e., the top electrode of the disk. This electrode is connected to the center pin of the electrical connector.

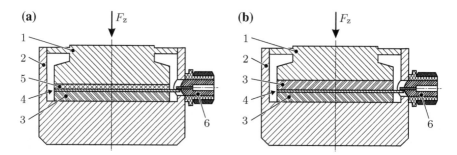

Fig. 9.3 Internal structure of simple piezoelectric force sensors of cylindrical shape with **a** one piezoelectric disk and **b** two piezoelectric disks [33]; 1 top plate; 2 cylindrical housing; 3 piezoelectric disk; 4 electrode; 5 insulating layer; 6 electrical connector

Instead of an insulating layer, the setup in Fig. 9.3b contains another disk whose axis of piezoelectricity points in opposite direction to the bottom disk. Both disks are subjected to the same mechanical force. Since they are electrically in parallel due to the common electrode that is again connected to the center pin of the electrical connector, the resulting electric charges of both disks are added up. If the piezoelectric disks feature equal material properties, the sensor sensitivity Q_S/F_z will double, which constitutes a great advantage concerning practical applications. Moreover, the second setup offers a higher sensor rigidity because common insulating materials usually have a lower elasticity than piezoelectric materials.

From the theoretical point of view, piezoelectric materials and, thus, piezoelectric force sensors should be applicable for measuring both compressive and tension forces without any preload. The brittleness of such materials and the difficulty to transmit tension forces to them, however, make a mechanical preload necessary. Of course, this preload has to exceed the applied tension forces because the piezoelectric force sensor might be damaged otherwise. We can achieve appropriate mechanical preloads either by a so-called dead weight or, more frequently, by a preloading bolt of high elasticity that is connected to the structure to be analyzed [33].

An often used setup of piezoelectric force sensors is shown in Fig. 9.4. This so-called *load washer sensor* (e.g., from Kistler Instrumente GmbH [49]) consists of the same components as the simple setup in Fig. 9.3b, but all components are ring-shaped. Again, the cylindrical housing is hermetically welded to the top plate under a controlled mechanical preload in z-direction, which is especially important when tension forces should be measured. If the load washer sensor is designed accordingly, we can also measure mechanical forces in x- or in y-direction. For this purpose, we need ring-shaped piezoelectric disks that are sensitive to the transverse shear mode in a specific direction. It is, therefore, possible to build up a compact triaxial piezoelectric force sensor consisting of six ring-shaped piezoelectric disks grouped into the three directions in space. While two disks have to be sensitive to the longitudinal mode, four disks have to be sensitive to the transverse shear mode, whereby two disks in x-direction and and two disks in y-direction, respectively. Note that such multiaxis force sensors will work correctly only if the mechanical preload is sufficient to transmit shear forces between top plate and piezoelectric disks. When friction between two subsequent sensor components is too small, shear forces will be partly transmitted.

Fig. 9.4 Internal structure of ring-shaped load washer sensor [33]; 1 housing; 2 top plate; 3 piezoelectric ring; 4 electrode; 5 electrical connector

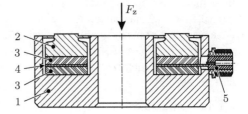

Fig. 9.5 Arrangement of
piezoelectric force sensors to
measure torque M_z [33];
arrows indicate sensitive axis
of individual sensor

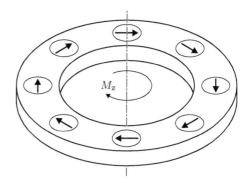

By a suitable arrangement of a few piezoelectric force sensors (e.g., load washer sensors) being sensitive to shear forces, it is possible to measure torques [33, 106]. The individual piezoelectric elements have to be aligned with their sensitive axis tangent to a circle (see Fig. 9.5). As a matter of fact, this arrangement requires a certain mechanical preload of the sensors because the torque gets transmitted through shear forces. If all sensors are connected electrically in parallel, the output voltage of a charge amplifier will be directly proportional to the acting torque.

One can also build up piezoelectric sensor devices allowing the measurement of three forces as well as three torques, i.e., F_x, F_y, F_z, M_x, M_y, and M_z. Such a sensor device is often termed *dynamometer* and usually consists of three or four individual multiaxis force sensors that are appropriately arranged between two rigid steel plates [33, 49]. However, when all forces and torques are different to zero, it will generally be impossible to determine the position of the force vector $\mathbf{F} = [F_x, F_y, F_z]^t$ in space. Besides, it should be noted that $\mathbf{F}$ produces an additional torque acting on the sensor device. This additional torque is, strictly speaking, independent of the external torque vector $\mathbf{M} = [M_x, M_y, M_z]^t$. A dynamometer, therefore, solely provides the resulting torque vector including external torques as well as torques stemming from $\mathbf{F}$.

According to Sect. 9.1.1, the utilized piezoelectric material greatly influences the sensitivity of piezoelectric force and torque sensors. Depending on the exploited mode of piezoelectric coupling, the sensor sensitivity is closely linked to a specific piezoelectric strain constant, e.g., d_{33} for the longitudinal mode. A high sensor sensitivity would be, thus, achieved by means of piezoceramic materials like PZT because these materials offer large values for d_{ij}. Another advantage of such materials would lie in the fact that one can fabricate piezoelectric elements of arbitrary shape. Despite the advantages of piezoceramic materials, piezoelectric force and torque sensors are mostly based on piezoelectric single crystals like quartz and tourmaline. Especially artificially grown quartz is often used since it has four key benefits over piezoceramics (see Sect. 3.6):

- better sensor linearity,
- practically free of hysteresis,

- better temperature behavior due to the comparatively weak pyroelectric effect of quartz, and
- higher electrical insulation resistance, which is important for quasi-static measurements.

A typical piezoelectric force sensor that exploits two quartz disks in longitudinal mode as transduction elements features the sensitivity $|Q_S/F| \approx 4\,\text{pC}\,\text{N}^{-1}$. The companies Kistler Instrumente GmbH [49] and Hottinger Baldwin Messtechnik (HBM) GmbH [42] are well-known manufacturers of piezoelectric force and torque sensors. The upper limits of the measuring range of commercially available piezoelectric force sensors start from less than 1 kN and go up to 1 MN. Piezoelectric torque sensors can be bought with upper limits of the measuring range from 1 Nm to 1 M Nm.

9.1.3 Pressure

Sensors for measuring pressure in fluids (i.e., liquids and gases) are often based on membranes being deflected due to pressure. Fundamentally, one can distinguish between two types of membrane-based pressure sensors, namely (i) absolute pressure sensors and (ii) differential pressure sensors (see Fig. 9.6a and b) [106]. Sensors for measuring the absolute pressure p_{abs} require a sealed chamber, which contains the reference pressure p_{ref}. Owing to the fact that p_{abs} always relates to $p_{ref} = 0$, the sealed chamber needs to be vacuumed and the reference side of the membrane is only exposed to vacuum. When a gas is filled into the sealed chamber, p_{ref} will naturally differ from zero.

In contrast to absolute pressure sensors, differential pressure sensors do not measure pressure relative to vacuum but to a selected reference value p_{ref}. The membranes of differential pressure sensors that measure the difference $p_{in} - p_{ref}$ between the input pressure p_{in} and p_{ref} have to withstand the fluid on both sides, which can especially be a problem in case of aggressive or corrosive liquids. If the ambient barometric pressure is used as p_{ref} (i.e., $p_{ref} = 101.325\,\text{kPa} \cong 1.01325\,\text{bar}$), differential pressure sensors will correspond to relative pressure sensors.

Irrespective of whether absolute pressure or differential pressure should be measured, various practical applications demand sensors that can measure statically. However, this is impossible with piezoelectric pressure sensors because of the nonideal properties of subsequent readout components such as charge amplifier circuits (cf. Sect. 9.1.5). A constant pressure will, consequently, not imply a constant output signal of these sensors. That is the reason why it does not make sense to clearly distinguish between absolute and differential piezoelectric pressure sensors. The latter represent merely a special sensor design with two ports allowing measurements of quasi-static and dynamic pressure differences. Even though quasi-static pressure measurements can be performed if the charge amplifier is reset just before data acquisition, piezoelectric pressure sensors are mostly applied for dynamic pressure measurements, e.g., sound pressure measurements.

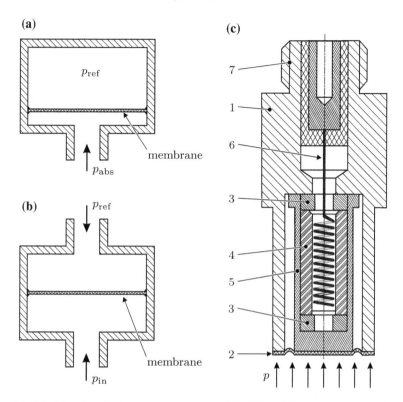

Fig. 9.6 Principle of **a** absolute pressure sensor and **b** differential pressure sensor; **c** internal structure of membrane-based piezoelectric pressure sensor [33]; 1 cylindrical housing; 2 membrane; 3 transfer plate; 4 piezoelectric elements; 5 preloading sleeve; 6 helical-shaped spring; 7 electrical connector

Figure 9.6c displays the internal structure of a typical membrane-based piezoelectric pressure sensor of cylindrical shape containing the main components membrane, preloading sleeve, transfer plates, piezoelectric elements, sensor housing, and electrical connector [33]. Preloading sleeve and sensor housing are made of electrically conductive materials like steel. The membrane with effective area A is hermetically welded under a light mechanical preload to the sensor housing. By means of the membrane, the fluid pressure p is converted into a mechanical force $F = pA$, which gets transmitted by the preloading sleeve onto the piezoelectric elements. Two transfer plates equalize mechanical stresses on the elements' end faces and should compensate temperature effects. Contrary to the simple piezoelectric force sensors in Fig. 9.3, these three to four piezoelectric elements exploit the transverse mode and, consequently, carry electric charges at mechanically unloaded element surfaces. The elements feature the shape of cylinder segments with a flattened inner surface, which is completely covered by thin electrodes. The cylindrical outer surface does not need to be coated with an electrode and is either directly contacting the preloading sleeve or through capacitive coupling. Therefore, this element surface exhibits the same electric potential as the sensor housing. The inner electrode is connected to the center of the electrical connector with the aid of a helical-shaped spring.

The metallic membranes of such piezoelectric pressure sensors are usually thinner than 0.1 mm. It is, thus, not surprising that the membrane represents a critical sensor component. On the one hand, the membrane should prevent the fluid from penetrating the sensor housing and should withstand aggressive substances as well as temperature changes. On the other hand, the membrane should be ideally elastic to ensure reliable sensor linearity over the entire measuring range. The necessary trade-offs lead to a large variety of membrane-based piezoelectric pressure sensors, which are commercially available. While low-pressure sensors require large membranes to obtain a reasonable sensitivity, the membranes of high-pressure sensors are small to avoid destruction. Since large membranes imply low eigenfrequencies, low-pressure sensors commonly provide smaller cutoff frequencies than high-pressure sensors.

The sensor housing of membrane-based piezoelectric pressure sensors is often equipped with a mounting thread for directly screwing them into objects. In case of high ambient temperature, an additional water cooling is sometimes used [33]. However, water cooling increases the necessary space for the sensor housing and may cause disturbing noise, which impairs the available measuring threshold. A further important point in the application of piezoelectric pressure sensors lies in their sensitivity to mechanical accelerations, especially if those accelerations are very pronounced and in the same frequency range as the pressure signal. This is due to the fact that all sensor components (e.g., membrane) in front of the piezoelectric elements act as a seismic mass just as it is exploited in piezoelectric acceleration sensors (see Sect. 9.1.4). If the pressure sensor is accelerated, a sensor output will emerge which superimposes with the aimed sensor signal stemming from pressure. By means of additional masses and piezoelectric elements exhibiting appropriate axis of piezoelectricity, we can, however, compensate accelerations in the sensor outputs.

Similar to piezoelectric force and torque sensors, the transduction elements in membrane-based piezoelectric pressure sensors are often made of artificially grown quartz. According to Sect. 9.1.1, the usage of the transverse length mode gives the possibility to increase sensor sensitivity with the aid of long and thin piezoelectric elements. Special quartz cuts as well as crystals of the CGG group enable, moreover, the operation of high-temperature pressure sensors without water cooling. However, the typical setup shown in Fig. 9.6c is not applicable for piezoelectric low-pressure sensors, which are used in the audible range and, thus, have to provide measuring thresholds far below 10 µbar in air. Such sensors are also named *piezoelectric microphones* [60]. Their setup corresponds to that of piezoelectric unimorph, piezoelectric parallel bimorph, or piezoelectric serial bimorph transducers (see Fig. 7.27 on p. 300). Instead of artificially grown quartz, piezoelectric microphones are mostly made of piezoceramic materials like PZT.

A hydrophone indicates a pressure sensor for underwater use. As the explanations in Sect. 8.1.1 reveal, the setup of piezoelectric hydrophones (e.g., membrane hydrophone) completely differs from Fig. 9.6c. Besides, piezoelectric hydrophones frequently utilize PVDF as transduction element since this piezoelectric material is mechanically flexible and can be fabricated as a thin film.

The companies Brüel & Kjær GmbH [13], Kistler Instrumente GmbH [49], and Onda Corporation [78] are well-known manufacturers of piezoelectric pressure sensors. Depending on the specific application and the medium in which the pressure should be measured, the commercially available sensors considerably differ in design, measuring threshold, measuring range as well as frequency range.

9.1.4 Acceleration

The measurement of mechanical accelerations is very important for various applications because it provides essential information about mechanical oscillations, vibrations and eigenfrequencies of the investigated structures. Accelerations are frequently related to the standard acceleration due to gravity, which is defined as $g_n = 9.80665 \, \text{m s}^{-2}$. While a *shock* indicates an impulse-like acceleration that can take values greater than $1000 \, g_n$, we often call periodical accelerations *vibrations*.

Commonly, acceleration sensors correspond to force sensors, which are equipped with an additional mass [106]. This so-called *seismic mass* with constant mass m_S will generate a mechanical force F acting on the sensor's transduction element if the sensor is exposed to an acceleration a. In case of a piezoelectric acceleration sensor, at least one piezoelectric element serves as transduction element. By neglecting the net weight of the piezoelectric elements, the resulting force on the elements becomes $F = m_S \cdot a$.

Figure 9.7a shows a typical setup of a piezoelectric acceleration sensor. The compression-type sensor of cylindrical shape consists of the main components seismic mass, preloading bolt, two piezoelectric rings, sensor housing, and electrical connector [33]. Again, the sensor housing and the seismic mass are made of electrically conductive materials like steel. They act as ground electrode for both piezoelectric rings, which exploit the longitudinal mode of piezoelectricity in opposite z-directions. Their common electrode is connected to the center pin of the electrical connector. The sensor housing should protect the sensor components from the environment and contains a mounting thread, which enables a mechanical link to the investigated structure. When the structure is accelerated by the value a_z in z-direction, the seismic mass exerts the force $F_z = m_S \cdot a_z$ on the piezoelectric disks.

Now, let us briefly study some fundamentals concerning the dynamic behavior of piezoelectric acceleration sensors. In simplified terms, a piezoelectric acceleration sensor can be interpreted as mechanical oscillator system with one degree of freedom (see Fig. 9.7b). The oscillator system comprises the seismic mass m_S, a spring with the spring rate κ_S describing the effective sensor rigidity, and a damper (dashpot) with the damping constant η_D. If we assume linear as well as time-invariant properties, the displacement u(t) of the seismic mass has to fulfill the differential equation [6]

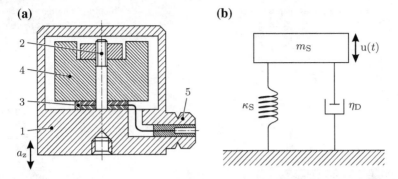

Fig. 9.7 a Internal structure of piezoelectric acceleration sensors of simple cylindrical shape [33]; 1 cylindrical housing; 2 preloading bolt; 3 piezoelectric rings; 4 seismic mass m_S; 5 electrical connector; **b** mechanical oscillator system with one degree of freedom consisting of seismic mass m_S, spring with spring rate κ_S and dashpot with damping constant η_D

$$m_S \frac{d^2 u(t)}{dt^2} + \eta_D \frac{du(t)}{dt} + \kappa_S u(t) = F_z(t) \tag{9.6}$$

with the external force $F_z(t)$ acting on the system. In case of a simple piezoelectric acceleration sensor as displayed in Fig. 9.7a, $F_z(t)$ denotes the force that is applied to the piezoelectric elements. The solution of this differential equation yields $u(t)$, which can be used to determine the velocity $v_z(t) = du(t)/dt$ and acceleration $a_z(t) = d^2 u(t)/dt^2$ of m_S. For a harmonic force $F_z(t) = \hat{F}_0 \sin(2\pi f t)$ with the force amplitude $\hat{F}_0$ and excitation frequency f, the solution of (9.6) takes the form

$$u(t) = \hat{u} \sin(2\pi f t - \varphi) \ . \tag{9.7}$$

The frequency-dependent displacement amplitude $\hat{u}(\omega)$ and phase angle $\varphi(\omega)$ computes as (angular frequency $\omega = 2\pi f$)

$$\hat{u}(\omega) = \frac{\hat{F}_0}{\sqrt{\left(\kappa_S - m_S \omega^2\right)^2 + (\eta_D \omega)^2}} \tag{9.8}$$

$$\varphi(\omega) = \arctan\left[\frac{\eta_D \omega}{\kappa_S - m_S \omega^2}\right] \ . \tag{9.9}$$

With a view to obtaining normalized results, it makes sense to introduce the displacement $u_0 = F_0/\kappa_S$ for a static force F_0, the angular frequency $\omega_0 = \sqrt{\kappa_S/m_S}$ of the undamped system (i.e., $\eta_D = 0$), and the dimensionless damping ratio $\xi_d = \eta_D/(2 m_S \omega_0)$. In doing so, the frequency-dependent ratio $\hat{u}(\omega)/u_0$ of displacements and $\varphi(\omega)$ become

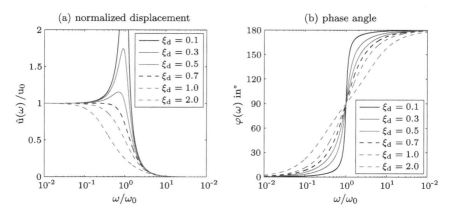

Fig. 9.8 a Ratio $\hat{u}(\omega)/u_0$ of displacements and **b** phase angle $\varphi(\omega)$ of oscillator system (see Fig. 9.7) with respect to normalized angular frequency ω/ω_0 for different damping ratios ξ_d

$$\frac{\hat{u}(\omega)}{u_0} = \frac{1}{\sqrt{\left(1 - \frac{\omega^2}{\omega_0^2}\right)^2 + \left(2\xi_d \frac{\omega}{\omega_0}\right)^2}} \tag{9.10}$$

$$\varphi(\omega) = \arctan\left[\frac{2\xi_d \frac{\omega}{\omega_0}}{1 - \left(\frac{\omega}{\omega_0}\right)^2}\right]. \tag{9.11}$$

Figure 9.8 depicts both quantities with respect to the normalized angular frequency ω/ω_0 for different values of ξ_d. As can be clearly observed, $\hat{u}(\omega)/u_0$ and $\varphi(\omega)$ change remarkably close to $\omega/\omega_0 = 1$. Especially in case of small damping ratios (e.g., $\xi_d = 0.1$), those changes are strongly pronounced. The global maximum of $\hat{u}(\omega)/u_0$ will, moreover, move toward lower frequencies if ξ_d is increasing. Hence, we can state that piezoelectric acceleration sensors should always be operated far below their resonance frequency because, otherwise, the sensor output will not reflect actually existing accelerations. This does not only refer to the acceleration amplitude but also to the phase angle. It should be noted in addition that a poor mechanical link of investigated structure and piezoelectric acceleration sensor, e.g., due to inadequate screwing, reduces the upper limit of the sensor's operating frequency.

Besides the compression-type sensor in Fig. 9.7a, several further designs of piezo-electric acceleration sensors are commercially available [33]. The internal structures of selected designs are shown in Fig. 9.9. The disadvantage of compression-type acceleration sensors lies in the fact that mechanical strains in the mounting surface of the investigated structure get transmitted directly to the piezoelectric elements. Owing to the transverse effect of piezoelectricity, such strains generate unwanted sensor outputs. We can avoid strain transmission by means of the so-called hanging design (see Fig. 9.9a) since the piezoelectric elements are preloaded against the inner face of sensor housing's top. Piezoelectric acceleration sensors in hanging design oftentimes serve as reference in calibrating other acceleration sensors.

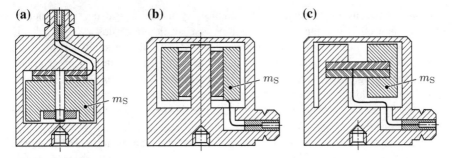

Fig. 9.9 Internal structure of selected piezoelectric acceleration sensors [33]; **a** hanging design; **b** acceleration sensor exploiting transverse shear mode of piezoelectricity; **c** acceleration sensor exploiting transverse length mode of two piezoelectric beams; seismic mass m_S

Figure 9.9b displays the internal structure of a piezoelectric acceleration sensor that exploits the transverse shear mode of piezoelectricity. In this design, both the piezoelectric transduction element and the seismic mass exhibit the shape of a hollow cylinder. The seismic mass is heated during sensor manufacturing. Through its shrinkage when cooling down, the piezoelectric cylinder gets radially preloaded against the cylindrical central stud of the sensor housing. Alternatively, three piezoelectric elements with three individual seismic masses are preloaded against a triangular-shaped center stud. This special sensor design is usually termed *DeltaShear* design and commercially distributed from the company Brüel & Kjær GmbH [13]. Compared to acceleration sensors being based on longitudinal mode of piezoelectricity, sensors exploiting the transverse shear mode offer a better thermal stability of the sensor sensitivity because the pyroelectric effect is weakly pronounced.

A further design of piezoelectric acceleration sensors is illustrated in Fig. 9.9c. Just as piezoelectric bimorph transducers (see Fig. 7.27 on p. 300), this kind of acceleration sensors exploits the transverse length mode of two piezoelectric beams, which are fixed at one end to the sensor housing. At the other end, the thin beams are equipped with the seismic mass m_S. If an acceleration acts on the sensor, m_S will exert a mechanical force on the piezoelectric bimorph. The force leads to a certain bending of the bimorph and, therefore, electric charges are electrostatically induced on the beam electrodes. The so-called *PiezoBeam* from the company Kister Instrumente GmbH [49] represents an alternative design, which does not need a seismic mass. In this design, the piezoelectric bimorph is mounted in its center to the sensor housing. PiezoBeam sensors measure exclusively accelerations orthogonal to the bimorph since angular accelerations will cause bendings in opposite directions with respect to the beam center. As a result, the electric charges on the beam electrodes cancel each other out.

Apart from special applications like high operating temperatures and in contrast to force as well as pressure sensors, piezoelectric acceleration sensors mostly exploit piezoceramic materials as transduction elements. This can be ascribed to the dynamic character of mechanical accelerations, e.g., harmonic vibrations. Consequently, we do not require quasi-static measurements. By means of piezoceramic materials, one can build up small as well as low-cost acceleration sensors with seismic masses $m_S < 1$ g and comparatively high sensor sensitivities. A small seismic mass goes hand in hand with high resonance frequencies, which constitutes a great advantage concerning practical applications like modal analysis of structures. It is also possible to build up compact triaxial piezoelectric acceleration sensors with piezoceramic materials. Special designs of such triaxial sensors need only one seismic mass.

The companies Brüel & Kjær GmbH [13], Kistler Instrumente GmbH [49], and Meggitt Sensing Systems [28] are well-known manufacturers of piezoelectric acceleration sensors. The upper limit of the operating frequency of commercially available sensors exceeds 10 kHz. To some extend, the sensor sensitivity Q_S/a is much greater than $10\,\mathrm{pC}g_n^{-1}$. One can also purchase piezoelectric acceleration sensors with integrated amplifier circuits (e.g., IEPE accelerometers from Kistler Instrumente GmbH) leading to the advantage that further amplifiers are not needed anymore.

9.1.5 Readout of Piezoelectric Sensors

A piezoelectric element that is equipped with electrodes can be interpreted as an electrical capacitance. If we apply a mechanical load (e.g., force) to the element, the electrodes of the capacitance will carry electric charges. This is a consequence of the changing polarization state inside the piezoelectric material (see Sect. 3.1). The larger the mechanical load, the more electric charges will be on the electrodes. Therefore, the amount of electric charges relates to the applied mechanical load, which should be measured by the piezoelectric sensor.

Let us assign C_S to the element's capacitance and Q_S to the electric charge on the electrodes. As it is the case for a capacitance, one can measure the electrical voltage $U_S = Q_S/C_S$ between the electrodes when the capacitance is charged, i.e., when the piezoelectric element is mechanically loaded. Since U_S is directly proportional to the applied mechanical load, it makes, thus, sense to measure this voltage. We can conduct such measurement by a so-called *electrometer amplifier*.

Alternatively to measuring U_S, it is possible to directly determine the electric charge by a so-called *charge amplifier*. In doing so, the electrodes of the piezoelectric element become virtually short-circuited, whereby the charges remain on the electrodes. In case of a real short-circuit, there will occur charge equalization, i.e., the electrodes do not carry electric charges anymore.

In the following, we will study both amplifier circuits (i.e., electrometer and charge amplifier) for reading out piezoelectric sensors. This includes advantages and disadvantages as well as commercially available products.

Electrometer Amplifier

An electrometer amplifier is an amplifier circuit featuring a very high insulation resistance at its input. It is possible to either amplify electric voltages or convert electric

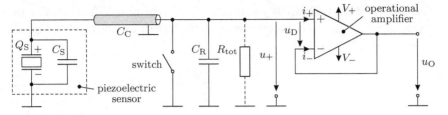

Fig. 9.10 Electrometer amplifier based on operational amplifier for piezoelectric sensor; argument time t omitted for compactness

charges into voltages. While in former times, such amplifiers were built up with electrometer tubes, they consist of transistors and/or operational amplifiers (opamp or op-amp) nowadays. In general, operational amplifiers comprise several transistors (e.g., field-effect transistors) and are used for measuring amplifiers [102]. Figure 9.10 shows the typical setup of electrometer amplifiers. Besides the operational amplifier and the piezoelectric sensor (e.g., force sensor) with electrical capacitance C_S, the setup contains a so-called range capacitor C_R and an electric switch to reset the measurement. The component C_C describes the capacitance of the cable, which connects piezoelectric sensor and amplifier. The capacitance of a cable of length l_C computes as $C_C = C'_C \cdot l_C$ with the capacitance C'_C per unit length.

To analyze the electrometer amplifier in Fig. 9.10, let us assume ideal components in a first step. Hence, C_S, C_C as well as C_R should be ideal capacities with an infinite resistor connected in parallel, i.e., the capacitances offer an infinite insulation resistance each and do not exhibit resistive losses. For the considered circuit, an ideal operational amplifier implies [59]:

- The input resistance between noninverting input (i.e., input $+$) and inverting input (i.e., input $-$) of the operational amplifier is infinite, i.e., the input currents $i_+(t)$ and $i_-(t)$ are zero.
- Its output resistance is zero.
- The amplification G_{OL} (open-loop voltage gain) of the differential input voltage $u_D(t) = u_+(t) - u_-(t)$, i.e., the potential difference between input $+$ and input $-$, is infinite. Because the amplifier output $u_O(t)$ takes always finite values, $u_D(t) = 0$ holds in stable operation mode, which is also called negative feedback.
- The operational amplifier provides output voltages $u_O(t)$ between negative supply voltage V_- and positive supply voltage V_+.
- The behavior of the operational amplifier does not depend on frequency f.

As mentioned previously, a piezoelectric sensor can be interpreted as a electrical capacitance C_S that carries the electric charge $Q_S(t)$ at its electrodes. The electric voltage $u_S(t)$ over C_S becomes $u_S(t) = Q_S(t)/C_S$. This relation will be, however, only valid if the setup does not contain further components, which is impossible since we always require cables as well as analysis units like analog-to-digital converters. In the present case, one has to consider both the cable capacitance C_C and the range capacitor C_R that are electrically connected in parallel to C_S. Consequently, the electric voltage $u_+(t)$ with respect to ground potential at the noninverting input of the operational amplifier reads as

$$u_+(t) = \frac{Q_S(t)}{C_S + C_C + C_R} \ . \tag{9.12}$$

Due to $u_D(t) = 0$ being valid for an ideal operational amplifier with negative feedback, the considered amplifier circuit exhibits the closed-loop voltage gain $G_{CL} = u_O(t)/u_+(t) = 1$. The output $u_O(t)$ of the electrometer amplifier, thus, corresponds to $u_+(t)$, i.e.,

$$u_O(t) = u_+(t) = \frac{Q_S(t)}{C_S + C_C + C_R} \ . \tag{9.13}$$

At this point, the question arises what (9.13) means for practical applications of the combination piezoelectric sensor and electrometer amplifier. On the one hand, it is possible to alter $u_O(t)$ for a given sensor charge $Q_S(t)$ by selecting an appropriate value of C_R, which constitutes an advantage concerning optimal use of the amplifier's output voltage range such as $\pm 10\,\mathrm{V}$. However, we can only determine $Q_S(t)$ from $u_O(t)$ if the capacity values of C_S, C_C, and C_R are known. This fact is a problem of electrometer amplifiers since $Q_S(t)$ directly relates to the aimed quantity, e.g., mechanical force $F_z \propto Q_S$ in (9.3). Of course, the combination of piezoelectric sensor and electrometer amplifier can also be calibrated by applying defined mechanical loads to the piezoelectric sensor like weights to a force sensor. When the configuration changes, e.g., due to a longer cable, the calibration result will not be suitable anymore. The capacitance per unit length of a common coaxial cable amounts $C_C' \approx 100\,\mathrm{pF\,m^{-1}}$. A cable of $l_C = 10\,\mathrm{m}$ can, therefore, exhibit a capacity value in the range of C_S. For instance, the capacity of a piezoceramic disk with diameter $d_S = 5\,\mathrm{mm}$ and thickness $t_S = 1\,\mathrm{mm}$ equals $C_S \approx 1\,\mathrm{nF}$.

Even though the combination of piezoelectric sensor and electrometer amplifier is perfectly calibrated, there exist further problems concerning quasi-static measurements. This stems from the nonideal properties of the piezoelectric sensor element, the connecting cable, the range capacitor as well as the operational amplifier. In reality, each capacitance (i.e., C_S, C_C, and C_R) suffers from a finite insulation resistance. Depending on the component quality, the insulation resistance can take values from more than $10\,\mathrm{T}\Omega$ down to a few $\mathrm{G}\Omega$ [33]. The insulation resistance between input $+$ and $-$ of the operational amplifier is also not infinite because of the nonzero leakage currents of the transistors at its input stage [59, 102]. As a result, the electrometer amplifier circuit including the piezoelectric sensor exhibits the total insulation resistance R_{tot}, which follows from the parallel connection of the individual resistances. The total capacitance $C_{tot} = C_S + C_C + C_R$ of the circuit gets discharged via R_{tot}; i.e., both the charge $Q_S(t)$ and the output voltage $u_O(t)$ decrease exponentially as time goes by. The smaller R_{tot}, the larger the discharge current and the shorter the discharge time will be.

The time constant $\tau_S = R_{tot}C_{tot}$ specifies the time after which $u_O(t)$ has decreased to 36.8% of its initial value. After the time $5\tau_S$, C_{tot} is almost completely discharged. Besides quasi-static measurements with piezoelectric sensors, the parameter τ_S is important for dynamic measurements because it also specifies the lower cutoff frequency f_l

$$f_l = \frac{1}{2\pi\tau_S} = \frac{1}{2\pi R_{tot}C_{tot}} \qquad (9.14)$$

of the combination sensor and electrometer amplifier. To avoid a noticeable influence on measured amplitude and phase angle, f_l should be much smaller than the lowest frequency in the measurement signal. We can, of course, increase τ_S and decrease f_l by choosing a higher value for the range capacitor C_R. However, in doing so, the amplifier output $u_O(t)$ for a given measurand will be automatically reduced which impairs the available signal-to-noise ratio.

Owing to the mentioned disadvantages, electrometer amplifiers are rarely used as readout electronics for piezoelectric sensors. An important exception to this represent piezoelectric acceleration sensors with integrated electrometer amplifiers like *Delta-Tron* from Brüel & Kjær GmbH [13], *Piezotron* from Kistler Instrumente GmbH [49], and *ISOTRON* from Meggitt Sensing Systems [28]. Such sensor/amplifier assemblies utilize the two-wire principle. They are usually powered by a constant electric current of 4 mA. The resulting resistance of the integrated electrometer amplifier varies with respect to the acting acceleration.

Charge Amplifier

The underlying principle of so-called charge amplifiers was firstly introduced by Kistler in 1950 [33]. Although the name *charge amplifier* suggests that electric charges get amplified, they are converted into a directly proportional electric voltage signal. Figure 9.11 displays the typical setup of charge amplifiers for piezoelectric sensors. Just as in case of electrometer amplifiers, the main component is an operational amplifier in stable operation mode [59, 102]. The range capacitor C_R serves a negative feedback; i.e., the output voltage $u_O(t)$ is capacitively led back to the inverting input of the operational amplifier. Moreover, the amplifier circuit contains the capacitance C_S stemming from the piezoelectric sensor and the cable capacitance C_C. These quantities are collected in Fig. 9.11 as total capacity $C_{tot} = C_S + C_C$ at the amplifier input. The electric switch in parallel to C_R allows again to reset the measurement.

With a view to analyzing the considered charge amplifier circuit, let us again assume ideal components in a first step. This refers to the properties of both the operational amplifier (e.g., $i_+ = i_- = 0$; see p. 423) and to the capacitances, which offer, thus, infinite insulation resistances. Irrespective of the fact that the open-loop voltage gain G_{OL} of an ideal operational amplifier is infinite, we will treat this amplification as finite number at the moment. The amplifier output $u_O(t)$ results from the differential input voltage $u_D(t)$ through $u_O(t) = G_{OL}u_D(t)$. Therewith, the voltage $u_C(t)$ over the range capacitor C_R has to fulfill

Fig. 9.11 Charge amplifier
for piezoelectric sensors;
total capacity C_{tot} includes
sensor capacitance C_S and
cable capacitance C_C;
argument time t omitted for
compactness

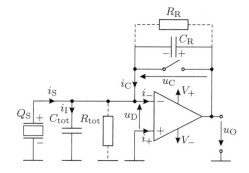

$$u_C(t) = u_O(t) + u_D(t) = u_O(t) + \frac{u_O(t)}{G_{OL}} = u_O(t)\left[1 + \frac{1}{G_{OL}}\right]. \qquad (9.15)$$

Since the electric input currents of an ideal operational amplifier are zero, Kirchhoff's
current law yields

$$i_S(t) - i_I(t) + i_C(t) = 0 \qquad (9.16)$$

with the current $i_S(t) = dQ_S(t)/dt$ generated from the electric charge $Q_S(t)$ on the
sensor's electrodes. Due to the assumed directions in Fig. 9.11, the currents $i_I(t)$
through C_{tot} and $i_C(t)$ through C_R become with (9.15)

$$i_I(t) = -C_{tot}\frac{du_D(t)}{dt} = -\frac{1}{G_{OL}}C_{tot}\frac{du_O(t)}{dt} \qquad (9.17)$$

$$i_C(t) = C_R\frac{du_C(t)}{dt} = \left[1 + \frac{1}{G_{OL}}\right]C_R\frac{du_O(t)}{dt}. \qquad (9.18)$$

By inserting both relations into (9.16), one obtains

$$i_S(t) = \frac{dQ_S(t)}{dt} = i_I(t) - i_C(t)$$
$$= -\frac{1}{G_{OL}}C_{tot}\frac{du_O(t)}{dt} - \left[1 + \frac{1}{G_{OL}}\right]C_R\frac{du_O(t)}{dt}. \qquad (9.19)$$

After integrating this equation and choosing the integration constant to be zero, which
is physically realized by resetting the amplifier circuit shortly before the measurement
starts, the amplifier output $u_O(t)$ finally results in

$$u_O(t) = -\frac{Q_S(t)}{\left[1 + \dfrac{1}{G_{OL}}\right]C_R + \dfrac{1}{G_{OL}}C_{tot}}. \qquad (9.20)$$

It should be noted that the amplifier output exhibits the opposite sign of the sensor charge $Q_S(t)$, i.e., a positive $Q_S(t)$ causes a negative $u_O(t)$. Because of this, the piezoelectric elements need to be installed appropriately with a view to achieving positive electrical outputs for positive mechanical inputs. For instance, in case of the disk-shaped piezoelectric element in Fig. 9.2, the mechanical force F_z and the axis of piezoelectricity have to point in opposite directions.

Ideal operational amplifiers satisfy $G_{OL} \to \infty$ and, therefore, (9.20) simplifies to

$$u_O(t) = -\frac{Q_S(t)}{C_R} = u_C(t) \; . \tag{9.21}$$

According to (9.21), neither the sensor capacitance C_S nor the cable capacitance C_C influence the amplifier output and the cable length l_C plays no role anymore. The connection between $u_O(t)$ and $Q_S(t)$ is solely specified by the range capacitor C_R, which justifies its name. This originates from the fact that the differential input voltage $u_D(t)$ equals zero for an ideal operational amplifier in stable operation mode. Since $u_D(t) = -u_-(t) = 0$ also corresponds to the voltages over the capacitances C_S and C_C, they are always discharged. The existing charges arise from the electric polarization of the piezoelectric sensor in case of mechanical loads. The property $u_D(t) = 0$ additionally ensures that a finite insulation resistance R_{tot} (e.g., of C_C) does not affect $u_O(t)$. As a result, $i_I(t) = 0$ holds and $i_S(t) = -i_C(t)$ follows from (9.16), which means that the electric charge of C_R and the piezoelectric sensor are equal but of opposite polarity. A charge amplifier, thus, continuously compensates charges on the sensor electrodes due to mechanical loads with equal charges in the range capacitor.

From the theoretical point of view, a charge amplifier comprising ideal components should enable true static measurements with piezoelectric sensors. We are, however, always confronted with nonideal components in reality. Especially the properties of the range capacitor C_R play a decisive role in this context. Its finite insulation resistance R_R discharges C_R. Consequently, the voltage $u_C(t)$ and the amplifier output $u_O(t)$ will change with respect to time if a constant mechanical load is applied to the piezoelectric element. Just as for electrometer amplifiers, we can define the time constant $\tau_S = R_R C_R$ and lower cutoff frequency $f_1 = 1/(2\pi\tau_S)$. By considering only the nonideal range capacitor, $u_O(t)$ will be close to zero after $5\tau_S$ for a constant mechanical load. It is, therefore, not surprising that true static measurements call for $\tau_S \to \infty$ and $f_1 = 0$. Nevertheless, an additional resistor R_f with resistance values much smaller than R_R is sometimes switched in parallel to the range capacitor for dynamic measurements [33]. In doing so, τ_S decreases and f_1 increases which implies that signal components with frequencies $\ll f_1$ do not arise in $u_O(t)$. Modern commercially available charge amplifiers enable the selection between different values of R_f.

Besides the nonideal range capacitor, the actual properties of the operational amplifier affect the performance of the charge amplifier. In particular, we have to consider three influencing factors, namely (i) the input current $i_-(t)$, (ii) the open-loop voltage gain G_{OL}, and (iii) the offset voltage U_{OS} at the amplifier input [59, 102].

The currents $i_+(t)$ and $i_-(t)$ originate from leakage currents of the transistors at the input stage of operational amplifiers. Depending on the transistor type (i.e., bipolar or field-effect transistors), these currents range from a few fA up to μA and strongly vary with temperature. As a matter of fact, $i_-(t)$ changes the charge state of C_R. Because G_{OL} takes finite values from 10^4 to 10^7, the differential input voltage $u_D(t)$ is not zero. This causes a current $i_I(t)$ through the nonideal sensor capacitance C_S and the nonideal cable capacitance C_C. The resulting current changes the charge state of C_R equally to $i_-(t)$. The same applies to U_{OS}, which is in the order of μV to mV, since this voltage also generates currents through the finite isolation resistances of C_S and C_C. Therefore, we can conclude that all three influencing factors change the charge state of C_R. Such change directly alters the amplifier output and becomes visible as drift, which leads to positive or negative saturation of $u_O(t)$ after a certain time. That is why one should reset charge amplifiers before starting measurements with piezoelectric sensors and also after a long measuring time.

The application of charge amplifiers for piezoelectric sensors offers several advantages over electrometer amplifiers:

- Since the voltage across sensor and cable is rather small, their capacitances as well as insulation resistances have a comparatively little impact on the output of charge amplifiers.
- The virtual short-circuit at the amplifier input prevents voltage peaks of the piezoelectric elements due to sudden mechanical loads.
- For ideal components, the output voltage $u_O(t)$ is directly proportional to the electric charge $Q_S(t)$ on the sensor electrodes and, thus, to the measurand. Note that this simple relation also represents a very good approximation for nonideal components of the charge amplifier circuit.
- One can connect several piezoelectric sensors of equal sensitivity in parallel to a single charge amplifier. In case of an electrometer amplifier, the parallel connection of several sensors requires an extensive calibration procedure.

On these grounds, it seems only natural that charge amplifiers are widely used for reading out piezoelectric sensors. The companies Brüel & Kjær GmbH [13] and Kistler Instrumente GmbH [49] are well-known manufacturers of charge amplifiers. To some extend, the sold charge amplifiers can be computer-controlled and include analog-to-digital converters, which gives the possibility for further digital processing.

9.2 Determination of Plate Thickness and Speed of Sound

Ultrasonic waves and, thus, piezoelectric ultrasonic transducers are widely used for nondestructive testing like acoustic microscopy [63, 94, 114], weld inspection [50] as well as material characterization (see Sect. 5.1.2). Thereby, geometric dimensions and material properties have to be known, e.g., sample thickness and speed of sound. If either one of the parameters is unknown, this parameter can be identified by simple time-of-flight measurements of ultrasound. Here, we will study a special ultrasonic measuring system enabling simultaneous determination of plate thickness and speed of sound inside the plate. The underlying approach was published by Kiefer et al. [48] and should demonstrate typical signal processing steps (e.g., Wiener filtering) as well as capabilities of ultrasonic measuring systems in material characterization. The measurement principle that is based on the through-transmission mode of longitudinal waves will be detailed in Sect. 9.2.1. Afterward, we model the investigated plates from the system point of view as a transmission line in the time and frequency domain. Section 9.2.3 deals with coded excitation signals of the ultrasonic transducers as well as requirements to achieve the aimed spatial resolution. The long duration of coded excitation signals demands pulse compression by appropriate filters, which is explained in Sect. 9.2.4. Finally, we will discuss experimental results that were obtained by a realized ultrasonic measuring system.

9.2.1   Measurement Principle

In the following, let us assume a homogenous as well as flat solid plate of thickness d_P in z-direction and of large extensions in the xy-plane. The plate material shall feature the wave propagation velocity c_P for longitudinal waves that is termed speed of sound (SOS). As mentioned above, we can directly determine d_P if c_P is known or c_P if d_P is known with the aid of time-of-flight (TOF) measurements of ultrasonic waves. In doing so, it makes sense to place the investigated plate in an appropriate coupling medium for ultrasonic waves, e.g., water. Because the acoustic impedance Z_{aco} of coupling medium and plate material differ, ultrasonic waves will be reflected at their interfaces. When an immersion transducer emits an ultrasonic wave propagating in z-direction toward the plate and is operated in pulse-echo mode, we will be able to compute the plate thickness d_P or the speed of sound c_P from the simple mathematical relation $t_R = 2d_P/c_P$. The expression t_R denotes the time difference between the reflected pulses from the front edge and rear edge of the plate. However, by additionally exploiting multiple reflections (reveberations) inside the plate, one can determine both parameters (i.e., d_P as well as c_P) simultaneously [43, 92]. The evaluation of these reflections is not only possible in pulse-echo mode but also in through-transmission mode, which represents a special case of the pitch-catch mode requiring two axially aligned ultrasonic transducers (cf. Fig. 7.1 on p. 262). Without limiting the generality, we will study the simultaneous determination of d_P and c_P for this through-transmission mode. An alternative approach that exploits an ultrasonic annular array is proposed in [55, 111].

Figure 9.12 depicts the considered configuration of the ultrasonic transducers. While the left transducer serves as ultrasonic transmitter, the right one is used as a receiver of ultrasonic waves. In principle, the simultaneous determination of plate thickness and SOS in the plate is based on two different measurements. The first measurement is conducted without plate between transmitter and receiver. From this reference measurement, we can immediately deduce the SOS $c_W = l/t_W$ inside the coupling medium when both the geometric transducer distance l and the TOF t_W

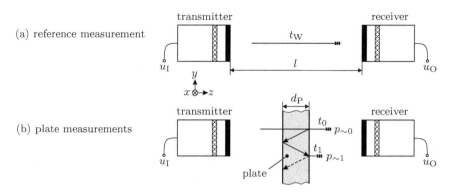

Fig. 9.12 Considered configuration of ultrasonic transmitter and receiver for **a** reference measurement and **b** plate measurements [48, 92]; vertical components of ultrasonic rays indicate time delays between sound pressure waves; argument t omitted for compactness

of the ultrasonic waves from transmitter to receiver are known. Alternatively, c_W also results from evaluating the difference of TOFs for two different geometric distances between transmitter and receiver, which can be arranged by a high-precision positioning system.

For the second measurement, the investigated plate has to be placed between transmitter and receiver (see Fig. 9.12b). In this case, there occurs a directly transmitted ultrasonic wave as well as waves stemming from multiple reflections inside the plate. At the receiver, the entire sound pressure wave $p_\sim(t)$ follows from

$$p_\sim(t) = \sum_{i=0}^{\infty} p_{\sim i}(t) \ . \tag{9.22}$$

The expressions $p_{\sim 0}(t)$ and $p_{\sim i}(t)$ $\forall i \in \mathbb{N}_+$ stand for the directly transmitted wave and the ith multiple reflection, respectively. $p_{\sim 0}(t)$ reaches the receiver after the TOF t_0, whereas the first multiple reflection $p_{\sim 1}(t)$ arrives after the TOF t_1. The TOFs t_0 and t_1 for plane wave propagation in z-direction are given by

$$t_0 = \frac{l - d_P}{c_W} + \frac{d_P}{c_P} \tag{9.23}$$

$$t_1 = \frac{l - d_P}{c_W} + \frac{3d_P}{c_P} \ . \tag{9.24}$$

Hence, the time difference $t_R = t_1 - t_0$ between the TOFs corresponds to $2d_P/c_P$ since the ultrasonic waves have to additionally travel through the plate twice for the first multiple reflection. This time difference, of course, also holds for successive multiple reflections inside the plate, i.e., $t_{i+1} - t_i = t_R$. The combination of (9.23) $t_W = l/c_W$ and $t_R = 2d_P/c_P$ finally yields [92]

$$d_P = c_W \left[t_W - t_0 + \frac{t_R}{2} \right] \tag{9.25}$$

$$c_P = c_W \left[1 + \frac{2(t_W - t_0)}{t_R} \right] \tag{9.26}$$

for the plate thickness and the SOS in the plate. Consequently, it is possible to determine both parameters simultaneously when the quantities c_W, t_W, t_0, and t_R are known. Due to the fact that (9.25) as well as (9.26) exclusively contain time differences, there do not arise measurement deviations from systematic time delays within the ultrasonic transducers and additional electronic components.

9.2.2 Transmission Line Model for Plate

To obtain a deeper understanding of the measurement principle and further important points like the required spatial resolution, let us regard the investigated plate from the system point of view. The plate between transmitter and receiver can be modeled as a three-layer problem [9, 60]. The middle layer corresponds to the elastic plate, while the outer layers represent the surrounding coupling medium. If ultrasonic waves impinge perpendicular onto the interface coupling medium/plate, mechanical waves will be generated inside the plate. At the rear interface plate/coupling medium, these mechanical waves will be partially converted again to ultrasonic waves propagating toward the receiver. The directly transmitted waves and the resulting multiple reflections exhibit a certain time difference. It makes sense to treat the three-layer configuration as a transmission line. In doing so, we can neglect attenuation within the coupling medium and the plate because the measurement principle is solely based on TOFs. Below, the transmission line will be modeled in the time and frequency domain.

Transmission Line in Time Domain

An incident wave gets partially transmitted as well as reflected at each interface of coupling medium and plate. As already detailed, the incident wave is, thus, not only directly transmitted through the plate but also after $2i$ reflections inside the plate. If the directly transmitted ultrasonic wave $p_{\sim 0}(t)$ reaches the receiver at t_0, the ith multiple reflection $p_{\sim i}(t)$ will arrive at the time

$$t_i = t_0 + i\, t_R \quad \forall i \in \mathbb{N} \tag{9.27}$$

with the TOF $t_R = 2d_P/c_P$ that a mechanical wave needs to propagate back and forth inside the plate. In the time domain, the resulting transmission line can be fully characterized by the discrete impulse response

$$h_P(t) = \sum_{i=0}^{\infty} a_i\, \delta(t - t_i) , \tag{9.28}$$

where $\delta(\cdot)$ stands for the Dirac delta distribution. The expression

$$a_i = q_{WP} \, q_{PW} \, r_{PW}^{2i} \tag{9.29}$$

is the amplitude factor for directly transmitted waves (i.e., $i = 0$) and ith multiple reflections. Here, q_{WP} and q_{PW} indicate the transmission coefficients at the interfaces coupling medium/plate and plate/coupling medium, respectively.[1] With the acoustic impedances $Z_{aco;W}$ and $Z_{aco;P}$ of coupling medium and plate, these transmission coefficients compute as (cf. (2.139, p. 38))

$$q_{WP} = \frac{2Z_{aco;P}}{Z_{aco;W} + Z_{aco;P}} \tag{9.30}$$

$$q_{PW} = \frac{2Z_{aco;W}}{Z_{aco;W} + Z_{aco;P}} \; . \tag{9.31}$$

The remaining variable r_{PW} explains the reflection coefficient at the interface plate/coupling medium and takes the form

$$r_{PW} = \frac{Z_{aco;W} - Z_{aco;P}}{Z_{aco;W} + Z_{aco;P}} \; . \tag{9.32}$$

When the parameters c_W and l_W are known, it can be stated that the impulse response $h_P(t)$ of the transmission line includes all information (i.e., t_0 and t_R), which is required to simultaneously determine d_P and c_P.

Figure 9.13a depicts the calculated impulse response $h_P(t)$ of a steel plate with thickness $d_P = 3\,\mathrm{mm}$ that is immersed in water. Thereby, the wave propagation velocities and acoustic impedances of water and steel were assumed to take the values $c_W = 1485\,\mathrm{m\,s^{-1}}$, $c_P = 5850\,\mathrm{m\,s^{-1}}$, $Z_{aco;W} = 1.49 \times 10^6\,\mathrm{N\,s\,m^{-3}}$, and $Z_{aco;P} = 45.63 \times 10^6\,\mathrm{N\,s\,m^{-3}}$. As expected, $h_P(t)$ decreases exponentially and shows a spacing of t_R between two successive Dirac impulses. The normalization of the abscissa ensures that the curve progression only depends on the ratio d_P/c_P.

Transmission Line in Frequency Domain

Even though the impulse response $h_P(t)$ includes the entire information of the transmission line (i.e., the plate), one should additionally take a closer look at its complex-valued transfer function $\underline{H}_P(f)$ in the frequency domain. This is especially advisable because $\underline{H}_P(f)$ reveals decisive aspects, which facilitate a reasonable choice of the transmitter excitation. Before $\underline{H}_P(f)$ of the plate is deduced, let us consider a Dirac comb $III(t)$ (also known as impulse train) that will be quite similar to the impulse response $h_P(t)$ of the plate if $a_i = 1 \; \forall\, i \in \mathbb{N}$ holds. In the time domain, the Dirac comb reads as

$$III(t) = \sum_{i=-\infty}^{\infty} \delta(t - i\, t_R) \; . \tag{9.33}$$

---

[1]To avoid confusions with the time t, the transmission coefficients are named q.

By applying the Fourier transform, the Dirac comb becomes (frequency f)

$$\text{III}(f) = \mathcal{F}\{\text{III}(t)\} = \frac{1}{t_R} \sum_{i=-\infty}^{\infty} \delta\left(f - \frac{i}{t_R}\right) \tag{9.34}$$

in the frequency domain and, therefore, remains a Dirac comb [110]. While the periodic spacing between two successive Dirac impulses amounts t_R in the time domain, the periodic spacing in the frequency domain equals $1/t_R$. In contrast, the complex-valued transfer function $\underline{H}_P(f)$ of the plate results in

$$\underline{H}_P(f) = \mathcal{F}\{h_P(t)\} = q_{WP}\, q_{PW}\, e^{-j2\pi f d_P/c_P} \sum_{i=0}^{\infty}\left(r_{PW}^2\, e^{-j2\pi f 2d_P/c_P}\right)^i. \tag{9.35}$$

Owing to the fact that $\left|r_{PW}^2\right| < 1$ is always satisfied for a solid plate being immersed in a liquid, the series converges and $\underline{H}_P(f)$ simplifies to [12]

$$\underline{H}_P(f) = \frac{q_{WP}\, q_{PW}\, e^{-j2\pi f d_P/c_P}}{1 - r_{PW}^2\, e^{-j2\pi f 2d_P/c_P}} = \frac{q_{WP}\, q_{PW}}{e^{j2\pi f d_P/c_P} - r_{PW}^2\, e^{-j2\pi f d_P/c_P}}, \tag{9.36}$$

which is a periodic function in the frequency domain. Just as for the Dirac comb, the spacing of the maxima in the magnitude $\left|\underline{H}_P(f)\right|$ equals $1/t_R$. Since $h_P(t)$ gradually decreases, there do not, however, appear Dirac impulses in $\left|\underline{H}_P(f)\right|$. The resulting maxima have a nonvanishing frequency width different to zero instead. This is also demonstrated in Fig. 9.13b, which displays $\left|\underline{H}_P(f)\right|$ for the steel plate of thickness $d_P = 3\,\text{mm}$ that was also considered for the impulse response $h_P(t)$. Due to the normalization of the abscissa, the curve progression depends again on the ratio d_P/c_P.

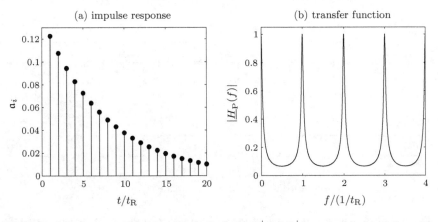

Fig. 9.13 **a** Impulse response $h_P(t)$ and **b** transfer function $\left|\underline{H}_P(f)\right|$ (magnitude) of steel plate with thickness $d_P = 3\,\text{mm}$ being immersed in water; x-axis normalized to $t_R = 1.026\,\mu\text{s}$ and $1/t_R = 975\,\text{kHz}$, respectively

9.2.3 Excitation Signal of Transmitter

The excitation signal $u_I(t)$ of the ultrasonic transmitter constitutes a decisive part of the considered measuring system, which comprises transmitter, investigated plate, receiver as well as an appropriate pulse compression approach (see Fig. 9.14). Under the assumption of a linear system, we are able to combine the impulse responses $h_{trans}(t)$ and $h_{rec}(t)$ of transmitter and receiver. The combined impulse response $h_T(t)$ of the transducer pair is defined as (temporal convolution $*$)

$$h_T(t) = h_{trans}(t) * h_{rec}(t) \ . \tag{9.37}$$

By using the impulse response $h_P(t)$ of the plate, the output $u_O(t)$ of the ultrasonic receiver becomes then

$$u_O(t) = u_I(t) * h_T(t) * h_P(t) \tag{9.38}$$

in the time domain and after applying the Fourier transform

$$\underline{U}_O(f) = \underline{U}_I(f) \cdot \underline{H}_T(f) \cdot \underline{H}_P(f) \ . \tag{9.39}$$

The expressions $\underline{U}_I(f)$ and $\underline{H}_T(f)$ denote the frequency spectrum of the excitation signal and the transfer function of the transducer pair, respectively. To facilitate the following explanations, let us introduce the so-called *interrogation signal* $g_T(t) = u_I(t) * h_T(t)$ that is generated from the transducer pair and the excitation signal, i.e., without plate [48]. The receiver output for the plate being placed between transmitter and receiver takes, thus, the form

$$u_O(t) = g_T(t) * h_P(t) \quad \xrightarrow{\mathcal{F}} \quad \underline{U}_O(f) = \underline{G}_T(f) \cdot \underline{H}_P(f) \tag{9.40}$$

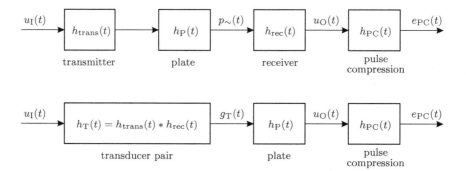

Fig. 9.14 Two equivalent structural diagrams of considered ultrasonic measuring system [48]; impulse responses $h_i(t)$; excitation signal $u_I(t)$; incident sound pressure wave $p_\sim(t)$ at receiver; receiver output $u_O(t)$; output signal $e_{PC}(t)$ after pulse compression with $h_{PC}(t)$; interrogation signal $g_T(t)$

with the frequency spectrum $\underline{G}_T(f)$ of the interrogation signal. Consequently, the receiver output corresponds to a filtered version of the interrogation signal, whereby the plate serves as filter. On the other hand, we can also state that the receiver output represents a filtered version of the plate behavior, whereby the interrogation signal serves as filter. Since the plate is the investigated object, it makes sense to prefer the second view. This means that both the excitation signal and the transducer behavior need to be appropriate if the desired information (i.e., plate thickness and SOS) should be deduced from the receiver output. The typical bandpass behavior of transmitter and receiver (cf. Fig. 7.46 on p. 329) is, however, specified by the available ultrasonic transducers. Therefore, one has to concentrate on the excitation signal of the transmitter.

Hereinafter, let us detail the requirements for the interrogation signal concerning simultaneous determination of plate thickness and SOS. We will also study coded transducer excitation as well as reasonable conditioning of the excitation signal to enhance the system's performance.

Requirements for Interrogation Signal

As mentioned above, we have to figure out the time difference t_R for determining plate thickness d_P and SOS c_P. In case of the considered ultrasonic measuring system, this implies requirements for the interrogation signal $g_T(t)$, which refer to two signal properties, namely its (i) bandwidth and (ii) energy. Let us start with the bandwidth B_g of $g_T(t)$. It is well known from ultrasonic imaging and radar imaging that the achievable spatial resolution is inversely proportional to B_g, e.g., [19, 67]. Hence, a high spatial resolution calls for a large bandwidth. This also applies to the considered ultrasonic measuring system. As has been shown in Sect. 9.2.2, the plate's transfer function $\underline{H}_P(f)$ is periodic in frequency. The receiver output should contain at least one entire period because, otherwise, we cannot identify $1/t_R$ and, thus, the quantity t_R. That is the reason why the bandwidth of the interrogation signal has to fulfill the condition

$$B_g \geq \frac{1}{t_R} \quad \longrightarrow \quad B_g \geq \frac{c_P}{2d_P}, \tag{9.41}$$

which will be violated especially in the case of a thin plate exhibiting a high SOS. Since the bandwidth of transmitter and receiver is always limited, one should choose not only broadband transducers but also an excitation signal $u_I(t)$ that exploits the available transducer bandwidth best possible.

The second requirement for $g_T(t)$ concerns its signal energy. Generally speaking, a high energy of the excitation signal improves the SNR values that are obtained from an ultrasonic measuring system [67]. Signal energy will rise if the amplitude $\hat{u}_I$, the duration T_u of $u_I(t)$, and/or its bandwidth B_u are increased. The same behavior, of course, applies to the interrogation signal, i.e., to $\hat{g}_T$, T_g as well as B_g. Unfortunately, an increased amplitude can lead to unwanted nonlinearities (e.g., in wave propagation) or may even damage the ultrasonic transmitter. It is, thus, recommended to extend duration and/or bandwidth. The so-called *time–bandwidth product TB* rates the energy of a signal [16, 67]. Whereas the bandwidth of a sinusoidal signal vanishes, a pulse-shaped signal owns a short duration. In both cases, TB takes small values, e.g., ≈ 1 for a pulse-shaped harmonic signal offering the property $B \approx 1/T$. The lower limit of TB arises for purely amplitude-modulated signals like a Gaussian pulse [10]. To achieve a high time–bandwidth product, one has to use phase or frequency modulation. The resulting excitation signals are commonly named *coded excitation signals* since their long duration demands an appropriate decoding process to achieve reasonable resolutions for TOF measurements [48]. Such decoding processes are usually referred to as *pulse compression* (see Sect. 9.2.4).

Coded Excitation

Coded signals result from phase or frequency modulation of a signal that was originally of sinusoidal shape. In doing so, the phase or frequency of the original signal is modified systematically. As a matter of fact, the choice of this modification determines the properties of the coded signal. Phase modulation can be based on code sequences like Barker codes, Golay codes, and Gold codes [67, 80, 92]. Such binary sequences always feature two defined states, e.g., 0 and 1 or -1 and 1. Due to the fact that Gold codes provide outstanding correlation properties, they are often exploited in telecommunication and satellite navigation. The autocorrelation of a Gold sequence shows a distinct maximum, while the cross-correlation of two differing Gold sequences equals almost zero. That is the reason why Gold codes should also be well suited for TOF measurements in ultrasonic measuring systems.

Now, let us detail a concrete example of phase modulation that is based on Gold codes. The selected Gold sequence s_{Gold} of order 6 has a length of $N_{Gold} = 63$ elements. Depending on the value of the single sequence element, the phase of the sinusoidal signal comprising M_s cycles gets altered. If $s_{Gold} = -1$, the phase will be shifted by $180°$; otherwise, i.e., $s_{Gold} = 1$, the phase remains unchanged. Therewith, we obtain a coded signal that contains N_{Gold} blocks with M_s sinusoidal cycles each. Figure 9.15a depicts a cutout of both the Gold sequence $s_{Gold}(t)$ and the resulting coded signal $s_{phase}(t)$ for $M_s = 1$ and the frequency $f = 2\,\text{MHz}$. As expected and demonstrated in Fig. 9.15b, the phase modulation has a remarkable impact on the frequency spectrum $\underline{S}_{phase}(f)$ of $s_{phase}(t)$. In contrast to a sinusoidal time signal, the spectral magnitude $\left|\underline{S}_{phase}(f)\right|$ of the coded signal becomes wide because of the conducted phase modulation.

The second kind of coded signals originates from frequency modulation; i.e., a variation of the instantaneous frequency $f_s(t)$ with respect to time. Such frequency modulation can be conducted again with the aid of code sequences. A frequency modulated signal will often be called chirp signal when $f_s(t)$ changes continuously with time. Without limiting the generality, we will exclusively consider linear chirp signals, which implies a linear variations of $f_s(t)$. One can distinguish between linear up-chirp signals and linear down-chirp signals. For an up-chirp signal, $f_s(t)$ increases from the lowest frequency f_{min} to the highest frequency f_{max} over time, whereas $f_s(t)$ decreases from f_{max} to f_{min} for a down-chirp signal. In complex notation, a linear chirp signal $\underline{s}_{chirp}(t)$ providing the bandwidth B_{chirp} is defined as [68]

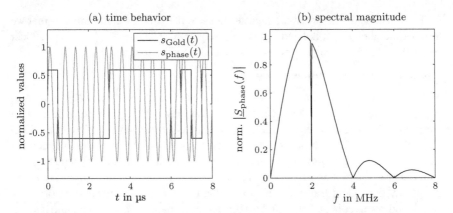

Fig. 9.15 a Cutout of Gold sequence $s_{\text{Gold}}(t)$ and phase modulated signal $s_{\text{phase}}(t)$ with respect to time t; **b** normalized spectral magnitude $\left|\underline{S}_{\text{phase}}(f)\right|$ of $s_{\text{phase}}(t)$

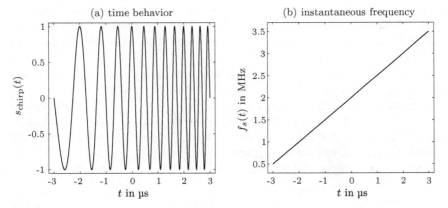

Fig. 9.16 a Time behavior of linear up-chirp signal $s_{\text{chirp}}(t)$; **b** instantaneous frequency $f_s(t)$ of $s_{\text{chirp}}(t)$; parameters of chirp signal $f_c = 2\,\text{MHz}$, $B_{\text{chirp}} = 3\,\text{MHz}$, and $T_{\text{chirp}} = 6\,\mu\text{s}$

$$\underline{s}_{\text{chirp}}(t) = e^{j2\pi(f_c t + F_{\text{chirp}} t^2/2)} \quad \text{for} \quad -\frac{T_{\text{chirp}}}{2} \leq t \leq \frac{T_{\text{chirp}}}{2} \qquad (9.42)$$

with the center frequency f_c, the so-called chirp rate $F_{\text{chirp}} = \pm B_{\text{chirp}}/T_{\text{chirp}}$ and the duration T_{chirp} of the signal, respectively. F_{chirp} takes positive values for up-chirp signals and negative values for down-chirp signals. The actual chirp signal $s_{\text{chirp}}(t)$ is given by the real part of $\underline{s}_{\text{chirp}}(t)$, i.e., $s_{\text{chirp}}(t) = \Re\{\underline{s}_{\text{chirp}}(t)\}$. Figure 9.16a and b display the time behavior of a linear up-chirp signal and the corresponding instantaneous frequency $f_s(t)$. The selected parameters amount $f_c = 2\,\text{MHz}$, $B_{\text{chirp}} = 3\,\text{MHz}$, and $T_{\text{chirp}} = 6\,\mu\text{s}$.

For the considered ultrasonic measuring system, coded excitation means that we apply either a phase modulated signal or a frequency modulated signal as transducer excitation $u_I(t)$. This signal represents the coded excitation signal.

Conditioning of Excitation Signal

From the theoretical point of view, a Dirac impulse has a flat amplitude spectrum and, therefore, contains all frequencies. We can exploit the entire bandwidth of an ultrasonic transducer when such impulse is utilized as transducer excitation. However, if the spatial resolution of an ultrasonic measuring system is of concern, it might be desirable to make better use of the transducer bandwidth by adjusting the transducer excitation appropriately. The fundamental idea lies in compensating the frequency-dependent transfer behavior of the ultrasonic transducers, i.e., of transmitter and receiver [48]. This can be achieved by enhancing spectral components in the excitation signal that are attenuated by the transducers. Consequently, the interrogation signal will offer a higher bandwidth than a short pulse and, thus, one is able to improve the spatial resolution of an ultrasonic measuring system. The underlying *signal conditioning* comprises three steps, namely (i) defining a conditioning filter, (ii) designing a conditioned transducer behavior, and (iii) deducing the required excitation signal. Below, let us explain these three steps, which are mainly motivated by the work of Oelze [74].

As a starting point, we assume a hypothetic transducer, the so-called *conditioned transducer*, that features the desired bandwidth of the transducer pair. The conditioned transducer with the impulse response $h_C(t)$ will generate the interrogation signal $g_C(t)$ when excited by the excitation signal $u_I(t)$. Now, the question arises which excitation signal will produce the same interrogation signal from the actually existing transducer pair being specified by the impulse response $h_T(t)$. This conditioned excitation signal $u_{IC}(t)$, of course, has to fulfill the convolution equivalence

$$u_{IC}(t) * h_T(T) \overset{!}{=} u_I(t) * h_C(T) = g_{TC}(t) \tag{9.43}$$

in the time domain and

$$\underline{U}_{IC}(f) \cdot \underline{H}_T(f) \overset{!}{=} \underline{U}_I(f) \cdot \underline{H}_C(f) = \underline{G}_{TC}(f) \tag{9.44}$$

in the frequency domain with the frequency spectra $\underline{U}_I(f)$ of $u_I(t)$, $\underline{H}_C(f)$ of $h_C(t)$, and $\underline{H}_T(f)$ of $h_T(t)$. From there, we can directly solve for the frequency spectrum $\underline{U}_{IC}(f)$ of $u_{IC}(t)$ through

$$\underline{U}_{IC}(f) = \underline{U}_I(f) \frac{\underline{H}_C(f)}{\underline{H}_T(f)} = \underline{U}_I(f) \cdot \underline{\Psi}(f) \ . \tag{9.45}$$

The expression $1/\underline{H}_T(f)$ represents the so-called inverse filter in the frequency domain and $\underline{\Psi}(f)$ describes a possible conditioning filter. Owing to the fact that ultrasonic transducers typically exhibit bandpass behavior, $1/\underline{H}_T(f)$ is unbounded which causes problems in (9.45). To achieve a stable deconvolution, one should, therefore, apply a Wiener filter instead of the inverse filter [59]. For the given situation, the filter $\underline{\Psi}_W(f)$ reads as

$$\underline{\Psi}_W(f) = \frac{\underline{H}_C(f) \cdot \underline{H}_T^*(f)}{\left|\underline{H}_T(f)\right|^2 + \beta_W \left|\underline{H}_T(f)\right|^{-2}}, \tag{9.46}$$

where the noise-to-signal ratio $NSR(f)$ is assumed to be $\beta_W \left|\underline{H}_T(f)\right|^{-2}$. The factor β_W is used to estimate the spectral energy density of the noise, and $\underline{H}_T^*(f)$ stands for the complex conjugate of $\underline{H}_T(f)$. Note that this conditioning filter tries to correct the phase difference between $\underline{H}_C(f)$ and $\underline{H}_T(f)$. However, since the subsequently applied pulse compression is very sensitive to phase mismatches, such correction might degrade the quality of the coded excitation signal. By leaving the phase spectrum of $\underline{U}_I(f)$ unchanged, we are, moreover, able to design pulse compression filters on the basis of the original excitation signal $u_I(t)$. It is, thus, advisable to adapt the complex-valued filter from (9.46) to the real-valued version

$$\Psi_W(f) = \frac{\left|\underline{H}_C(f) \cdot \underline{H}_T^*(f)\right|}{\left|\underline{H}_T(f)\right|^2 + \beta_W \left|\underline{H}_T(f)\right|^{-2}}, \tag{9.47}$$

which exclusively alters spectral magnitudes in $\underline{U}_I(f)$. Therewith, the frequency spectrum $\underline{U}_{IC}(f)$ of the conditioned excitation signal results from

$$\underline{U}_{IC}(f) = \underline{U}_I(f) \frac{\left|\underline{H}_C(f) \cdot \underline{H}_T^*(f)\right|}{\left|\underline{H}_T(f)\right|^2 + \beta_W \left|\underline{H}_T(f)\right|^{-2}}. \tag{9.48}$$

The inverse Fourier transform finally leads to the conditioned excitation signal $u_{IC}(t)$ in the time domain.

In a next step, let us take a closer look at the conditioned transducer. The impulse response of an ultrasonic transducers equals approximately a Gaussian pulse (cf. Fig. 7.23 on p. 292). On this account, one should also model the impulse response $h_C(t)$ of the conditioned transducer by such a pulse, which mathematically takes the form [10]

$$h_C(t) = e^{-t^2/(2\sigma_C^2)} \cos(2\pi f_c t) \tag{9.49}$$

with the carrier frequency f_c and the parameter σ_C defining the duration of signal envelope. While f_c equals the center frequency of the transducer, σ_C determines the transducer bandwidth. According to the modulation theorem, the frequency spectrum $\underline{H}_C(t)$ of $h_C(t)$ will correspond to the frequency spectrum of its envelope $e^{-t^2/(2\sigma_C^2)}$ when the envelope's spectrum is shifted by f_c. This means that we only have to inspect the envelope's spectrum

$$\mathcal{F}\left\{e^{-t^2/(2\sigma_C^2)}\right\} = \sqrt{2\pi}\sigma_C e^{-2(\pi f \sigma_C)^2} \tag{9.50}$$

to figure out the -6 dB bandwidth $B_C^{-6\text{dB}}$ of $\underline{H}_C(t)$. For the Gaussian pulse in (9.49), $B_C^{-6\text{dB}}$ becomes

$$B_C^{-6\text{dB}} = \frac{\sqrt{2\ln 2}}{\pi \sigma_C}. \tag{9.51}$$

Consequently, it is possible to modify the bandwidth of the conditioned transducer by choosing σ_C appropriately.

With a view to determining a suitable excitation signal, one has to design the conditioned transducer in advance, i.e., its impulse response $h_C(t)$. Guided by the actually existing transducer pair in the experiments, the center frequency f_c equals 2.29 MHz. Theoretically, the transducer bandwidth can be chosen arbitrarily between 0 and $2f_c$. Here, let us set the bandwidth $B_C^{-6\text{dB}}$ to $1.1 f_c$, which amounts 2.52 MHz. This conditioned transducer was used to adapt a linear up-chirp signal featuring the duration $T_{\text{chirp}} = 150\,\mu\text{s}$ and the same center frequency as the transducer pair. In order to obtain the lowest side lobe level for the conditioned transducer after pulse compression (see Sect. 9.2.4), the chirp bandwidth was chosen to be $B_{\text{chirp}} = 1.14 B_C^{-6\text{dB}} = 2.87$ MHz [88]. The time–bandwidth product TB of the linear up-chirp signal equals approximately 430. Figure 9.17a shows this chirp signal, which is referred to as unconditioned excitation signal $u_I(t)$ in the following. By applying the real-valued filter $\Psi_W(f)$ from (9.47) as conditioning filter with the empirically determined parameter $\beta_W = 500$, one obtains the conditioned excitation signal $u_{IC}(t)$ in Fig. 9.17b. It is not surprising that the signal amplitudes of $u_{IC}(t)$ take higher values at the beginning and the end because $\Psi_W(f)$ compensates the transducer's bandpass behavior.

Both the unconditioned excitation signal $u_I(t)$ and the conditioned excitation signal $u_{IC}(t)$ were convolved with the experimentally acquired impulse response $h_T(t)$ of the transducer pair. To compare the resulting unconditioned interrogation signal $g_T(t)$ and conditioned interrogation signal $g_{TC}(t)$, let us regard the resulting spectral magnitudes $\left|\underline{G}_T(f)\right|$ and $\left|\underline{G}_{TC}(f)\right|$ in Fig. 9.18a. We can clearly see that $g_{TC}(t)$ owns a larger bandwidth than $g_T(t)$. Therefore, the spatial resolution of the ultrasonic measuring system should be better in case of a conditioned transducer excitation.

9.2.4　Pulse Compression

Even though coded excitation signals provide a high time–bandwidth product, their signal duration demands additional signal processing steps for the considered ultrasonic measuring system. This is demonstrated in Fig. 9.18b, which shows the measured receiver output $u_O(t)$ as well as its envelope for the conditioned excitation signal $u_{IC}(t)$ and a steel plate of 3 mm thickness. To determine plate thickness and

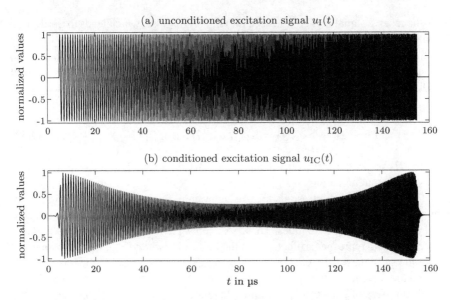

Fig. 9.17   a Unconditioned excitation signal $u_\mathrm{I}(t)$ of ultrasonic transmitter; **b** conditioned excitation signal $u_\mathrm{IC}(t)$ resulting from real-valued Wiener filter $\Psi_\mathrm{W}(f)$ (see (9.47))

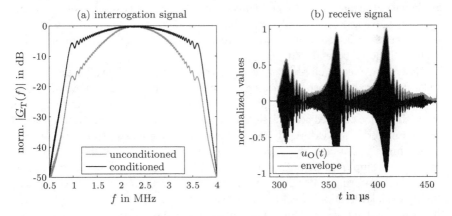

Fig. 9.18   a Spectral magnitudes $\left|\underline{G}_\mathrm{T}(f)\right|$ and $\left|\underline{G}_\mathrm{TC}(f)\right|$ of unconditioned interrogation signal $g_\mathrm{T}(t)$ and conditioned interrogation signal $g_\mathrm{TC}(t)$; **b** measured receiver output $u_\mathrm{O}(t)$ and envelope for coded transmitter excitation (i.e., conditioned chirp signal) and steel plate of 3 mm thickness

SOS simultaneously, we have to know the TOF t_R between two subsequent multiple reflections. Since t_R is much smaller than the signal duration, the multiple reflections cause a disturbing interference pattern in $u_\mathrm{O}(t)$. However, by applying appropriate decoding procedures that are commonly referred to as pulse compression, t_R can be figured out from $u_\mathrm{O}(t)$. The main idea of pulse compression lies in eliminating the phase spectrum of a frequency modulated or phase modulated signal [48]. In doing so, the receive signal gets compressed into a short impulse, which leads to the desired spatial resolution of the ultrasonic measuring system. After compression of $u_O(t)$ with the compression filter exhibiting the impulse response $h_{PC}(t)$, one obtains the pulse compression waveform $e_{PC}(t) = u_O(t) * h_{PC}(t)$ (see Fig. 9.14). The required TOF t_R can finally be identified through the envelope $e_{ev}(t)$ of $e_{PC}(t)$ that results from (cf. (7.69, p. 292))

$$e_{ev}(t) = |e_{PC}(t) + j\mathcal{H}\{e_{PC}(t)\}| \tag{9.52}$$

where $\mathcal{H}\{\cdot\}$ stands for the Hilbert operator. The time signal $e_{ev}(t)$ is termed compression output hereafter.

For the considered ultrasonic measuring system, the compression outputs do not only consist of a main peak, the so-called main lobe, at the signal arrival time (e.g., t_0) but also of undesired side lobes. These side lobes stem from the pulse compression procedure and represent artifacts. Both the half-pulse width T_P of the main lobe and the ratio of highest side lobe to main lobe in dB, termed the side lobe level (SLL), are quality characteristics of the compression output. When $T_P < t_R$ is not satisfied, the main lobes of two subsequent multiple reflections will overlap in $e_{ev}(t)$. On the other hand, a high value of SLL (e.g., SLL $= -15\,$dB) may yield an unusable compression output because side lobes might be mistaken for main lobes. Each situation is, therefore, accompanied by problems concerning the identification of t_R. Owing to this fact, we need sufficiently low values for both quantities (i.e., T_P and SLL) to determine plate thickness and SOS in a reliable way, especially if a thin plate with high SOS should be analyzed.

Now, let us take a closer look at pulse compression that is well established in radar imaging and ultrasonic imaging [19, 67]. There exist various filters for pulse compression like matched filters, mismatched filters as well as Wiener filters. Not surprisingly, the selected compression filter $h_{PC}(t)$ is decisive for the performance of pulse compression and, thus, strongly depends on the particular application. When a linear chirp signal is used as excitation signal, the fractional Fourier transform (FrFT) may also be employed for pulse compression [20, 38]. However, for the considered ultrasonic measuring system, this special kind of Fourier transform does not provide higher spatial resolution and a lower SLL than conventional compression filters. That is the reason why we will concentrate here on matched, mismatched, and Wiener filters. Without limiting the generality, the transmitter input is thereby supposed to be an unconditioned linear up-chirp signal. The same filters can be, of course, designed for any kind of coded signal.

Matched Filter

Pulse compression is often conducted by means of matched filters, which are also named correlation filters or conjugate filters. To explain the idea of matched filters, let us assume an arbitrary time signal $s(t)$ being transmitted through a channel with the known impulse response $h(t)$. For this signal, the matched filter becomes $h_{PC}^M(t) = s^*(-t)$ in the time domain and

$$\underline{H}_{PC}^M(f) = \underline{S}^*(f) = |\underline{S}(f)| \, e^{-j \arg\{\underline{S}(f)\}} \tag{9.53}$$

in the frequency domain with the frequency spectrum $\underline{S}(f)$ of $s(t)$ [75]. The operator $\arg\{\cdot\}$ leads to the argument (phase) of a complex-valued quantity. If the channel neither distorts nor delays $s(t)$, the channel's transfer function will be $\underline{H}(f) = 1$ for all frequencies and, consequently, the channel output will coincide with the input. Because the matched filter is the complex conjugate of $\underline{S}(f)$, it eliminates then the phase $\arg\{\underline{S}(f)\}$. Therefore, the frequency spectrum of the filter output takes the form

$$\underline{S}(f) \cdot \underline{H}(f) \cdot \underline{H}_{PC}^M(f) = |\underline{S}(f)|^2 . \tag{9.54}$$

In case of the considered ultrasonic measuring system, the channel's transfer function comprises the transfer behavior of both the transducer pair and the plate between them, i.e., $\underline{H}(f) = \underline{H}_T(f) \cdot \underline{H}_P(f)$. When we apply the matched filter to the receive signal $\underline{U}_O = \underline{U}_I(f) \cdot \underline{H}(f)$ for pulse compression, the frequency spectrum $\underline{E}_{PC}(f)$ of the pulse compression waveform $e_{PC}(t)$ will result with $\underline{H}_{PC}^M(f) = \underline{U}_I^*(f)$ in

$$\begin{aligned} \underline{E}_{PC}(f) &= \underline{U}_O(f) \cdot \underline{H}_{PC}^M(f) \\ &= \underline{U}_I(f) \cdot \underline{H}(f) \cdot \underline{H}_{PC}^M(f) \\ &= |\underline{U}_I(f)|^2 \cdot |\underline{H}(f)| \, e^{j \arg\{\underline{H}(f)\}} . \end{aligned} \tag{9.55}$$

Accordingly, the phase characteristic of $\underline{E}_{PC}(f)$ is solely specified by $\underline{H}(f)$. The present channel consisting of transducer pair and plate introduces a time delay between $u_I(t)$ and $u_O(t)$, which corresponds to a linear phase of $\underline{H}(f)$. Since $\underline{E}_{PC}(f)$ contains $\arg\{\underline{H}(f)\}$, one is, thus, able to recover the time delay by means of the matched filter. In other words, this kind of pulse compression should allow TOF measurements. However, any other linear or nonlinear phase distortions of the entire transmission system will also remain in the pulse compression waveform. Such phase distortions are mainly generated by the ultrasonic transducers. To mitigate the resulting deviations in TOF measurements, it makes, therefore, sense to design the matched filter on basis of the interrogation signal $g_T(t)$ of the ultrasonic measuring system, i.e., $h_{PC}^M(t) = g_T^*(-t)$. The pulse compression will then eliminate disturbing phase distortions because they are included in $g_T(t)$. The problem herein lies in the fact that we have to know $g_T(t)$ in advance. Both these options are used in practical scenarios.

Mismatched Filter

As previously mentioned, the parameters T_P and SLL are decisive quality characteristics of the compression output $e_{ev}(t)$ for the considered ultrasonic measuring system. Generally speaking, if T_P is reduced, the SLL will increase and vice versa. With a view to illustrating this behavior, let us regard a rectangular function representing an ideal bandpass in the frequency domain. A rectangular function with infinitely sharp edges yields a sinc function in the time domain [12]. Whereas such a function offers a small T_P, its SLL equals ≈ -13 dB, i.e., a rather high value (see Fig. 9.19b). When

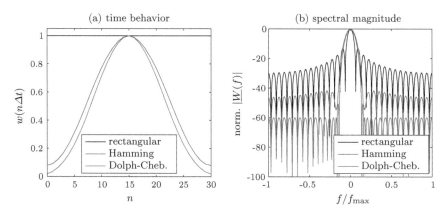

Fig. 9.19 **a** Discrete-time functions $w(n\Delta t)$ of rectangular, Hamming and Dolph–Chebyshev window with length of 31 samples; Dolph–Chebyshev designed for SLL = 60 dB; sample index n; sampling time Δt; **b** corresponding spectral magnitudes $|\underline{W}(f)|$ in frequency domain; frequency axis normalized to maximum f_{max}

the compression output coincides with a sinc function, the ultrasonic measuring system will, consequently, feature an outstanding spatial resolution, but the separation of two subsequent multiple reflections in $e_{ev}(t)$ might be problematic due to side lobes.

To reduce the SLL, we have to conduct smoothing of the band edges. In practice, edge smoothing can be achieved either by applying an appropriate filter in the frequency domain or by windowing, which means that a suitable window function is used in the time domain [48]. Actually, there exist various time-limited, even as well as real-valued window functions $w(t)$ like the Hamming window (see Fig. 9.19), whose frequency spectrum $W(f)$ is also real-valued and exhibits narrow main lobes as well as low side lobes [39]. Due to the duality of the Fourier transform (see Fig. 9.20), the same holds for a frequency-limited, real-valued, and even window functions $w(f)$ in the frequency domain. The resulting time signals $W(t)$ exhibit again narrow main lobes as well as low side lobes. Here, we are dealing with linear chirp signals of a high time–bandwidth product (i.e., $TB \gg 1$) that have spectral magnitudes with a shape of the signal's envelope [68]. Edge smoothing can, thus, be conducted either in the time or frequency domain by a suitable window. However, it is usually more convenient for practical reasons (e.g., filter implementation) to perform windowing in the time domain. Because the Dolph–Chebyshev window is optimal to maintain spatial resolution while reducing the SLL, we will exploit this window for the considered ultrasonic measuring system.

From the system point of view, windowing by $w(t)$ can be applied to each part of the transmission line, e.g., to the transducer excitation $u_I(t)$ before transmitting. If the windowing is performed in the course of pulse compression that relates to $u_I(t)$, the compression filter will change to

Fig. 9.20 Duality of Fourier transform; Fourier transform $\mathcal{F}$ from time to frequency domain; inverse Fourier transform $\mathcal{F}^{-1}$ from frequency to time domain

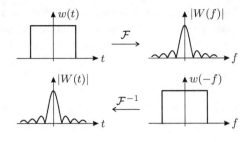

$$h_{\text{PC}}^{\text{MM}}(t) = u_{\text{I}}^*(-t) \cdot w(t) \tag{9.56}$$

in the time domain. The resulting filter with the impulse response $h_{\text{PC}}^{\text{MM}}(t)$ is commonly named mismatched filter. By additionally applying $w(t)$ to $u_{\text{I}}(t)$ or $g_{\text{I}}(t)$, one can reduce the SLL in the compression output $e_{\text{ev}}(t)$ further.

Wiener Filter

As an alternative to matched and mismatched filters, pulse compression can be carried out with deconvolution. The considered ultrasonic measuring system aims at reconstructing the impulse response $h_{\text{P}}(t)$ of the investigated plate from the receiver output $u_{\text{O}}(t)$ to determine plate thickness and SOS. By utilizing the interrogation signal $g_{\text{T}}(t) = u_{\text{I}}(t) * h_{\text{T}}(t)$, the receiver output is given by $u_{\text{O}}(t) = g_{\text{T}}(t) * h_{\text{P}}(t)$ (cf. (9.40)). For this situation, the conventional deconvolution that is also named inverse filtering reads as

$$h_{\text{P}}(t) = u_{\text{O}}(t) * g_{\text{T}}^{-1}(t) \quad \overset{\mathcal{F}}{\rightarrow} \quad H_{\text{P}}(f) = \frac{U_{\text{O}}(f)}{G_{\text{T}}(f)} \tag{9.57}$$

in the time and frequency domain, respectively. The expressions $H_{\text{P}}(f)$, $G_{\text{T}}(f)$ and $U_{\text{O}}(f)$ stand again for the corresponding frequency spectra. According to Sect. 9.2.3, ultrasonic transducers typically show bandpass characteristic, which also arises in $g_{\text{T}}(t)$. The deconvolution by means of the inverse filter, therefore, becomes unstable (cf. (9.45)). However, when a Wiener filter is exploited instead, we will achieve a stable deconvolution with best possible noise suppression. The Wiener filter for pulse compression takes the form (cf. (9.46))

$$H_{\text{PC}}^{\text{W}}(f) = \frac{G_{\text{T}}^*(f)}{|G_{\text{T}}(f)|^2 + 1/\text{SNR}(f)} \tag{9.58}$$

where $\text{SNR}(f)$ rates the frequency-dependent signal-to-noise ratio[2] of $G_{\text{T}}(f)$. As requested for pulse compression, the numerator eliminates the phase spectrum of the interrogation signal because $G_{\text{T}}(f) \cdot G_{\text{T}}^*(f) = |G_{\text{T}}(f)|^2$ holds. The Wiener filter

---

[2] The signal-to-noise ratio $\text{SNR}(f)$ is the reciprocal of the noise-to-signal ratio $\text{NSR}(f)$, i.e., $\text{SNR}(f) = 1/\text{NSR}(f)$.

behaves similar to the inverse filter for noiseless spectral components but like the matched filter where signal energy is low. Consequently, the bandwidth of the filtered signal increases as it is the case in inverse filtering. A well-designed Wiener filter, therefore, improves the resolution of a imaging system compared to conventional matched filtering.

The difficulty arising when implementing a Wiener filter lies in the fact that we have to know $1/\mathrm{SNR}(f) = \mathrm{NSR}(f)$ or at least an appropriate estimate hereof. Since the actual $\mathrm{SNR}(f)$ is usually unknown, it is commonly supposed to be constant over frequency. The relation

$$1/\mathrm{SNR}_0 = \mathrm{NSR}_0 = \beta_\mathrm{W} \left| \underline{G}_\mathrm{T}(f) \right|^2_{\mathrm{max}} \qquad (9.59)$$

with the factor $\beta_\mathrm{W} = 10^{-2}$ and the maximum $\left| \underline{G}_\mathrm{T}(f) \right|_{\mathrm{max}}$ of the spectral magnitudes represents an estimate, which is often used in practical applications [41, 71]. For the realized ultrasonic measuring system, $1/\mathrm{SNR}_0 = \mathrm{NSR}_0$ was determined empirically.

The Wiener filter tends to produce a pulse compression waveform $e_\mathrm{PC}(t)$ featuring approximately a rectangular shape of the spectral magnitudes $\left| \underline{E}_\mathrm{PC}(f) \right|$. Sharp edges of $\left| \underline{E}_\mathrm{PC}(f) \right|$ will, however, be accompanied by a high SLL of the resulting compression output $e_\mathrm{ev}(t)$, which leads to problems concerning separation of multiple reflections. Hence, one should perform edge smoothing with the aid of an appropriate window function $w(f)$ in the frequency domain [48]. For this purpose, it is advisable to apply again a frequency-shifted as well as stretched Dolph–Chebyshev window that overlaps exactly with the bandwidth of the excitation signal $u_\mathrm{I}(t)$. The adapted Wiener filter for pulse compression finally becomes

$$\underline{H}_\mathrm{PC}^\mathrm{WW}(f) = \frac{w(f) \cdot \underline{G}_\mathrm{T}^*(f)}{\left| \underline{G}_\mathrm{T}(f) \right|^2 + \mathrm{NSR}_0} \qquad (9.60)$$

in the frequency domain. The window function leads to a band-limited as well as stable deconvolution. Owing to this fact, we will be able to omit NSR_0 in (9.60) when $w(f)$, especially its bandwidth, is chosen appropriately. The resulting pulse compression filter could then be interpreted as band-limited inverse filter.

9.2.5   Experiments

In this subsection, the main parts of the realized measurement setup will be explained. We will, furthermore, discuss the resulting axial point spread functions of the realized ultrasonic measuring system for different transmitter excitations and pulse compression filters. At the end, measurement results (i.e., d_P and c_P) for various plate thicknesses as well as plate materials are shown and compared to reference values.

Fig. 9.21 Main parts of measurement setup consisting of piston-type ultrasonic transmitter and receiver (Olympus V306) being immersed in water; investigated plates were fixed by special mount

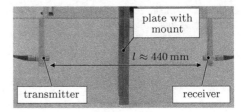

Measurement Setup

Figure 9.21 illustrates the main parts of the measurement setup, which was realized at the Chair of Sensor Technology (Friedrich-Alexander-University Erlangen-Nuremberg). Two identical piston-type transducers (Olympus V306 [77]) that were immersed in water served as ultrasonic transmitter and ultrasonic receiver. The piezoelectric transducers with the active element diameter $2R_T = 12.7$ mm have a nominal center frequency of $f_c = 2.25$ MHz and provide the -6 dB bandwidth $B_s^{-6dB} = 1.38$ MHz in pulse-echo mode, i.e., the fractional bandwidth equals 61.5%. In order to avoid near field effects, the geometric distance l between both transducers was chosen to exceed the near field distance $N_{near} \approx R_T^2/\lambda_{aco} \approx 60$ mm. By means of a power amplifier, the ultrasonic transmitter was excited with a signal amplitude $\hat{u}_I$ up to 50 V. After the generated sound pressure waves have propagated through water path as well as investigated flat plate, they reach the ultrasonic receiver. The receiver output $u_O(t)$ was acquired by a digital storage oscilloscope (Tektronix DPO 7104C [100]) at the sampling frequency of 200 MHz. Averaging of 50 recorded waveform ensured a reasonable signal-to-noise ratio of $u_O(t)$.

As discussed in Sect. 9.2.1, the simultaneous determination of plate thickness d_P and SOS c_P inside the plate requires a reference measurement without plate. The reference measurement needs to be done only once before investigating plates and yields the SOS c_W in water as well as the TOF t_W of ultrasonic waves propagating from transmitter to receiver. For this purpose, the ultrasonic receiver was mounted on a linear translation axes (Physik Instrumente M-531.DG [81]) that allows precise variation of the geometric distance l between both transducers. By evaluating the time difference Δt_W for distinct changes Δl of l, one can measure the SOS inside water with the aid of $c_W = \Delta l/\Delta t_W$ in a reliable manner. It is, of course, convenient to apply the same excitation signals and signal processing steps for the reference measurement as for the subsequent plate measurements. In the present case, the resulting quantities amounted $c_W = 1488$ m s^{-1} and $t_W = 294.5$ μs, which leads to the geometric distance $l \approx 440$ mm of the ultrasonic transducers, i.e., $l \gg N_{near}$ holds.

Axial Point Spread Function

The point spread function (PSF) denotes generally a decisive quantity of imaging systems since it provides information about the achievable spatial resolution of the system [14, 40, 99]. For the considered ultrasonic measuring system, the axial PSF will equal the waveform resulting from signal processing when only the directly

Fig. 9.22 Normalized axial point spread functions for pulse-shaped transmitter excitation and for conditioned chirp excitation signal $u_{IC}(t)$ after pulse compression with mismatched filter (MM) $h_{PC}^{MM}(t) = u_I(-t) \cdot w(t)$ and adapted Wiener filter (WW) $\underline{H}_{PC}^{WW}(f)$

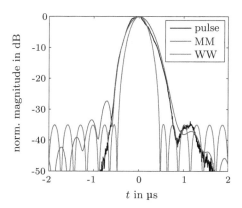

transmitted signal $p_{0\sim}(t)$ arises at the receiver, i.e., without a plate between transmitter and receiver. This corresponds to the compression output $e_{ev}(t)$ of an infinitely thin plate. Owing to the fact that the compression output does not contain any multiple reflections, we can directly deduce both the achievable spatial resolution T_P and the SLL from $e_{ev}(t)$.

Figure 9.22 shows three axial point spread functions of the realized measurement setup. The first PSF refers to the envelope of the system's impulse response $h_T(t)$ (i.e., the impulse response of the transducer pair) that results from pulse-shaped transmitter excitation, which is conventionally exploited for ultrasonic measurements. While the second PSF relates to pulse compression by means of the mismatched filter $h_{PC}^{MM}(t) = u_I(-t) \cdot w(t)$ (cf. (9.56)), the adapted Wiener filter $\underline{H}_{PC}^{WW}(f)$ was applied for the third PSF (cf. (9.60)). In both cases, side lobe level reduction was conducted with the aid of a Dolph–Chebyshev window $w(t)$ and $w(f)$ targeting SLL $= -35$ dB. Moreover, the conditioned chirp signal $u_{IC}(t)$ served as transmitter excitation for the second and third PSF. The comparison of the three PSFs reveals that the adapted Wiener filter leads to the narrowest main lobe and, thus, to the best spatial resolution of the ultrasonic measuring system. This result can be ascribed to the filter characteristic of $\underline{H}_{PC}^{WW}(f)$, which corresponds to an inverse filter within the bandwidth of the excitation signal. Apart from that, the Wiener filter provides the lowest SLL due to the utilized Dolph–Chebyshev window $w(f)$. In contrast to the mismatch filter that was designed on basis of the unconditioned chirp signal $u_I(t)$, the resulting PSF of the Wiener filter is symmetric because $\underline{H}_{PC}^{WW}(f)$ was designed on basis of the interrogation signal $g_T(f)$.

Besides the mentioned combinations of excitation signal and pulse compression filter for the realized ultrasonic measuring system, there exist various other combinations that, of course, yield different PSFs and, consequently, different values for T_P and SLL. This follows from the performed conditioning of the transmitter excitation (see Sect. 9.2.3). Since the real-valued conditioning filter $\Psi_W(f)$ (cf. (9.47)) exclusively modifies the spectral magnitudes of the unconditioned chirp signal, pulse compression filters can be designed on basis of both conditioned and unconditioned

Table 9.1 Comparison of spatial resolution T_P and highest side lobe level SLL for different combinations of excitation signal (i.e., either unconditioned excitation $u_I(t)$ or conditioned excitation $u_{IC}(t)$) and pulse compression filter; values refer to measured PSFs

Excitation	Compression filter	T_P/δ_P	SLL in dB
Pulse	–	1	−33.9
$u_I(t)$	Matched: $h_{PC}^M(t) = u_I(-t)$	1.06	−28.0
	Mismatched: $h_{PC}^{MM}(t) = u_I(-t) \cdot w(t)$	1.32	−34.1
	Adapted Wiener: $\underline{H}_{PC}^{WW}(f)$	0.80	−34.8
$u_{IC}(t)$	Matched: $h_{PC}^M(t) = u_{IC}(-t)$	0.66	−7.5
	Mismatched: $h_{PC}^{MM}(t) = u_{IC}(-t) \cdot w(t)$	0.86	−14.5
	Mismatched: $h_{PC}^{MM}(t) = u_I(-t)$	0.83	−14.5
	Mismatched: $h_{PC}^{MM}(t) = u_I(-t) \cdot w(t)$	1.09	−27.3
	Mismatched: $h_{PC}^{MM}(t) = g_T(-t)$	0.95	−29.4
	Mismatched: $h_{PC}^{MM}(t) = g_T(-t) \cdot w(t)$	0.99	−30.6
	Adapted Wiener: $\underline{H}_{PC}^{WW}(f)$	0.80	−34.8

excitation signal. Table 9.1 contains the measured values for selected combinations. Note that the spatial resolutions T_P have been normalized to $\delta_P = 0.59\,\mu s$, which represents the spatial resolution achievable by pulse-shaped excitation of the employed ultrasonic transducers.

In a first step, let us take a closer look at the expected spatial resolution in the measured PSFs. The unconditioned chirp signal $u_I(t)$ with the mismatched filter $h_{PC}^{MM}(t)$ exhibits the worst spatial resolution, whereas the conditioned chirp signal $u_{IC}(t)$ with the matched filter $h_{PC}^M(t)$ enables the best resolution. From there, one could conclude that the second combination should be optimal for the realized ultrasonic measuring system. However, the very high SLL of −7.5 dB causes remarkable problems in separating multiple reflections from directly transmitted signals, which is absolutely necessary to determine plate thickness and SOS. It is, furthermore, noticeable that $u_{IC}(t)$ in combination with mismatched filtering does not substantially improve the spatial resolution although the obtained resolutions are always better than in case of $u_I(t)$. Nevertheless, the SNR value of the compression output increases compared with pulse-shaped excitation since the time–bandwidth product TB takes high values. The table entries also demonstrate that regardless of matched or mismatched filtering, $u_I(t)$ always offers a better SLL in the PSF than $u_{IC}(t)$. This fact originates from the opposite behavior of conditioning filter $\Psi_W(f)$ and window function $w(t)$ for SLL reduction. While $\Psi_W(f)$ tends to enhance the band edges, $w(t)$ is designed for smoothing them. As a last aspect, it should be mentioned that pulse compression by the adapted Wiener filter $\underline{H}_{PC}^{WW}(f)$ is almost independent of the excitation signal because such filter acts as an inverse filter in the frequency band of $w(f)$. Compared to the other combinations of excitation signal and pulse compression filter, $\underline{H}_{PC}^{WW}(f)$ produces the best values for both T_P and SLL.

Fig. 9.23 Normalized
compression output $e_{ev}(t)$
for three steel plates of
different thicknesses d_P;
pulse compression conducted
with mismatched fil-
ter (MM) $h_{PC}^{MM}(t) = u_I(-t) \cdot w(t)$
and adapted Wiener
filter (WW) $\underline{H}_{PC}^{WW}(f)$

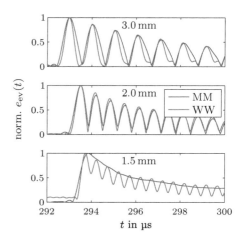

Measurement Results

The realized measurement setup was utilized to analyze flat plates of different thicknesses and materials, e.g., steel. Before the determined values for plate thickness d_P
and SOS c_P are compared to reference values, let us discuss the resulting compression outputs $e_{ev}(t)$ of steel plates differing in thickness. Figure 9.23 depicts $e_{ev}(t)$
for the nominal plate thicknesses 3.0, 2.0, and 1.5 mm. Just as in Fig. 9.22, the
conditioned chirp signal $u_{IC}(t)$ served as transmitter excitation. Pulse compression
was conducted either through the mismatched filter $h_{PC}^{MM}(t) = u_I(-t) \cdot w(t)$ or the
adapted Wiener filter $\underline{H}_{PC}^{WW}(f)$. For simultaneous determination of d_P and c_P, it is
indispensable that directly transmitted waves as well as multiple reflections cause
pronounced maxima in $e_{ev}(t)$. Both compression filters enable separation of these
maxima for $d_P = 3.0$ mm as well as $d_P = 2.0$ mm. However, in case of the thinnest
steel plate (i.e., $d_P = 1.5$ mm), the mismatched filter does not allow maxima identification and, therefore, d_P as well as c_P cannot be calculated. By contrast, the compression output resulting from the adapted Wiener filter contains separable maxima.

To actually measure plate thickness d_P and SOS c_P by means of the realized ultrasonic measuring system, we have to figure out the TOFs t_0 and t_R in the compression
output $e_{ev}(t)$. The maxima in the normalized $e_{ev}(t)$, which contain this information,
were found algorithmically. With a view to avoiding side lobe detection, normalized
maxima were only taken into account if they exceeded 0.15. The TOF t_R was estimated by computing the mean value of the time differences between all successive
maxima. Together with the parameters c_W and t_W, the aimed quantities d_P as well
as c_P can finally be determined simultaneously from (9.25) and (9.26).

Figure 9.24a and b show the obtained results for d_P and c_P, respectively. Overall,
seven plates of the dimension 250 mm × 150 mm × d_P were investigated. The plates
were made of steel, aluminum or poly(methyl methacrylate) (PMMA). Again, the
mismatched filter $h_{PC}^{MM}(t) = u_I(-t) \cdot w(t)$ and the adapted Wiener filter $\underline{H}_{PC}^{WW}(f)$
were applied for pulse compression. The given percentage values relate to the relative

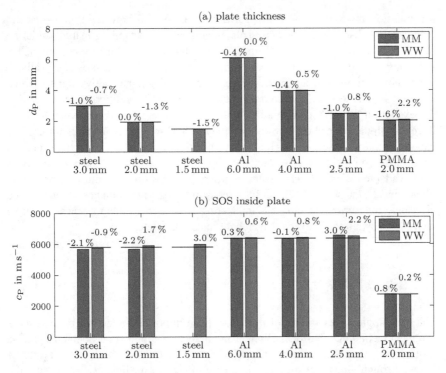

Fig. 9.24 Comparison of determined **a** plate thickness d_P and **b** SOS c_P inside plate for different plate materials and plate thicknesses; pulse compression conducted with mismatched filter (MM) $h_{PC}^{MM}(t) = u_I(-t) \cdot w(t)$ and adapted Wiener filter (WW) $\underline{H}_{PC}^{WW}(f)$; relative deviations from reference values as percentage values above bares

deviations from reference values, which are displayed as horizontal lines. While a micrometer screw served as reference for d_P, the reference values for c_P were determined by standard TOF measurements (see Fig. 5.8d on p. 138) with a piezo-electric contact transducer that generates longitudinal waves inside the investigated plates. The contact transducer (Olympus V112 [77]) with the active element diameter $2R_T = 6.4$ mm offers a nominal center frequency of $f_c = 10$ MHz and, thus, provides small wavelengths. Consequently, the relative deviations from the reference values can be interpreted as relative measurement error of the ultrasonic measuring system. With the exception of the thinnest steel plate (i.e., $d_P = 1.5$ mm), the combination of conditioned chirp excitation $u_{IC}(t)$ and mismatched filter leads to reliable measurement results for d_P as well as c_P. The Wiener filter additionally allows determination for this steel plate because maxima in $e_{ev}(t)$ can still be separated (cf. Fig. 9.23). For both compression filters, the maximum relative deviations of d_P and c_P from the reference take values within $\pm 3\%$.

In summary, we are able to simultaneously determine plate thickness and speed of sound by the considered ultrasonic measuring system that is based on the through-transmission mode, i.e., the investigated plate is located between transmitter and receiver. The characterization of thin plates calls for an outstanding spatial resolution of the measuring system, which goes hand in hand with a high center frequency and a large bandwidth of the used ultrasonic transducers. To exploit the available bandwidth best possible and achieve a reasonable SNR in the receive signals, linear up-chirp signals served as coded transmitter excitation. By additionally conditioning the excitation signal before sending, one can enhance spectral components at the lower and upper edge of the transducer bandwidth. However, owing to their duration, coded excitation signals always require pulse compression approaches. In the present case, pulse compression was conducted by matched filters, mismatched filters as well as adapted Wiener filters. As the comparison of the different pulse compression filters reveals, the adapted Wiener filter leads to the best performance of the realized ultrasonic measuring system. It is interesting to note that plate thicknesses can be measured down to 60% of the wavelength inside the plate.

9.3   Fluid Flow

The metrological registration of fluid flow is an important branch of process measurement technology. Various technical and industrial applications demand a precise acquisition of mass flow rates and volumetric flow rates of fluids (i.e., liquids or gases) through pipes. For instance, it is very important for industry to measure the oil quantity, which is flowing through pipelines and for drivers to be aware of the obtained fuel at petrol stations. Moreover, the amount of consumed water represents a decisive quantity for private households. Owing to the great variety of applications and large value range of flow rates, there exists a remarkable number of different measurement principles [30, 106]. Basically, one can distinguish between direct and indirect measurement methods for mass flow rates and volumetric flow rates. As the names suggest, direct methods provide immediately the mass flow rates or volumetric flow rates through pipes, while indirect methods require the pipe cross section to determine the desired quantities. Direct measurement methods include displacement flow meters and coriolis flow meters. In contrast, turbine flow meters, vortex shedding flow meters, magnetic flow meters as well as ultrasonic flow meters belong to the group of indirect measurement methods.

In the following, let us concentrate on ultrasonic flow meters since these flow meters usually exploit piezoelectric transducers for transmitting and receiving ultrasonic waves. Before the fundamental measurement principles of ultrasonic flow meters are studied in Sect. 9.3.2, we will discuss decisive physical quantities in fluid flow measurements. Section 9.3.3 addresses typical arrangements of transmitters and receivers for ultrasonic flow meters. Finally, a modeling approach will be presented which enables efficient simulation of clamp-on ultrasonic flow meters.

9.3.1 Fundamentals of Fluid Flow Measurements

In this subsection, let us explain decisive physical quantities in fluid flow measurements as well as fundamental relations for ultrasonic flow meters. The mass flow rate $\dot{m}_F(t)$ (unit kg s^{-1}) represents such a physical quantity because it measures the mass $m_F(t)$ of a fluid, which passes per unit of time t. Therefore, $\dot{m}_F(t)$ is defined as

$$\dot{m}_F(t) = \lim_{\Delta t \to 0} \frac{\Delta m_F(t)}{\Delta t} = \frac{dm_F(t)}{dt} . \tag{9.61}$$

If we assume a uniform fluid of constant density ϱ_0 that flows in one direction through a circular pipe (inner cross section A_P) with the flow rate $v_F(r, t)$ depending on both radial position r inside the pipe and time, $\dot{m}_F(t)$ will result from

$$\dot{m}_F(t) = \varrho_0 \int_{A_P} v_F(r, t) \, dA . \tag{9.62}$$

Besides the mass flow rate, the volumetric flow rate $\dot{V}_F(t)$ (volume flow rate; unit m^3 s^{-1}) denotes a decisive physical quantity in fluid flow measurements. It is given by

$$\dot{V}_F(t) = \lim_{\Delta t \to 0} \frac{\Delta V_F(t)}{\Delta t} = \frac{dV_F(t)}{dt} = \int_{A_P} v_F(r, t) \, dA \tag{9.63}$$

and, thus, measures the fluid volume inside the pipe, which passes per unit of time with $v_F(r, t)$. For a uniform fluid of constant density, there exists a simple connection of volumetric flow rate and mass flow rate according to $\dot{m}_F(t) = \varrho_0 \dot{V}_F(t)$.

Common ultrasonic flow meters do not provide the spatially resolved flow rate $v_F(r, t)$ of a fluid along the pipe cross section. Actually, one can mostly just determine the average flow rate $\bar{v}_F(t)$, which indicates the average value of $v_F(r, t)$ along the path of propagating ultrasonic waves (see Sect. 9.3.2). Before this fact is discussed in detail, it makes sense to introduce as a further quantity the average area velocity $\bar{v}_A(t)$ of the pipe cross section A_P. The volumetric flow rate $\dot{V}_F(t)$ and $\bar{v}_A(t)$ are linked by

$$\dot{V}_F(t) = \bar{v}_A(t) \, A_P . \tag{9.64}$$

By inserting (9.63), $\bar{v}_A(t)$ then becomes

$$\bar{v}_A(t) = \frac{\dot{V}_F(t)}{A_P} = \frac{1}{A_P} \int_{A_P} v_F(r, t) \, dA = \frac{2}{R_{Pi}^2} \int_0^{R_{Pi}} v_F(r, t) \, r \, dr \tag{9.65}$$

with the inner radius R_{Pi} of the circular pipe. In contrast to the average area velocity, the average flow rate $\bar{v}_F(t)$ being defined as

$$\bar{v}_F(t) = \frac{1}{R_{\text{Pi}}} \int\limits_0^{R_{\text{Pi}}} v_F(r, t)\,dr \tag{9.66}$$

does not directly lead to the volumetric flow rate $\dot{V}_F(t)$ within the pipe. To handle this, one should take a closer look at the ratio k_v of both velocities, which takes the form

$$k_v = \frac{\bar{v}_A(t)}{\bar{v}_F(t)} = \frac{2}{R_{\text{Pi}}} \frac{\int\limits_0^{R_{\text{Pi}}} v_F(r, t)\,r\,dr}{\int\limits_0^{R_{\text{Pi}}} v_F(r, t)\,dr} \tag{9.67}$$

and is always smaller than 1. When k_v is known, we can calculate $\dot{V}_F(t)$ from $\bar{v}_F(t)$ with the aid of

$$\dot{V}_F(t) = k_v \bar{v}_F(t)\,A_P\ . \tag{9.68}$$

The problem lies, however, in the fact that the ratio k_v strongly depends on the flow profile of the fluid [36, 62]. In the case of an inline ultrasonic flow meter (see Sect. 9.3.3), one can calibrate the measurement system in advance since the arrangement at the measuring point (transducers and pipe) is well known. The aim is to determine a relation between measured average flow rate $\bar{v}_F(t)$ and volumetric flow rates $\dot{V}_F(t)$. By contrast, we usually do not exactly know the arrangement in case of clamp-on ultrasonic flow meters that represent a noninvasive configuration. It should be, nevertheless, possible to estimate k_v if the flow profile inside the pipe is known. Because the flow profile itself depends on the average flow rate $\bar{v}_F(t)$ of the fluid, we have to calculate k_v iteratively during flow measurement [84]. The actual existing flow profile gets approximated while doing so. In this context, the so-called *Reynolds number Re*, which denotes an important dimensionless quantity in fluid mechanics [25], plays a significant role. For the flow inside a circular pipe, the Reynolds number is defined as[3]

$$Re = \frac{\varrho_0 \bar{v}_F D_{\text{Pi}}}{\eta} \tag{9.69}$$

with the inner diameter $D_{\text{Pi}} = 2R_{\text{Pi}}$ of the pipe and the dynamic viscosity η of the fluid.

Table 9.2 contains the underlying mathematical relations of the spatially resolved flow rate $v_F(r)$ as well as the resulting ratio k_v for three different characteristic flow profiles that are often used to approximate the actual flow profile inside a circular

---

[3]For compactness, the argument time t is omitted hereinafter.

Table 9.2 Fluid velocity $v_F(r)$ inside pipe with respect to radial position r; ratio $k_v = \overline{v}_A/\overline{v}_F$ of average area velocity $\overline{v}_A$ and average flow rate $\overline{v}_F$; inner radius R_{Pi} of pipe; maximum v_{max} of $v_F(r)$; parameter n_v depends on Reynolds number Re and differs for the velocity profiles

Velocity profile	Fluid velocity $v_F(r)$	Ratio k_v
Power law	$v_{max}\left[1 - \dfrac{r}{R_{Pi}}\right]^{\frac{1}{n_v}}$	$\dfrac{2n_v}{2n_v + 1}$
Logarithmic	$v_{max}\left[1 + n_v \ln\left(1 - \dfrac{r}{R_{Pi}}\right)\right]$	$\dfrac{1 - \frac{3}{2}n_v}{1 - n_v}$
Parabolic	$v_{max}\left[1 - \left(\dfrac{r}{R_{Pi}}\right)^{2n_v}\right]$	$\dfrac{2n_v + 1}{2(n_v + 1)}$

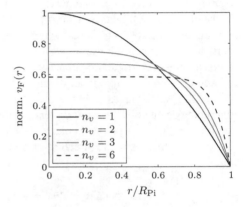

Fig. 9.25 Normalized parabolic velocity profiles $v_F(r)$ inside pipe of inner radius R_{Pi} with respect to parameter n_v (see Table 9.2); each velocity profile features identical volumetric flow rate $\dot{V}_F$, i.e., identical average area velocity $\overline{v}_A$

pipe. The flow profiles are named (i) power law velocity profile, (ii) logarithmic velocity profile, and (iii) parabolic velocity profile [36]. Each of these velocity profiles depends on the Reynolds number Re since the parameter n_v in the mathematical relations is a function of Re. Obviously, n_v modifies the three velocity profiles in a different way and, thus, it does not make any sense to compare them for the same parameter value.

The parabolic velocity profile with $n_v = 1$ describes the flow profile for laminar flow inside a pipe, which means that the fluid flows in parallel layers without any disruption between the layers [25]. A laminar flow will occur if $Re < 2300$ holds, whereas a turbulent flow is present for $Re > 4000$. Such a turbulent flow implies chaotic changes in pressure as well as velocity and corresponds to a parabolic velocity profile with $n_v > 5$. When a fluid flow exhibits a Reynolds number from 2300 to 4000, it is usually named transition flow. Figure 9.25 displays normalized parabolic velocity profiles inside a pipe for different values of n_v.

9.3.2   *Measurement Principles of Ultrasonic Flow Meters*

In this subsection, we will explain the underlying measurement principles of transit time as well as Doppler ultrasonic flow meters. These two types of flow meters are frequently employed in technical applications that use ultrasound for flow measurements of fluids, i.e., liquids and gases. Moreover, we will briefly study speckle-tracking ultrasonic flow meters, which enable determination of velocity profiles inside pipes. Further types of ultrasonic flow meters such as the tag cross-correlation flow meter are discussed in, e.g., [3, 62].

Transit Time Flow Meters

Ultrasonic flow meters are mostly based on the so-called *windfall effect* for propagating sound waves, i.e., one makes use of the fact that sound waves travel faster in the direction of flow than against [30, 36]. The utilization of this windfall effect is particularly suitable in fluids, which contain only a small number of scattering particles for incident ultrasonic waves. For illustration of the measuring principle, let us assume a configuration consisting of two ultrasonic transducers (T1 and T2) as well as a circular pipe (inner diameter D_{Pi}) through which a homogeneous fluid is flowing. As displayed in Fig. 9.26a, the connecting line of both transducers and the pipe axis enclose the angle β_F and, therefore, this connecting line exhibits the angle $\alpha_F = 90° - \beta_F$ perpendicular to the pipe axis. According to the windfall effect, the sound velocities in the direction of flow and against will become (average flow rate $\bar{v}_F$)

$$c_{down} = c_F + \bar{v}_F \cos \beta_F = c_F + \bar{v}_F \sin \alpha_F = c_F + v_\parallel \qquad (9.70)$$

$$c_{up} = c_F - \bar{v}_F \cos \beta_F = c_F - \bar{v}_F \sin \alpha_F = c_F - v_\parallel \qquad (9.71)$$

if the sound velocity in the resting fluid (i.e., without any flow; $\bar{v}_F = 0$) amounts c_F. The expressions c_{down} and c_{up} denote the downstream and upstream sound velocity, respectively. The terms $\pm\bar{v}_F \cos \beta_F = \pm\bar{v}_F \sin \alpha_F = \pm v_\parallel$ in (9.70) and (9.71) indicate the flow-induced changes in these sound velocities (see Fig. 9.26b). Not surprisingly, the difference of c_{down} and c_{up} will take its maximum for a given average flow rate when $\beta_F = 0°$ (i.e., $\alpha_F = 90°$) holds.

In order to acquire $\bar{v}_F$, we have to determine the sound velocities c_{down} as well as c_{up}. This can be done by evaluating the transit times t_{down} and t_{up} of sound waves propagating in the direction of flow and against, respectively. The transit time t_{down} will arise when T1 serves as transmitter and T2 as receiver. In the other case (i.e., transmitter T2 and receiver T1), it is possible to figure out t_{up}. For the considered configurations, the sound waves cover in both directions the geometric distance

$$L_F = \frac{D_{Pi}}{\sin \beta_F} = \frac{D_{Pi}}{\cos \alpha_F} \qquad (9.72)$$

(a) **(b)**

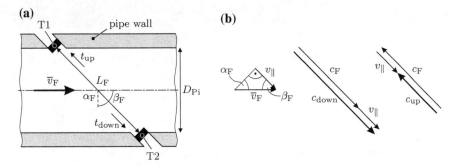

Fig. 9.26 **a** Measurement principle of transit time ultrasonic flow meters; transit times t_{down} and t_{up} differ in downstream and upstream direction; ultrasonic transducers T1 and T2; **b** component $v_\parallel$ of average flow rate $\bar{v}_F$ causes difference of sound velocities c_{down} and c_{up} in downstream and upstream direction

inside the fluid. Consequently, the transit times $t_{\text{down;up}} = L_F/c_{\text{down;up}}$ of the propagating sound waves result from[4]

$$t_{\text{down}} = \frac{D_{Pi}}{c_{\text{down}} \cos \alpha_F} + t_{\text{delay}} \quad \text{and} \quad t_{\text{up}} = \frac{D_{Pi}}{c_{\text{up}} \cos \alpha_F} + t_{\text{delay}} . \tag{9.73}$$

Here, t_{delay} stands for a constant time delay, which does not depend on $\bar{v}_F$ and comprises sound propagation times within the ultrasonic transducers as well as further delays of the obtained electrical signals. By computing the time difference Δt_F of both transit times, t_{delay} cancels out and, therefore, does not affect the identified value of $\bar{v}_F$. The time difference takes the form

$$\Delta t_F = t_{\text{up}} - t_{\text{down}} = \frac{D_{Pi}}{c_{\text{up}} \cos \alpha_F} - \frac{D_{Pi}}{c_{\text{down}} \cos \alpha_F}$$
$$= \frac{D_{Pi}}{\cos \alpha_F} \frac{c_{\text{down}} - c_{\text{up}}}{c_{\text{down}} c_{\text{up}}} , \tag{9.74}$$

which can be directly converted to the phase difference $\Delta \varphi_F = 2\pi f \Delta t_F$ in case of a known ultrasonic frequency f. If c_{down} and c_{up} are replaced by (9.70) and (9.71), (9.74) will read as

$$\Delta t_F = \frac{D_{Pi}}{\cos \alpha_F} \frac{2\bar{v}_F \sin \alpha_F}{c_F^2 - \bar{v}_F^2 \sin^2 \alpha_F} . \tag{9.75}$$

Because $\bar{v}_F \ll c_F$ is usually fulfilled in ultrasound-based flow measurements, one can simplify this relation to

---

[4]The following mathematical relations are given exclusively for the angle α_F.

$$\Delta t_F \approx \frac{2 D_{Pi} \, \overline{v}_F \tan \alpha_F}{c_F^2} \tag{9.76}$$

without causing significant deviations. In doing so, the average flow rate $\overline{v}_F$ is given by

$$\overline{v}_F = \frac{\Delta t_F \, c_F^2}{2 D_{Pi} \tan \alpha_F} . \tag{9.77}$$

If the delay time t_{delay} is known or comparatively small, we can also exploit the sum Σt_F of both transit times to determine $\overline{v}_F$. This sum becomes

$$\Sigma t_F = t_{up} + t_{down} = \frac{D_{Pi}}{c_{up} \cos \alpha_F} + \frac{D_{Pi}}{c_{down} \cos \alpha_F}$$

$$= \frac{D_{Pi}}{\cos \alpha_F} \frac{c_{down} + c_{up}}{c_{down} \, c_{up}} \tag{9.78}$$

and after replacing c_{down} as well as c_{up}

$$\Sigma t_F = \frac{D_{Pi}}{\cos \alpha_F} \frac{2 c_F}{c_F^2 - \overline{v}_F^2 \sin^2 \alpha_F} = \frac{2 L_F \, c_F}{c_F^2 - \overline{v}_F^2 \sin^2 \alpha_F} . \tag{9.79}$$

Just as in (9.75), it is possible to neglect the expression $\overline{v}_F^2 \sin^2 \alpha_F$ since $\overline{v}_F \ll c_F$ holds. The resulting relation $\Sigma t_F \approx 2 L_F / c_F$ in combination with the time difference Δt yields

$$\overline{v}_F = \frac{c_F}{\sin \alpha_F} \frac{\Delta t_F}{\Sigma t_F} \tag{9.80}$$

for the average flow rate. In contrast to (9.77), this equation requires neither the inner diameter D_{Pi} of the pipe in which the fluid flows nor the geometric distance L_F between the ultrasonic transducers, but the time delay t_{delay} has to be known. Nevertheless, the sum Σt_F will lead to the sound velocity c_F of the fluid when L_F is known. This can be particularly useful to consider the temperature dependency of c_F in the framework of determining $\overline{v}_F$.

Figure 9.27 displays typical receive signals of transit time ultrasonic flow meters for pulse-shaped electrical excitation and an average flow rate $\overline{v}_F \neq 0$. While u_{rec}^{down} refers to the output of transducer T2, u_{rec}^{up} represents the output of transducer T1. As can be clearly seen, a certain time difference Δt_F between both receive signals occurs. Especially in case of low flow rates, $\overline{v}_F$ and, thus, the time difference Δt_F will be small, e.g., in the range of a few nanoseconds [7, 84]. On the other hand, a large flow rate may produce strong differences of u_{rec}^{down} and u_{rec}^{up} regarding their amplitudes as well as signal shape [52]. For these reasons, one should apply cross-correlation approaches [59] and interpolation techniques [82, 108] to precisely determine Δt_F. Further improvements can be achieved by means of coded transducer excitation signals, such as chirp signals, Barker codes, and Gold codes [48, 67]. However, the

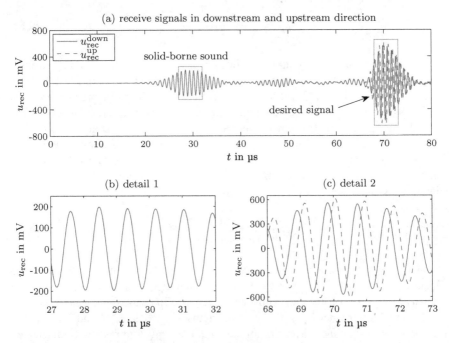

Fig. 9.27 Results of FE simulations for receive signals of transit time ultrasonic flow meter in clamp-on configuration; **a** entire receive signals $u_{\mathrm{rec}}^{\mathrm{down}}$ and $u_{\mathrm{rec}}^{\mathrm{up}}$ in downstream and upstream direction; **b** detail 1 shows receive signals due to solid-borne sound propagating in pipe wall; **c** detail 2 shows desired signals offering time difference Δt_{F} between $u_{\mathrm{rec}}^{\mathrm{down}}$ and $u_{\mathrm{rec}}^{\mathrm{up}}$ for determination of $\overline{v}_{\mathrm{F}}$

available electrical energy and computing power as well as the utilized ultrasound transducers primarily specify the possible measuring range of transit time ultrasonic flow meters.

Besides conventional transit time ultrasonic flow meters consisting of one measuring section (i.e., two ultrasonic transducers), there exist various other implementations that require two or even more measuring sections. The so-called *sing-around ultrasonic flow meter* is based on two measuring sections (i.e., four ultrasonic transducers), which are operated simultaneously as well as permanently [30, 36]. One of these sections captures the downstream direction, while the other one the upstream direction. Due to the different sound velocities c_{down} and c_{up}, the resulting transit times t_{down} and t_{up} will differ again. If a pulse-shaped transducer excitation is applied, we can distinguish between two sing-around methods, namely (i) fixed measurement time and (ii) fixed number of signal circulations. In each case, the receiver triggers the transmitter within one measuring section, which means that the transmitter will generate a pulse-shaped sound wave when a sound wave reaches the receiver. The number n_{down} and n_{up} of trigger events in downstream and upstream direction indirectly relate to the average flow rate $\overline{v}_{\mathrm{F}}$. As the name suggests, the trigger events are counted during a fixed measurement time for the first sing-around method. By means of

$$\overline{v}_F = \frac{c_F}{\sin \alpha_F} \frac{n_{\text{down}} - n_{\text{up}}}{n_{\text{down}} + n_{\text{up}}} , \tag{9.81}$$

one finally obtains $\overline{v}_F$. Owing to the integer discretization, the accuracy strongly depends on the number of trigger events that are considered for a single measurement. A large number and, therefore, a long measurement time will lead to precise results for $\overline{v}_F$. However, a long measurement time prevents dynamic measurements of $\overline{v}_F$. The second sing-around method is based on a fixed number of trigger events, i.e., $n_{\text{down}} = n_{\text{up}}$. Of course, the time intervals T_{down} and T_{up} to reach this number will differ in downstream and upstream direction because the effective sound velocities and, consequently, the transit times do not coincide. The connection of $\overline{v}_F$ and the time intervals is given by

$$\overline{v}_F = \frac{c_F}{\sin \alpha_F} \frac{T_{\text{up}} - T_{\text{down}}}{T_{\text{up}} + T_{\text{down}}} . \tag{9.82}$$

A further type of transit time ultrasonic flow meter exploits the so-called lambda locked loop principle [60]. Thereby, we require again at least one measuring section in downstream direction and one in upstream direction. Since the sound velocities c_{down} and c_{up} will differ when $\overline{v}_F \neq 0$, the wavelengths $\lambda_{\text{down}} = c_{\text{down}}/f$ and $\lambda_{\text{up}} = c_{\text{up}}/f$ of the generated sound waves do not coincide for a given ultrasonic frequency f. The lambda locked loop principle aims at a constant wavelength in downstream and upstream direction. This condition can only be satisfied if the ultrasonic frequencies f_{down} and f_{up} in both flow directions differ, i.e.,

$$\lambda_0 = \lambda_{\text{down}} = \frac{c_F + \overline{v}_F \sin \alpha_F}{f_{\text{down}}} = \lambda_{\text{up}} = \frac{c_F - \overline{v}_F \sin \alpha_F}{f_{\text{up}}} \tag{9.83}$$

where $\lambda_0 = c_F/f$ stands for the wavelength without flow. As a result, we obtain an identical phase difference $\Delta\varphi_F$ in the measuring sections between transmitted and received signal. For the practical implementation, it makes, thus, sense to vary f_{down} as well as f_{up} according to the phase differences in the measuring sections. When equal values for $\Delta\varphi_F$ are achieved, the average flow rate will follow from

$$\overline{v}_F = \frac{\lambda_0}{2 \sin \alpha_F} \left(f_{\text{down}} - f_{\text{up}} \right) . \tag{9.84}$$

Generally speaking, transit time ultrasonic flow meters are utilized in several applications, where the mass flow rate or the volumetric flow rate of fluids has to be acquired over a wide measuring range of flow velocities. Depending on the investigated medium, such ultrasonic flow meters typically operate at frequencies ranging from 40 kHz up to a few megahertz [3, 62]. Rather small operating frequencies are chosen for gases due to sound attenuation, while the ultrasonic transducers feature center frequencies of ≈ 1 MHz for liquids. Especially in case of multipath configurations, which means that more than one measuring section is available for determining the flow velocity, transit time ultrasonic flow meters offer a relative measurement error smaller than 1%.

Doppler Flow Meters

Instead of the windfall effect, Doppler ultrasonic flow meters exploit the *Doppler effect* inside flowing fluids [30, 36]. Such ultrasonic flow meters will be especially appropriate if the fluid contains many scattering particles for incident ultrasonic waves. Before the idea behind Doppler ultrasonic flow meters is detailed, we will briefly repeat the fundamentals of the Doppler effect for sound waves. Basically, one can distinguish between three different standard scenarios: (i) stationary transmitter and moving receiver, (ii) moving transmitter and stationary receiver, and (iii) stationary transmitter and stationary receiver for a moving reflector (see Fig. 9.28). Let us suppose for each scenario that the transmitter generates sound waves with the frequency f_T in a fluid featuring the sound velocity c_F. In case of scenario (i), the resulting wavelength λ_T also arises at the receiver, which is assumed to move toward the transmitter with the velocity given by the vector $\mathbf{v}_0$. Due to this movement, the effective sound velocity of the incident waves changes to $c_F + \mathbf{v}_0 \cdot \mathbf{e}_{RT}$ and, therefore, the relation (wavelength λ_R at receiver)

$$\lambda_T = \frac{c_F}{f_T} = \lambda_R = \frac{c_F + \mathbf{v}_0 \cdot \mathbf{e}_{RT}}{f_R} \tag{9.85}$$

holds with the unit vector $\mathbf{e}_{RT}$ pointing from receiver to transmitter. By rewriting (9.85), the frequency f_R of the sound waves at the receiver becomes

$$f_R = f_T \left(1 + \frac{\mathbf{v}_0 \cdot \mathbf{e}_{RT}}{c_F} \right) . \tag{9.86}$$

Since the transmitter moves toward the receiver with the velocity vector $\mathbf{v}_0$ in scenario (ii), the wavelength λ_R of the incident sound waves at the stationary receiver reads as (unit vector $\mathbf{e}_{TR}$ pointing from transmitter to receiver)

$$\lambda_R = \lambda_T - \frac{\mathbf{v}_0 \cdot \mathbf{e}_{TR}}{f_T} . \tag{9.87}$$

Inserting this equation as well as $\lambda_T = c_F/f_T$ in $f_R = c_F/\lambda_R$ leads to

$$f_R = f_T \left(1 - \frac{\mathbf{v}_0 \cdot \mathbf{e}_{TR}}{c_F} \right)^{-1} \tag{9.88}$$

for the sound frequency at the receiver. If $|\mathbf{v}_0 \cdot \mathbf{e}_{TR}| \ll c_F$ is satisfied, (9.88) can be simplified to

$$f_R \approx f_T \left(1 + \frac{\mathbf{v}_0 \cdot \mathbf{e}_{TR}}{c_F} \right) . \tag{9.89}$$

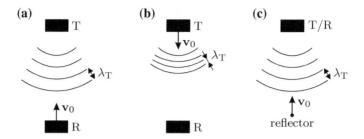

Fig. 9.28 Standard scenarios concerning Doppler effect; **a** stationary transmitter and moving receiver; **b** moving transmitter and stationary receiver; **c** stationary transmitter and receiver for moving reflector; transmitter T and receiver R; velocity vector $\mathbf{v}_0$

According to (9.86) and (9.89), f_R is solely influenced by the relative velocity between transmitter and receiver. If they move toward each other, the receive frequency f_R will be higher and, otherwise, smaller than the transmit frequency f_T.

Scenario (iii) relates to a stationary transmitter and receiver, which are located at the same position. In case of a reflector moving toward the transmitter–receiver combination with the velocity vector $\mathbf{v}_0$, the reflector observes the same frequency as given in (9.86). The reflector itself serves as source of sound waves and can, thus, be treated as a transmitter that moves toward the receiver. The sound frequency f_R of reflected sound waves at the receiver results then from the link of (9.86) and (9.88), i.e.,

$$f_R = f_T \frac{1 + \frac{\mathbf{v}_0 \cdot \mathbf{e}_{PT}}{c_F}}{1 - \frac{\mathbf{v}_0 \cdot \mathbf{e}_{PR}}{c_F}} \qquad (9.90)$$

with the unit vectors $\mathbf{e}_{PT}$ pointing from moving reflector to stationary transmitter and $\mathbf{e}_{PR}$ ($\hat{=}\,\mathbf{e}_{PT}$) pointing from moving reflector to stationary receiver, respectively. Under the assumption of $|\mathbf{v}_0 \cdot \mathbf{e}_{PT}| \ll c_F$, this equation takes the form

$$f_R \approx f_T \left(1 + \frac{2\mathbf{v}_0 \cdot \mathbf{e}_{PT}}{c_F} \right) . \qquad (9.91)$$

The velocity of the reflector has, consequently, the double impact on the receive frequency as a moving transmitter or moving receiver without reflector.

Now, let us consider an actual realization of a Doppler ultrasonic flow meter, which is displayed in Fig. 9.29. The central axis of both ultrasonic transducers and the pipe axis encloses the angle β_F. A scattering particle shall be moving along the pipe axis with the flow velocity $v_F = \|\mathbf{v}_F\|_2$. If the transmitter generates a continuous sound wave of frequency f_T, the sound frequency f_P at the moving scattering particle will become (cf. (9.86))

$$f_P = f_T \left(1 + \frac{\mathbf{v}_F \cdot \mathbf{e}_{PT}}{c_F} \right) = f_T \left(1 - \frac{v_F \cos \beta_F}{c_F} \right) \qquad (9.92)$$

with the unit vector $\mathbf{e}_{PT}$ pointing from particle to stationary transmitter. Again, this scattering particle can be treated as source of sound waves. The resulting sound frequency f_R at the stationary receiver is finally given by (cf. (9.90))

$$f_R = f_T \frac{1 + \frac{\mathbf{v}_F \cdot \mathbf{e}_{PT}}{c_F}}{1 - \frac{\mathbf{v}_F \cdot \mathbf{e}_{PR}}{c_F}} = f_T \frac{1 - \frac{v_F \cos \beta_F}{c_F}}{1 + \frac{v_F \cos \beta_F}{c_F}} \tag{9.93}$$

where $\mathbf{e}_{PR}$ denotes the unit vector pointing from moving particle to stationary receiver. Owing to the fact that $v_F \ll c_F$ is always fulfilled for common fluid flows being measured with Doppler ultrasonic flow meters, this equation simplifies to

$$f_R \approx f_T \left(1 - \frac{2 v_F \cos \beta_F}{c_F} \right) . \tag{9.94}$$

The velocity of the scattering particle reads then as

$$v_F = \frac{c_F (f_T - f_R)}{2 \cos \beta_F f_T} . \tag{9.95}$$

When the quantities c_F, f_T, f_R, and β_F are known, we can, therefore, calculate the flow velocity. In doing so, it is supposed that the particles move uniformly with the flow, which sometimes constitutes a problem in practical applications. Besides, (9.95) is based on the assumption that scattering particles exclusively exist along the pipe axis. As a matter of course, scattering particles also arise apart from this axis. Because the angles β_F for the ultrasonic transducer differ for such particle locations (see Fig. 9.29), the mathematical link between v_F, f_T, and f_R changes. According to Fig. 9.25, the particle velocities depend, moreover, on the radial position r inside the pipe. Those are the reasons why one should use a transducer combination, which features a limited spatial extension of the measuring volume around the pipe axis. Similar to transit time ultrasonic flow meters, one has to know the velocity profile $v_F(r)$ inside the pipe with a view to determining the mass flow rate $\dot{m}_F$ and the volumetric flow rate $\dot{V}_F$ from the flow velocity along the pipe axis.

The studied Doppler ultrasonic flow meter requires two ultrasonic transducers (cf. Fig. 9.29). While one transducer serves as transmitter of continuous sound waves that commonly exhibit constant frequency, the other transducer is utilized as receiver. If we apply a pulse-shaped excitation, it will be possible to build up a Doppler ultrasonic flow meter with only one transducer operating in pulse-echo mode [3]. Just as in case of two transducers, we can evaluate the frequency shift between transmitted and received pulse to determine the flow velocity. However, a single transducer exhibits a sustained measuring volume, which is usually larger than that of a transmitter–receiver combination. Due to the large measuring volume, scattering particles at different radial positions and, thus, with different velocities contribute to the receiver output. As a result, we obtain a broadband frequency spectrum, which cannot be assigned to a single flow velocity of the fluid.

Fig. 9.29 Measurement principle of Doppler ultrasonic flow meters; moving scattering particle causes difference of transmit frequency f_T and receive frequency f_R for stationary ultrasonic transmitter and receiver

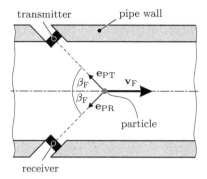

In general, typical Doppler ultrasonic flow meters operate at frequencies of a few megahertz and provide a relative measurement error of the flow velocity smaller than 5% [30]. They can be used for flow velocities of liquids in the range of 0.2–10 m s^{-1}. The frequency difference $\Delta f = f_T - f_R$ between transmit and receive frequency normally accounts 200 Hz to 10 kHz. The optimum concentration of the scattering particles in the liquid is in the order of 0.01% [3]. Their size should be approximately in the range of 30–100 μm. Apart from industrial applications, Doppler ultrasonic flow meters are employed in medical diagnostics for blood flow measurements [45].

Speckle-Tracking Flow Meters

To determine volumetric flow rates as well as mass flows inside a pipe, one has to know the velocity profile $v_F(r)$ for standard realizations of transit time ultrasonic flow meters and Doppler ultrasonic flow meters. Speckle-tracking ultrasonic flow meters do not require the knowledge of $v_F(r)$ because they can be used to reconstruct this quantity [60, 62]. Just as Doppler ultrasonic flow meters, speckle-tracking flow meters exploit reflections of incident sound waves at scattering particle, which are supposed to move uniformly with the flow inside the pipe. The underlying measurement principle is based on the pulse-echo mode of an ultrasonic transducer. By exciting the transducer with a pulse-shaped electrical signal, we are able to determine the radial position r of the scattering particles from the time-of-flight of the resulting echo. The cross-correlation of the echo signals for successive transducer excitations yields the position change of the scattering particles. In case of a sufficient concentration of scattering particles, we can, thus, deduce the spatially resolved flow velocity of the fluid, i.e., the velocity profile inside the pipe.

Speckle-tracking ultrasonic flow meters offer a comparatively low relative measurement error in the range of 1%. However, such flow meters cannot be utilized for small flow velocities like 0.2 m s^{-1} since the scattering particles commonly do not move with the liquid anymore.

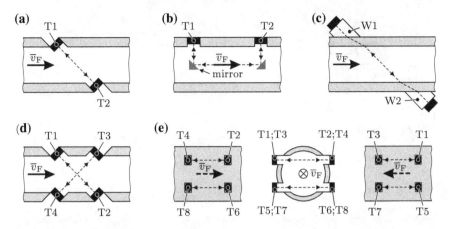

Fig. 9.30 Selected transducer arrangements of ultrasonic flow meters [30, 36]; **a** single-path invasive; **b** single-path invasive with acoustic mirrors; **c** single-path noninvasive; **d** two paths invasive;
e four paths invasive; ultrasonic transducers Tx; ultrasonic wedge transducers Wx

9.3.3 Arrangement of Ultrasonic Transducers

There exist many different arrangements of transmitters and receivers for ultrasonic flow meters, in particular for transit time ultrasonic flow meters [30, 36, 62].
Figure 9.30 depicts selected arrangements for such flow meters, which are frequently
utilized in technical applications. Mainly, one can distinguish between single-path
and multipath arrangements as well as invasive and noninvasive configurations.
Because multipath arrangements exploit more than one measuring section (e.g., four
sound paths), the determination of the flow velocity is more reliable and does not
heavily depend on the velocity profile $v_F(r)$ inside the pipe [69, 116]. However, the
operation of several measuring sections within ultrasonic flow meters requires a large
number of ultrasonic transducers and is, furthermore, accompanied by an increasing effort of both control and readout electronics. In the following, let us discuss
the difference between invasive and noninvasive configurations of ultrasonic flow
meters.

Invasive Configurations

In case of invasive configurations of ultrasonic flow meters that are also named
inline ultrasonic flow meters, the ultrasonic transducers themselves and/or associated
components are in contact with the flowing fluid [3]. Depending on the geometric
circumstances, this contact does not influence the fluid flow or it can be intrusive
disturbing the fluid flow (see Fig. 9.31a and b). Invasive configurations of ultrasonic
flow meters always require special pipe sections. If the pipe sections are intrusive, they
may cause various additional problems in practical applications. For example, the
flowing fluid can be swirled at the measuring point and, therefore, the velocity profile
will be modified locally which alters the measured average flow rate $\overline{v}_F$ for transit time

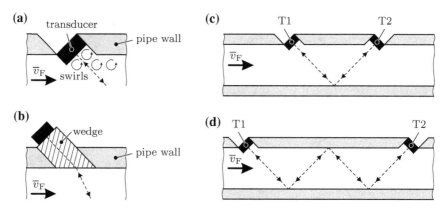

Fig. 9.31 a Intrusive and **b** nonintrusive transducer configurations of invasive ultrasonic flow meters; **c** V path, i.e., one reflection at inner pipe wall; **d** W path, i.e., three reflections at pipe wall; ultrasonic transducers T1 and T2 [3]

flow meters. Although inline ultrasonic flow meters that exploit acoustic mirrors (see Fig. 9.30b) represent an intrusive configuration, they often facilitate evaluating the difference Δt_F of transit times in upstream and downstream direction. This is a consequence of the enlarged transducer distance L_F, which linearly increases Δt_F for a given value of $\overline{v}_F$ in case of $\alpha_F = 90°$. Note that we are also able to enlarge L_F by using one (V path; see Fig. 9.31c), two, three (W path; see Fig. 9.31d), or even more acoustic reflections at the inner pipe wall and by increasing the angle α_F. Each of these methods for enlarging L_F leads to a frequently unwanted geometric extension of the measuring point as well as sometimes to a remarkably reduced SNR in the receive signals.

Figure 9.32a shows the commercially available inline ultrasonic flow meter HYDRUS from the company Diehl Metering [23]. This low-cost transit time flow meter is equipped with a pipe section containing acoustic mirrors and can be applied for recording water consumption in private households over a wide measuring range. The low energy consumption of the HYDRUS flow meter allows a battery lifetime of up to 16 years.

Noninvasive Configurations

In contrast to invasive configurations, there does not occur any contact between fluid and ultrasonic transducer for noninvasive configurations of ultrasonic flow meters. This will constitute a great advantage when the flowing fluid exhibits a wide temperature range and would damage the transducers. However, since the transducers are not in direct contact with the flowing fluid but have to be appropriately coupled to the outer pipe wall, remarkable sound energy does not reach the receiver due to reflections at the interface fluid/inner pipe wall. Such reflections reduce the receive signals and, consequently, the achievable SNR. This fact needs to be taken into account, especially when further acoustic reflections at the inner pipe wall are used for enlarging the effective transducer distance L_F.

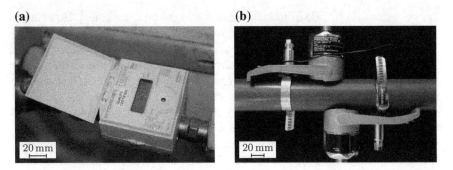

Fig. 9.32 a Inline ultrasonic flow meter HYDRUS from company Diehl Metering; **b** clamp-on ultrasonic flow meter from company Endress+Hauser; ultrasonic wedge transducers mounted on pipe

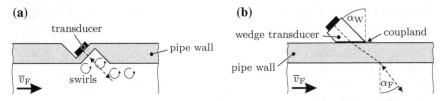

Fig. 9.33 a Intrusive and **b** nonintrusive configurations of noninvasive ultrasonic flow meters [3]; clamp-on ultrasonic flow meters represent well-known nonintrusive configuration

Just as in case of inline ultrasonic flow meters, we can distinguish between intrusive and nonintrusive configurations [3]. The primary difference between intrusive ultrasonic flow meters in invasive and noninvasive configurations lies in the position of the ultrasonic transducers. While the transducers are in contact with the fluid for invasive configurations, we have to mount the transducers on designated places at the outer pipe wall for noninvasive configurations. The specific shaping of the inner pipe wall (see Fig. 9.33a) can cause, again, swirls in the flowing fluid and, thus, measurement errors of the average flow rate $\overline{v}_F$ if an appropriate calibration procedure was not conducted in advance. Besides, we need a special pipe section at the measuring point of noninvasive ultrasonic flow meters in intrusive configuration.

The so-called *clamp-on ultrasonic flow meters* represent a very well-known and advantageous noninvasive configuration. They do not demand a special pipe section but can theoretically be installed at any location along the pipe. Hence, one has neither to open nor to modify an existing pipe system for inserting a pipe section at the measuring point [30, 36]. Clamp-on ultrasonic flow meters typically use ultrasonic wedge transducers that consist of a piezoelectric element being mounted on a wedge featuring the angle α_W (see Fig. 9.33b). By utilizing such wedge transducers as transmitters and receivers, one obtains a fluid angle $\alpha_F \neq 0°$, which is required for the measurement principle of transit time as well as Doppler ultrasonic flow meters. To avoid wave refractions and reflections at the interface wedge/outer pipe wall, the wedge could be made of the same material as the pipe, e.g., steel. However, in doing so, we cannot reduce wave reflections at the interface inner pipe wall/fluid. According to the fundamental law of refraction for acoustics (see Sect. 2.3.4), a large fluid angle α_F for extending Δt_F calls, moreover, for an even larger wedge angle α_W (i.e., $\alpha_W > \alpha_F$) because the propagation velocity of longitudinal waves in the fluid is smaller than in the pipe wall.

Apart from problems concerning technical feasibility, large wedge angles go hand in hand with a reduced wave transmission at the interface inner pipe wall/fluid. Those are the reasons why the wedge material should not coincide with the pipe material. If a special wedge material like plastic of low sound velocity is used instead, we can, however, exploit mode conversions at the interfaces wedge/outer pipe wall and inner pipe wall/fluid [3, 84]. Starting from the propagating longitudinal wave in the wedge, the wave gets converted into a transverse wave propagating in the pipe wall. At the interface inner pipe wall/fluid, the transverse wave gets converted again into a longitudinal wave propagating in the fluid. Such ultrasonic wedge transducers provide primarily three advantages for clamp-on ultrasonic flow meters.

- When the propagation velocities of longitudinal waves in the wedge material approximately corresponds to the fluid, α_W will be similar to α_F. Therefore, we do not need a large value of α_W, which facilitates technical realization as well as coupling of wedge transducers.
- The impedance mismatch at the interfaces inner pipe wall/fluid is much lower since the acoustic impedance Z_{aco} of the pipe material is smaller for transverse waves than for longitudinal waves. As a result, we can transmit much more sound energy at these interfaces, which leads to higher receive signals.
- The third advantage lies in the possibility of generating mechanical Lamb waves in the pipe wall (cf. Fig. 9.39), which yield a large axial extension of the sound field in the fluid. This so-called wide beam makes transducer positioning easier but also reduces the receiver outputs [30, 70].

Despite the great advantages of clamp-on ultrasonic flow meters, both their installation and commissioning are crucial points for a proper operation [95]. As mentioned above, clamp-on ultrasonic flow meters can be theoretically installed at any location along the pipe. However, when the ultrasonic transducers are mounted a short distance after a pipe elbow or a change of the pipe cross section, the introduced turbulences inside the flowing fluid will influence the measured average flow rate $\overline{v}_F$. Just as the other ultrasonic flow meters, clamp-on ultrasonic flow meters require, thus, a sufficient inlet path to minimize unwanted disturbances in the flow profile. Another important aspect refers to the actual measuring point. Only if parameters such as the pipe inner diameter are known or an adequate calibration can be conducted, it will be possible to determine $\overline{v}_F$ correctly. Furthermore, the ultrasonic wedge transducers have to be perfectly aligned at the measuring point. Small axial and radial misalignments of the transducers can alter the measured average flow rate significantly [32, 64]. Proper coupling of the transducers to the outer pipe wall is, of course, a further decisive point for clamp-on ultrasonic flow meters. Coupling can be ensured by means of a thin layer of, e.g., an appropriate liquid, epoxy resins, or silicon grease between wedge transducer and cleaned outer pipe wall. Finally, one always has to keep in mind that the pipe also directly transmits mechanical waves from transmitter to receiver [7, 84]. This so-called solid-borne sound does not depend on the fluid velocity and, therefore, represents unwanted components in the receive signals (see Fig. 9.27b). Let us assume that those signal components arise at the same time as the desired signals (see Fig. 9.27c) being propagating through the fluid. In such a case, the determination of $\overline{v}_F$ might be impossible because we are not able to identify the time difference Δt_F of transit times in upstream and downstream direction anymore. Note that solid-borne sound can also cause problems for other types of ultrasonic flow meters, e.g., inline ultrasonic flow meters.

Figure 9.32b illustrates a picture of two ultrasonic transducers from the Prosonic Flow W series from the company Endress+Hauser [29] that are mounted on a pipe. Each cylindrical transducer contains a wedge made of plastics, which is equipped with a piezoceramic disk. By evaluating the difference of transit times, an extern analysis unit yields the average flow rate and, consequently, the sought-after quantities mass flow rate as well as volumetric flow rate. Recently, the company Bürkert [17] presented the product FLOWave as a modification of conventional clamp-on ultrasonic flow meters. Instead of wedge transducers, interdigital transducers are attached to the outer pipe wall for transmitting and receiving waves.

9.3.4 Modeling of Clamp-on Transit Time Ultrasonic Flow Meters in Frequency–Wavenumber Domain

Notwithstanding the advantages of clamp-on transit time ultrasonic flow meters[5] (CTU), the wave propagation from transmitter to receiver is much more complicated than for invasive configurations of ultrasonic flow meters. This fact originates from refraction effects taking place at the interfaces ultrasonic wedge transducer/pipe wall as well as pipe wall/fluid. Such refraction effects alter the dominant fluid angle α_F^{dom}, which indicates the dominant direction of propagation for plane sound waves within the fluid. If the flow profile is ideal and the parameters inner pipe diameter D_{Pi}, time difference Δt_F, and sound velocity c_F of the fluid are exactly known, the calculated average flow rate $\overline{v}_F$ will exclusively depend on α_F^{dom} (cf. (9.77)). In the majority of cases, the expected fluid angle α_F^0 is determined by means of the fundamental law of refraction for acoustics. Angle deviations $\Delta\alpha_F = \alpha_F^{dom} - \alpha_F^0$ between expected and dominant fluid angle result in the relative systematic measurement error ϵ_v [85, 87]

$$\epsilon_v = \frac{\overline{v}_F|_{\alpha_F=\alpha_F^0}}{\overline{v}_F|_{\alpha_F=\alpha_F^{dom}}} - 1 = \frac{\tan\alpha_F^{dom}}{\tan\alpha_F^0} - 1 \qquad (9.96)$$

---

[5]For compactness, clamp-on transit time ultrasonic flow meter is abbreviated as CTU flow meter hereinafter.

of the calculated average flow rate with respect to the true value. For instance, a slight angle deviation of 0.1° from the expected fluid angle $\alpha_F^0 = 25°$ will already yield a relative systematic measurement error of 0.46%. Therefore, it is of utmost importance to accurately predict α_F^{dom} for practical applications of CTU flow meters, especially when one requires a high precision of the measured flow rate.

In principle, we are able to identify the dominant fluid angle α_F^{dom} of a CTU flow meter with the aid of measurements and numerical simulations. For the measurements, it is recommended to acquire the generated sound field in the fluid for a representative configuration of the flow meter; i.e., a setup comprising ultrasonic wedge transducer, pipe wall as well as fluid [84]. Both hydrophone and Schlieren measurements can provide the decisive sound field information (see Sect. 8.1). Special signal processing techniques like a modified Hough transform finally lead to the orientation of each wavefront inside a spatially resolved wave packet [86, 87]. As a matter of course, these orientations are closely linked to α_F^{dom}.

Several approaches to determine α_F^{dom} by means of numerical simulations are based on the finite element (FE; see Chap. 4) method. FE simulations additionally enable predicting the flow-dependent electrical output of the receiver in case of CTU flow meters. However, reliable simulation results call for precise material parameters of all components, e.g., piezoelectric element and pipe. Conventional FE simulations are, moreover, accompanied by a remarkable computational effort because we have to discretize both time and the spatial domain including all components of the CTU flow meter as well as the fluid. That was the reason why Bezděk et al. [7, 8] proposed a hybrid simulation approach combining the FE method with a particular boundary integral method, the so-called *Helmholtz integral ray-tracing method* (HIRM). While they apply the FE method for the ultrasonic wedge transducer and pipe wall, the HIRM is exploited to efficiently calculate the sound propagation within the fluid, which would take plenty of time if conventional FE simulations were used. Such a hybrid approach allows simulations for three-dimensional models of CTU flow meters and, thus, facilitates product development. When one is primarily interested in the dominant fluid angle α_F^{dom}, the coupled FEM-HIRM scheme will deliver, however, information (e.g., spatially resolved sound field) that is not required but takes most of the computation time.

Besides FE-based simulations, there exist various analytical as well as semi-analytical approaches, which can be exploited to determine α_F^{dom} for CTU flow meters. In the following, we will briefly discuss a few approaches from the literature. Montegi et al. [70] suggested a method that provides the spatial frequency representation (i.e., wave number spectrum) of the transmitted beam in the fluid. The generated beam of the ultrasonic wedge transducer is modeled in the spatial frequency domain. By applying the transmission model from Oliner [76], they describe the filter effect of the pipe wall for CTU flow meters. The combination of the wave number spectra for both the transducer's beam and the pipe wall yields the overall transmission coefficient as a function of the horizontal wave number. From that, it is possible to calculate the angle deviations $\Delta\alpha_F$ between expected and dominant fluid angle. Funck et al. [32] presented an extended modeling version by additionally introducing a coordinate transform. The authors also recognized filter effects of the pipe wall and, moreover, deduced appropriate excitation signals for the ultrasonic transmitter. Both modeling approaches exclusively concentrate on the spatial frequency domain and, thus, consider neither the transducer's transfer behavior in the frequency domain nor in the time domain. In contrast to these approaches, Wöckel et al. [113] aim at predicting the output signal of CTU flow meters in the time domain. Since they suppose only plane wave propagation resulting from geometric acoustics, the spatial filter effect of the pipe wall cannot be considered which represents a certain oversimplification for describing such flow meters. However, this model takes the complex-valued frequency spectra of the electrical transducer excitation as well as the transfer behavior of ultrasonic transmitter and receiver into account. By evaluating the inverse Fourier transform, it is then possible to compute the receive signal of a CTU flow meter in the time domain.

As a conclusion of the previous explanations, it makes sense to combine the spatial frequency domain and the conventional frequency domain referring to time. When we use such a combination, the modeling will be carried out in the so-called *frequency–wavenumber domain*. Below, let us detail an efficient modeling approach in the frequency–wavenumber domain for CTU flow meters. The semi-analytical approach was developed within the framework of the doctoral thesis of Ploß [84] and published in [85].

General Idea of the Modeling Approach

The general idea of the modeling approach is based on the *angular spectrum method* originating from Fourier optics [34]. According to this method, one can decompose each wave field into an angular spectrum of plane waves. Each plane wave is assumed to travel in a unique direction given by the wave vector

$$\mathbf{k} = k_x \mathbf{e}_x + k_y \mathbf{e}_y + k_z \mathbf{e}_z \tag{9.97}$$

with the components k_i. The magnitude $\|\mathbf{k}\|_2$ of the wave vector is linked to the frequency f of the propagating wave and its wave propagation velocity c through (wave number k)

$$\|\mathbf{k}\|_2 = k = \frac{2\pi f}{c} . \tag{9.98}$$

Because it is commonly sufficient to model CTU flow meters in the two-dimensional space, we can restrict the following calculation steps to the xy-plane [84, 85]. Figure 9.34 depicts the considered setup of a CTU flow meter, which contains two equal ultrasonic wedge transducers being mounted on the outer pipe wall on opposite sides. For this setup, the fundamental law of refraction for acoustics (cf. Sect. 2.3.4) leads to the relation

$$\underbrace{\frac{c}{\sin \alpha}}_{\text{general}} = \underbrace{\frac{c_{\mathrm{l,W}}}{\sin \alpha_{\mathrm{W}}}}_{\text{wedge}} = \underbrace{\frac{c_{\mathrm{t,P}}}{\sin \alpha_{\mathrm{t,P}}^0}}_{\text{pipe}} = \underbrace{\frac{c_{\mathrm{F}}}{\sin \alpha_{\mathrm{F}}^0}}_{\text{fluid}} = c_{\mathrm{ph}}^0 \tag{9.99}$$

Fig. 9.34 Considered
two-dimensional setup of
CTU flow meter; ultrasonic
wedge transducers contain
piezoelectric element for
generating and receiving
ultrasonic waves; angles with
superscript 0 refer to
expected orientations of
wave propagation according
to fundamental law of
refraction for acoustics [84]

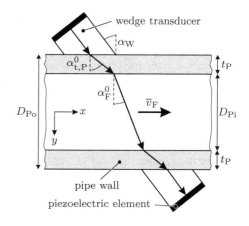

between expected directions of plane wave propagation and wave propagation veloc-
ity in the components of the CTU flow meter. The angles α_W, $\alpha_{t,P}^0$, and α_F^0 refer to
the wedge transducers, the pipe wall, and the fluid, respectively. Note that here, the
wave propagation in both wedge and fluid is supposed to consist exclusively of longitudinal waves (index l), while only transverse waves (index t) should be present in
the pipe wall. The expression c_{ph}^0 stands for the so-called design phase velocity of
the CTU flow meter, i.e., the desired phase velocity.

The phase velocity c_{ph} denotes an appropriate parameter to describe refraction of
waves at interfaces of different media since the interface conditions depend on c_{ph}.
For the considered setup, c_{ph} is linked to the component k_x of **k** in x-direction by

$$k_x = \frac{2\pi f}{c_{ph}} \, . \tag{9.100}$$

This means that if c_{ph} changes, k_x and, consequently, the direction of plane wave
propagation (i.e., angle $\alpha = \arctan(k_x/k_y)$) will be altered because

$$k_y = \sqrt{k^2 - k_x^2} = \sqrt{\left(\frac{2\pi f}{c}\right)^2 - k_x^2} \tag{9.101}$$

has always to be fulfilled.

Owing to the finite-sized transducer's apertures, the wave propagation within a
CTU flow meter is not limited to a single plane wave propagating under a defined
angle. Instead, one obtains a certain range of angles α and, thus, phase velocities c_{ph}.
Depending on the frequency f and wave propagation medium (e.g., pipe wall),
each c_{ph} is directly connected to an angle α by means of (9.100) and (9.101).

To apply the angular spectrum method, the considered CTU flow meter should be
split up into five components, namely (i) ultrasonic transmitter, (ii) first pipe wall,
(iii) fluid, (iv) second pipe wall, and (v) ultrasonic receiver. Under the assumption

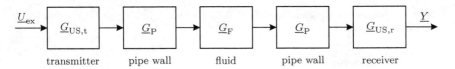

Fig. 9.35 Complex-valued transfer functions $\underline{G}_i(f, c_{ph})$ of components for modeling considered CTU flow meter in frequency–wavenumber domain; arguments f and c_{ph} omitted for compactness

of a linear time-invariant system, we can assign an individual complex-valued transfer function $\underline{G}_i(f, c_{ph})$ to the components, which depends on frequency f as well as phase velocity c_{ph}. From the system point of view, the output $\underline{Y}(f, c_{ph})$ of the ultrasonic receiver results then from (cf. Fig. 9.35)

$$\underline{Y}(f, c_{ph}) = \underline{U}_{ex}(f, c_{ph}) \cdot \underline{G}_{US,t}(f, c_{ph}) \cdot \underline{G}_P(f, c_{ph}) \cdot \underline{G}_F(f, c_{ph}) \cdot$$
$$\cdot \underline{G}_P(f, c_{ph}) \cdot \underline{G}_{US,r}(f, c_{ph}) , \tag{9.102}$$

whereby $\underline{U}_{ex}(f, c_{ph})$ represents the electrical excitation of the ultrasonic transmitter in the frequency–wavenumber domain. The transfer functions $\underline{G}_{US,t}(f, c_{ph})$, $\underline{G}_P(f, c_{ph})$, $\underline{G}_F(f, c_{ph})$, and $\underline{G}_{US,r}(f, c_{ph})$ rate the transfer behavior of ultrasonic transmitter, first as well as second pipe wall, fluid and ultrasonic receiver in the frequency–wavenumber domain, respectively. Since we are able to directly convert $\underline{G}_i(f, c_{ph})$ into $\underline{G}_i(f, \alpha)$, each complex-valued transfer function can be interpreted as frequency-dependent directivity pattern.

Motivated by an actually existing CTU flow meter that was exploited to verify the modeling approach, let us suppose the quantities as listed in Table 9.3. For this arrangement, the fundamental law of refraction for acoustics yields the expected angles $\alpha_{t,P}^0 = 52.00°$ within the pipe wall and $\alpha_F^0 = 21.91°$ inside the fluid as well as the design phase velocity $c_{ph}^0 = 3980\,\mathrm{m\,s^{-1}}$. Hereinafter, the electrical excitation and all complex-valued transfer functions of the CTU flow meter will be studied separately.

Excitation Signal

In order to generate an ultrasonic wave within the fluid, the piezoelectric element of one ultrasonic wedge transducer has to be excited by an electrical signal $u_{ex}(t)$ that changes over time t. There are various possibilities regarding the shape of $u_{ex}(t)$, ranging from simple bipolar square bursts to tailored arbitrary signals such as coded signals (see Sect. 9.2.3). Without limiting the generality, let us assume a bipolar square burst as excitation signal that consists of $n_{burst} = 3$ burst periods and features the frequency $f_{ex} = 2\,\mathrm{MHz}$. Figure 9.36a and b display the excitation signal in the time domain and the magnitude $|\underline{U}_{ex}(f)| = |\mathcal{F}\{u_{ex}(t)\}|$ of the resulting frequency spectrum. Due to the rather short signal duration of $u_{ex}(t)$, the frequency $f_{ex,max} = 1.94\,\mathrm{MHz}$ of the maximum spectral magnitude $U_{ex,max}$ does not coincide with f_{ex}. However, when n_{burst} is increased, the difference between $f_{ex,max}$ and f_{ex} will be

Table 9.3 Decisive quantities of considered CTU flow meter (see Fig. 9.34) for modeling in frequency–wavenumber domain

Component	Variable	Value
Ultrasonic wedge transducers		
Center frequency	f_c	2 MHz
Bandwidth	B_{US}^{-6dB}	1 MHz
Diameter of piezoelectric disk	D_{piezo}	20 mm
Wedge angle	α_W	38°
Material density of wedge	ϱ_W	1270 kg m^{-3}
Propagation velocity of longitudinal waves	$c_{l,W}$	2450 m s^{-1}
Steel pipe		
Outer diameter	D_{Po}	90 mm
Wall thickness	t_P	2 mm
Inner diameter	D_{Pi}	86 mm
Material density	ϱ_P	7897 kg m^{-3}
Propagation velocity of longitudinal waves	$c_{l,P}$	5729 m s^{-1}
Propagation velocity of transverse waves	$c_{t,P}$	3136 m s^{-1}
Fluid (water)		
Material density	ϱ_F	1000 kg m^{-3}
Propagation velocity of longitudinal waves	c_F	1485 m s^{-1}

reduced. The bandwidth $B_{ex}^{-6dB} = 0.79$ MHz refers to the frequency range, in which the signal's spectral magnitude $|U_{ex}(f)|$ stays above $U_{ex,max}/2$.

As mentioned above, the modeling approach for CTU flow meters demands the electrical excitation in the frequency–wavenumber domain, i.e., $U_{ex}(f, c_{ph})$. We have, therefore, to extend the frequency spectrum of $u_{ex}(t)$ by the phase velocity c_{ph}. Because $U_{ex}(f)$ does not depend on c_{ph}, this step leads to

$$U_{ex}(f, c_{ph}) = U_{ex}(f) \quad \forall \quad c_{ph} . \tag{9.103}$$

Ultrasonic Wedge Transducer

The transfer function $G_{US}(f, c_{ph})$ of an ultrasonic transducer in the frequency–wavenumber domain results from combining both the effective frequency-dependent directivity pattern $\Gamma_{t,az}(f, \alpha_{t,P})$ and the electroacoustic transfer function $H_{EA}(f)$ of the transducer. In a first step, let us discuss the expression $\Gamma_{t,az}(f, \alpha_{t,P})$ that is defined as

$$\Gamma_{t,az}(f, \alpha_{t,P}) = \Gamma_{t,geo}(f, \alpha_{t,P}) \cdot P_t(\alpha_{t,P}) , \tag{9.104}$$

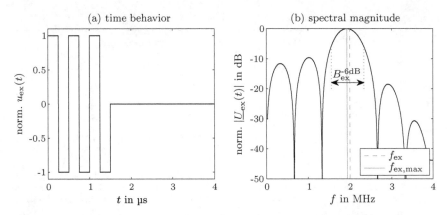

Fig. 9.36 a Time behavior of transducer's excitation signal $u_{ex}(t)$ consisting of $n_{burst} = 3$ burst periods; **b** resulting normalized spectral magnitude $|\underline{U}_{ex}(f)|$ of $u_{ex}(t)$; excitation frequency $f_{ex} = 2\,\mathrm{MHz}$; frequency $f_{ex,max} = 1.94\,\mathrm{MHz}$ of maximum $U_{ex,max}$; bandwidth $B_{ex}^{-6dB} = 0.79\,\mathrm{MHz}$

whereby $\Gamma_{t,geo}(f, \alpha_{t,P})$ and $P_t(\alpha_{t,P})$ stand for the geometric directivity pattern and the point source directivity of the wedge transducer in the xy-plane (azimuthal plane), respectively. The geometric directivity pattern depends on various geometric parameters of the wedge transducer and computes as [24, 115]

$$\Gamma_{t,geo}(f, \alpha_{t,P}) = \frac{D_{piezo}}{\cos \alpha_W} \mathrm{sinc}\left[\frac{D_{piezo}}{\cos \alpha_W} \frac{f}{c_{t,P}} \left(\sin \alpha_{t,P} - \frac{c_{t,P}}{c_{1,W}} \sin \alpha_W \right) \right] \quad (9.105)$$

with the sinc function $\mathrm{sinc}(x) = \sin(\pi x)/(\pi x)$. The inspection of (9.105) reveals that increasing the diameter D_{piezo} of the piezoelectric disk, the wedge angle α_W or frequency f yields a geometric directivity, which is more pronounced. Hence, one approaches the assumption of a single plane wave as stated by the fundamental law of refraction (cf. (9.99)). This is proven in Fig. 9.37 showing $\Gamma_{t,geo}(f, \alpha_{t,P})$ for two different frequencies.

The point source directivity $P_t(\alpha_{t,P})$ can significantly affect the direction of the dominant wave propagation in the pipe wall. As a result, there arise remarkable deviations between the angle $\alpha_{t,P}^{dom}$ of this direction and the expected angle $\alpha_{t,P}^0$ from the fundamental law of refraction. In case of a CTU flow meter, we are interested in the point source directivity of transverse waves that are generated in the pipe wall. Due to the thin fluid coupling layer being usually present between wedge transducer and pipe wall, shear forces can hardly be transmitted. It is, therefore, sufficient to take only point forces acting perpendicular to the interface wedge/pipe wall into account. According to [54, 115], the point source directivity for such arrangement becomes

$$P_t(\alpha_{t,P}) = \frac{4 \sin \alpha_{t,P} \cdot \cos \alpha_{t,P}}{N_1 + N_2} \sqrt{\left(\frac{c_{t,P}}{c_{1,P}} \right)^2 - \sin^2 \alpha_{t,P}} \quad (9.106)$$

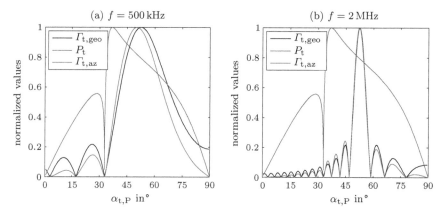

Fig. 9.37 Normalized geometric directivity pattern $\Gamma_{\mathrm{t,geo}}$, point source directivity P_{t}, and directivity pattern $\Gamma_{\mathrm{t,az}}$ of ultrasonic wedge transducer for **a** $f = 500\,\mathrm{kHz}$ and **b** $f = 2\,\mathrm{MHz}$

with

$$N_1 = 4\sin^2\alpha_{\mathrm{t,P}} \cdot \cos\alpha_{\mathrm{t,P}} \sqrt{\left(\frac{c_{\mathrm{t,P}}}{c_{\mathrm{l,P}}}\right)^2 - \sin^2\alpha_{\mathrm{t,P}}} + \left(1 - 2\sin^2\alpha_{\mathrm{t,P}}\right)^2 \tag{9.107}$$

$$N_2 = \frac{\varrho_{\mathrm{w}}}{\varrho_{\mathrm{P}}} \sqrt{\left(\frac{c_{\mathrm{t,P}}}{c_{\mathrm{l,P}}}\right)^2 - \sin^2\alpha_{\mathrm{t,P}}} \left[\frac{\left(1 - 2\left(\frac{c_{\mathrm{t,P}}}{c_{\mathrm{l,P}}}\right)^2 \sin^2\alpha_{\mathrm{t,P}}\right)^2}{\sqrt{\left(\frac{c_{\mathrm{t,P}}}{c_{\mathrm{l,w}}}\right)^2 - \sin^2\alpha_{\mathrm{t,P}}}} + \right.$$

$$\left. + 4\left(\frac{c_{\mathrm{t,w}}}{c_{\mathrm{t,P}}}\right)^4 \sin^2\alpha_{\mathrm{t,P}} \sqrt{\left(\frac{c_{\mathrm{t,P}}}{c_{\mathrm{t,w}}}\right)^2 - \sin^2\alpha_{\mathrm{t,P}}} \right]. \tag{9.108}$$

In contrast to the geometric directivity pattern, $P_{\mathrm{t}}(\alpha_{\mathrm{t,P}})$ exclusively depends on material parameters and is not a function of frequency. Figure 9.37 depicts the point source directivity for the considered setup as well as the obtained combined directivity pattern $\Gamma_{\mathrm{t,az}}(f, \alpha_{\mathrm{t,P}})$ for two different frequencies. One can clearly observe that $P_{\mathrm{t}}(\alpha_{\mathrm{t,P}})$ influences $\Gamma_{\mathrm{t,az}}(f, \alpha_{\mathrm{t,P}})$, especially for comparatively low frequencies, e.g., $f = 500\,\mathrm{kHz}$. Figure 9.38a shows the resulting frequency-dependent deviation

$$\Delta\alpha_{\mathrm{t,P}}(f) = \alpha_{\mathrm{t,P}}^{\mathrm{dom}}(f) - \alpha_{\mathrm{t,P}}^0 \tag{9.109}$$

between dominant angle $\alpha_{\mathrm{t,P}}^{\mathrm{dom}}(f)$ and expected angle $\alpha_{\mathrm{t,P}}^0$ for the studied CTU flow meter due to $\Gamma_{\mathrm{t,az}}(f, \alpha_{\mathrm{t,P}})$. For high frequencies, $\Delta\alpha_{\mathrm{t,P}}(f)$ will take small values because of the geometric directivity pattern $\Gamma_{\mathrm{t,geo}}(f, \alpha_{\mathrm{t,P}})$. However, $\Delta\alpha_{\mathrm{t,P}}(f)$ is rather large for low frequencies.

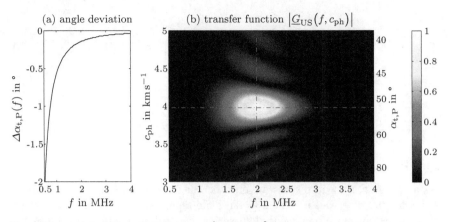

Fig. 9.38 a Angle deviation $\Delta\alpha_{t,P}(f) = \alpha_{t,P}^{\text{dom}}(f) - \alpha_{t,P}^{0}$ inside pipe wall between dominant angle $\alpha_{t,P}^{\text{dom}}(f)$ of sound radiation and expected angle $\alpha_{t,P}^{0}$ with respect to frequency f; **b** normalized transfer function $|\underline{G}_{\text{US}}(f, c_{\text{ph}})|$ (magnitude) of ultrasonic wedge transducer in frequency–wavenumber domain; dotdashed blue lines indicate transducer's excitation frequency $f_{\text{ex}} = 2\,\text{MHz}$ and design phase velocity $c_{\text{ph}}^{0} = 3980\,\text{m s}^{-1}$ corresponding to expected angle $\alpha_{t,P} = 52.00°$

As mentioned at the beginning, the transfer function $\underline{G}_{\text{US}}(f, c_{\text{ph}})$ of an ultrasonic transducer in the frequency–wavenumber domain also depends on its electroacoustic transfer behavior $\underline{H}_{\text{EA}}(f)$. This transfer function is connected to the electroacoustical impulse response $h_{\text{EA}}(t)$ of the transducer through the relation $\underline{H}_{\text{EA}}(f) = \mathcal{F}\{h_{\text{EA}}(t)\}$. There exist various measurement techniques to determine $h_{\text{EA}}(t)$ and $\underline{H}_{\text{EA}}(f)$, e.g., hydrophone measurements (see Chap. 8). For the sake of simplicity, let us suppose here a Gaussian distribution for $\underline{H}_{\text{EA}}(f)$, namely

$$\underline{H}_{\text{EA}}(f) = e^{-\left(\frac{f-f_c}{\sigma_\tau}\right)^2} \quad \text{with} \quad \sigma_\tau = \frac{B_{\text{US}}^{\text{-6dB}}}{\sqrt{2\ln(2)}}. \tag{9.110}$$

The final transfer function $\underline{G}_{\text{US}}(f, \alpha_{t,P})$ of the ultrasonic transducer reads as

$$\underline{G}_{\text{US}}(f, \alpha_{t,P}) = \Gamma_{t,\text{az}}(f, \alpha_{t,P}) \cdot \underline{H}_{\text{EA}}(f) \tag{9.111}$$

and applies, of course, to both transmitter and receiver because they are assumed to exhibit identical behavior, i.e., $\underline{G}_{\text{US}}(f, \alpha_{t,P}) = \underline{G}_{\text{US,t}}(f, \alpha_{t,P}) = \underline{G}_{\text{US,r}}(f, \alpha_{t,P})$. It is again possible to convert $\underline{G}_{\text{US}}(f, \alpha_{t,P})$ into $\underline{G}_{\text{US}}(f, c_{\text{ph}})$ with the aid of (9.100) and (9.101).

The resulting transfer function $|\underline{G}_{\text{US}}(f, c_{\text{ph}})|$ (magnitude) of the considered ultrasonic wedge transducer can be seen in Fig. 9.38b. As expected, the transducer's radiation behavior is limited in the frequency and wave number domain. Consequently, the generated wave packet after pulsed transducer excitation will be also limited in time as well as space.

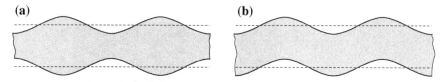

Fig. 9.39 Example of **a** symmetrical Lamb wave modes s_n and **b** antisymmetric Lamb wave modes a_n; dashed line indicate plate in original state

Pipe Wall Transmission

As a next step, we will discuss the influence of the pipe on CTU flow meters. In doing so, the pipe wall is treated as flat plate representing the intermediate layer of the three-layered system wedge half-space/pipe wall/fluid half-space. Before this configuration is studied in detail, let us take a brief look at the sound transmission through a plate being immersed in a fluid. Sound transmission through such an immersed plate was already subject to numerous publications, e.g., [18, 83]. It is well known that Lamb wave modes can be excited in a plate for specific combinations of frequency f and phase velocity c_{ph} within the plate. These modes represent waves guided along a plane plate of finite thickness and infinite extent otherwise. Mainly, one distinguishes between symmetrical and antisymmetric Lamb wave modes (see Fig. 9.39), which are usually termed s_n and a_n, respectively. The index n stands for the order of the Lamb wave mode, e.g., $n = 0$ indicates the zero-order modes. For the simple configuration of a free plate, the possible combinations of f and c_{ph} for the different Lamb wave modes (i.e., s_n and a_n) result from the solution to the so-called *Rayleigh–Lamb frequency equations* [90]. However, the pipe wall of a CTU flow meter calls for an alternative approach due to two reasons: (i) the considered structure consists of three different materials and (ii) we are not only interested in generated Lamb waves but in the quantitative transfer behavior of the pipe wall for various combinations (f, c_{ph}). The so-called *Global Matrix Method* (GMM) enables the calculation of the desired transfer behavior for arbitrary layered structures [61].

Figure 9.40a displays the transfer function $\left|\underline{G}_{\mathrm{P}}(f, c_{\mathrm{ph}})\right|$ (magnitude) of the pipe wall in the frequency–wavenumber domain, which results from the GMM for the considered CTU flow meter. The horizontal axis was rescaled for the wall thickness $t_{\mathrm{P}} = 2\,\mathrm{mm}$ of the steel pipe. As expected from the theoretical point of view, most energy will be transmitted through the pipe wall in the vicinity of Lamb wave modes (e.g., a_1). The resonance frequencies $f_{\mathrm{res,P}}$ of the pipe wall are specified by the intersections of design phase velocity $c_{\mathrm{ph}}^0 = 3980\,\mathrm{m\,s^{-1}}$ and Lamb wave modes in $\left|\underline{G}_{\mathrm{P}}(f, c_{\mathrm{ph}})\right|$. In the relevant frequency range (i.e., $f \in [0.5, 4.0]\,\mathrm{MHz}$), the Lamb wave modes s_0, a_1, and s_1 arise at 1.21, 2.49, and 3.76 MHz, respectively. Note that the transfer function $\underline{G}_{\mathrm{P}}(f, c_{\mathrm{ph}})$ and, therefore, these resonance frequencies strongly depend on the wall thickness t_{P}.

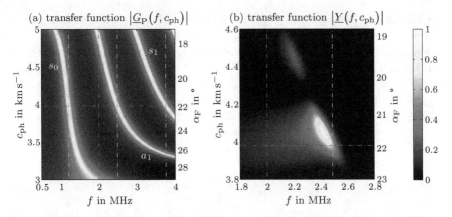

Fig. 9.40 a Normalized transfer function $\left|\underline{G}_{\mathrm{P}}(f, c_{\mathrm{ph}})\right|$ (magnitude) of pipe wall in frequency–wavenumber domain; Lamb wave modes s_0, a_1 and s_1; **b** normalized overall transfer function $\left|\underline{Y}(f, c_{\mathrm{ph}})\right|$ (magnitude) of considered CTU flow meter in frequency–wavenumber domain; dotdashed blue lines indicate transducer's excitation frequency $f_{\mathrm{ex}} = 2\,\mathrm{MHz}$, design phase velocity $c_{\mathrm{ph}}^{0} = 3980\,\mathrm{m\,s^{-1}}$ and expected fluid angle $\alpha_{\mathrm{F}} = 21.91°$; dotdashed green lines indicate resonance frequencies $f_{\mathrm{res,P}}$ of pipe wall at 1.21, 2.49 and 3.79 MHz

Fluid

The remaining transfer function $\underline{G}_{\mathrm{F}}(f, c_{\mathrm{ph}})$ of the CTU flow meter in the frequency–wavenumber domain refers to the sound propagation inside the fluid. This transfer function is influenced by both the flow-dependent measurement effect and the attenuation of propagating sound waves. Let us start with the flow-dependent measurement effect. According to the explanations in Sect. 9.3.2, there arise different sound velocities in upstream and downstream direction, which are here indicated as $c_{\mathrm{F,up}}$ and $c_{\mathrm{F,down}}$, respectively. Due to the differing sound velocities, the time-of-flights $t_{\mathrm{F},i}$ of the sound waves propagating from transmitter to receiver also vary in the upstream and downstream direction. We can directly convert the time difference $\Delta t_{\mathrm{F}} = t_{\mathrm{F,up}} - t_{\mathrm{F,down}}$ between both time-of-flights (transit times) into a phase difference $\Delta\phi_{\mathrm{F}}(f, \alpha_{\mathrm{F}})$ by relating Δt_{F} to the period duration $T = 1/f$. The phase difference, of course, depends on the angle α_{F} of the sound waves propagating inside the fluid (cf. Fig. 9.34). By additionally inserting (9.77), $\Delta\phi_{\mathrm{F}}(f, \alpha_{\mathrm{F}})$ takes the form

$$\Delta\phi_{\mathrm{F}}(f, \alpha_{\mathrm{F}}) = \frac{2\pi\,\Delta t_{\mathrm{F}}}{T} = 2\pi\,\Delta t_{\mathrm{F}}\,f = \frac{4\pi D_{\mathrm{Pi}}\tan\alpha_{\mathrm{F}}}{c_{\mathrm{F}}^{2}}\overline{v}_{\mathrm{F}}\,f \qquad (9.112)$$

with the sound velocity c_{F} of the fluid and its average flow rate $\overline{v}_{\mathrm{F}}$ through the pipe. It makes sense to split $\Delta\phi_{\mathrm{F}}(f, \alpha_{\mathrm{F}})$ into phase differences for upstream and downstream direction that become

$$\frac{\Delta\phi_{\mathrm{F}}(f, \alpha_{\mathrm{F}})}{2} = \Delta\phi_{\mathrm{F,up}}(f, \alpha_{\mathrm{F}}) = -\Delta\phi_{\mathrm{F,down}}(f, \alpha_{\mathrm{F}}) = \frac{2\pi D_{\mathrm{Pi}}\,\overline{v}_{\mathrm{F}}\,f}{c_{\mathrm{F}}\sqrt{c_{\mathrm{ph}}^{2} - c_{\mathrm{F}}^{2}}} \qquad (9.113)$$

by substituting $\alpha_F = \arcsin(c_F/c_{ph})$ in (9.112) and after some mathematical treatment.

As already discussed in Sect. 2.3.5, sound propagation in fluids is always accompanied by certain absorption mechanisms. The resulting attenuation depends on the fluid and the frequency f of the sound waves. To incorporate this attenuation in the modeling procedure, one should introduce a frequency-dependent factor $\Psi_F(f, \alpha_F)$, which stems from the frequency-dependent attenuation coefficient $\alpha_{at}(f)$ (see (2.143, p. 39)) of the fluid and the geometric distance of sound propagation. For the considered CTU flow meter, $\Psi_F(f, \alpha_F)$ can be approximated by

$$\Psi_F(f, \alpha_F) \approx e^{-\frac{\alpha_{at}(f)D_{Pi}}{\cos \alpha_F}} \tag{9.114}$$

with the inner diameter D_{Pi} of the pipe.

The combination of the part $\Delta\phi_{F,i}(f, \alpha_F)$ originating from the flow-dependent measurement effect with $\Psi_F(f, \alpha_F)$ yields the fluid's transfer function

$$\underline{G}_{F,i}(f, \alpha_F) = \Psi_F(f, \alpha_F)\, e^{j\Delta\phi_{F,i}(f,\alpha_F)} \tag{9.115}$$

in the frequency–wavenumber domain for upstream and downstream direction, respectively. If the flow rate $\bar{v}_F$ of the fluid is zero, both transfer functions will coincide, i.e., $\underline{G}_{F,up}(f, \alpha_F) = \underline{G}_{F,down}(f, \alpha_F) = \underline{G}_F(f, \alpha_F)$. Just as the other transfer functions, $\underline{G}_F(f, \alpha_F)$ can easily be converted to $\underline{G}_F(f, c_{ph})$.

Resulting System Response

After all transfer functions of the considered CTU flow meter have been determined, we can evaluate the output $\underline{Y}(f, c_{ph})$ as well as $\underline{Y}(f, \alpha_F)$ of the ultrasonic receiver representing the system response in the frequency–wavenumber domain. Without limiting the generality, let us assume that the average flow rate $\bar{v}_F$ of the fluid through the pipe is zero. Hence, the phase difference $\Delta\phi_F(f, \alpha_F)$ is also zero; i.e., one can apply $\underline{G}_F(f, \alpha_F)$ for upstream and downstream direction.

Figure 9.40b shows the obtained magnitude $|\underline{Y}(f, c_{ph})|$, which reveals two important aspects for the considered CTU flow meter [84, 85]. The first aspect refers to the maximum in $|\underline{Y}(f, c_{ph})|$ that mainly specifies the dominant frequency f_{dom} and the dominant phase velocity c_{ph}^{dom}. Neither f_{dom} nor c_{ph}^{dom} coincide with the excitation frequency $f_{ex} = 2\,\text{MHz}$ and the design phase velocity $c_{ph}^0 = 3980\,\text{m s}^{-1}$. Since there is a distinct connection of phase velocity and angle, the dominant fluid angle α_F^{dom} for sound propagation, thus, also differs from the expected angle α_F^0. If this angle deviation is not taken into account in the analysis unit of the CTU flow meter, we will be inevitably confronted with remarkable measurement errors of the calculated flow rates.

The second aspect refers to the additional maximum in $|\underline{Y}(f, c_{ph})|$ at $(f, c_{ph}) = (2.2\,\text{MHz}, 4500\,\text{m s}^{-1})$ being located far away from c_{ph}^0. This maximum is caused by the synergy of the first side lobe in the transducer's effective directivity pattern $\Gamma_{t,az}(f, \alpha_{t,P})$ at $\alpha_{t,P} = 45°$ (cf. Figs. 9.37b and 9.38b) and the Lamb wave

Fig. 9.41 Normalized
angular spectrum $\Gamma(\alpha_F)$ for
considered CTU flow meter
with respect to fluid
angle α_F; expected fluid
angle $\alpha_F^0 = 21.91°$;
dominant fluid
angle $\alpha_F^{dom} = 21.57°$

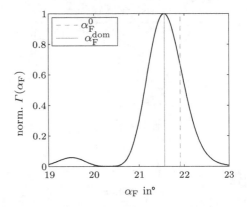

mode a_1 of the pipe wall (cf. Fig. 9.40a). Consequently, substantial sound energy is
transmitted for combinations (f, c_{ph}) close to the additional maximum.

Besides the conducted qualitative observations, the complex-valued system
response $\underline{Y}(f, \alpha_F)$ can be exploited to predict the *angular spectrum* for propagating waves inside the fluid as well as the *receive signal* of the considered CTU flow
meter. As a matter of fact, the dominant angle α_F^{dom} does not exclusively depend on
the peak of $|\underline{Y}(f, \alpha_F)|$ but on each frequency containing sound energy. It is possi-
ble to approximate the angular spectrum $\Gamma(\alpha_F)$ by performing an integration of the
energy $\propto |\underline{Y}|^2$ over the relevant frequencies $f \in [f_{min}, f_{max}]$, i.e.,

$$\Gamma(\alpha_F) = \int_{f_{min}}^{f_{max}} \left(|\underline{Y}(f, \alpha_F)|\right)^2 df . \tag{9.116}$$

The resulting angular spectrum for the considered CTU flow meter is displayed
in Fig. 9.41. The progress of $\Gamma(\alpha_F)$ can be interpreted as a directivity plot, which
provides a maximum that corresponds to the dominant fluid angle α_F^{dom}. In the present
case, α_F^{dom} takes the value $21.57°$ and, thus, exhibits a significant deviation from the
expected angle $\alpha_F^0 = 21.91°$ by $\Delta\alpha_F = -0.34°$. According to (9.96), such angle
deviation always leads to a relative systematic measurement error of $\epsilon_v = -1.7\%$
for the average flow rate, regardless of the actually value.

The receive signal $u_{rec}(t)$ of the CTU flow meter is an electrical output, which
depends only on time t in the time domain and, consequently, only on frequency f in
the frequency domain. That is the reason why we have to get rid of the argument α_F
in the system response $\underline{Y}(f, \alpha_F)$ with a view to calculating $u_{rec}(t)$. It, therefore,
makes sense to conduct an integration of $\underline{Y}(f, \alpha_F)$ over all relevant fluid angles $\alpha_F \in$
$[\alpha_{F,min}, \alpha_{F,max}]$; i.e., the complex-valued frequency spectrum $\underline{U}_{rec}(f)$ of the receive
signal is obtained from

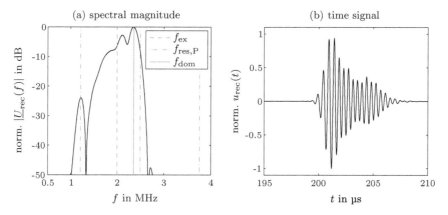

Fig. 9.42 **a** Normalized spectral magnitude $|\underline{U}_{rec}(f)|$ of receive signal for considered CTU flow meter; excitation frequency $f_{ex} = 2\,\text{MHz}$; dominant frequency $f_{dom} = 2.35\,\text{MHz}$; relevant resonance frequencies $f_{res,P}$ of pipe wall at 1.21, 2.49 and 3.79 MHz; **b** normalized receive signal $u_{rec}(t)$ in time domain

$$\underline{U}_{rec}(f) = \int_{\alpha_{F,min}}^{\alpha_{F,max}} \underline{Y}(f, \alpha_F)\, d\alpha_F \ . \tag{9.117}$$

Figure 9.42a depicts the resulting spectral magnitude $|\underline{U}_{rec}(f)|$ of the receive signal for the considered CTU flow meter. Let us take a closer look at the dominant frequency $f_{dom} = 2.35\,\text{MHz}$, which indicates the maximum $U_{rec,max}$ of $|\underline{U}_{rec}(f)|$. Not surprisingly, f_{dom} strongly deviates from the excitation frequency $f_{ex} = 2\,\text{MHz}$ of the ultrasonic transmitter. The maximum seems to be shifted toward the second pipe wall resonance at $f_{res,P} = 2.49\,\text{MHz}$ that refers to Lamb wave mode a_1. Furthermore, $|\underline{U}_{rec}(f)|$ contains an additional peak at 1.21 MHz and a notch at 1.33 MHz. While the additional peak originates from Lamb wave mode s_0 of the pipe wall, the notch results from the frequency spectrum $\underline{U}_{ex}(f)$ of the excitation signal (cf. Fig. 9.36b).

In a last step, it is possible to compute the receive signal $u_{rec}(t)$ of the considered CTU flow meter in the time domain (see Fig. 9.42b) by applying the inverse Fourier transform to $\underline{U}_{rec}(f)$, i.e., $u_{rec}(t) = \mathcal{F}^{-1}\{\underline{U}_{rec}(f)\}$. This step concludes the modeling procedure starting from the electrical excitation signal $u_{ex}(t)$ of the ultrasonic transmitter and ending with the electrical output signal $u_{rec}(t)$ of the ultrasonic receiver.

Experimental Verification

To verify the modeling approach for CTU flow meters in the frequency–wavenumber domain, let us finally compare predicted relative systematic measurement errors ϵ_v for the average flow rate with measurements, as observed on a water flow rig. In doing so, a constant volume flow of $\dot{V}_F = 20\,\text{L s}^{-1}$ water was prescribed whereby a high-precision Coriolis flow meter served as reference measurement device

[84, 85]. Owing to the constant $\dot{V}_\mathrm{F}$, the average flow rate $\bar{v}_\mathrm{F}$ in the steel pipe also remains constant. The entries in Table 9.3 correspond to the material properties and most of the geometric quantities of the realized experimental setup. The only exceptions are the pipe dimensions (i.e., outer and inner diameter as well as wall thickness) at the measuring point, which take the values $D_\mathrm{Po} = 88.9\,\mathrm{mm}$, $D_\mathrm{Pi} = 85.1\,\mathrm{mm}$, and $t_\mathrm{P} = 1.9\,\mathrm{mm}$. For the wall thickness being actually present, the relevant Lamb wave modes within the pipe arise at the frequencies 1.27, 2.62 and 3.96 MHz.

The frequency–wavenumber modeling approach was exploited for calculating the dominant fluid angle $\alpha_\mathrm{F}^\mathrm{dom}$ that is required to determine the average flow rate $\bar{v}_\mathrm{F}$ through the pipe (cf. (9.77)). The angle deviation $\Delta\alpha_\mathrm{F}$ between $\alpha_\mathrm{F}^\mathrm{dom}$ and the expected angle α_F^0 was then used to predict ϵ_v through (9.96). The measured average flow rate was calculated on basis of α_F^0 and the time difference Δt_F between sound wave propagation in upstream and downstream direction of the utilized CTU flow meter. By comparing the measurement result for $\bar{v}_\mathrm{F}$ with the true value, we obtain the relative systematic error of the conducted measurement.

In Fig. 9.43, one can see measured as well as predicted values for the relative systematic measurement error ϵ_v with respect to the excitation frequency f_ex of the ultrasonic transmitter. Each frequency refers to a bipolar square burst signal consisting of $n_\mathrm{burst} = 3$ burst periods. As the comparison of the curve progressions clearly demonstrates, measured and predicted values coincide very well in a wide frequency range around the center frequency $f_\mathrm{c} = 2\,\mathrm{MHz}$ of the ultrasonic transducers. In both cases, the relative systematic measurement error of the average flow rate $\bar{v}_\mathrm{F}$ takes almost values in the range of $\pm 3\%$, which is too large for various industrial applications of flow meters. Besides, it is possible to deduce further findings for CTU flow meters from the results of Fig. 9.43. For instance, excitation frequencies close to one of the pipe wall resonances $f_\mathrm{res,P}$ do not always imply a small value of ϵ_v. By exploiting the presented modeling approach in frequency–wavenumber domain, we are, however, able to figure out appropriate transducer excitations that lead to a small systematic measurement error of a CTU flow meter [84]. The modeling approach also allows the optimization of the transducer design regarding the expected pipe wall properties.

Fig. 9.43 Measured and predicted systematic measurement error ϵ_v of actually realized CTU flow meter; pipe wall thickness $t_\mathrm{P} = 1.9\,\mathrm{mm}$ at measuring point; relevant resonance frequencies $f_\mathrm{res,P}$ of pipe wall at 1.27 and 2.62 MHz

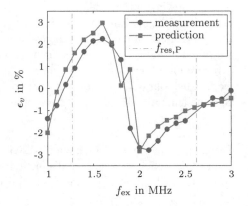

9.4   Cavitation Sensor for Ultrasonic Cleaning

Ultrasonic cleaning is widely used in practical applications for cleaning of objects like surgical instruments, parts of precision mechanics, optical lenses, dentures, jewelry as well as assembled printed circuit boards. Especially in case of irregularly shaped object surfaces, ultrasonic cleaning constitutes an excellent choice since traditional cleaning methods such as grinding are hardly feasible.

During ultrasonic cleaning, the object to be cleaned gets immersed in an appropriate cleaning liquid in which an ultrasonic field of high sound intensity is present. Such an ultrasonic field leads to nucleations of cavitation bubbles, which subsequently can collapse close to the object surface. If the process parameters (e.g., cleaning duration) are chosen carefully, the resulting mechanical forces will release dirt particles from the object surface and, therefore, the immersed object will be cleaned. However, it is impossible to generate an evenly distributed ultrasonic field of high intensity inside the cleaning liquid due to the exploitation of standing waves and the presence of the object to be cleaned. We are, consequently, confronted with a varying cavitation activity and cleaning efficiency along the object surface. That is the reason why a special cavitation sensor has been developed at the Chair of Sensor Technology (Friedrich-Alexander-University Erlangen-Nuremberg) [93, 97, 98]. The mechanically flexible cavitation sensor is based on a ferroelectret material and can be directly applied to curved surfaces of objects.

This section deals with the realized cavitation sensor. We will start with fundamentals of acoustic cavitation and ultrasonic cleaning. Afterward, conventional approaches for measuring cavitation activity are explained. Section 9.4.3 details then the setup of the realized cavitation sensor and Sect. 9.4.4 its characterization. At the end, selected experimental results will be presented.

9.4.1   Fundamentals of Acoustic Cavitation and Ultrasonic Cleaning

In the following, let us briefly discuss the fundamentals of acoustic cavitation and ultrasonic cleaning. This includes the nucleation of cavities inside a liquid as well as the dynamics of cavitation bubbles because the bubble collapse is exploited in ultrasonic cleaning. Such cavities can be generated either by high sound fields which is referred to as acoustic cavitation or by intense pulsed laser radiation. Owing to the fact that ultrasonic cleaning is commonly based on high sound fields inside a cleaning tank, we will, however, concentrate exclusively on acoustic cavitation.

Nucleation of Cavities

The nucleation or initial formation of a cavity inside a liquid constitutes the starting point of cavitation. Basically, a distinction is made between *homogeneous nucleation* and *heterogeneous nucleation* [109]. Homogeneous nucleation can occur in a homogeneous liquid that contains neither impurities nor gas bubbles. If the negative pressure of a propagating sound wave takes higher values than the liquid's tensile strength, the liquid will be locally torn and, consequently, a cavity will be formed. As stated by Temperley, the van der Waals forces have to be exceeded which implies sound pressure amplitudes $> 10^8$ MPa in water [26, 104]. However, in real systems and practical experiments, nucleation of cavities already exists at much lower sound pressure amplitudes. This so-called heterogeneous nucleation arises at impurities inside the liquid as well as at the interfaces liquid/solid and liquid/gas.

Regardless of whether homogeneous nucleation or heterogeneous nucleation takes place in the liquid, the resulting cavitation bubbles are filled with vapor. According to the classical nucleation theory, cavitation bubbles must overcome a certain energy barrier. The energy in the system comprising cavitation bubble as well as surrounding liquid can be altered by two counteracting factors during bubble formation. When a cavitation bubble is generated in a liquid, the Gibbs free energy $\mathcal{G}_B$ of the system will be reduced because vapor exhibits a lower energy density than liquid. However, $\mathcal{G}_B$ also increases by an amount proportional to the bubble surface due to the formation of the bubble/liquid interface. For a spherical cavitation bubble with radius R_B, the change $d\mathcal{G}_B$ of Gibbs free energy reads as [65]

$$d\mathcal{G}_B = \underbrace{4\pi R_B^2 \, \gamma_{surf}}_{surface} - \underbrace{\frac{4}{3}\pi R_B^3 \left(p_{v,sat} - p\right)}_{volume} \tag{9.118}$$

where γ_{surf} stands for the surface energy density in $J\,m^{-2}$. The expressions p and $p_{v,sat}$ denote the local pressure inside the liquid and the saturation vapor pressure,[6] respectively. The critical radius R_{crit} of a spherical cavitation bubble follows from the first-order derivative of (9.118) and finally computes as

$$R_{crit} = \frac{2\gamma_{surf}}{p_{v,sat} - p} . \tag{9.119}$$

Note that for this bubble radius, $d\mathcal{G}_B$ reaches its maximum, which becomes

$$d\mathcal{G}_{B,max} = \frac{16\pi\gamma_{surf}^3}{3\left(p_{v,sat} - p\right)^2} . \tag{9.120}$$

If the actual bubble radius R_B is smaller than R_{crit}, cavitation bubbles will tend to shrink, whereas for $R_B > R_{crit}$, cavitation bubbles will tend to expand. Therefore, $d\mathcal{G}_{B,max}$ corresponds to the activation energy that is required for nucleation of a cavity. The activation energy will be exceeded when sound waves feature sufficiently high amplitudes.

---

[6]The saturation vapor pressure equals the pressure at which liquid and vapor are in phase equilibrium for a given temperature ϑ.

Fig. 9.44 Heterogeneous
nucleation of cavitation
bubble of radius R_B at flat
wall; bubble filled with vapor
and surrounded by liquid;
contact angle Θ_B between
bubble and wall

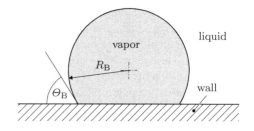

In case of heterogeneous nucleation, the formed cavitation bubbles are not spherical anymore and, thus, the activation energy gets altered. For instance, if a cavitation
bubble is formed at a flat wall (see Fig. 9.44), the change $d\mathcal{G}_B$ of Gibbs free energy
will take the form [65]

$$
d\mathcal{G}_B = \left[4\pi R_B^2 \gamma_{\text{surf}} - \frac{4}{3}\pi R_B^3 \left(p - p_{\text{v,sat}} \right) \right] \underbrace{\frac{(1 + \cos \Theta_B)^2 (2 - \cos \Theta_B)}{4}}_{=\lambda_{\text{cav}}} \tag{9.121}
$$

with the contact angle Θ_B of cavitation bubble and wall. Here, the quantity λ_{cav}
describes a scaling factor between homogeneous and heterogeneous nucleation. The
critical radius R_{crit} for heterogeneous nucleation corresponds again to (9.118), but
the resulting maximum change $d\mathcal{G}_{B,\text{max}}$ is given by

$$
d\mathcal{G}_{B,\text{max}} = \frac{16\pi \gamma_{\text{surf}}^3}{3\left(p_{\text{v,sat}} - p \right)^2} \lambda_{\text{cav}} . \tag{9.122}
$$

Not surprisingly, when the cavitation bubble does not touch the wall, $\Theta_B = 0$ as well
as $\lambda_{\text{cav}} = 1$ will be satisfied and heterogeneous becomes homogeneous nucleation.

As the comparison of (9.120) and (9.122) reveals, $d\mathcal{G}_{B,\text{max}}$ is always smaller for
heterogeneous than for homogeneous nucleation since $\lambda_{\text{cav}} < 1 \; \forall \Theta_B \neq 0$ holds. In
other words, heterogeneous nucleation arises at lower sound pressure amplitudes than
homogeneous nucleation. Due to the fact that liquids are contaminated with impurities in practical situations and contain, moreover, fluid/solid interfaces in ultrasonic
cleaning, heterogeneous nucleation is the dominating mechanism for the formation
of cavitation bubbles.

Cavitation Bubble Dynamics

After nucleation, the size of the cavitation bubbles will change dynamically when
there exists a sound field within the surrounding liquid. This includes bubble growth
and bubble shrinking as well as bubble oscillations that are commonly referred to
as stable cavitation. Moreover, cavitation bubble dynamics also comprises bubble
collapse, the so-called inertial cavitation.

One can find various differential equations in the literature to describe the dynamic
behavior of cavitation bubbles. Hereinafter, let us briefly detail the fundamental

idea behind those equations. In doing so, we assume a single cavitation bubble and an incompressible surrounding liquid. In the initial equilibrium state, the gas pressure p_{B0} inside the cavitation bubble of radius R_{B0} results from [53]

$$p_{B0} = p_0 + \frac{2\,T_{surf}}{R_{B0}}\,. \tag{9.123}$$

The expressions p_0 and T_{surf} stand for the hydrostatic pressure within the surrounding liquid and the surface tension on the bubble surface, respectively. By applying an additional sound pressure field $p_\sim(t)$, the pressure within the liquid changes to $p(t) = p_0 + p_\sim(t)$. Hence, the time-dependent pressure $p_W(t)$ that acts on the bubble wall reads as

$$p_W(t) = p(t) + \frac{2\,T_{surf}(t)}{R_B(t)} = p_0 + p_\sim(t) + \frac{2\,T_{surf}(t)}{R_B(t)} \tag{9.124}$$

with $T_{surf}(t)$ and $R_B(t)$ also depending on time. As a matter of fact, the bubble radius will change if $p_W(t)$ does not coincide with the gas pressure $p_B(t)$. Let us assume a constant gas quantity inside the cavitation bubble, i.e., diffusion processes are neglected, and moderate sound pressure amplitudes $\hat{p}_\sim$. Then, cavitation bubbles shrink in case of positive sound pressure values, i.e., $p_\sim(t) > 0$. On the other hand, $p_\sim(t) < 0$ is accompanied by a growth of the cavitation bubbles. A propagating sound pressure wave, therefore, alters the bubble size (see Fig. 9.45).

Every change in the size of a cavitation bubble generates a certain fluid flow in the surrounding liquid. The time-dependent velocity $v_L(r, t)$ of this fluid flow at the radial distance r from the bubble center is calculated as

$$v_L(r, t) = \left[\frac{R_B(t)}{r}\right]^2 \frac{dR_B(t)}{dt} \tag{9.125}$$

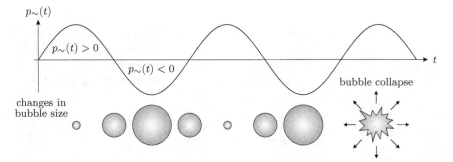

Fig. 9.45 Size of cavitation bubbles changes periodically with respect to time t through sinusoidal sound pressure wave $p_\sim(t)$; bubble shrinks for $p_\sim(t) > 0$ and grows for $p_\sim(t) < 0$; high sound pressure amplitudes cause bubble collapse generating shock waves in surrounding liquid

and tends to zero for $r \to \infty$. Overall, the kinetic energy $\mathcal{E}_{kin}(t)$ of the fluid flow becomes (density ϱ_L of the liquid) [53]

$$\mathcal{E}_{kin}(t) = \frac{\varrho_L}{2} \int_{R_B(t)}^{\infty} v_L(r, t)^2 \cdot 4\pi r^2 dr = 2\pi \varrho_L R_B(t)^3 \dot{R}_B^2 \qquad (9.126)$$

with $\dot{R}_B = dR_B(t)/dt$, which represents the velocity of the bubble wall. The change in the bubble size implies, however, an additional work $W_B(t)$ since the bubble wall has to move against the pressure difference $p_W(t) - p_B(t)$. This work takes the form

$$\frac{dW_B(t)}{dt} = [p_W(t) - p_B(t)] 4\pi R_B(t)^2 \dot{R}_B \qquad (9.127)$$

per unit time and has to be compensated by reducing the kinetic energy of the fluid flow, i.e.,

$$\frac{dW_B(t)}{dt} = -\frac{d\mathcal{E}_{kin}(t)}{dt}. \qquad (9.128)$$

Inserting (9.124), (9.126), and (9.127) in (9.128) leads to[7]

$$\varrho_L \left[R_B \ddot{R}_B + \frac{3}{2} \dot{R}_B^2 \right] + p + \frac{2 T_{surf}}{R_B} - p_B = 0 \qquad (9.129)$$

with $\ddot{R}_B = d^2 R_B(t)/dt^2$. This differential equation is known as *Noltingk–Neppiras equation* [73]. By additionally considering the dynamic viscosity η_L of the surrounding liquid, we arrive at the so-called *Rayleigh–Plesset equation* [58]

$$\varrho_L \left[R_B \ddot{R}_B + \frac{3}{2} \dot{R}_B^2 \right] + p + \frac{2 T_{surf}}{R_B} + \frac{4\eta_L}{R_B} \dot{R}_B - p_B = 0. \qquad (9.130)$$

It is reasonable to assume adiabatic changes in state, which means that heat exchanges between gas and surrounding liquid do not occur. According to this assumption, the gas pressure $p_B(t)$ inside the cavitation bubble exclusively depends on its current radius $R_B(t)$ as well as on the radius R_{B0} and the gas pressure p_{B0} in the initial equilibrium state. The mathematical link is given by [53]

$$p_B(t) = p_{B0} \left[\frac{R_{B0}}{R_B(t)} \right]^{3\kappa} \qquad (9.131)$$

where κ represents the adiabatic exponent of the enclosed gas. By combining this relation with (9.123) and replacing p_B in (9.130), one finally obtains the *RPNNP*

---

[7]For compactness, the argument time t is omitted.

equation[8] [58]

$$\varrho_L \left[R_B \ddot{R}_B + \frac{3}{2} \dot{R}_B^2 \right] + p + \frac{2\,T_{\text{surf}}}{R_B} + \frac{4\eta_L}{R_B} \dot{R}_B - \left[p_0 + \frac{2\,T_{\text{surf}}}{R_{B0}} \right] \left[\frac{R_{B0}}{R_B} \right]^{3\kappa} = 0 \,.$$

(9.132)

This equation describes movements of the bubble wall with sufficient precision in case of moderate wall velocities [72]. However, if strongly nonlinear bubble oscillations arise, the assumption of an incompressible surrounding liquid will lead to large deviations after the first bubble collapse. Extended versions like the so-called *Herring-Trilling equation* [105] and *Gilmore equation* [58] consider the compressibility of the surrounding liquid in addition.

Now, let us take a look at the numerical solution of the RPNNP Eq. (9.132) for an applied sound pressure $p_\sim(t)$. In doing so, sinusoidal sound pressure waves are assumed to propagate in water (density $\varrho_L = 1000\,\text{kg m}^{-3}$; $c_L = 1484\,\text{m s}^{-1}$) at a typical frequency of $f_{\text{ex}} = 30\,\text{kHz}$ in ultrasonic cleaning. The bubble radius and the gas pressure in the initial equilibrium state equal $R_{B0} = 10\,\mu\text{m}$ and $p_{B0} = 10^5\,\text{Pa}$, respectively. The adiabatic exponent κ of the enclosed gas was set to 1.0. The surface tension T_{surf} on the bubble wall was neglected which will be permitted if the bubble radius is not too small.

Figure 9.46a, b, and c depict the numerical solutions for three sound pressure amplitudes $\hat{p}_\sim$, namely 10, 30, and 80 kPa. The top and bottom panels show the current bubble radius $R_B(t)$ and the resulting wall velocity $\dot{R}_B(t)$, respectively. It can be clearly observed that the cavitation bubble expands during negative sound pressure and shrinks during positive sound pressure for low values of $\hat{p}_\sim$, i.e., for 10 and 30 kPa. As expected, the greater $\hat{p}_\sim$, the larger the variations in $R_B(t)$ and $\dot{R}_B(t)$ will be. Both quantities show almost a sinusoidal progression (see Fig. 9.46a, b) and, therefore, overtones are only weakly pronounced. However, in case of comparatively large sound pressure amplitudes like $\hat{p}_\sim = 80\,\text{kPa}$, $R_B(t)$ as well as $\dot{R}_B(t)$ strongly deviate from sinusoidal progression (see Fig. 9.46c). The bubble shrinks to less than $0.5 R_{B0}$ and grows to more than R_{B0} several times in a row during positive sound pressure. These bubble oscillations are accompanied by high velocities of the bubble wall. Moreover, there arise remarkable overtones (e.g., at $2 f_{\text{ex}}$ and $3 f_{\text{ex}}$) in the frequency spectra of $R_B(t)$ and $\dot{R}_B(t)$ since the bubble oscillations are of higher frequency than the excitation frequency [56].

When the sound pressure amplitude is increased further, the bubble oscillations will also increase and $R_B(t)$ can grow to a multiple of R_{B0}; e.g., a cavitation bubble can exhibit a radius of more than 100 μm for $f_{\text{ex}} = 30\,\text{kHz}$. In such cases, the bubble oscillations comprise frequency components lower than the excitation frequency f_{ex} in addition because this bubble growth usually takes more than one excitation period. The frequency spectra of $R_B(t)$ and $\dot{R}_B(t)$, consequently, do not only contain pronounced components at f_{ex} and its overtones at, e.g., $2 f_{\text{ex}}$ but also so-called subharmonic components at $f_{\text{ex}}/2$, $f_{\text{ex}}/3$, etc., as well as so-called

---

[8]The abbreviation RPNNP stands for Rayleigh–Plesset–Noltingk–Neppiras–Poritsky.

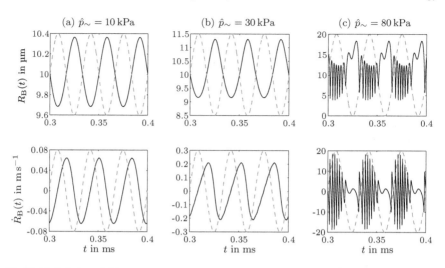

Fig. 9.46 Solution of RPNNP equation for sound pressure amplitudes **a** $\hat{p}_\sim = 10\,\text{kPa}$, **b** $\hat{p}_\sim = 30\,\text{kPa}$, and **c** $\hat{p}_\sim = 80\,\text{kPa}$ at excitation frequency $f_{ex} = 30\,\text{kHz}$; initial bubble radius $R_{B0} = 10\,\mu\text{m}$; time-dependent bubble radius $R_B(t)$ in top panels; time-dependent wall velocity $\dot{R}_B(t)$ of bubble in bottom panels; gray dotted line illustrates normalized sound pressure wave $p_\sim(t)$ in surrounding water

ultraharmonic components at $3f_{ex}/2$, $5f_{ex}/2$, etc., [53, 56]. It should be mentioned that bubble oscillations are in general larger for low than for high excitation frequencies. This circumstance directly follows from the longer period of negative pressure for low values of f_{ex}. That is the reason why lower excitation frequencies are accompanied by a greater cavitation activity than higher excitation frequencies.

An oscillating cavitation bubble itself acts as sound source, which generates sound pressure waves according to $p_\sim(t) \propto \ddot{R}_B(t)$. As a result, there arise sound pressure waves in the surrounding liquid due to the excitation at f_{ex} as well as sound pressure waves originating from bubble oscillations. If the sound pressure signal is measured in the surrounding liquid and this signal includes pronounced overtones, subharmonic and ultraharmonic components, we can, thus, expect that the liquid contains oscillating cavitation bubbles.

As a matter of fact, cavitation bubbles will tend to collapse when we do not have stable cavitation in the liquid, i.e., stable bubble oscillations. Note that a cavitation bubble usually does not vanish completely during the collapse (inertial cavitation), but the bubble size is enormously reduced, e.g., to one-twentieth of the diameter before collapse [53]. Afterward, the small cavitation bubble can unite with other bubbles and grow again. The inertial cavitation represents an extremely fast process, whereby the wall velocity $\dot{R}_B(t)$ of the bubble can reach values greater than the sound velocity c_L of the surrounding liquid [21]. Therefore, a bubble collapse generates a very short sound pressure pulse in the form of a shock wave, which leads to a broadband noise in the frequency spectrum of the measured sound pressure signal [109]. To this end, we can state that the measured sound pressure signal in the surrounding

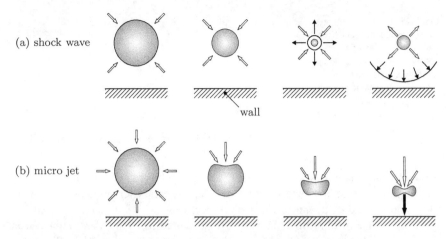

Fig. 9.47 Two effects of inertial cavitation near to wall, i.e., object surface; **a** shock wave exhibiting high acoustic intensity; **b** microjet impinging at very high speed on wall [97]

liquid contains characteristic frequency components as well as broadband noise in case of a pronounced cavitation activity, which includes bubble oscillations and bubble collapse. Since the broadband noise is also present for frequencies smaller than 20 kHz, people commonly perceive cavitation activity as noise.

Besides a shock wave, an extremely short flash of light is emitted during inertial cavitation. This phenomenon is usually named *sonoluminescence* [22]. The emitted light flashes of a typical duration up to a few hundred picoseconds exhibit a peak radiation intensity of the order of a few milliwatts.

Ultrasonic Cleaning

Ultrasonic cleaning is based on inertial cavitation of bubbles nearby the object surface, which has to be cleaned. In principle, one can distinguish between two different effects of the inertial cavitation near to a wall representing the object surface. While the first effect is the generation of an acoustic shock wave during bubble collapse, the second effect refers to the formation of a microjet (see Fig. 9.47) [97]. The generated shock wave exhibits high acoustic intensity acting on the object surface. In case of the microjet formation, the cavitation bubble near to the object surface loses its spherical shape due to differences in the flow conditions around the bubble [57]. The bubble area facing away from the object becomes invaginated. In the further course, a microjet is developed that impinges at a very high speed of more than $100 \, \text{km s}^{-1}$ on the object surface [60].

Shock waves as well as micro jets have a certain impact on the object surface because both effects are accompanied by a local energy input. Hammitt [37] figured out that a material-dependent energy barrier has to be exceeded in order to remove particles from the object surface. Not surprisingly, only the energy portion exceeding this so-called damaging threshold contributes to the surface erosion, which is caused by inertial cavitation. Fortes-Patella et al. [31] suggested a material-dependent

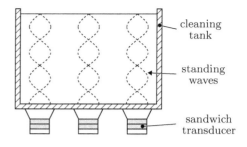

Fig. 9.48 Cleaning tank with piezoelectric sandwich transducers attached to bottom; transducers generate standing sound waves of high intensity inside cleaning liquid [53]

parameter that enables estimating the damaging threshold and, therefore, rating the cavitation resistance of a material. For example, this parameter amounts $\approx 4\,J\,mm^{-3}$ for aluminum and $\approx 30\,J\,mm^{-3}$ for steel, respectively.

In the course of ultrasonic cleaning, the objects to be cleaned are immersed in cleaning liquids like aqueous solutions or organic cleaning agents [53]. Since dissolved gases lower the cavitation activity, it is recommended to degas the cleaning liquid in advance and to heat it up to 80 °C during the cleaning process. The liquid is filled into a so-called cleaning tank that is usually made of stainless steel. With a view to achieving pronounced cavitation activities inside the cleaning liquid through acoustic waves, we need sound pressure waves of remarkable amplitudes because cavitation bubbles have to be formed and inertial cavitation must take place. That is the reason why standing sound waves of high intensity (i.e., up to $5\,W\,cm^{-2}$) are commonly generated inside the cleaning liquid (see Fig. 9.48). In doing so, there arises a good cleaning effect at the antinodes but a little cleaning effect close to the nodes of the standing waves. A uniform cleaning requires, thus, movements and rotations of the object to be cleaned during the cleaning process. It should be also noted that the immersed object influences the formation of standing waves inside the cleaning liquid.

The standing sound waves are usually generated by several piezoelectric sandwich transducers, which are attached to the bottom of the cleaning tank (see Fig. 9.48) or immersed into the cleaning liquid as encapsulated unit. To avoid noise pollution in immediate surroundings and to facilitate inertial cavitation inside the liquid, the operating frequency f_{ex} of typical ultrasonic cleaning systems lies between 20 and 40 kHz [60, 97]. Higher operating frequency (e.g., 100 kHz) can, however, be helpful if gentle cleaning is desired. Depending on the degree of contamination of the object to be cleaned, the sound intensity and the used cleaning liquid, the cleaning duration ranges from a few seconds to one minute [53]. On grounds of efficiency, the resonance frequency f_r of the utilized piezoelectric sandwich transducers should coincide with the operating frequency of the ultrasonic cleaning system.

9.4.2   Conventional Measurements of Cavitation Activity

As mentioned above, inertial cavitation implies broadband noise in the sound pressure signal, which arises in the surrounding liquid. Furthermore, short flashes of light are emitted during bubble collapse. It is not surprising that both effects are exploited in practical applications to evaluate cavitation activity, i.e., inertial cavitation as well as bubble oscillations. Here, we will concentrate exclusively on sound pressure measurements, in particular on ultrasonic measurements.

Fundamentally, one can distinguish between active and passive cavitation detection if sound pressure measurements are utilized [109]. In the framework of active cavitation detection, the area of cavitation gets treated with ultrasound by an appropriate ultrasonic transducer. The scattered ultrasonic waves contain information about the existing cavitation activity because processes like bubble oscillations and bubble collapses affect those waves. When a focused ultrasonic transducer (e.g., a linear array) is operated in pulse-echo mode, we can additionally localize the cavitation area [103]. In such a case, active cavitation detection is also termed active cavitation imaging or active cavitation mapping. The problem of active cavitation detection lies, however, in the fact that this kind of detection is technically feasible only during the intermission of the ultrasonic source, which generates inertial cavitation. Otherwise, the scattered ultrasonic waves become completely covered by the high-intensity ultrasound generating cavitation and, thus, we are not able to acquire them in a reliable way. That is the reason why active cavitation detection is hardly applicable for ultrasonic cleaning.

In contrast to active cavitation detection, passive cavitation detection does not require an ultrasonic transducer for providing an additional sound field. The utilized transducer solely serves as receiver for the sound waves, which are generated from inertial cavitation and bubble oscillations [4]. Passive cavitation detection can also be applied during the operation of the ultrasonic source being responsible for inertial cavitation. Owing to this fact, passive cavitation detection should be applicable for ultrasonic cleaning. However, similar to active cavitation detection, we have to deal with the problem that the receive signals comprise pronounced spectral components at the fundamental frequency f_{ex} of the excitation as well as its overtones due to nonlinear sound propagation. Subharmonic components (e.g., $f_{ex}/2$) and ultraharmonic components (e.g., $3f_{ex}/2$) pointing out the existence of oscillating cavitation bubble become, nevertheless, visible [109]. Besides, the frequency spectra of the receive signals contain the broadband noise, which is typical for inertial cavitation. An appropriate analysis of these frequency spectra enables measuring bubble oscillations as well as inertial cavitation by means of passive cavitation detection. When special imaging techniques like beam forming are used, the spatial resolution of the measurement can be improved, e.g., [35]. Passive cavitation detection is then also named passive cavitation imaging or passive cavitation mapping.

9.4.3 Realized Sensor Array

Even though there exist several approaches for measuring cavitation activity (cf. Sect. 9.4.2), the conventional measurement techniques are suitable only to a limited extent in ultrasonic cleaning. This can be mainly ascribed to the fact that objects

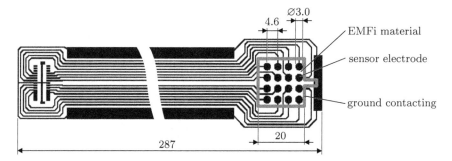

Fig. 9.49 Geometric structure of realized sensor array and signal lines [93]; array consists of 16 elements arranged in 4 rows and 4 columns; geometric dimensions in mm

to be cleaned influence the sound field in the cleaning liquid and, therefore, the cavitation activity will change spatially which will also alter the cleaning efficiency along object surfaces. Consequently, reliable investigations in ultrasonic cleaning demand a cavitation sensor that can be attached directly onto an object surface and allows spatial resolved measurements of the cavitation activity.

Below, we will discuss a special cavitation sensor, which was developed within the framework of the doctoral thesis of Strobel [97]. The underlying approach and obtained results were also published in [93, 98]. The realized sensor array represents a device for passive cavitation detection and is based on the mechanically flexible ferroelectret material electromechanical film (EMFi; see Sect. 3.6.3) from the company Emfit Ltd [27]. To achieve a pronounced piezoelectric coupling, the material type EMFi-HS of thickness $\approx 70\,\mu$m was utilized as active sensor material. Owing to its low mechanical stiffness and material density, the EMFi material should provide a comparable large frequency bandwidth for operating in cleaning liquids (cf. Fig. 7.39b on p. 322). The mechanical flexibility of this material constitutes, furthermore, a great advantage since it can be attached directly onto the surface of an object, even if the surface is curved.

Figure 9.49 displays the geometric dimensions of the realized sensor array containing 16 elements of $2R_T = 3.0$ mm diameter each, which are arranged in 4 rows and 4 columns. Both the element diameter and the lateral spacing between them were chosen according to the typical frequencies ranging from 20 to 40 kHz in ultrasonic cleaning. A further design criterion was the mechanical and electrical crosstalk of the array elements. For the maximum frequency of 40 kHz, the wavelength λ of the generated ultrasonic waves equals 37.1 mm in water. The geometric distance between neighboring sound pressure minimum (i.e., node) and maximum (i.e., antinode) amounts then $\lambda/4 = 9.3$ mm. Consequently, the chosen lateral spacing of 4.6 mm between two array elements guarantees that a single wave train can be sufficiently resolved in both time and spatial domain.

A schematic cross-sectional view of a single element of the realized sensor array is illustrated in Fig. 9.50. The active sensor component, a square-shaped EMFi material with an edge length of 20 mm, was fixed onto the bottom electrodes of the array

Fig. 9.50 Schematic cross-sectional view of single element of sensor array; copper (Cu) electrode at bottom of EMFi material and common aluminum (Al) electrode at top

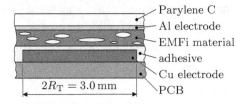

Parylene C
Al electrode
EMFi material
adhesive
Cu electrode
$2R_T = 3.0\,\text{mm}$
PCB

elements by a highly viscous and solvent-free adhesive. The bottom electrodes as well as the signal lines are made of copper. Electrodes and signal lines are, moreover, part of a flexible printed circuit board (PCB), which allows bending of the sensor array. While the sensor signals can be measured separately at these circular element electrodes of 3.0 mm diameter, a single top aluminum layer of 20 nm thickness that was vapor-deposited on the EMFi material serves as common ground for all sensor elements. To protect sensor array as well as signal lines from damaging environmental conditions due to cavitation effects, an additional coating with a Parylene C layer of 40 μm thickness is used. This polymer coating material is optically transparent, features an excellent chemical resistance, and enables a uniform surface covering [15, 112]. The Parylene C layer was formed by a chemical vapor deposition (CVD) process.

9.4.4 Characterization of Sensor Array

In the following, the metrological and simulation-based characterization of the realized sensor array will be detailed. This includes directivity pattern, resonance frequency, achievable sensitivity, and signal-to-noise ratio of the array elements. Moreover, we will study mechanical as well as electrical crosstalk between the sensor components.

Directivity Pattern

The directivity pattern describes the spatial sensitivity of the sensor array with respect to the angle of incident sound pressure waves. Concerning the practical application of the sensor array, the sensitivity of the array elements should not depend on the angle of incidence. To verify this, let us treat the individual array elements as piston-type transducers with element radius $R_T = 1.5\,\text{mm}$. According to Sect. 7.2.1, piston-type transducers will feature almost a spherical directivity pattern in the half-space if the product of wave number $k = 2\pi/\lambda$ and element radius is smaller than one, i.e., $kR_T < 1$. The higher the frequency f, the smaller the acoustic wavelength λ and the larger k will be. In the present case, the realized sensor array operates in water ($c_W \approx 1500\,\text{m s}^{-1}$) and should provide sound field information for frequencies up to 100 kHz, which leads to the maximum wave number $k \approx 420$. Hence, the condition $kR_T < 1$ always holds and the array elements may theoretically offer a spherical directivity pattern. This behavior was also confirmed by experiments as well as FE simulations [97].

Resonance Frequency

The mechanical resonance frequency f_r of a piezoelectric sensor denotes an essential parameter. From the theoretical point of view, the sensor sensitivity remains constant for frequencies smaller than f_r and decreases by $-20\,\mathrm{dB}/\mathrm{decade}$ for frequencies $f > f_r$. Here, let us a regard a simple analytical model to calculate f_r of a single array element. Basically, f_r is influenced by the mechanical mass m_S and mechanical compliance n_S of the sensor material. In case of a homogeneous and disk-shaped sensor element (thickness t_S) that can oscillate uniformly and freely in space, the resonance frequency in thickness direction reads as

$$f_r = \frac{1}{2\pi\sqrt{m_S \cdot n_S}} = \frac{1}{2\pi\sqrt{M_S \cdot N_S}} \tag{9.133}$$

with the so-called area density $M_S = m_S/A_S$ and the surface-related compliance $N_S = n_S \cdot A_S$. The expression $A_S = R_T^2 \pi$ stands for the base area of the disk. As shown in Fig. 9.50, the EMFi material of the realized sensor array is one-sided clamped at its bottom area. The effective mass of the EMFi material is, thus, reduced to one-third [51].

Even though the realized sensor array exhibits a rather complicated setup, we can neglect the adhesive layer at the bottom as well as the aluminum electrodes at the top of the EMFi material due to their low thickness. However, the same does not apply to the Parylene C coating because its thickness is comparable to the EMFi material. A further important point for the sensor's resonance frequency is the propagation medium of the received sound pressure waves, which has to be taken into account for estimating f_r. In the present case, the sensor array operates in a cleaning liquid that is very similar to water. While one can neglect the influence of air as wave propagation medium, water represents a heavy load for the realized sensor array. Not surprisingly, f_r will be remarkable reduced when the sensor array operates in water. For a piston-type transducer, the area density M_W of water oscillating with the active transducer surface results in [51]

$$M_W = \varrho_W \frac{8R_T}{3\pi} \tag{9.134}$$

where ϱ_W is the equilibrium density of water. The oscillating volume can be interpreted as a cylinder with height $8R_T/3\pi$ and the base area $R_T^2\pi$.

Table 9.4 contains the decisive quantities of the EMFi material (M_E and N_E), the Parylene C coating (M_P and N_P), and water for calculating f_r. Thereby, the area density M_i and surface-related compliance N_i are derived from

$$M_i = \varrho_i \cdot t_i \quad \text{and} \quad N_i = \frac{E_i}{t_i} \tag{9.135}$$

Table 9.4 Thickness t_i, equilibrium density ϱ_i, Young's modulus E_i, and resulting area density M_i as well as surface-related mechanical compliance N_i for element diameter $2R_T = 3.0$ mm; entries '−' not meaningful or not required for calculating resonance frequency f_r of array element; material parameters from [15, 51]

Layer	t_i μm	ϱ_i kg m^{-3}	E_i 10^6 N m^{-2}	M_i kg m^{-2}	N_i 10^{-12} m^3 N^{-1}
EMFi material	70	330	≈ 2.0	0.023	35
Parylene C coating	40	1289	≈ 3000	0.052	0.013
Water	−	998	−	1.271	−

with the equilibrium density ϱ_i, Young's modulus E_i, and the layer thickness t_i, respectively. Because the compliances of EMFi material and Parylene C coating differ by more than three orders of magnitude, we are able to neglect N_P. It can also be seen that the area density M_W dominates the other values, i.e., $M_W \gg M_E + M_P$. That is the reason why the resonance frequency of the sensor array is mainly determined by water.

Overall, the resonance frequency of a single array element can be approximated through

$$f_r = \frac{1}{2\pi\sqrt{(M_E/3 + M_P + M_W) \cdot N_E}} . \tag{9.136}$$

For the given layer thicknesses and material parameters, this equation leads to $f_r = 23$ kHz. When air serves as wave propagation medium (i.e., $M_W \approx 0$), the resonance frequency of an array element equals ≈ 300 kHz and, thus, coincides very well with the measurement results presented in Fig. 7.39a on p. 322.

Actual measurements in a water tank revealed that the maximum sensitivity of the realized sensor array for incident sound pressure waves arises at ≈ 40 kHz, which is much higher than the approximated resonance frequency of 23 kHz. This deviation follows, on the one hand, from uncertainties of the supposed material parameters. Besides, we assume a uniform mechanical oscillation for the analytical model, which is not fulfill (cf. Fig. 9.52). As a consequence, the effective area density decreases and, therefore, f_r takes higher values. By adjusting the relevant material parameters appropriately, coupled FE simulations provide a similar behavior and resonance frequency as the realized sensor array [97].

Sensor Sensitivity and Signal-to-Noise Ratio

The frequency-resolved sensitivity $B_S(f)$ in V Pa^{-1} of the realized sensor array was measured in a water tank. In doing so, an ultrasonic transmitter generated approximately plane sound pressure waves that impinge perpendicular to the sensor array. By comparing the electrical outputs of the array elements and a reference hydrophone, it was possible to determine the aimed quantity. The sensor sensitivity stays almost constant up to the cutoff frequency of 33 kHz and equals $B_S(f) = 8.7\,\mu$V Pa^{-1}. For higher frequencies, $B_S(f)$ increases until the resonance frequency $f_r \approx 40\,\text{kHz}$ is reached and strongly decreases for $f > f_r$. This behavior could be also proven by the results of coupled FE simulations [97]. When one uses the common reference value $p_{ref} = 1\,\mu\text{Pa}$ for sound pressure waves in water, the sensor sensitivity will become $-221.1\,\text{dB re }1\,\text{V}\,\mu\text{Pa}^{-1}$. To improve the sensor sensitivity in conventional sound field measurements with one array element, a low-noise amplifier (Reson VP1000 [101]) was exploited that provides a constant gain factor of 32 dB in the relevant frequency band. The overall sensitivity $B_{S,sys}(f)$ of the combination sensor array and amplifier takes the values $350\,\mu\text{V Pa}^{-1}$ and $-189.1\,\text{dB re }1\,\text{V}\,\mu\text{Pa}^{-1}$ below 33 kHz.

In general, the SNR results from the ratio of the root mean square (RMS) value of wanted system output that is noiseless to the noise signal at the system's output [59, 60]. For the combination of realized sensor array and amplifier, the SNR in dB is given by

$$\text{SNR} = 20 \log_{10}\left(\frac{U_{S,sys}}{U_{noise}}\right) \cong L_p - L_{noise} \tag{9.137}$$

where $U_{S,sys}$ and U_{noise} denote RMS values for both wanted output signal of amplifier and resulting noise signal at its output, respectively. The expression L_p stands for the sound pressure level of the wanted signal, and L_{noise} is the equivalent acoustic noise level

$$L_{noise} = 20 \log_{10}\left(\frac{\sqrt{\int_{f_{min}}^{f_{max}} U_{noise,f}(f)^2\,df}}{B_{S,sys}(f)\,p_{ref}}\right) \tag{9.138}$$

of the overall system. The noise voltage spectral density $U_{noise,f}(f)$ in $\text{V}/\sqrt{\text{Hz}}$ was measured at the amplifier output by the vector signal analyzer Keysight HP89441A [47]. In the relevant frequency band ranging from 0 to 100 kHz, the equivalent acoustic noise level amounts $L_{noise} = 115\,\text{dB}$. Preliminary investigations indicated that the sensor array should provide a resolution limit of 100 Pa to detect cavitation in the cleaning liquid. With the resulting sound pressure level $L_p = 160\,\text{dB}$, the SNR value of the combination sensor array and amplifier equals 45 dB. Consequently, the operational capability of a single array element including the amplifier Reson VP1000 could be confirmed.

Mechanical Crosstalk

Crosstalk always constitutes a decisive aspect for array systems. For the considered piezoelectric sensor array, we have to deal with mechanical as well as electrical crosstalk between neighboring array elements. When the mechanical crosstalk should be measured directly, it would be necessary that sound pressure waves impinge exclusively on a single array element but do not arise at the remaining elements of the sensor array. As a matter of fact, this requirement cannot be satisfied in practical situations. That is the reason why the array elements should be operated as transmitter;

(a)   **(b)**

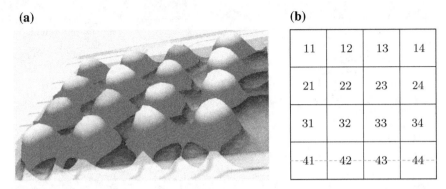

11	12	13	14
21	22	23	24
31	32	33	34
41	42	43	44

Fig. 9.51 **a** Measured spatially resolved displacement amplitudes û of realized sensor array; array elements excited by 100 V_{pp} at $f = 27.2$ kHz; **b** numbering of array elements; signal lines border on top elements, i.e., 11 to 14; dashed line shows scan line of laser Scanning vibrometer in Fig. 9.52

i.e., the sensor array exploiting the direct piezoelectric effect transforms to an actuator array, which is based on the inverse piezoelectric effect [93]. Because the electrome-chanical coupling factor of mechanical and electrical energy is identical in both directions of energy flow (see Sect. 3.5), high sensor sensitivities directly imply high actuator deformations and vice versa. In other words, one should be able to rate the mechanical crosstalk by electrically exciting a single element and measuring the resulting deformations of all array elements, which can be done by a laser scanning vibrometer. With a view to minimizing disturbing coupling effects due to propagating sound waves, the measurements were performed in air.

Before we study mechanical crosstalk, let us regard the mechanically displace-ments of the realized sensor array if it operates as actuator array. Figure 9.51a shows the resulting spatially resolved displacement amplitudes û, which were measured by the laser Scanning vibrometer Polytec PSV-300 [89]. Thereby, the array elements were simultaneously excited by the voltage 100 V_{pp} at the frequency $f = 27.2$ kHz. Note that this frequency was also used as excitation frequency for ultrasonic cleaning in the considered cleaning bath. As can be clearly observed, the displacement ampli-tudes of the array elements coincide very well. Moreover, the array elements oscillate in phase, which is especially important for time-resolved sound pressure measure-ments. The maximum relative deviation of the element's displacement amplitudes is smaller than 10 %, and, thus, the difference of the element sensitivities B_S should be below 1 dB. These deviations mainly originate from inhomogeneities of the EMFi material and slight differences of the realized array structure.

Figure 9.52 displays both measured and simulated values for û along a horizon-tal line ranging from array element 41 to 44 (see Fig. 9.51b). Since in contrast to Fig. 9.51a, only the array element 41 was excited by 100 V_{pp} at 27.2 kHz, it should be possible to rate the mechanical crosstalk by means of this measurement. Within the excited element, the coupled FE simulations correspond well to the measurement results. There arise, however, rather large deviations between measurements and

Fig. 9.52 Measured and
simulated displacement
amplitudes $\hat{u}$ along
horizontal line ranging from
array element 41 to 44 (see
Fig. 9.51b); array element 41
excited by voltage $100\,V_{pp}$ at
frequency 27.2 kHz

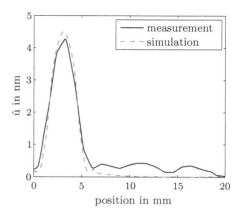

simulations outside the excited element, i.e., at the array elements 42, 43, and
44. Interestingly, the measured displacement amplitudes take the highest values
in the center of those array elements. Such behavior, of course, does not emerge
from mechanical crosstalk inside the realized sensor array but stems from electrical
crosstalk between neighboring signal lines, which exhibit a length >200 mm for
practical reasons (see Fig. 9.49). An isolated evaluation of the mechanical crosstalk
inside the realized sensor array requires, strictly speaking, numerical simulations. In
the present case, coupled FE simulations yielded displacement amplitudes <0.1 nm
outside the excited array element, which equals for the maximum amplitude 4.5 nm a
crosstalk attenuation of 33 dB [93]. According to further FE simulations, such a high
attenuation does not affect the performance (e.g., directivity pattern) of the realized
sensor array because the individual elements are mechanically decoupled.

Electrical Crosstalk

As just mentioned, the electrical crosstalk of the realized sensor array seems to
dominate its mechanical crosstalk. In the present case, electrical crosstalk is caused
by the coupling capacitances between the different array components. This does not
only refer to the coupling capacitances C_{CE} between the individual array elements
but also to the coupling capacitances C_{CL} between the signal lines. Owing to the
fact that it is hardly possible to measure the individual capacitances separately, FE
simulations were exploited for this task [97]. The aimed capacitances follow from
applying variable voltages to the sensor electrodes and evaluating the fundamental
relation $C = Q/U$. As the simulation results revealed, C_{CE} takes much smaller values
than the intrinsic capacitance C_S of a single array element. This is a consequence
of the thin EMFi material and the comparatively large lateral distance of 4.6 mm
between the centers of two neighboring array elements (cf. Fig. 9.49). C_{CE} and C_S
amount 0.07 and 1.48 pF, respectively. Hence, the crosstalk attenuation calculates
to 27 dB, which represents again a quite high value.

Now, let us discuss the electrical crosstalk between the utilized signal lines of lateral extension 0.3 mm, whereby the lateral distance (from center to center) between

two neighboring signal lines equals 1.0 mm. In contrast to the realized sensor array, the signal lines are not equipped with a common ground layer. As a result, capacitive coupling occurs on both top and bottom side of the signal lines. Especially if the signal lines are immersed in water, which is the case here, the coupling capacitance C_{CL} between two signal lines will take large values because water exhibits a high relative electric permittivity of $\varepsilon_r \approx 80$. According to FE simulations, the maximum of C_{CL} will amount 16.1 pF when the signal lines are completely immersed in water. This value is only slightly reduced for two signal lines that are not neighboring. The coupling capacitances of the signal lines, therefore, dominate the intrinsic capacitance C_S of a single array element.

To suppress electrical crosstalk due to coupling capacitances between the sensor components, especially between the signal lines, each array element was connected to a separate charge amplifier circuit. By means of these charge amplifiers, which contain operational amplifiers, both the bottom array electrodes and the signal lines are forced to ground potential (see Sect. 9.1.5). Consequently, the coupling capacitances C_{CE} as well as C_{CL} become short-circuited and, ideally, they do not influence the sensor performance anymore. Measurements demonstrated that the used charge amplifier circuits yield more than 50 dB attenuation of electric crosstalk in the relevant frequency band. Electric crosstalk is, thus, negligible for the combination of realized sensor array and charge amplifier circuits.

9.4.5 Experimental Results

Finally, let us discuss some experimental results that were achieved by the realized sensor array. We will study the connection of cavitation activity and amplified electrical element output. In doing so, the sensor array was fixed to a cylindrically shaped body and placed in a cleaning liquid. The sensor output will also be verified with regard to the cleaning effect by a special test layer.

Frequency-resolved Sound Pressure Amplitudes

As mentioned in Sect. 9.4.1, cavitation activity is closely linked to the spectral sound pressure amplitudes $\left|\underline{P}_{\sim}(f)\right|$ in the surrounding liquid. It is, thus, meaningful to measure $\left|\underline{P}_{\sim}(f)\right|$ by means of the realized sensor array. Figure 9.53a illustrates the experimental setup that was used for this task. The sensor array was fixed to a cylindrically shaped steel body of diameter 15 mm (see Fig. 9.53b). This is possible because the realized sensor array including the signal lines is mechanically flexible. The steel body together with the sensor array was placed in a cylindrically shaped and optically transparent tank featuring an inner diameter of 140 mm. A steel membrane was attached to the base of the tank. At the bottom side, the membrane is equipped with a special piezoelectric sandwich transducer serving as ultrasonic source. The resonance frequency f_r of the utilized sandwich transducer lies at 27.2 kHz.

The tank was filled with demineralized water, which served as cleaning liquid. Its temperature was kept constant at 50 °C with the aid of an external infrared source.

(a)   **(b)**

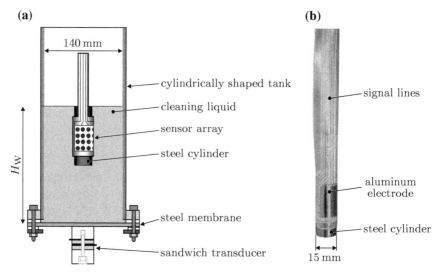

Fig. 9.53 a Schematic of experimental setup for measurements with realized sensor array being immersed in cleaning liquid; liquid level $H_W = 200\,\text{mm}$; **b** sensor array was fixed to cylindrically shaped steel body of 15 mm diameter [98]

By electrically exciting the ultrasonic sandwich transducer, one can generate sound pressure waves inside the tank. In order to obtain high sound pressure amplitudes that are required for initiating cavitation activities, standing waves were produced. The formation of such standing waves strongly depends on the liquid level H_W inside the tank and the frequency f_{ex} of the excited sound pressure waves [11]. Due to the transducer's resonance frequency f_r and the desire for six maxima along tank's cylinder axis, the liquid level was set to $H_W = 200\,\text{mm}$.

Now, let us consider two different scenarios, namely a low and a high excitation voltage U_{ex} of the sandwich transducer. While $U_{ex} = 100\,\text{V}_{pp}$ represents a low excitation voltage, $U_{ex} = 500\,\text{V}_{pp}$ denotes a high excitation voltage. For both scenarios, the excitation frequency f_{ex} should coincide with the resonance frequency of the transducer, i.e., $f_{ex} = f_r$. The experimental investigations revealed that there do not arise cavitation bubbles inside the cleaning bath for low excitation voltage and, consequently, cavitation activity is not present [97]. In contrast, several cavitation bubbles could be observed for $U_{ex} = 500\,\text{V}_{pp}$. The resulting cavitation activity could also be recognized as broadband noise in the audible range.

Figure 9.54a and b depict for both scenarios the resulting spectral magnitude $\left|\underline{U}_{SC}(f)\right|$ of the measured element output, which has been amplified by the charge amplifier. Thereby, the selected array element was located in the top maximum of the standing waves. In case of the low excitation voltage $U_{ex} = 100\,\text{V}_{pp}$ (see Fig. 9.54a), $\left|\underline{U}_{SC}(f)\right|$ contains a pronounced maximum at the fundamental frequency corresponding to f_{ex} as well as overtones at $2f_{ex}$, $3f_{ex}$ etc. For the high excitation voltage $U_{ex} = 500\,\text{V}_{pp}$ (see Fig. 9.54b), these spectral components increase. We are,

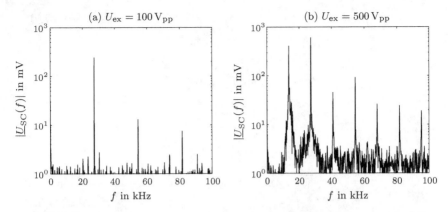

Fig. 9.54 Spectral magnitude $|\underline{U}_{SC}(f)|$ of amplified element output for **a** low excitation voltage $U_{ex} = 100\,\text{V}_{pp}$ and **b** high excitation voltage $U_{ex} = 500\,\text{V}_{pp}$ of sandwich transducer at excitation frequency $f_{ex} = 27.2\,\text{kHz}$

moreover, confronted with an additional subharmonic component at $f_{ex}/2$ and ultraharmonic components at $3/2\,f_{ex}$, $5/2\,f_{ex}$, etc. of appreciable magnitudes, especially at $f_{ex}/2$. According to Sect. 9.4.1, such spectral components result from bubble oscillations and, thus, prove the existence of cavitation bubbles. Because the broadband noise in $|\underline{U}_{SC}(f)|$ remarkably increases for high excitation voltage compared with low excitation voltage, we can also conclude that these bubbles implode (inertial cavitation) which goes hand in hand with arising cavitation activity. Therefore, the recorded spectral magnitude contains the expected information. The realized sensor array seems to be applicable for cavitation analysis in ultrasonic cleaning.

Cavitation Measurements

The cavitation activity does not only depend on the sound field intensity but, among other things, on the concentration of dissolved gases inside the cleaning liquid [21, 91]. If demineralized water serves as cleaning liquid, the concentration of dissolved oxygen will play an important role. Even though gas bubbles represent cavitation nuclei (cf. Sect. 9.4.1), an increasing concentration of dissolved oxygen is accompanied by a reduced cavitation activity and, thus, leads to a reduced cleaning effect. To test whether the realized sensor array enables such observations, the progress of individual spectral components in the recorded spectral magnitude $|\underline{U}_{SC}(f)|$ of a selected array element was rated with respect to two quantities. The first quantity refers to the excitation voltage U_{ex} of the sandwich transducer, whereas the second quantity is the concentration of dissolved oxygen inside the cleaning liquid. Again, the steel cylinder together with the realized sensor array was placed in the cylindrically shaped tank.

Figure 9.55a and b show the resulting progress of the measured spectral components at $f_{ex}/2$, f_{ex} as well as $2\,f_{ex}$ in $|\underline{U}_{SC}(f)|$ for two different oxygen concentrations, namely 3.5 and 7.0 mgO_2/L. The current oxygen concentration inside the cleaning liquid was determined with the aid of a electrochemical oxygen

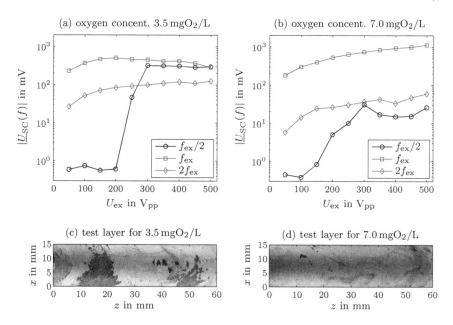

Fig. 9.55 Components at $f_{ex}/2$ (subharmonic), f_{ex} (fundamental frequency), and $2f_{ex}$ (second harmonic) in spectral magnitude $\left|\underline{U}_{SC}(f)\right|$ of amplified element output for oxygen concentration **a** $3.5\,mgO_2/L$ and **b** $7.0\,mgO_2/L$ with respect to excitation U_{ex} of sandwich transducer at $f_{ex} = 27.2\,kHz$; resulting test layer after cleaning time of $30\,s$ for oxygen concentration **c** $3.5\,mgO_2/L$ and **d** $7.0\,mgO_2/L$; dark areas indicate cleaned areas

meter and titrimetric tests [106]. Not surprisingly, the considered spectral components get altered by U_{ex}. The subharmonic component $\left|\underline{U}_{SC}(f_{ex}/2)\right|$ exhibits the strongest change, while the modification of the spectral component $\left|\underline{U}_{SC}(f_{ex})\right|$ at the excitation frequency is least pronounced. Especially for the oxygen concentration of $3.5\,mgO_2/L$, the subharmonic component increases dramatically between $U_{ex} = 200\,V_{pp}$ and $U_{ex} = 300\,V_{pp}$. However, such a big change does not arise for the oxygen concentration of $7.0\,mgO_2/L$. In case of the transducer excitation $U_{ex} = 500\,V_{pp}$, the spectral component $\left|\underline{U}_{SC}(f_{ex}/2)\right|$ takes much smaller values for high than for low oxygen concentrations. Since this spectral component represents an unambiguous indicator for oscillations of existing cavitation bubbles, cavitation activity and, thus, the cleaning effect should be very small for the oxygen concentration $7.0\,mgO_2/L$. In contrast, one can expect a good cleaning effect for the low oxygen concentration. Just as the subharmonic component, the broadband noise in $\left|\underline{U}_{SC}(f)\right|$ (cf. Fig. 9.54b) increases remarkably for $3.5\,mgO_2/L$. Hence, it is reasonable to assume that inertial cavitation occurs frequently.

To prove the findings, which were deduced from the measurements by the realized sensor array, the actual cleaning effect was rated in addition. For this purpose, the steel cylinder was coated with a special test layer. Figure 9.56 displays the setup of the used test layer, which consists of nickel, tin, and copper [97]. The metal

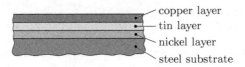
copper layer
tin layer
nickel layer
steel substrate

Fig. 9.56 Schematic cross-sectional view of test layer comprising nickel, tin, and copper layer; test layer exhibits thickness of $\approx 1.5\,\mu m$

layers were formed successively by electroplating. While the nickel layer serves as bonding agent, the tin layer reduces the adhesion between copper and nickel. The copper layer represents the top layer and is, therefore, in permanent contact with the cleaning liquid. Copper was applied because it features an average resistance against cavitation and can be optically distinguished from other metals such as tin. Altogether, the thickness of the test layer equals $\approx 1.5\,\mu m$.

Figure 9.55c and d depict photos of the test layer for the considered oxygen concentrations after a cleaning time of 30 s. The sandwich transducer was excited by $U_{ex} = 500\,V_{pp}$ at $f_{ex} = 27.2\,kHz$. For the low oxygen concentration of $3.5\,mgO_2/L$, one can recognize a remarkable cleaning effect since the bright copper layer was removed in large areas. As a matter of fact, these dark areas coincide with the maxima of the standing waves inside the cleaning bath. However, for the high oxygen concentration of $7.0\,mgO_2/L$, the test layer remains almost unchanged and, thus, the cleaning effect is negligible after a cleaning time of 30 s. These observations coincide with the expectations from the corresponding spectral component $\left| \underline{U}_{SC}(f_{ex}) \right|$ in Fig. 9.55a and b.

In summary, the realized sensor array consisting of 4×4 individual elements can be attached onto the object surface, which should be cleaned in an ultrasonic cleaning bath, even if the surface is curved. We are able to detect cavitation bubbles in a spatially resolved manner by analyzing subharmonic and/or ultraharmonic spectral components of the measured element output. When there arise high magnitudes for these spectral components as well as broadband noise in the resulting spectral magnitudes, cavitation bubbles will implode and a remarkable cleaning effect on the object surface will be present.

References

1. Antlinger, H., Clara, S., Beigelbeck, R., Cerimovic, S., Keplinger, F., Jakoby, B.: A differential pressure wave-based sensor setup for the acoustic viscosity of liquids. IEEE Sens. J. **16**(21), 7609–7619 (2016)
2. Arnau, A.: Piezoelectric Transducers and Applications, 2nd edn. Springer, Berlin (2008)
3. Asher, R.C.: Ultrasonic Sensors. Institute of Physics Publishing, Bristol (1997)
4. Atchley, A.A., Frizzell, L.A., Apfel, R.E., Holland, C.K., Madanshetty, S., Roy, R.A.: Thresholds for cavitation produced in water by pulsed ultrasound. Ultrasonics **26**(5), 280–285 (1988)

5. Beigelbeck, R., Antlinger, H., Cerimovic, S., Clara, S., Keplinger, F., Jakoby, B.: Resonant pressure wave setup for simultaneous sensing of longitudinal viscosity and sound velocity of liquids. Meas. Sci. Technol. **24**(12) (2013)

6. Beitz, W., Küttner, K.H.: Dubbel Handbook of Mechanical Engineering. Springer, Berlin (1994)

7. Bezděk, M.: A boundary integral method for modeling sound waves in moving media and its application to ultrasonic flowmeters. Ph.D. thesis, Friedrich-Alexander-University Erlangen-Nuremberg (2006)

8. Bezděk, M., Landes, H., Rieder, A., Lerch, R.: A coupled finite-element, boundary-integral method for simulating ultrasonic flowmeters. IEEE Trans. Ultrason. Ferroelectr. Freq. Control **54**(3), 636–646 (2007)

9. Blackstock, D.T.: Fundamentals of Physical Aocustics. Wiley, New York (2000)

10. Blahut, R.E., Miller, W., Wilcox, C.H.: Radar and Sonar: Part I. Springer, Berlin (1991)

11. Blevins, R.D.: Formulas for Natural Frequency and Mode Shape. Krieger Publishing Company, Malabar (1995)

12. Bronstein, I.N., Semendjajew, K.A., Musiol, G., Mühlig, H.: Handbook of Mathematics, 6h edn. Springer, Berlin (2015)

13. Brüel & Kjær: Product portfolio (2018). http://www.bksv.com

14. Buzug, T.M.: Computed Tomography, 6th edn. Springer, Berlin (2008)

15. Chen, P.J., Rodger, D.C., Humayun, M.S., Tai, Y.C.: Unpowered spiral-tube parylene pressure sensor for intraocular pressure sensing. Sens. Actuators A: Phys. **127**(2), 276–282 (2006)

16. Chiao, R.Y., Hao, X.: Coded excitation for diagnostic ultrasound: a system developer's perspective. IEEE Trans. Ultrason. Ferroelectr. Freq. Control **52**(2), 160–170 (2005)

17. Christian Bürkert GmbH & Co. KG: Manufacturer of transit time ultrasonic flow meters (2018). Homepage: https://www.burkert.com/en/

18. Claeys, J.M., Leroy, O.: Reflection and transmission of bounded sound beams on half-spaces and through plates. J. Acoust. Soc. Am. **72**(2), 585–590 (1982)

19. Cook, C.E.: Pulse compression - key to more efficient radar transmission. Proc. IRE **48**(3), 310–316 (1960)

20. Cowell, D.M.J., Freear, S.: Separation of overlapping linear frequency modulated (LFM) signals using the fractional Fourier transform. IEEE Trans. Ultrason. Ferroelectr. Freq. Control **57**(10), 2324–2333 (2010)

21. Crum, L.A.: Rectified diffusion. Ultrasonics **22**(5), 215–223 (1984)

22. Crum, L.A., Mason, T.J., Reisse, J.L., Suslick, K.S.: Sonochemistry and Sonoluminescence. Kluwer Academic Publishers, Dordrecht (1999)

23. Diehl Metering GmbH: Manufacturer of transit time ultrasonic flow meters (2018). Homepage: http://www.diehl.com/en/diehl-metering.html

24. Ditri, J.J., Rose, J.L.: Excitation of guided waves in generally anisotropic layers using finite sources. J. Appl. Mech. Trans. ASME **61**(2), 330–338 (1994)

25. Durst, F.: Fluid Mechanics: An Introduction to the Theory of Fluid Flows. Springer, Berlin (2008)

26. Eisenmenger, W., Köhler, M., Pecha, R., Wurster, C.: Neuartige Methode zur Messung der Zerreißspannung von Wasser. In: Proceedings of Fortschritte der Akustik (DAGA), pp. 574–575 (1997)

27. Emfit Ltd: Manufacturer of electro-mechanical films (2018). https://www.emfit.com

28. Endevco as part of Meggitt Sensing Systems: Product portfolio (2018). https://endevco.com

29. Endress+Hauser AG: Manufacturer of transit time ultrasonic flow meters (2018). http://www.endress.com

30. Fiedler, O.: Strömungs- und Durchflußmeßtechnik. Oldenbourgh Verlag München (1992)

31. Fortes-Patella, R., Reboud, J., Archer, A.: Cavitation erosion mechanism: numerical simulation of the interaction between pressure waves and solid boundaries. In: Proceedings of CAV, pp. 1–8 (2001)

32. Funck, B., Mitzkus, A.: Acoustic transfer function of the clamp-on flowmeter. IEEE Trans. Ultrason. Ferroelectr. Freq. Control **43**(4), 569–575 (1996)

33. Gautschi, G.: Piezoelectric Sensorics. Springer, Berlin (2002)
34. Goodman, J.W.: Introduction to Fourier Optics, 3rd edn. Roberts & Company Publishers, Englewood (2005)
35. Gyöngy, M., Coussios, C.C.: Passive cavitation mapping for localization and tracking of bubble dynamics. J. Acoust. Soc. Am. **128**(4), EL175–180 (2010)
36. Gätke, J.: Akustische Strömungs- und Durchflußmessung. Akademie, Berlin (1991)
37. Hammitt, F.G.: Observations on cavitation damage in a flowing system. J. Basic Eng. **85**(3), 347–356 (1963)
38. Harput, S., Evans, T., Bubb, N., Freear, S.: Diagnostic ultrasound toothiimaging using fractional fourier transform. IEEE Trans. Ultrason. Ferroelectr. Freq. Control **58**(10), 2096–2106 (2011)
39. Harris, F.J.: On the use of windows for harmonic analysis with the discrete fourier transform. Proc. IEEE **66**(1), 51–83 (1978)
40. Hecht, E.: Optics, 5th edn. Pearson, London (2016)
41. Honarvar, F., Sheikhzadeh, H., Moles, M., Sinclair, A.N.: Improving the time-resolution and signal-to-noise ratio of ultrasonic NDE signals. Ultrasonics **41**(9), 755–763 (2004)
42. Hottinger Baldwin Messtechnik (HBM) GmbH: Product portfolio (2018). https://www.hbm.com
43. Hsu, D.K., Hughes, M.S.: Simultaneous ultrasonic velocity and sample thickness measurement and application in composites. J. Acoust. Soc. Am. **92**(2), 669–675 (1992)
44. Ilg, J., Rupitsch, S.J., Lerch, R.: Impedance-based temperature sensing with piezoceramic devices. IEEE Sens. J. **13**(6), 2442–2449 (2013)
45. Jensen, J.A.: Estimation of Blood Velocities Using Ultrasound. Cambridge University Press, Cambridge (1996)
46. Keiji K.K., Gordon II, J.G.: The oscillation frequency of a quartz resonator in contact with liquid. Analytica Chimica Acta **175**(C), 99–105 (1985)
47. Keysight Technologies Inc.: Product portfolio (2018). http://www.keysight.com
48. Kiefer, D.A., Fink, M., Rupitsch, S.J.: Simultaneous ultrasonic measurement of thickness and speed of sound in elastic plates using coded excitation signals. IEEE Trans. Ultrason. Ferroelectr. Freq. Control **64**(11), 1744–1757 (2017)
49. Kistler Instrumente GmbH: Product portfolio (2018). https://www.kistler.com
50. Krautkrämer, J., Krautkrämer, H.: Werkstoffprüfung mit Ultraschall. Springer, Berlin (1986)
51. Kressmann, R.: New piezoelectric polymer for air-borne and water-borne sound transducers. J. Acoust. Soc. Am. **109**(4), 1412–1416 (2001)
52. Kupnik, M., Krasser, E., Gröschl, M.: Absolute transit time detection for ultrasonic gas flowmeters based on time and phase domain characteristics. In: Proceedings of International IEEE Ultrasonics Symposium (IUS), pp. 142–145 (2007)
53. Kuttruff, H.: Phyik und Technik des Ultraschalls. S. Hirzel, Stuttgart (1988)
54. Kühnicke, E.: Elastische Wellen in geschichteten Festkörpersystemen. TIMUG (2001)
55. Kümmritz, S., Wolf, M., Kühnicke, E.: Simultaneous determination of thicknesses and sound velocities of layered structures. Tech. Messen **82**(3), 127–134 (2015)
56. Lauterborn, W.: Numerical investigation of nonlinear oscillations of gas bubbles in liquids. J. Acoust. Soc. Am. **59**(2), 283–293 (1976)
57. Lauterborn, W., Hentschel, W.: Cavitation bubble dynamics studied by high speed photography and holography: Part I. Ultrasonics **23**(6), 260–268 (1985)
58. Leighton, T.: The Acoustic Bubble. Academic Press, New York (1994)
59. Lerch, R.: Elektrische Messtechnik, 7th edn. Springer, Berlin (2016)
60. Lerch, R., Sessler, G.M., Wolf, D.: Technische Akustik: Grundlagen und Anwendungen. Springer, Berlin (2009)
61. Lowe, M.J.: Matrix techniques for modeling ultrasonic waves in multilayered media. IEEE Trans. Ultrason. Ferroelectr. Freq. Control **42**(4), 525–542 (1995)
62. Lynnworth, L.C.: Ultrasonic Measurements for Process Control. Academic Press, Boston (1989)

63. Maev, G.: Advances in Acoustic Microscopy and High Resolution Imaging. Wiley-VCH, Weinheim (2012)
64. Mahadeva, D.V., Baker, R.C., Woodhouse, J.: Further studies of the accuracy of clamp-on transit-time ultrasonic flowmeters for liquids. IEEE Trans. Instrum. Meas. **58**(5), 1602–1609 (2009)
65. Maris, H.J.: Introduction to the physics of nucleation. Comptes Rendus Physique **7**(9–10), 946–958 (2006)
66. Martin, S.J., Frye, G.C., Wessendorf, K.O.: Sensing liquid properties with thickness-shear mode resonators. Sens. Actuators: A. Phys. **44**(3), 209–218 (1994)
67. Misaridis, T., Jensen, J.A.: Use of modulated excitation signals in medical ultrasound. Part I: basic concepts and expected benefits. IEEE Trans. Ultrason. Ferroelectr. Freq. Control **52**(2), 177–190 (2005)
68. Misaridis, T., Jensen, J.A.: Use of modulated excitation signals in medical ultrasound. Part II: design and performance for medical imaging applications. IEEE Trans. Ultrason. Ferroelectr. Freq. Control **52**(2), 192–206 (2005)
69. Moore, I.P., Brown, G.J., Stimpson, B.P.: Ultrasonic transit-time flowmeters modelled with theoretical velocity profiles: methodology. Meas. Sci. Technol. **11**(12), 1802–1811 (2000)
70. Motegi, R., Takeuchi, S., Sato, T.: Widebeam ultrasonic flowmeter. In: Proceedings of International IEEE Ultrasonics Symposium (IUS), pp. 331–336 (1990)
71. Neal, S.P., Speckman, P.L., Enright, M.A.: Flaw signature estimation in ultrasonic nondestructive evaluation using the Wiener filter with limited prior information. IEEE Trans. Ultrason. Ferroelectr. Freq. Control **40**(4), 347–353 (1993)
72. Neppiras, E.A.: Acoustic cavitation. Phys. Rep. **61**(3), 159–251 (1980)
73. Noltingk, B.E., Neppiras, E.A.: Cavitation produced by ultrasonics. Proc. Phys. Soc. Sect. B **63**(9), 674–685 (1950)
74. Oelze, M.L.: Bandwidth and resolution enhancement through pulse compression. IEEE Trans. Ultrason. Ferroelectr. Freq. Control **54**(4), 768–781 (2007)
75. Ohm, J., Lüke, H.D.: Grundlagen der digitalen und analogen Signalübertragung. Springer, Berlin (2015)
76. Oliner, A.A.: Microwave network methods for guided elastic waves. IEEE Trans. Microwave Theory Tech. **17**(11), 812–826 (1969)
77. Olympus Corporation: Product portfolio (2018). https://www.olympus-ims.com
78. Onda Corporation: Product portfolio of hydrophones (2018). http://www.ondacorp.com
79. O'Sullivan, C.K., Guilbault, G.G.: Commercial quartz crystal microbalances - theory and applications. Biosens. Bioelectr. **14**(8–9), 663–670 (1999)
80. Peng, Q., Zhang, L.Q.: High-resolution ultrasound displacement measurement using coded excitations. IEEE Trans. Ultrason. Ferroelectr. Freq. Control **58**(1), 122–133 (2011)
81. Physik Instrumente (PI) GmbH & Co. KG: Product portfolio (2018). https://www.physikinstrumente.com/en/
82. Pinton, G.F., Trahey, G.E.: Continuous delay estimation with polynomial splines. IEEE Trans. Ultrason. Ferroelectr. Freq. Control **53**(11), 2026–2035 (2006)
83. Plona, T.J., Pitts, L.E., Mayer, W.G.: Ultrasonic bounded beam reflection and transmission effects at a liquid/solid-plate/liquid interface. J. Acoust. Soc. Am. **59**(6), 1324–1328 (1976)
84. Ploß, P.: Untersuchung von Clamp-on-Ultraschalldurchflussmessgeräten im k-Raum. Ph.D. thesis, Friedrich-Alexander-University Erlangen-Nuremberg (2017)
85. Ploß, P., Rupitsch, S.J.: Modeling of clamp-on ultrasonic flow meters in the wavenumber domain for prediction of flow measurement errors. IEEE Trans. Ultrason. Ferroelectr. Freq. Control (2018). Submitted
86. Ploß, P., Rupitsch, S.J., Fröhlich, T., Lerch, R.: Identification of acoustic wave orientation for ultrasound-based flow measurement by exploiting the Hough transform. Procedia Eng. **47**, 216–219 (2012)
87. Ploß, P., Rupitsch, S.J., Lerch, R.: Extraction of spatial ultrasonic wave packet features by exploiting a modified Hough transform. IEEE Sens. J. **14**(7), 2389–2395 (2014)

88. Pollakowski, M., Ermert, H., Bernus, L., Schmeidl, T.: The optimum bandwidth of chirp signals in ultrasonic applications. Ultrasonics **31**(6), 417–420 (1993)

89. Polytec GmbH: Product portfolio (2018). http://www.polytec.com

90. Rose, J.L.: Ultrasonic Waves in Solid Media. Cambridge University Press, Cambridge (1999)

91. Rozenberg, L.: Physical Principles of Ultrasonic Technology. Springer, Berlin (1973)

92. Rupitsch, S.J., Glaser, D., Lerch, R.: Simultaneous determination of speed of sound and sample thickness utilizing coded excitation. In: Proceedings of International IEEE Ultrasonics Symposium (IUS), pp. 711–714 (2012)

93. Rupitsch, S.J., Lerch, R., Strobel, J., Streicher, A.: Ultrasound transducers based on ferro-electret materials. IEEE Trans. Dielectr. Electr. Insul. **18**(1), 69–80 (2011)

94. Rupitsch, S.J., Zagar, B.G.: Acoustic microscopy technique to precisely locate layer delamination. IEEE Trans. Instrum. Meas. **56**(4), 1429–1434 (2007)

95. Sanderson, M.L., Yeung, H.: Guidelines for the use of ultrasonic non-invasive metering techniques. Flow Meas. Instrum. **13**(4), 125–142 (2002)

96. Sauerbrey, G.: Verwendung von Schwingquarzen zur Wägung dünner Schichten und zur Mikrowägung. Zeitschrift für Physik **155**(2), 206–222 (1959)

97. Strobel, J.: Werkzeuge zur Charakterisierung der Kavitation in Ultraschall-Reinigungsbädern. Ph.D. thesis, Friedrich-Alexander-University Erlangen-Nuremberg (2009)

98. Strobel, J., Rupitsch, S.J., Lerch, R.: Ferroelectret sensor for measurement of cavitation in ultrasonic cleaning systems. Tech. Messen **76**(11), 487–495 (2009)

99. Szabo, T.L.: Diagnostic Ultrasound Imaging: Inside Out, 2nd edn. Academic Press, Amsterdam (2014)

100. Tektronix, Inc.: Product portfolio (2018). https://www.tek.com

101. Teledyne Marine: Product portfolio (2018). http://www.teledynemarine.com

102. Tietze, U., Schenk, C., Gamm, E.: Electronic Circuits - Handbook for Design and Application. Springer, Berlin (2008)

103. Ting, D., Yuan, Y., Supin, W., Mingxi, W.: Spatial-temporal dynamics of cavitation bubbles induced by pulsed hifu thrombolysis within a vessel and parameters optimization for cavitation enhancement. In: Proceedings of International IEEE Ultrasonics Symposium (IUS) (2016)

104. Trevena, D.H.: Cavitation and the generation of tension in liquids. J. Phys. D: Appl. Phys. **17**(11), 2139–2164 (1984)

105. Trilling, L.: The collapse and rebound of a gas bubble. J. Appl. Phys. **23**(1), 14–17 (1952)

106. Tränkler, H.R., Reindl, L.M.: Sensortechnik - Handbuch für Praxis und Wissenschaft. Springer, Berlin (2014)

107. Vellekoop, M.J.: Acoustic wave sensors and their technology. Ultrasonics **36**(1–5), 7–14 (1998)

108. Viola, F., Walker, W.F.: A spline-based algorithm for continuous time-delay estimation using sampled data. IEEE Trans. Ultrason. Ferroelectr. Freq. Control **52**(1), 80–93 (2005)

109. Wan, M., Feng, Y., ter Haar, G.: Cavitation in Biomedicine. Springer, Berlin (2015)

110. Wolf, D.: Signaltheorie: Modelle und Strukturen. Springer, Berlin (1999)

111. Wolf, M., Kühnicke, E., Kümmritz, S., Lenz, M.: Annular arrays for novel ultrasonic measurement techniques. J. Sens. Sens. Syst. **5**(2), 373–380 (2016)

112. Wolgemuth, L.: Assessing the performance and suitability of parylene coating. Med. Device Diagn. Ind. **22**(8), 42 (2000)

113. Wöckel, S., Steinmann, U., Auge, J.: Signal processing for ultrasonic clamp-on-sensor-systems. Tech. Messen **81**(2), 86–92 (2014)

114. Wüst, M., Eisenhart, J., Rief, A., Rupitsch, S.J.: System for acoustic microscopy measurements of curved structures. Tech. Messen **84**(4), 251–262 (2017)

115. Wüstenberg, H.: Untersuchungen zum Schallfeld von Winkelprüfköpfen für die Materialprüfung mit Ultraschall. Ph.D. thesis, Technische Universität Berlin (1972)

116. Zhao, H., Peng, L., Takahashi, T., Hayashi, T., Shimizu, K., Yamamoto, T.: Ann based data integration for multi-path ultrasonic flowmeter. IEEE Sens. J. **14**(2), 362–370 (2014)

Chapter 10
Piezoelectric Positioning Systems and Motors

As already stated, piezoelectric elements (in particular piezoceramic elements) enable an efficient conversion of electrical energy into mechanical energy. They provide high mechanical forces and a high dynamic performance. Piezoelectric elements are, moreover, nonwearing and feature a high rigidity. On those grounds, such elements should be ideally suited as active components in various drives. In this chapter, we will concentrate on piezoelectric positioning systems and motors. Such devices contain piezoelectric actuators, which consist of one or more piezoelectric elements. Figure 10.1 illustrates four different actuator structures that are often employed, namely (a) piezoelectric stack actuators, (b) piezoelectric bimorph actuators, (c) piezoelectric trimorph actuators, and (d) the so-called macrofiber composite (MFC [25]) actuators. While piezoelectric stack actuators comprise several piezoelectric elements being stacked, bimorph and trimorph actuators contain only two piezoelectric bars (cf. Fig. 7.27 on p. 300). In contrast to bimorph actuators, trimorph actuators are equipped with a thick metallic layer between the piezoelectric bars. MFC actuators belong to the group of piezoelectric composite transducers (see Sect. 7.4.3), which contain either thin piezoceramic plates or a large number of stripes or fibers that are skillfully contacted. To obtain robust as well as mechanically flexible devices, the active components (i.e., the piezoelectric elements) of the piezoelectric composite transducers are commonly surrounded by appropriate passive materials, e.g., polymers [22]. Not surprisingly, piezoelectric composite transducers are not restricted to actuator applications but can also be used as sensors.

Apart from their structure, the considered actuators in Fig. 10.1 differ significantly in the achievable mechanical forces and displacements in working direction, e.g., [20, 25]. Basically, we always have to choose a compromise between force and displacement; i.e., a piezoelectric actuator features either high forces or large displacements. For instance, the stacking of several elements in case of piezoelectric stack actuators yields high forces in working direction. The available displacements are, however, small in comparison with the tip displacements of piezoelectric bimorph and trimorph actuators. As a matter of fact, the large displacement of bimorph and

© Springer-Verlag GmbH Germany, part of Springer Nature 2019   511
S. J. Rupitsch, *Piezoelectric Sensors and Actuators*, Topics in Mining, Metallurgy and Materials Engineering, https://doi.org/10.1007/978-3-662-57534-5_10

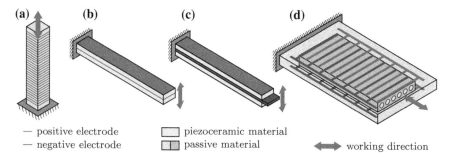

— positive electrode   ☐ piezoceramic material
— negative electrode   ☐ passive material   ⬌ working direction

Fig. 10.1 Typical piezoelectric actuator structures for positioning systems and motors; **a** piezoelectric stack actuator; **b** piezoelectric bimorph actuator; **c** piezoelectric trimorph actuator; **d** macrofiber composite (MFC) actuator

trimorph actuators is achieved at the expense of the reachable force and rigidity. This also applies to MFC actuators.

The chapter starts with the fundamentals of piezoelectric stack actuators as well as the effect of mechanical prestress on the stack performance. Preisach hysteresis modeling from Chap. 6 will be applied to describe the large-signal behavior of prestressed stack actuators. Section 10.2 deals with so-called amplified piezoelectric actuators, which provide relatively large mechanical displacements by converting mechanical forces into displacements. The conversion is performed with the aid of special metallic hinged frames. In Sect. 10.3, the applicability of piezoelectric trimorph actuators for positioning tasks will be demonstrated. For this purpose, model-based hysteresis compensation is conducted. At the end of the chapter, a brief overview of piezoelectric motors will be given which includes selected examples of linear as well as rotary motors.

10.1   Piezoelectric Stack Actuators

Piezoelectric stack actuators are often utilized in practical applications because these actuators enable much larger strokes than single piezoelectric elements. The section starts with the fundamentals of piezoelectric stack actuators as well as typical setups. Since piezoelectric stack actuators should be mechanically prestressed in practical applications, we will take a closer look at effects of such prestress on the resulting electrical and mechanical quantities in Sect. 10.1.2. Moreover, Preisach hysteresis modeling is exploited to predict the electrical as well as mechanical large-signal behavior of a stack actuator in case of prestress.

10.1.1   Fundamentals

Actuators that are based on piezoceramic materials provide an efficient conversion of electrical energy into mechanical energy and large operating frequencies. However, the achieved deflections of such materials for common excitation voltages are rather small. With a view to increasing the available stroke of a piezoelectric device for practical applications, it makes sense to stack several piezoceramic elements. The resulting piezoelectric device is usually termed piezoelectric stack actuator (PSA). Depending on the operating direction, we can distinguish between longitudinal PSAs and shear PSAs [6]. While longitudinal PSAs exploit the longitudinal mode of piezo-electricity (i.e., the d_{33}-effect), shear PSAs are often based on the transverse shear mode of piezoelectricity (i.e., the d_{15}-effect).

Figure 10.2a illustrates the conventional setup of a longitudinal PSA, which con-sists of a large number of polarized piezoceramic disks that are equipped with two electrodes each, i.e., a positive and a negative electrode. The electric polarization **P** of the piezoceramic disks points from the positive to the negative electrode. As a matter of fact, the PSA demands an electrical link of all positive electrodes and all negative electrodes, respectively. We can reduce the resulting wiring effort by stacking the disks appropriately. This means that either the positive or the negative electrode of two neighboring disks should border on each other. If a conductive adhesive is used for connecting the disks, the wiring effort will then be minimal. The manufacturing costs of such longitudinal PSAs are, however, extremely high because they have to be handmade [24]. Therefore, the conventional setup is only occasionally utilized in practical applications.

The so-called *multilayer stack actuators* (see Fig. 10.2b) represent an alternative to the conventional setup of longitudinal PSAs [8]. They can be fabricated in large numbers by a multistage manufacturing process starting with an unfired layer of the piezoceramic material. The typical layer thickness amounts 50–100 μm. By means of screen printing, the piezoceramic layer gets equipped with a thin metallic film, which serves as inner electrode of the multilayer stack actuator. Commonly, more than 100 piezoceramic layers including the printed electrodes are laminated to a block. This is done at elevated temperature and under a certain mechanical stress. During the subsequent process steps, the multilayer block is tailored, fired, and sintered just as in case of piezoceramic single elements. The block is, moreover, equipped with external

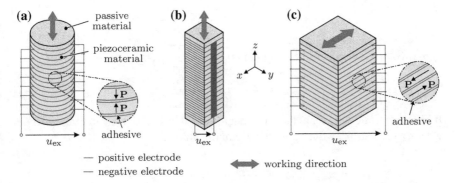

Fig. 10.2 Principle setup of **a** conventional longitudinal PSAs, **b** multilayer stack actuators, and **c** shear PSAs; **P** indicates direction of electric polarization

electrodes, which connect the inner electrodes in an appropriate manner. Finally, the multilayer stack actuator has to be polarized with a view to activating piezoelectric coupling. Such stack actuators are used in high-speed switching applications (e.g., injection system in diesel engines [6]) as well as for precision positioning.

Besides of longitudinal PSAs, one can also fabricate shear PSAs. In contrast to longitudinal PSAs, it is, however, hardly possible to build up a multilayer stack actuator because the required direction of electric polarization cannot be created by the inner electrodes. That is the reason why shear PSAs comprise individual piezoceramic elements, which are equipped with a positive and a negative electrode. Again, the wiring effort can be minimized by a conductive adhesive and an appropriate element stacking. Figure 10.2c depicts a shear PSA consisting of piezoceramic elements, which are alternately polarized in positive and negative x-direction. Therefore, the shear PSA provides comparatively large displacements between its lower and upper end in this direction. If we additionally utilize a shear PSA that contains piezoceramic elements being alternately polarized in positive and negative y-direction, a PSA combination will be achieved which allows displacements in both directions, i.e., in x- and in y-direction. For instance, such shear PSA combinations can be used in scanning microscopes. There also exists an alternative design of shear PSAs, the so-called multilayer pseudo-shear actuator, which exploits the transverse mode of piezoelectricity, i.e., the d_{31}-effect [31]. By alternately using a stiff conductive adhesive at the left and right end of the piezoceramic plates, the overall block behaves like a shear PSA.

To demonstrate the general operation principle of piezoelectric stack actuators, let us assume a mechanically unloaded (i.e., $T_3 = 0$) longitudinal PSA consisting of $n_{disk} = 100$ cylindrical piezoceramic disks. Each disk shall feature the diameter $d_S = 10\,\text{mm}$ and thickness $t_S = 0.5\,\text{mm}$. By neglecting the adhesive layer between the disks, the stack exhibits, thus, the overall length $l_{stack} = n_{disk} \cdot t_S = 50\,\text{mm}$. Furthermore, we assume a typical piezoelectric strain constant $d_{33} = 4 \cdot 10^{-10}\,\text{m V}^{-1}$ of a piezoceramic material, an excitation voltage of $u_{ex;stack} = 500\,\text{V}$, and purely linear material behavior. The assumptions lead to the mechanical strains of piezoelectric disk $S_{3;disk}$ and stack $S_{3;stack}$

$$S_{3;disk} = S_{3;stack} = d_{33}E_3 = d_{33}\frac{u_{ex;stack}}{t_S} = 4 \cdot 10^{-4} \qquad (10.1)$$

in 3-direction. Therefore, the longitudinal PSA offers a stroke of

$$z_{stroke} = S_{3;stack} \cdot l_{stack} = n_{disk} \cdot S_{3;disk} \cdot t_S = 20\,\mu\text{m} \ . \qquad (10.2)$$

If a piezoceramic cylinder of the same length is used instead (i.e., cylinder length $l_S = 50\,\text{mm}$), such stroke will require the excitation voltage

$$u_{ex;cylinder} = \frac{S_{3;cylinder} \cdot l_S}{d_{33}} = \frac{z_{stroke}}{d_{33}} = 50\,\text{kV} \ , \qquad (10.3)$$

which is n_{disk} times higher than the excitation voltage $u_{ex;stack}$ of a stack actuator with the same performance, i.e., $u_{ex;cylinder} = n_{disk} \cdot u_{ex;stack}$. When we increase the number n_{disk} of piezoceramic disks and reduce their thickness t_S accordingly that the stack length l_{stack} stays constant, the obtained stroke will be improved further. This simple example already reveals the great advantage of piezoelectric stack actuators. However, one also has to consider that a stack actuator behaves electrically like the parallel connection of n_{disk} single piezoceramic disks. The capacitance C_{disk} of a single disk is given by

$$C_{disk} = \frac{\varepsilon_{33}^T A_S}{t_S} = \frac{\varepsilon_{33}^T d_S^2 \pi}{4 t_S} \qquad (10.4)$$

with the electric permittivity ε_{33}^T for constant mechanical stress. Therewith, the total capacitance C_{stack} of the longitudinal PSA results in

$$C_{stack} = n_{disk} \cdot C_{disk} = \frac{n_{disk} \varepsilon_{33}^T d_S^2 \pi}{4 t_S} = \frac{n_{disk}^2 \varepsilon_{33}^T d_S^2 \pi}{4 l_{stack}} . \qquad (10.5)$$

Compared to the capacitance $C_{cylinder} = \varepsilon_{33}^T A_S / l_{stack}$ of the piezoceramic cylinder with the same geometric dimensions, C_{stack} takes values that are n_{disk}^2 times as large, i.e., $C_{stack} = n_{disk}^2 \cdot C_{cylinder}$. Although the excitation voltage $u_{ex;stack}$ is relatively small, the parallel connection of n_{disk} capacitances C_{disk} increases the current consumption of the PSA remarkably. If the electric current of a single disk is i_{disk}, the current i_{stack} of the entire stack becomes $i_{stack} = n_{disk} \cdot i_{disk}$. It seems only natural that both the increasing total capacitance and the increasing current consumption have to be taken into account when designing the control electronics for a PSA.

With regard to practical applications, one should always bear in mind that PSAs are very sensitive to tensile forces. On the one hand, this is due to the adhesive layer between the piezoceramic elements in conventional setups. On the other hand, piezoceramic materials itself should generally not be loaded with tensile forces because these materials exhibit low tensile strength. Hence, tensile forces cannot only damage conventional PSAs but also multilayer stack actuators. Even if there do not act external tensile forces on the PSA, the inner forces in case of electrical excitation may cause damages. That is the reason why PSAs (especially longitudinal PSAs) are commonly mechanically prestressed in practical applications [6]. For the longitudinal PSAs in Fig. 10.2a and b, mechanical prestress means a mechanical force acting in negative z-direction on the top end, i.e., parallel to the working direction. We can generate the required prestress either with the aid of a suitable PSA housing or through an external preloading.

The permitted range of the excitation voltage u_{ex} represents a further important point concerning the practical application of PSAs. In order to avoid partial or full depolarization of the involved piezoceramic materials, PSA must not be excited by large negative voltages. To some extent, the permitted range is exclusively limited to positive voltages, i.e., $u_{ex} \geq 0\,\mathrm{V}$. Exceptions to this are shear PSAs, which can usually be operated symmetrically around zero up to a few hundred volts. The permitted voltage range for conventional longitudinal PSAs often goes up to $u_{ex} = +1000\,\mathrm{V}$.

The company PI Ceramic GmbH [20] is a well-known manufacturer of longitudinal and shear PSAs. Depending on the stack length l_{stack}, commercially available longitudinal PSAs in the conventional setup provide strokes of more than $30\,\mu m$. The blocking forces can reach values $>50\,kN$ in longitudinal direction. Note that the blocking force will correspond to the maximum mechanical force when actuator displacements are completely prevented. Multilayer stack actuators allow strokes $>30\,\mu m$ and blocking forces greater than $3000\,N$. The maximum displacement of commercially available shear PSAs typically amounts $10\,\mu m$. However, due to the setup of shear PSAs, the permitted maximum of the shear load hardly exceeds $200\,N$.

10.1.2 Effect of Mechanical Prestress on Stack Performance

Figure 10.3 depicts the investigated longitudinal piezoelectric stack actuator PICA P-010.20P of cylindrical shape, which was manufactured by the company PI Ceramic GmbH [20]. The PSA consists of $n_{disk} = 50$ single disks made of the ferroelectrically soft material PIC255. Owing to the fact that the disks are polarized and stacked in 3-direction, the stack actuator mainly elongates in 3-direction, which also represents the working direction. Consequently, it makes sense to restrict the further investigations to the 3-direction. This includes the decisive physical quantities (e.g., mechanical strain), which means that we consider solely their 3-components. The most important specifications of the PSA are listed in Table 10.1.

Fig. 10.3 Piezoelectric stack actuator PICA P-010.20P manufactured by PI Ceramic GmbH; total length l_{stack} and cross section $A_S = d_S^2\pi/4$; electrical excitation u_{ex}; applied mechanical prestress $T_3 = F_3/A_{disk}$

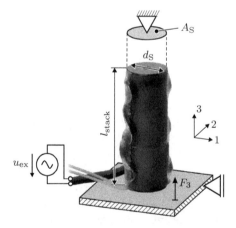

Table 10.1 Specifications of investigated longitudinal piezoelectric stack actuator PICA P-010.20P

Thickness t_S of a single disk	$\approx 0.5\,\mathrm{mm}$
Diameter d_S of a single disk	$10\,\mathrm{mm}$
Disk material	PIC255
Amount n_{disk} of single disks	50
Total length l_{stack} of the stack actuator	$31\,\mathrm{mm}$
Maximum permissible mechanical prestress $T_{3;\mathrm{max}}$	$30\,\mathrm{MPa}$
Electrical excitation voltage u_{ex}	$0\ldots 1\,\mathrm{kV}$

Let us start with the maximum achievable values for both the electric polarization P_{max}^{+} and the mechanical strain[1] S_{max}^{+} of the PSA in the unipolar working area (see Fig. 6.1 on p. 196) state of the actuator is unknown, we will exclusively quantify maximum changes of those quantities, i.e., $\Delta P_{\mathrm{max}}^{+}$ and $\Delta S_{\mathrm{max}}^{+}$. Just as in Chap. 6, a Sawyer–Tower circuit and a linear variable differential transformer were utilized to acquire ΔP and ΔS, respectively. The PSA was exposed to a mechanical prestress $T_3 = F_3/A_{\mathrm{disk}}$ in thickness direction through a tension–compression testing machine. Firstly, T_3 was stepwise increased from $0\,\mathrm{MPa}$ to $30\,\mathrm{MPa}$ in steps of $2.5\,\mathrm{MPa}$ and, secondly, stepwise reduced again to the mechanically unloaded case (i.e., $T_3 = 0\,\mathrm{MPa}$). In order to ensure that all transient phenomena within the actuator have decayed, the electrical excitation was applied after a waiting time of $5\,\mathrm{min}$. The sinusoidal excitation voltage (frequency $f = 0.1\,\mathrm{Hz}$) featured an amplitude of $\hat{u}_{\mathrm{ex}} = 500\,\mathrm{V}$ and an offset of $U_{\mathrm{off}} = +500\,\mathrm{V}$; i.e., the PSA was operated in the permissible range (see Table 10.1). This leads to the maximum electric field intensity $E = 2\,\mathrm{kV\,mm^{-1}}$ within a single disk.

Figure 10.4a and b show the obtained results for $\Delta P_{\mathrm{max}}^{+}$ and $\Delta S_{\mathrm{max}}^{+}$ of the PSA with respect to $T_3^{\pm}$. Interestingly, both quantities rise with increasing mechanical prestress T_3^{+} and drop with decreasing prestress T_3^{-} in the investigated value range. Such behavior is mainly attributable to the fact that several domains within the ferroelectric material switch to the ferroelastic intermediate stage due to applied prestresses [34]. As a consequence, the global electric polarization of the PSA reduces, but the amount of domains that can be aligned in parallel to the applied electric field increases. Therefore, the changes of electric polarizations $\Delta P_{\mathrm{max}}^{+}$ and mechanical strains $\Delta S_{\mathrm{max}}^{+}$ also rise in the considered range of mechanical prestresses (cf. Fig. 6.21c on p. 236). It has to be noted that a further increase of the prestress would, however, drastically reduce $\Delta P_{\mathrm{max}}^{+}$ as well as $\Delta S_{\mathrm{max}}^{+}$ since the domains stay in the ferroelastic intermediate stage. In other words, the applied electric field is no longer capable to align domains within the ferroelectric material. Besides, there occurs a certain hysteresis for increasing prestress T_3^{+} and decreasing prestress T_3^{-} in $\Delta P_{\mathrm{max}}^{+}(T_3)$ as well as $\Delta S_{\mathrm{max}}^{+}(T_3)$, i.e., $\Delta P_{\mathrm{max}}^{+}(T_3^{+}) \neq \Delta P_{\mathrm{max}}^{+}(T_3^{-})$ and $\Delta S_{\mathrm{max}}^{+}(T_3^{+}) \neq \Delta S_{\mathrm{max}}^{+}(T_3^{-})$. The reason for this lies in altered domain configurations within the ferroelectric material for increasing and decreasing prestresses.

---

[1] The given mechanical strains of the PSA always relate to its total length l_{stack}.

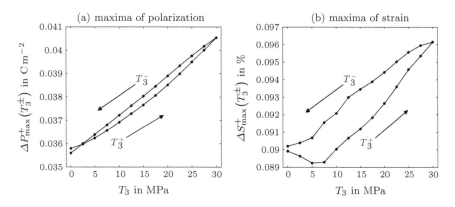

Fig. 10.4 Variation of **a** maximum electric polarization $\Delta P_{max}^+(T_3)$ and **b** maximum mechanical strain $\Delta S_{max}^+(T_3)$ versus applied mechanical prestress T_3; increasing prestress T_3^+ and decreasing prestress T_3^-; piezoelectric stack actuator PICA P-010.20P operating in unipolar working area

10.1.3   Preisach Hysteresis Modeling for Prestressed Stack

As the results in Fig. 10.4 demonstrate, ΔP_{max}^+ and ΔS_{max}^+ of the PSA strongly depend on the applied mechanical prestress T_3. The values (especially ΔP_{max}^+) are altered up to almost 15% in the investigated range of T_3. Accordingly, it seems only natural that the underlying large-signal behavior of the electric polarization and mechanical strain also varies. With a view to studying this effect more in detail, additional measurements were conducted for the actuator operating in the unipolar working area. In particular, a unipolar electrical excitation signal of sinusoidal shape was utilized which features decreasing amplitudes and a frequency of $f = 0.1$ Hz. The mechanical prestress T_3^+ was again stepwise increased from 0 MPa to 30 MPa in steps of 2.5 MPa. After a waiting time of 5 min, the PSA was excited with the unipolar electrical input sequence. Figure 10.5a and b show the collected measurements for the electric polarization $\Delta P_{meas}(t, T_3)$ and mechanical strain $\Delta S_{meas}(t, T_3)$ with respect to time t as well as applied prestress T_3. Similar to the previous experiments, the strong dependence on T_3 becomes apparent. This can also be seen in the resulting hysteresis curves $\Delta P_{meas}(E, T_3)$ and $\Delta S_{meas}(E, T_3)$ in Fig. 10.5c and d.

To incorporate the applied mechanical prestress T_3 in Preisach hysteresis modeling for the PSA, it is advisable to proceed in the same manner as in Sect. 6.6.5. This means that we introduce a weighting distribution $\mu_{DAT}(\alpha, \beta, T_3)$ for the elementary switching operators of the generalized Preisach hysteresis operator $\mathcal{H}_G$, which additionally depends on T_3. In a first step, the entire parameter set of $\mu_{DAT}(\alpha, \beta, T_3)$ is identified for $\Delta P(E, T_3)$ as well as $\Delta S(E, T_3)$ separately in case of the mechanically unloaded PSA, i.e., $T_3 = 0$. Subsequently, selected parameters have to be modified with respect to T_3. In contrast to the piezoceramic disk in Sect. 6.6.5, it is here sufficient to solely alter the parameter B because the considered PSA only operates in the unipolar working area. As the comparison of measured and simulated hysteresis

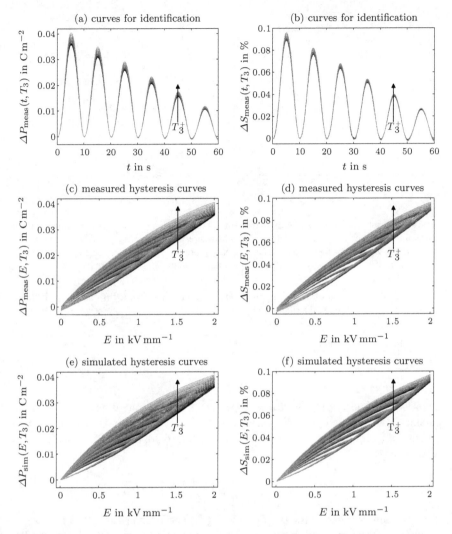

Fig. 10.5 **a** and **b** Measured curves $\Delta P_{\mathrm{meas}}(t, T_3)$ and $\Delta S_{\mathrm{meas}}(t, T_3)$ versus time t and increasing mechanical prestress T_3^+ for identifying parameters of Preisach hysteresis operator; **c** and **d** measured hysteresis curves $\Delta P_{\mathrm{meas}}(E, T_3)$ and $\Delta S_{\mathrm{meas}}(E, T_3)$; **e** and **f** simulated hysteresis curves $\Delta P_{\mathrm{sim}}(E, T_3)$ and $\Delta S_{\mathrm{sim}}(E, T_3)$; piezoelectric stack actuator PICA P-010.20P operating in unipolar working area

curves in Fig. 10.5c–f indicates, the generalized Preisach hysteresis model performs excellently. Even if the mechanical prestress changes, we will be able to realistically describe the large-signal behavior of the PSA in the unipolar working area.

The same investigations were conducted for a decreasing mechanical prestress T_3^-; i.e., T_3^- was stepwise reduced from 30 MPa to 0 MPa in steps of 2.5 MPa. Figure 10.6a and b contains the identified parameter values of $B_P(T_3)$ as well as $B_S(T_3)$ for

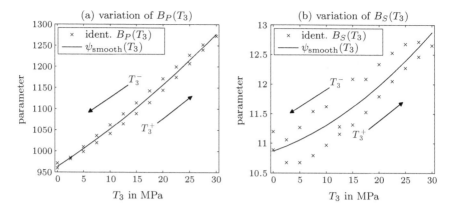

Fig. 10.6 Variation of parameter B as well as smoothing function $\psi_{\text{smooth}}(T_3)$ according to (10.6) versus applied mechanical prestress T_3 for Preisach hysteresis modeling; **a** $B_P(T_3)$ for electric polarization and **b** $B_S(T_3)$ for mechanical strain; increasing prestress T_3^+ and decreasing prestress T_3^-; piezoelectric stack actuator PICA P-010.20P operating in unipolar working area

Table 10.2 Parameters ς_i of smoothing function $\psi_{\text{smooth}}(T_3)$ in (10.6) for B_P and B_S, which are required to define weighting distribution $\mu_{\text{DAT}}(\alpha, \beta, T_3)$

		ς_1	ς_2	ς_3
Polarization	$B_P(T_3)$	962.8092	8.1108	0.0788
Strain	$B_S(T_3)$	10.8679	0.0319	0.0012

increasing and decreasing prestress, respectively. Due to smooth progression of the parameter values, they can also serve as data points of an appropriate smoothing function $\psi_{\text{smooth}}(T_3)$. Here, we use the function (cf. (6.30, p. 237))

$$\psi_{\text{smooth}}(T_3) = \varsigma_1 + \varsigma_2 \left(\frac{T_3}{1\,\text{MPa}} \right) + \varsigma_3 \left(\frac{T_3}{1\,\text{MPa}} \right)^2 . \tag{10.6}$$

The resulting values of ς_i for $B_P(T_3)$ and $B_S(T_3)$ are listed in Table 10.2.

For validation purpose, let us exploit the determined parameter set to predict polarizations and strains of the PSA for an electrical excitation signal, which was not considered during the identification procedure. Contrary to the identification signal, the unipolar excitation signal for validation features rising amplitudes. Figure 10.7a and b display the measured curves $\Delta P_{\text{meas}}(t, T_3)$ and $\Delta S_{\text{meas}}(t, T_3)$ with respect to t and T_3. The corresponding simulation results $\Delta P_{\text{sim}}(t, T_3)$ and $\Delta S_{\text{sim}}(t, T_3)$ are shown in Fig. 10.7c and d. Again, the comparison clearly points out that one is able to reliably predict the large-signal behavior of the PSA through Preisach hysteresis modeling, even in case of applied mechanical prestress. This is also confirmed by the normalized relative deviations ϵ_r between simulations and measurements, which always stay below 5% (see Fig. 10.7e and f). However, if the mechanical prestress is not taken into account, the modeling approach will yield relative deviations of more than 10% [34].

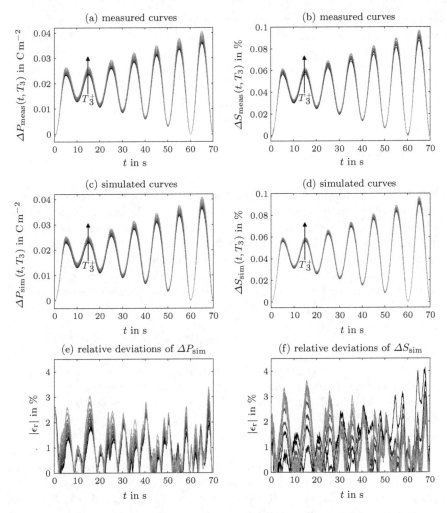

Fig. 10.7 a and **b** Measured curves $\Delta P_{\mathrm{meas}}(t, T_3)$ and $\Delta S_{\mathrm{meas}}(t, T_3)$ versus time t and increasing mechanical prestress T_3^+ for validating Preisach hysteresis model; **c** and **d** simulated curves $\Delta P_{\mathrm{sim}}(t, T_3)$ and $\Delta S_{\mathrm{sim}}(t, T_3)$ versus time t and increasing mechanical prestress T_3^+; **e** and **f** collected normalized relative deviations $|\epsilon_{\mathrm{r}}|$ in % (magnitude); piezoelectric stack actuator PICA P-010.20P operating in unipolar working area

10.2 Amplified Piezoelectric Actuators

Even though piezoelectric stack actuators provide huge blocking forces, excellent positioning accuracy as well as large operating frequencies, the available strokes are much smaller than those of electromagnetic actuators. This fact constitutes a considerable disadvantage for various practical applications like vibration sources. The so-called amplified piezoelectric actuators (APA) supply remedy because they provide large strokes by converting mechanical forces into displacements. The working principle and basic design of an APA will be explained in Sect. 10.2.1. Subsequently, we will discuss simulation results, which allow to deduce design criteria for amplified piezoelectric actuators. In Sect. 10.2.3, experimental results for different APA configurations will be compared to corresponding simulation results.

10.2.1 Working Principle

Amplified piezoelectric actuators always consist of piezoelectric elements and appropriate structures for converting mechanical forces into displacements. In the majority of cases, piezoelectric stack actuators serve as piezoelectric elements since the available strokes of such actuators are much higher than those of single piezoelectric elements, e.g., a disk. The conversion structure contains a certain amount of arms and is often diamond-shaped, which can be achieved either by the pitch angles of straight structure arms or by arms featuring the shape of suitable free-form curves, e.g., [11, 15, 17, 38]. Apart from the structure shape, the connection of the arms greatly influences the performance of the APA.

Figure 10.8a shows the typical setup of an APA containing a PSA and a metallic hinged frame for conversion. The hinged frame comprises four arms with two joints each. If the PSA expands due to electrical excitation, a mechanical force will be introduced to the hinged frame and its geometric dimension in x-direction will increase. This goes hand in hand with a reduction of the frame dimensions in y-direction. As a matter of fact, the geometric changes depend on the excitation signal, the geometric circumstances of stack actuator and frame as well as on their material properties.

In a first step, let us conduct a purely geometric consideration of the APA in Fig. 10.8a. Therefore, we assume ideal hinges and neglect acting forces as well as material properties. The symmetrical hinged frame comprising four nondeformable arms can then be reduced to a single arm of constant length l_A [14, 15]. While the left-end P_L of the arm moves only in y-direction, the right-end P_R moves only in

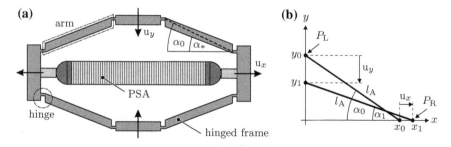

Fig. 10.8 a Typical setup of APA containing piezoelectric stack actuator (PSA) and closed hinged frame with four arms; **b** quantities for geometric consideration of single APA arm

x-direction. Figure 10.8b depicts the geometric circumstances for the initial state (i.e., without actuator excitation) and for the expanded state. In the initial and expanded state, the end positions of the arm are given by (x_0, y_0) and (x_1, y_1), respectively. The resulting effective pitch angles of the arm read as

$$\alpha_0 = \arctan\left(\frac{y_0}{x_0}\right) \quad \text{and} \quad \alpha_1 = \arctan\left(\frac{y_1}{x_1}\right). \tag{10.7}$$

Owing to the fact that l_A is assumed to remain constant, both states have to satisfy

$$l_A = \sqrt{x_0^2 + y_0^2} = \sqrt{x_1^2 + y_1^2} = \sqrt{(x_0 + u_x)^2 + (y_0 + u_y)^2} \tag{10.8}$$

with the displacement u_x of P_R in positive x-direction and u_y of P_L in positive y-direction. By solving for u_y, which represents the aimed quantity, (10.8) becomes

$$u_y = -y_0 \pm \sqrt{y_0^2 - u_x^2 - 2x_0 u_x}$$
$$= -l_A \sin \alpha_0 + \sqrt{(l_A \sin \alpha_0)^2 - u_x^2 - 2l_A \cos \alpha_0 u_x}. \tag{10.9}$$

This means that the displacement u_y of the hinged frame in y-direction depends exclusively on the actuator stroke u_x, the arm length l_A, and the pitch angle α_0 in the initial state. Because we modeled a quarter of the setup, u_x and u_y denote half of the actuator stroke and half of the entire frame displacement, respectively. It seems only natural that an APA should fulfill the condition $u_y > u_x$.

10.2.2 Numerical Simulations for Parameter Studies

The geometric considerations in the previous subsection neglect both the acting forces and the material properties of the APA components. It is not surprising that such an approach represents an oversimplification of the actual circumstances. To describe the behavior of an APA in a reliable way, we need, therefore, alternative three-dimensional approaches like elastostatic modeling, compliance-based modeling, or finite element (FE) analysis [13]. In the context of elastostatic modeling, the support reactions (i.e., forces and torques) at P_L and P_R are evaluated as function of the acting forces. These forces are given by the generated force of the PSA and the weight forces. The combination of support reactions and strain energy yields the displacements u_x and u_y of the APA. In case of compliance-based modeling, one has to introduce an elastic compliance tensor [s] for each component of the APA, i.e., for the structure arms and the PSA. From the acting forces and the resulting compliance tensor of the APA, we are again able to compute u_x as well as u_y. Although the elastostatic and compliance-based modeling approaches are rather simple, they exhibit serious weaknesses. Both modeling approaches do not allow to determine eigenfrequencies

Fig. 10.9 Schematic front
and side view of
three-dimensional FE model
of APA with sample
holder [14]; FE model
represents quarter of overall
structure; point of interest
(POI); geometric pitch
angle α_* of arm

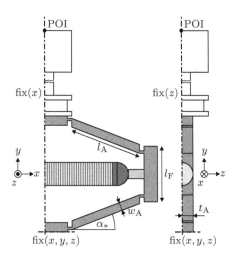

of the APA, which are of utmost importance concerning practical applications [14]. Moreover, the underlying calculation procedures require the acting force of the PSA. On these grounds, we will exclusively focus on the FE method hereinafter. In contrast to elastostatic and compliance-based modeling, the FE method can be used for several types of analysis, e.g., eigenfrequency analysis. The consideration of piezoelectric coupling (see Sect. 4.5.1) also enables the determination of u_x and u_y for a given electrical excitation of the PSA.

Figure 10.9 depicts a schematic front and side view of the three-dimensional FE model including characteristic geometric dimensions. The FE model was created according to the APA configurations that were built up at the Chair of Sensor Technology (Friedrich-Alexander-University Erlangen-Nuremberg). Besides of the PSA and the metallic hinged frame, the simulation model contains a sample holder made of acrylic glass. Owing to its symmetry, the entire setup can be reduced to a quarter, which remarkably reduces to computation effort and makes certain boundary conditions for the mechanical displacement necessary. The boundary conditions fix(x) and fix(z) imply that the displacements are zero in x- and z-direction, respectively. Because the ground plate of the realized APA is fixed, the FE model contains additionally fix(x, y, z) at its bottom.

The cylindrically shaped piezoelectric stack actuator P-010.40P from the company PI Ceramic GmbH [20] serves as piezoelectric element in the realized setup. The stack with diameter $d_S = 10$ mm and total length $l_{stack} = 58$ mm consists of 98 active disks that are made of the ferroelectrically soft material PIC255. They are equipped with electrodes and glued together. At the bottom and top end, the stack contains a ceramic plate as well as a steel plate.The active disks are alternately polarized

Fig. 10.10 Schematic front and side view of three-dimensional FE model of PSA [14]; FE model represents quarter of overall structure; **P** indicates direction of electric polarization

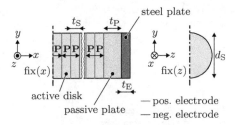

Table 10.3 Numbers and geometric dimensions of components for FE model of PSA in Fig. 10.10

	Number	Geometric dimensions
Passive ceramic plate	1	Thickness $t_P = 1.0$ mm
Active piezoelectric disk	49	Thickness $t_S = 0.56$ mm
Steel plate	1	Thickness $t_E = 0.5$ mm

in opposite directions. Figure 10.10 illustrates a schematic front and side view of the PSA in the FE model. Table 10.3 summarizes the number of components and geometric dimensions. Note that the adhesive layers and electrodes between the disks were not considered in the simulation. The material parameters of PIC255 were taken from the results of the inverse method in Table 5.3 on p. 160. For the metallic hinged frame that is made of tool steel, the decisive material parameters density, Young's modulus, and Poisson's ratio were assumed to be $\varrho_0 = 7800$ kg m^{-3}, $E_M = 210$ GPa, and $\nu_P = 0.28$.

As already mentioned, the numerical simulations were conducted for different configurations of the APA [14]. While the PSA remains unchanged for all configurations, the metallic frame was altered. This refers to the hinge design and its geometric dimensions as well as to the arm length l_A. In Fig. 10.11, one can see the two types of considered hinge designs with the characteristic dimensions w_H and l_H. Type A contains rounded cuts, whereas hinges of type B feature a rectangular shape. It is not surprising that both arm length and hinge influence the effective pitch angle α_0 of the arms and, therefore, the conversion of u_x into u_y, which is provided from the APA.

Table 10.4 contains the initial parameters of the hinged frame for the harmonic FE simulations, which were carried out in frequency range from 10 Hz to 3 kHz. Let us start with simulation results for a hinge of type B. Figure 10.12 displays the normalized simulated velocity amplitude $\hat{v}_y(f) = 2\pi f \hat{u}_y(f)$ in y-direction at the point of interest (POI) that corresponds to the top end of the sample holder (cf. Fig. 10.9). The first resonance in $\hat{v}_y(f)$ occurs at a frequency of $f_r \approx 300$ Hz. Until shortly before this resonance, $\hat{v}_y(f)$ rises linearly with the frequency since the displacement amplitude $\hat{u}_y(f)$ at the POI stays almost constant. If the arm length l_A of the hinged frame is reduced (e.g., $l_A = 15$ mm), f_r will increase which might be a major advantage. However, the available velocity amplitude below f_r will decrease

Fig. 10.11 Characteristic geometric dimensions of hinges of **a** type A and **b** type B

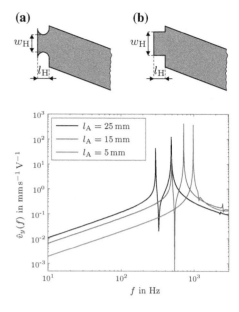

Fig. 10.12 Normalized simulated velocity amplitude $\hat{v}_y(f)$ of APA at POI for different arm lengths l_A; hinges of type B; geometric parameters listed in Table 10.4; normalization to amplitude $\hat{u}_{ex}$ of PSA excitation voltage

for shorter values of l_A. Similar studies can be performed for the other geometric parameters of the hinged frame. Table 10.4 summarizes the simulated influence of the geometric parameters on $\hat{v}_y(f)$ and f_r. Thereby, one parameter was increased, while the other parameters remain unchanged. The table entries demonstrate that similar to l_A, a larger length l_H of the hinge is accompanied by a higher $\hat{v}_y(f)$ and a lower f_r. The opposite behavior arises for the arm height w_A, the hinge height w_H, and the geometric pitch angle α_* of the arms. If these parameters are increased, $\hat{v}_y(f)$ will decrease and f_r will increase. The influence of frame thickness t_A as well as height l_F on both parameters is comparatively low. As a matter of fact, the material parameters of the PSA and the hinged frame also have a strong impact on the APA performance. Because these material parameters are predefined for the realized setup, their influence was not studied.

Guided by the parameter study, one can create different APA designs with regard to the practical application. In the present case, the APA should provide high velocities $\hat{v}_y(f)$ at the POI up to a frequency of $f = 80\,\text{Hz}$, which implies that f_r has to be much greater than 80 Hz. For this purpose, five metallic hinged frames were designed and fabricated at the Chair of Sensor Technology [14]. The closed frames (see Fig. 10.13) of constant frame thickness $2t_A = 10\,\text{mm}$ and constant arm height $w_A = 5\,\text{mm}$ differ in the geometric dimensions l_A, l_H, w_H and α_* as well as

Table 10.4 Initial geometric parameters of metallic hinged frame for FE simulations; simulated influence of individually increasing parameters on velocity amplitude $\hat{v}_y(f)$ below resonance and first resonance frequency f_r; ↑ and ↓ indicate strong increase and strong decrease, respectively; ↗ and ↘ indicate slight increase and slight decrease, respectively

Parameter	Initial value	$\hat{v}_y(f)$	f_r
l_A	25 mm	↑	↓
l_H	1 mm	↑	↓
w_A	5 mm	↓	↑
w_H	0.5 mm	↓	↑
α_*	1°	↑	↓
t_A	5 mm	↘	↗
l_F	30 mm	↘	↓

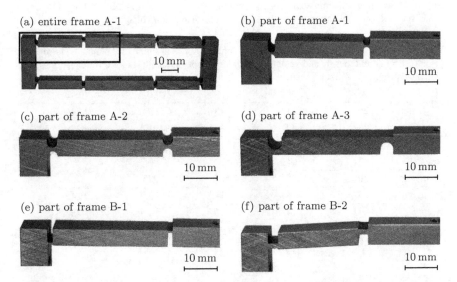

(a) entire frame A-1

(b) part of frame A-1

10 mm

10 mm

(c) part of frame A-2

(d) part of frame A-3

10 mm

10 mm

(e) part of frame B-1

(f) part of frame B-2

10 mm

10 mm

Fig. 10.13 Fabricated closed metallic frames for APA [14]; **a** entire frame A-1; part of **b** frame A-1, **c** frame A-2, **d** frame A-3, **e** frame B-1, and **f** frame B-2; decisive geometric dimensions are listed in Table 10.5

in the hinge designs. The metallic frames are designated as A-1, A-2, A-3, B-1, and B-2, whereby the letter stands for the hinge type. Table 10.5 lists the characteristic geometric dimensions of the individual frames. Figure 10.14 displays the obtained normalized simulation results of $\hat{v}_y(f)$ for the frames. To some extent, the frequency-resolved amplitudes show large differences. Due to the thin hinges of A-1 and B-1, these frames provide large velocities but exhibit small resonance frequencies. By contrast, the hinges of the frames A-2, A-3, and B-2 are stiffer which goes hand in hand with lower values of $\hat{v}_y(f)$ and higher values of f_r. In each case, f_r takes much higher values than 80 Hz. The achieved velocity amplitudes at 80 Hz and resonance frequencies are listed in Table 10.5.

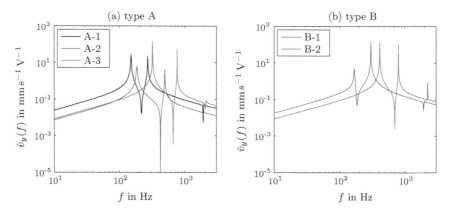

Fig. 10.14 Normalized simulated velocity amplitudes $\hat{v}_y(f)$ of APA at POI for hinges of **a** type A and **b** type B; normalization to amplitude $\hat{u}_{ex}$ of PSA excitation voltage

Table 10.5 Decisive geometric dimensions of hinged frames in Fig. 10.13; simulated normalized displacement amplitudes $\hat{u}_y$ and velocity amplitudes $\hat{v}_y$ of POI at 80 Hz for different frames; simulated frequency f_r of first resonance

Type	A-1	A-2	A-3	B-1	B-2
l_A in mm	26	31	29	34	24
l_H in mm	2.0	3.0	4.0	1.5	3.0
w_H in mm	0.5	1.0	1.5	0.5	1.5
α_*	0°	0°	0°	0°	3°
l_F in mm	30	25	25	25	26
$\hat{u}_y$ in nm V^{-1}	539	147	149	338	145
$\hat{v}_y$ in μm s^{-1} V^{-1}	271	74	75	170	73
f_r in Hz	151	188	321	168	412

10.2.3 Experimental Verification

With a view to verifying the simulation results, measurements were conducted in addition [14]. Figure 10.15 shows the realized APA with the sample holder. At the bottom end, the APA is equipped with a rigid adapter plate made of stainless steel. The displacements amplitudes $\hat{u}_y(f)$ and velocity amplitudes $\hat{v}_y(f)$ at the POI (i.e., at the top end of the sample holder) were acquired by the laser vibrometer OFV 303/3001 from the company Polytec GmbH [21]. In doing so, the PSA was excited harmonically by a sinusoidal voltage of the amplitude $\hat{u}_{ex} = 1.5$ V around the constant offset $U_{off} = +1.5$ V. Therefore, the excitation voltage varied between 0 and 3 V. The frequency range of the electrical excitation was chosen according to the simulations, i.e., from 10 Hz up to 3 kHz.

Fig. 10.15 Realized APA including piezoelectric stack actuator, closed metallic hinged frame (types A-1, A-2, A-3, B-1, or B-2), adapter plate as well as sample holder [14]

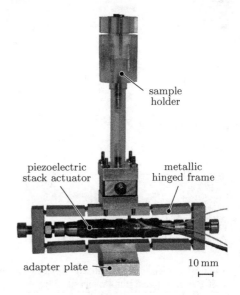

sample holder

piezoelectric stack actuator

metallic hinged frame

adapter plate

10 mm

Figure 10.16a and b display the normalized measurement results of $\hat{v}_y(f)$ for the hinges of type A and type B, respectively. As the comparison with Fig. 10.14 reveals, the measurements show a similar behavior as the simulation results. This means that the frames with the hinges A-1 and B-1 provide large velocity amplitudes at $f = 80\,\text{Hz}$, while the frames with the hinges A-2, A-3, and B-2 offer higher values of the resonance frequency f_r. For low excitation frequencies (i.e., $f < 40\,\text{Hz}$), the measurement results should be interpreted with caution since small velocities lead to noisy output signals of the laser vibrometer.

Table 10.6 contains for each frame the measurement values for $\hat{u}_y(80\,\text{Hz})$, $\hat{v}_y(80\,\text{Hz})$, and f_r. Although the basic behavior of simulations and measurements coincides very well, the differences in the absolute values are remarkable (cf. Table 10.5). In particular, the displacement and velocity amplitudes strongly deviate from each other. The deviations are mainly caused by three reasons. Firstly, the geometric circumstances of the metallic frames change slightly after inserting the PSA. This applies above all to the effective pitch angle α_0 of the frame arms, which influences the nonlinear relation between u_x and u_y (see (10.8)). Secondly, the conducted linear FE simulations do not consider prestresses inside PSA and hinged frame. Last but not least, the simple FE model of the stack actuator contains neither adhesive layers nor electrode layers between the almost 100 disks. Of course, both layers greatly affect the performance of the APA. It makes, however, hardly sense to take them into account since the computation effort of the FE simulations would increase remarkably. Moreover, we do not know their material behavior and geometric dimensions, in particular of the adhesive layers. A potential remedy would be a homogenized FE model of the PSA [32, 33]. Instead of the complex layer structure, the actuator would then consist of one homogeneous cylindrical-shaped material with fictive properties.

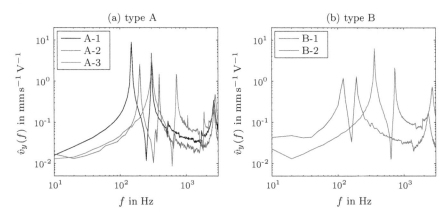

Fig. 10.16 Normalized measured velocity amplitudes $\hat{v}_y(f)$ of APA at POI for hinges of **a** type A and **b** type B; normalization to amplitude $\hat{u}_{ex}$ of PSA excitation voltage

Table 10.6 Measured normalized displacement amplitudes $\hat{u}_y$ and velocity amplitudes $\hat{v}_y$ of POI at 80 Hz for different frames; measured frequency f_r of first resonance

Type	A-1	A-2	A-3	B-1	B-2
$\hat{u}_y$ in nm V^{-1}	190	55	70	237	69
$\hat{v}_y$ in $\mu m\,s^{-1}\,V^{-1}$	101	31	37	133	38
f_r in Hz	150	200	300	120	360

These material properties can be identified by means of the inverse method (see Sect. 5.2), i.e., by an iterative adjustment of numerical simulations to measurements.

The practical use of the APA including the sample holder calls commonly for much higher displacement and velocity amplitudes than given in Table 10.6. That is the reason why the piezoelectric stack actuator has to be excited by AC voltages with amplitudes $\hat{u}_{ex} \gg 1\,V$. In the present case, the experiments were repeated with a sinusoidal excitation of $\hat{u}_{ex} = 250\,V$ and an offset of $U_{off} = +250\,V$ [14]. With a view to avoiding plastic deformations and mechanical damages of the APA, the excitation frequency f was kept below the first resonance. Figure 10.17a and b depict the measured velocity amplitudes $\hat{v}_y(f)$ at the POI for the hinges of type A and type B, respectively. Compared to the measurement results in Fig. 10.16a and b, the achieved velocity amplitudes are much higher. However, they do not coincide with the amplitudes that arise from multiplying the curves in Fig. 10.16 by the factor 250, i.e., 250 V/1 V. This circumstance can be ascribed to the nonlinear behavior of stack actuator and frame. The nonlinearity gets also visible in the total harmonic distortion *THD* relating the energy of the harmonics to the entire energy of a signal [12]. For the small excitation voltage $\hat{u}_{ex} = 1.5\,V$, *THD* is always lower than 5% in the considered frequency range, while it takes values up to 15% at $f = 80\,Hz$ for $\hat{u}_{ex} = 250\,V$.

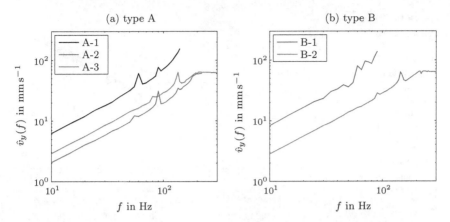

Fig. 10.17 Measured velocity amplitudes $\hat{v}_y(f)$ of APA at POI for hinges of **a** type A and **b** type B; amplitude of excitation voltage $\hat{u}_{ex} = 250\,\mathrm{V}$

Summing up, it can be stated that linear FE simulations allow predicting the frequency-dependent behavior of an APA qualitatively. As the simulation and measurement results demonstrated, one can build up a compact APA that exceeds velocity amplitudes of $50\,\mathrm{mms}^{-1}$ at an excitation frequency of 80 Hz. The required excitation voltages are in the range of a few hundred volts.

10.3 Piezoelectric Trimorph Actuators

Piezoelectric bending actuators like bimorph and trimorph actuators provide large mechanical deflections in short periods of time. Therefore, they are mostly utilized as mechanical switches in various applications, e.g., in circular knitting machines. However, the large deflections of those actuators might be also interesting for positioning tasks. In this section, let us verify the suitability of a piezoelectric trimorph actuator for such tasks. Section 10.3.1 deals with Preisach hysteresis modeling to describe the hysteretic behavior of the investigated actuator. Since positioning demands the precise knowledge of the electrical actuator excitation, model-based hysteresis compensation is exploited and characterized in Sect. 10.3.2.

10.3.1 Preisach Hysteresis Modeling for Trimorph

In Fig. 10.18, one can see the investigated trimorph actuator 427.0086.12F from the company Johnson Matthey Piezo Products GmbH [9]. This bending actuator containstwo piezoceramic layers (ferroelectric material M1100), both polarized in

Fig. 10.18 Piezoelectric
trimorph actuator
427.0086.12F manufactured
by Johnson Matthey
Catalysts GmbH; total
length l_{tri} and width w_{tri};
electrical excitation u_{ex}; tip
deflection x_{tip}

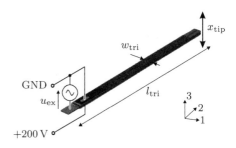

Table 10.7 Specifications of
investigated trimorph actuator
427.0086.12F

Thickness h_{layer} of a single piezoceramic layer	260 μm
Thickness h_{int} of the intermediate layer	240 μm
Total thickness h_{tri} of the trimorph actuator	780 μm
Width w_{tri} of the trimorph actuator	2.1 mm
Length l_{tri} of the trimorph actuator	49 mm
Maximum permissible excitation voltage u_{ex}	230 V

positive 3-direction. They are mechanically linked through an additional intermediate
layer, which does not exhibit piezoelectric properties and serves as central electrode.
The most important specifications of the trimorph actuator are listed in Table 10.7.
Due to the chosen electrical connection assignment (see Fig. 10.18), the actuator
will deflect in 3-direction if an electrical excitation signal $u_{ex} \neq 100$ V is applied.
Considering linear material behavior, bending in positive 3-direction arises for $u_{ex} >$
100 V, while an excitation signal $u_{ex} < 100$ V causes bending in negative 3-direction.

Hereinafter, we concentrate on tip deflections x_{tip} for the case that the investigated trimorph actuator is fixed at the other end. Since x_{tip} can reach values up
to 1 mm, an optical triangulation position sensor was used for nonreactive displacement measurements [4]. With regard to practical applications of the trimorph actuator
in positioning tasks, x_{tip} as function of u_{ex} represents the decisive transfer behavior.
To characterize this transfer behavior, a unipolar electrical excitation signal $u_{ex}(t)$
of sinusoidal shape was utilized which features decreasing amplitudes and a frequency of $f = 0.1$ Hz [34]. The results $x_{tip}(u_{ex})$ in Fig. 10.19 clearly indicate that
the investigated trimorph actuator shows strongly pronounced hysteresis. Even for
small excitation signals, the hysteresis will not be negligible if precise positioning is
required. It is for this reason very important to predict x_{tip} with respect to u_{ex}. Again,
let us exploit Preisach hysteresis modeling, in particular the generalized Preisach
hysteresis operator $\mathcal{H}_G$ (see Sect. 6.6). The model parameters (e.g., B and h_1) were
identified by means of adjusting simulations to measured hysteresis curves $x_{tip}(u_{ex})$.
In doing so, an offset of 100 V has to be added because the trimorph actuator
remains in its neutral position (i.e., $x_{tip} = 0$) for the excitation signal $u_{ex} = 100$ V.
As the comparisons of measurements and simulations in Fig. 10.19a and b (magnified detail) demonstrate, Preisach hysteresis modeling allows precise prediction of tip

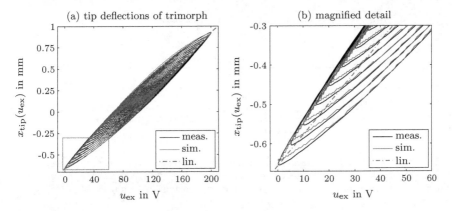

Fig. 10.19 a Measured and simulated hysteresis curves $x_{\mathrm{tip}}(u_{\mathrm{ex}})$ as well as linearization; **b** magnified detail of **a**; piezoelectric trimorph actuator 427.0086.12F

deflections. This holds for the entire working area of the investigated trimorph actuator. In contrast, the assumption of linear material behavior, which yields the linearized tip deflection x_{linear}

$$x_{\mathrm{linear}} = x_{\mathrm{meas,min}} + \frac{x_{\mathrm{meas,max}} - x_{\mathrm{meas,min}}}{u_{\mathrm{ex,max}} - u_{\mathrm{ex,min}}} u_{\mathrm{ex}} \qquad (10.10)$$

will lead to remarkable deviations between predicted results and measurements. Here, the expressions $x_{\mathrm{meas,min}}$ and $x_{\mathrm{meas,max}}$ stand for minimum and maximum tip deflections that are achieved through the excitation signals $u_{\mathrm{ex,min}}$ and $u_{\mathrm{ex,max}}$ in the considered working area, respectively.

10.3.2 Model-Based Hysteresis Compensation for Trimorph

The previous results have proven that Preisach hysteresis modeling is well suited to predict the hysteretic behavior of the investigated piezoelectric trimorph actuator. However, positioning tasks in practical applications demand knowledge of the electrical excitation signal u_{ex} to achieve the desired actuator's tip deflection x_{tar}. One has, therefore, to invert the generalized Preisach hysteresis operator leading to $\mathcal{H}_G^{-1}$. For this purpose, let us apply the same inversion procedure as in Sect. 6.8. Figure 10.20a illustrates the general approach of the underlying model-based hysteresis compensation: Starting from the desired target output $x_{\mathrm{tar}}(t)$ with respect to time t, we determine the electrical excitation signal $u_{\mathrm{inv}}(t)$ through $\mathcal{H}_G^{-1}$. The resulting tip deflections $x_{\mathrm{model}}(t)$ of the trimorph actuator are then measured and compared to $x_{\mathrm{tar}}(t)$. Besides, the investigated actuator was excited with the electrical signal $u_{\mathrm{linear}}(t)$, which results from inverting (10.10) and represents the case that the

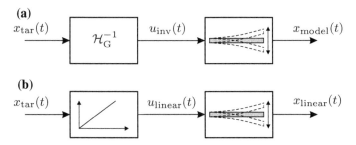

Fig. 10.20 Block diagram to achieve desired mechanical deflections $x_{\mathrm{tar}}(t)$ of the trimorph actuator for **a** model-based hysteresis compensation and **b** uncompensated case (i.e., linearization); determined quantities: $u_{\mathrm{inv}}(t)$ and $u_{\mathrm{linear}}(t)$; measured quantities: $x_{\mathrm{model}}(t)$ and $x_{\mathrm{linear}}(t)$

hysteretic behavior of the actuator is not taken into account (see Fig. 10.20b). For this excitation, the acquired tip deflections $x_{\mathrm{linear}}(t)$ are also compared to $x_{\mathrm{tar}}(t)$.

As already mentioned in Sect. 10.3.1, the parameters of the generalized Preisach hysteresis operator were identified on basis of sinusoidal excitations signal featuring decreasing amplitudes. With a view to evaluating the performance of the model-based hysteresis compensation for the piezoelectric trimorph actuator, a target quantity $x_{\mathrm{tar}}(t)$ should be chosen that remarkably differs from the one considered during parameter identification. Figures 10.21b and 10.22b display such target quantities containing several local minima and maxima as well as different slopes of the tip deflection with respect to time. The top panels (Figs. 10.21a and 10.22a) depict the applied excitation signals $u_{\mathrm{inv}}(t)$ and $u_{\mathrm{linear}}(t)$ of the investigated trimorph actuator resulting from model-based hysteresis compensation and linearization (see Fig. 10.20), respectively. Although the chosen target quantities are of completely other shape as the identification signal, $\mathcal{H}_{\mathrm{G}}^{-1}$ yields normalized relative deviations $|\epsilon_{\mathrm{r}}|$ (magnitude) between $x_{\mathrm{model}}(t)$ and $x_{\mathrm{tar}}(t)$ that always stay below 5%. In contrast, the normalized relative deviations for the linearization approach partially exceed 15%, which confirms the relevance of model-based hysteresis compensation for the trimorph actuator in positioning tasks. Nevertheless, especially steep changes in $x_{\mathrm{tar}}(t)$ followed by a constant value are accompanied by large relative deviations ϵ_{r}. This fact can be ascribed to the creep behavior of the actuator that is not taken into account even if we conduct generalized Preisach hysteresis modeling [34].

In various practical applications, positioning actuators have to move exclusively between two positions at constant speed. For the piezoelectric trimorph actuator, it is, thus, desired that the slope $\partial x_{\mathrm{tar}}(t)/\partial t$ (i.e., velocity) of the tip deflections remains constant between the two positions. Consequently, the target tip deflection $x_{\mathrm{tar}}(t)$ of the actuator with respect to time corresponds to a triangular waveform. Figure 10.23a shows such a tip deflection oscillating between the positions ± 0.580 mm with the frequency $f = 0.1$ Hz. To obtain the desired tip deflections $x_{\mathrm{tar}}(t)$ of the investigated trimorph actuator, we have to consider its hysteric behavior through model-based hysteresis compensation (see Fig. 10.20a). By means of the inverted generalized Preisach hysteresis operator $\mathcal{H}_{\mathrm{G}}^{-1}$, the measured tip deflections $x_{\mathrm{model}}(t)$ oscillate

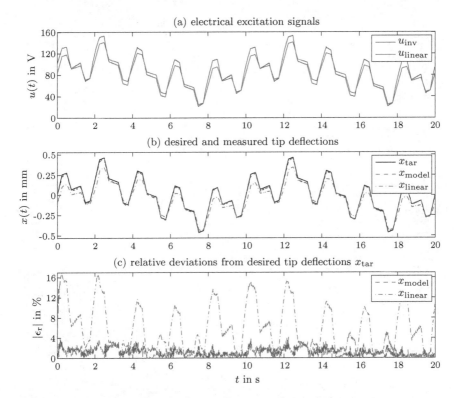

Fig. 10.21 a Electrical excitation signals $u_{inv}(t)$ and $u_{linear}(t)$ for model-based hysteresis compensation and uncompensated case, respectively; **b** desired tip deflections $x_{tar}(t)$ and measured ones $x_{model}(t)$ as well as $x_{linear}(t)$; **c** normalized relative deviations $|\epsilon_r|$ in % (magnitude); piezoelectric trimorph actuator 427.0086.12F

between $+0.584$ mm and -0.573 mm. The normalized relative deviations $|\epsilon_r|$ (magnitude) in Fig. 10.23b reveal that $x_{model}(t)$ and $x_{tar}(t)$ do not only coincide well for maximum as well as minimum tip deflections but also in between [34, 35]. On the other hand, when we neglect the hysteretic behavior of the trimorph actuator meaning linearization according to Fig. 10.20b, the resulting tip deflections $x_{linear}(t)$ will remarkably differ from $x_{tar}(t)$. For instance, $x_{linear}(t)$ oscillates between $+0.476$ mm and -0.565 mm. While the maximum of $|\epsilon_r|$ for model-based compensation is less than 3%, assuming linear behavior can lead to values greater than 12%.

Actually, the tip deflections x_{tip} of the investigated trimorph actuator exhibit a certain frequency dependence. If resonance phenomena are not considered, an increasing frequency f of the electrical excitation signal u_{ex} will reduce the achievable tip deflection. Figure 10.23c illustrates this behavior with the aid of hysteresis curves $x_{tip}(u_{ex}, f)$, which result from sinusoidal excitation signals featuring different frequencies. For the excitation voltage $u_{ex} = 200$ V, the maximum tip deflection decreases from 1.003 mm for 0.01 Hz to 0.842 mm for 10 Hz. As a matter of course,

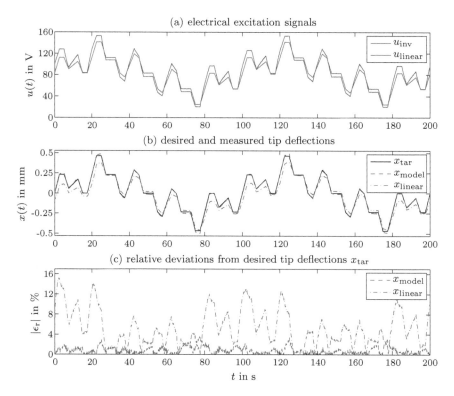

Fig. 10.22 **a** Electrical excitation signals $u_{inv}(t)$ and $u_{linear}(t)$ for model-based hysteresis compensation and uncompensated case, respectively; **b** desired tip deflections $x_{tar}(t)$ and measured ones $x_{model}(t)$ as well as $x_{linear}(t)$; **c** normalized relative deviations $|\epsilon_r|$ in % (magnitude); piezoelectric trimorph actuator 427.0086.12F

we have to consider the frequency-dependent behavior in Preisach hysteresis modeling for the trimorph actuator. Similar to Sect. 6.6.4, the excitation frequency can be incorporated in modeling through varying only a few parameters. Here, it is sufficient to exclusively alter the model parameter h_2 with respect to f [34]. Figure 10.23d depicts the identified values for $h_2(f)$ that can serve as data points for the smoothing function (cf. (6.29, p. 233))

$$\psi_{smooth}(f) = \varsigma_1 + \varsigma_2 \cdot f^{\varsigma_3} \tag{10.11}$$

with the function parameters ς_i. On basis of this smoothing function, the generalized Preisach hysteresis operator was inverted, which is required for model-based hysteresis compensation of actuator deflections. Figure 10.23e contains the target tipdeflection $x_{tar}(t)$ of triangular shape with a frequency of 10 Hz as well as the

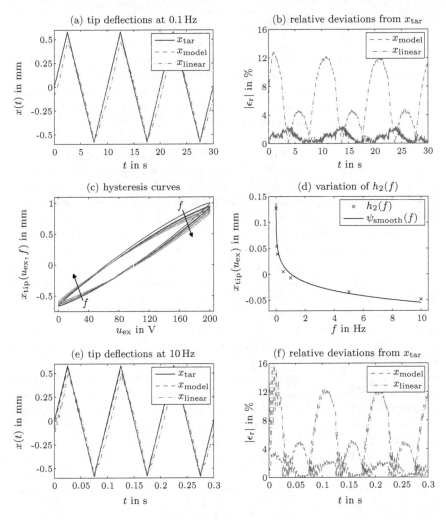

Fig. 10.23 a Desired tip deflections $x_{\text{tar}}(t)$ (triangular; frequency $f = 0.1\,\text{Hz}$) and measured ones $x_{\text{model}}(t)$ as well as $x_{\text{linear}}(t)$; **b** normalized relative deviations $|\epsilon_{\text{r}}|$ in % (magnitude); **c** measured hysteresis curves $x_{\text{tip}}(u_{\text{ex}}, f)$ with respect to excitation frequency f; **d** variation of model parameter $h_2(f)$ as well as smoothing function $\psi_{\text{smooth}}(f)$ according to (10.11) with the parameters $\varsigma_1 = 0.5042$, $\varsigma_2 = -0.5030$, and $\varsigma_3 = 0.0457$; **e** desired tip deflections $x_{\text{tar}}(t)$ (triangular; frequency $f = 10\,\text{Hz}$) and measured ones $x_{\text{model}}(t)$ as well as $x_{\text{linear}}(t)$; **f** normalized relative deviations $|\epsilon_{\text{r}}|$ in % (magnitude); piezoelectric trimorph actuator 427.0086.12F

measured quantities $x_{\text{model}}(t)$ and $x_{\text{linear}}(t)$. Due to the steep changes of the electrical excitation, there occur high-frequency mechanical vibrations of the actuator tip that are not covered by phenomenological Preisach hysteresis modeling. However, once again, the model-based hysteresis compensation provides much more reliable tip deflections of the investigated trimorph actuator than the linearization (see Fig. 10.23f).

10.4   Piezoelectric Motors

A piezoelectric motor is a device that converts electrical energy into mechanical energy. In doing so, piezoelectric motors perform either translational or rotational motions. Depending on the movement type, they are designated as *linear piezoelectric motors* or *rotary piezoelectric motors*. Figure 10.24 depicts the principle components of piezoelectric motors. The components can be grouped into the vibrator and the slider [27]. While the vibrator is composed of a piezoelectric driving element and an elastic vibrator piece, the slider contains a friction coating as well as an elastic sliding piece. The motion of vibrator and slider against each other leads to the mechanical output of the piezoelectric motor. Since the arising component movements are often in the ultrasonic range, piezoelectric motors are also termed piezoelectric ultrasonic motors [37].

In contrast to electromagnetic motors, one can easily build up efficient piezoelectric motors with sizes smaller than $1\,\mathrm{cm}^{-3}$. This stems from the fact that the conversion of electrical into mechanical energy does not depend on the size of piezoelectric motors. The conversion efficiency is predominantly determined by the utilized piezoelectric material and the basic motor design. As a result, we can achieve a large ratio of mechanical power to motor weight. In general, linear and rotary piezoelectric motors provide low translation velocities and slow rotational speeds, respectively [23, 27, 28]. Further advantages of piezoelectric motors over electromagnetic motors lie in the simple structure, easy production process, high retention forces, and negligible impacts of external magnetic fields on the motor performance. However, piezoelectric motors demand a high-frequency power supply and mostly offer less durability due to friction wear. Besides, the ratio of available force to generated velocity decreases with increasing velocities of linear piezoelectric motors. The same holds for the ratio of available torque to generated rotational speed for increasing rotational speeds of rotary piezoelectric motors. Nevertheless, since the advantages of piezoelectric motors often outweigh their disadvantages, they are used in various applications, e.g., as space-saving and efficient drives in camera lenses [1, 18].

Fig. 10.24 Principle components of piezoelectric motors that convert electrical inputs into mechanical outputs

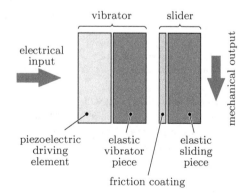

It is not surprising that the piezoelectric elements strongly influence the performance of piezoelectric motors. Due to this fact, piezoelectric motors are mostly based on piezoceramic materials because such materials offer high electromechanical coupling factors (see Sect. 3.6). To avoid excessive heat development during operation, especially in case of motors that exploit the resonance mode, one should use ferroelectrically hard PZT materials. Heat-induced depolarization can be prevented if the piezoceramic material additionally exhibits a high Curie temperature ϑ_C.

There can be found many different designs of piezoelectric motors in the literature. In the following, we will briefly discuss selected examples of linear piezoelectric motors (see Sect. 10.4.1) and rotary piezoelectric motors (see Sect. 10.4.2).

10.4.1 Linear Piezoelectric Motors

The selected examples of linear piezoelectric motors include inchworm, stepper as well as slip-stick motors.

Inchworm Motors

Inchworm motors can be considered as one the oldest categories of piezoelectric motors [3]. The name is justified by the underlying motion sequence, which is reminiscent of the movement of an inchworm. Basically, an inchworm motor contains two clamp actuators, a feed actuator, and two end plates and a slider (see Fig. 10.25) [5]. Because of the required strokes, longitudinal PSAs usually serve as clamp and feed actuators. A single motion sequence of an inchworm motor comprises six steps (see Fig. 10.26).

Fig. 10.25 Principle setup of piezoelectric inchworm motor; slider can be moved in positive and negative y-direction

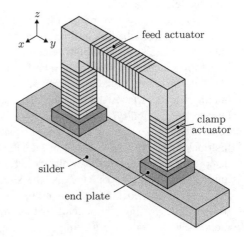

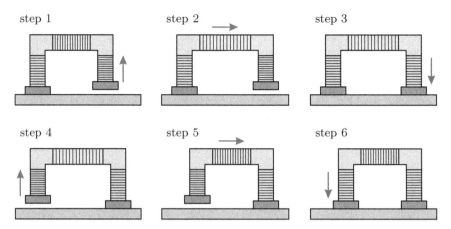

Fig. 10.26 Single motion sequence of inchworm motor; red arrows indicate direction of current actuator expansion and contraction

- Step 1: Releasing the right clamping by contraction of the right clamp actuator.
- Step 2: Forward motion by expansion of the feed actuator.
- Step 3: Expansion of the right clamp actuator, i.e., activation of the right clamping.
- Step 4: Releasing the left clamping by contraction of the left clamp actuator.
- Step 5: Contraction of the feed actuator.
- Step 6: Expansion of the left clamp actuator, i.e., activation of the left clamping.

After step 6, the motion sequence starts again with step 1. If the center of the feed actuator is fixed in space, the slider will move from the right to the left. By exchanging right through left and left through right in the motion sequence, the slider will move from the left to the right.

In general, inchworm motors offer high positioning accuracy, high rigidity, and a travel distance, which is infinite from the theoretical point of view. Even though the deployed PSAs provide high velocities, the complicated motion sequence results in greatly reduced travel speeds. The friction-type connection between end plates and slider in conventional inchworm motors leads, furthermore, to limited feed and holding forces. That is the reason why many efforts have been recently made to replace the friction-type connection by a positive-locking connection, which can be realized by equipping both the end plates and the slider with an appropriate interlocking [2, 19]. As a matter of fact, the lateral distance between adjacent teeth determines the minimal travel distance. From there, it makes sense to use a tight interlocking. If the end plates and the slider are made of silicon, one is able to fabricate the interlocking by anisotropic etching. A more robust interlocking can be obtained by using components that are made of hardened steel. In this case, the fabrication of a tight interlocking calls for laser ablation methods.

Fig. 10.27 Principle setup of piezoelectric stepper motor; slider can be moved in positive and negative y-direction

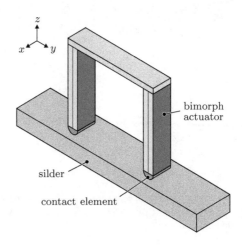

bimorph actuator

silder

contact element

Stepper Motors

The working principle of piezoelectric stepper motors is quite similar to that of inchworm motors. The main difference between both motor types lies in the feed actuator, which is not required for piezoelectric stepper motors [16, 26]. Stepper motors are commonly based on either piezoelectric bimorph actuators or combinations of longitudinal and shear PSAs. Here, let us describe the working principle for piezoelectric bimorph actuators. Figure 10.27 displays a fundamental setup, which consists of two piezoelectric serial bimorph actuators with contact elements and a slider that gets mechanically pressed against at least one actuator. Both piezoelectric bimorph actuators are connected at their top end to the housing. The two piezoelectric bars inside a single bimorph actuator are polarized in opposite direction and have to be controlled separately. In Fig. 10.28, one can see a single motion sequence of a piezoelectric serial bimorph actuator as well as the sinusoidal excitation voltages $u_{\mathrm{ex;A}}(t)$ and $u_{\mathrm{ex;B}}(t)$ of both piezoelectric bars. It is possible to distinguish between three excitation scenarios.

- $u_{\mathrm{ex;A}}(t) = u_{\mathrm{ex;B}}(t)$: The bimorph actuator keeps its shape. If both voltages are negative, the bimorph actuator will expand (state I in Fig. 10.28a). If both voltages are positive, the bimorph actuator will contract (state III in Fig. 10.28a).
- $u_{\mathrm{ex;A}}(t) > u_{\mathrm{ex;B}}(t)$: The bimorph actuator bends to the right because the left bar expands and the right bar contracts (state II in Fig. 10.28a).
- $u_{\mathrm{ex;A}}(t) < u_{\mathrm{ex;B}}(t)$: The bimorph actuator bends to the left because the left bar contracts and the right bar expands (state IV in Fig. 10.28a).

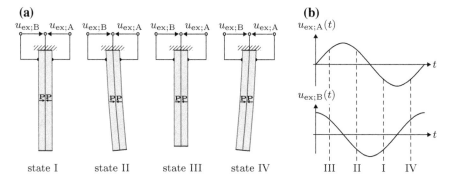

Fig. 10.28 a Motion sequence of single piezoelectric serial bimorph actuator; direction of electric polarization **P**; **b** electrical excitation signals $u_{ex;A}(t)$ and $u_{ex;B}(t)$ for elliptical motion of bimorph tip

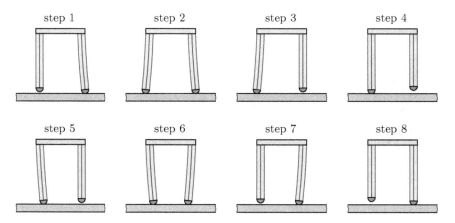

Fig. 10.29 Single motion sequence of piezoelectric stepper motor; slider moves from left to right

When $u_{ex;A}(t)$ and $u_{ex;B}(t)$ have identical amplitude and frequency but exhibit a phase difference of 90°, the lower end of the piezoelectric bimorph actuator performs an elliptical motion. By exciting both bimorph actuators (i.e., the four piezoelectric bars) appropriately, the slider of the piezoelectric stepper motor moves due to this elliptical motion. Figure 10.29 shows the motion sequence comprising eight steps for a slider movement from the left to the right. If the piezoelectric bimorph actuators are regarded as legs, the resulting motion sequence will reminiscent of a walking human.

Slip-Stick Motors

Piezoelectric slip-stick motors exploit the inertial principle, i.e., the inertia of moving objects [7, 36]. In Fig. 10.30a, one can see a possible setup and the underlying working principle of such a motor. The setup contains a longitudinal PSA, a drive shaft, and a moving part. While the left-hand side of the PSA is fixed, its right-hand

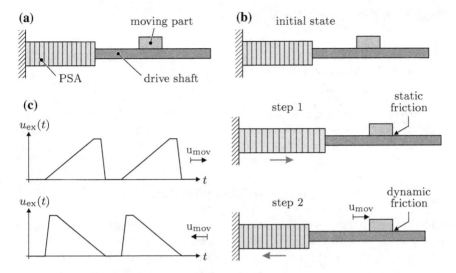

Fig. 10.30 **a** Possible setup of piezoelectric slip-stick motor; **b** initial state and steps of single motion sequence; displacement u_{mov} of moving part; red arrow indicates expansion and contraction of longitudinal PSA; **c** electrical excitation voltage $u_{ex}(t)$ of PSA for shifts of moving part to right and left, respectively

side is attached to the drive shaft. The moving part, which represents the slider of the piezoelectric motor, is located on the drive shaft. A single motion sequence of the considered slip-stick motor comprises two steps.

- Step 1: Slow expansion of the PSA.
- Step 2: Fast contraction of the PSA.

The steps can be achieved by exciting the PSA with the electrical voltages $u_{ex}(t)$ as shown at the top of Fig. 10.30c. During step 1, the static friction between moving part and drive shaft leads to a slow shift of both components (see Fig. 10.30b). Due to the fast PSA contraction during step 2, the static friction between moving part and drive shaft converts into dynamic friction. This stems from the fact that the inertia force acting on the moving part exceeds the static friction force. As a result, the moving part alters its position by u_{mov}. In the present case, the moving part is shifted to the right. If the speed of PSA expansion and contraction is exchanged (i.e., fast expansion and slow contraction), which results from $u_{ex}(t)$ at the bottom of Fig. 10.30c, the moving part will be shifted to the left. The attainable velocity of the moving part mainly depends on the duration of the slow PSA deformation. When this duration is too short, the accelerations will yield high inertia forces exceeding the static friction force between moving part and drive shaft. Consequently, the moving part will remain at the same position.

10.4.2   Rotary Piezoelectric Motors

The selected examples of rotary piezoelectric motors include standing wave, traveling wave as well as so-called Kappel motors.

Standing Wave Motors

As the name already suggests, standing wave motors are based on the formation of standing waves. Before we discuss rotary piezoelectric motors that exploit standing waves, let us regard a one-dimensional standing wave from the mathematical point of view. Such a wave can be expressed as

$$u_S(x, t) = \hat{u}_S \cos(kx) \cos(\omega t) \tag{10.12}$$

with the displacement amplitude $\hat{u}_S$, the wave number k, the position x, the angular frequency ω, and the time t, respectively. The resulting displacement $u_S(x, t)$ exhibits fixed nodes at which $u_S(x, t) = 0$ holds as well as fixed antinodes at which $u_S(x, t) = \hat{u}_S$ holds.

In the context of rotary piezoelectric motors, one makes use of shifts of vibrator against slider due to standing waves [23, 27]. Figure 10.31a depicts a simple setup of a rotary standing wave motor. It comprises a longitudinal PSA that is spatially fixed at one end, a plunger being connected to the other end of the PSA, and a rotor. According to the definition in Fig. 10.24, the combination of PSA and plunger represents the vibrator, while the rotor is the slider. To generate rotational movements of the rotor, we need a slight angle Θ_M between the central axis of the plunger and the surface normal of the rotor surface, where the plunger is pressed to the rotor. If the PSA expands, the arising contact area will move along the rotor surface. In case of a sufficient static friction between plunger and rotor, this plunger movement yields a rotational movement of the rotor. Not surprisingly, large rotational movements can be reached when the longitudinal PSA excites a standing wave that offers antinodes at the plunger tip.

The superposition of the longitudinal PSA stroke and the movement along the rotor surface results in an almost elliptical motion of the plunger tip with respect to a fixed coordinate system. Figure 10.31b illustrates a possible tip motion with the characteristic points A and B. From A to B, the plunger tip contacts the rotor. That is the reason why the tip motion corresponds to the rotor surface. Since there does not exist any contact from B to A, plunger and rotor do not affect each other. The plunger tip undergoes, thus, an elliptical motion. Although the tip does not contact the rotor from B to A, the rotor continues rotating, which is a consequence of the nonzero rotor's moment of inertia I_R. A small value of I_R is accompanied by strong fluctuations of the rotor movement. Of course, the driven load also influences I_R and, therefore, the uniformity of the rotor movement.

By using only one vibrator (i.e., one longitudinal PSA with plunger) as shown in Fig. 10.31a, we are restricted to a single direction of rotation. When the piezo-electric motor contains a second vibrator, which is appropriately arranged, it is possible to introduce rotational movements in the other direction of rotation. For the

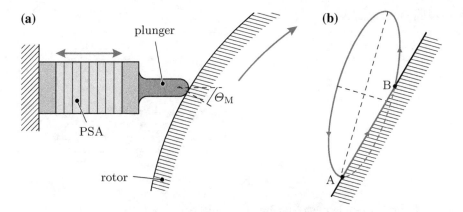

Fig. 10.31 a Simple setup of rotary piezoelectric standing wave motor with single longitudinal PSA; red arrows indicate direction of movements; **b** magnified plunger's tip motion at contact area with characteristic points A and B [27]

given example, this can be achieved by placing the second vibrator on the opposite side of the rotor. There also exist rotary piezoelectric motors, which are based on a butterfly-shaped vibrator that contains both vibrators [37]. In further motor designs, the piezoelectric vibrators directly act on the front surface of the rotor.

Traveling Wave Motors

Just as in case of standing wave motors, let us start with a one-dimensional traveling wave from the mathematical point of view. The location-dependent as well as time-dependent displacement $u_T(x, t)$ of such a wave can be expressed as

$$u_T(x, t) = \hat{u}_T \cos(kx - \omega t) \tag{10.13}$$

with the displacement amplitude $\hat{u}_T$. In contrast to a standing wave, a traveling wave contains neither fixed nodes nor fixed antinodes. However, by using basic trigonometric relations, we are able to convert (10.13) into

$$u_T(x, t) = \hat{u}_T \cos(kx) \cos(\omega t) + \hat{u}_T \cos\left(kx - \frac{\pi}{2}\right) \cos\left(\omega t - \frac{\pi}{2}\right)$$
$$= \hat{u}_T \cos(kx) \cos(\omega t) + \hat{u}_T \sin(kx) \sin(\omega t) . \tag{10.14}$$

Both terms of the sum represent standing waves, whose phases differ by $\pi/2$ from each other in space and time. Thus, a traveling wave can be generated by superimposing two standing waves. Note that this is not limited to the fundamental wave but also holds for the nth harmonic $u_{T;n}(x, t)$ of a traveling wave, which is given by

$$u_{T;n}(x, t) = \hat{u}_{T;n} \cos(nkx - \omega t) . \tag{10.15}$$

Fig. 10.32 Principle of
piezoelectric traveling wave
motors; slider movement is
opposite to propagation
direction of surface wave

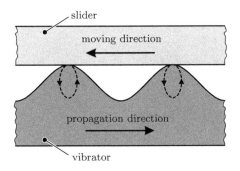

The decomposition into two standing waves becomes then

$$u_{T;n}(x, t) = \hat{u}_{T;n} \cos(nkx) \cos(\omega t) + \hat{u}_{T;n} \sin(nkx) \sin(\omega t) . \qquad (10.16)$$

From there, it should be possible to build up a piezoelectric motor that exploits
traveling waves by means of two standing waves. This fact is decisive because we can
easily generate standing waves inside a structure of finite size through piezoelectric
actuators.

Figure 10.32 shows the underlying principle of common piezoelectric traveling
wave motors. Let us assume an elastic wave, which propagates from the left to the
right on the vibrator surface that faces the slider. Such traveling wave corresponds
to a surface wave (Rayleigh wave) and, therefore, comprises longitudinal as well as
transverse waves. A surface particle of the vibrator undergoes an elliptical motion
in counterclockwise direction. The contact areas between vibrator and slider arise at
the positive local maxima of the propagating surface wave. When the static friction
between vibrator and slider at these areas is sufficient, the longitudinal part of the
elliptical motion will lead to a slider movement against the direction, in which the
surface wave propagates. Not surprisingly, the contact mechanism is decisive for the
operational characteristic (e.g., rotational speed) of traveling wave motors [29].

In Fig. 10.33a, one can see a well-known practical implementation of a rotary
piezoelectric motor that is based on traveling waves. This so-called *Sashida motor*
consists of a piezoceramic ring, an elastic ring, a slider, and a rotor [23, 27]. While
the elastic ring is linked to the piezoelectric ring, the slider is linked to the rotor.
The piezoceramic ring contains 16 active areas, which are polarized in either positive or negative thickness direction (see Fig. 10.33b). The active areas are grouped
into the two parts A and B, each with eight elements being contacted by common
electrodes. In peripheral direction, the active areas feature the geometric dimension $\lambda_T/2$, whereby λ_T stands for the wavelength of the resulting surface wave on
the piezoceramic ring. The spacings (i.e., areas without electrodes) between the two
parts amount $3/4\lambda_T$ and $\lambda_T/4$, respectively. With the aid of a single part, we can generate a standing wave along the piezoelectric and elastic ring. Relating to the ring's
circumference, this standing wave corresponds to the ninth harmonic. The spacings
of the part A and B lead to the spatial phase shift $90°/9 = 10°$ of both standing

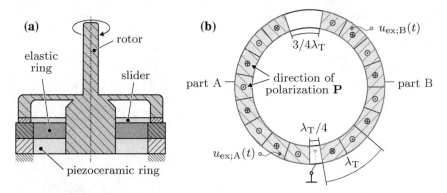

Fig. 10.33 a Cross-sectional view of Sashida motor; **b** top view of piezoceramic ring containing 16 active areas with different directions of electric polarization **P** [23, 27]; wavelength λ_T of traveling wave along ring; electrical excitation $u_{ex;A}(t)$ and $u_{ex;B}(t)$ of part A and B, respectively

waves. According to (10.16), it will, thus, be possible to generate a traveling wave along the rings if one part is electrically excited by $\hat{u}_{ex} \sin(\omega t)$ and the other part by $\hat{u}_{ex} \cos(\omega t)$. Exchanging the excitation signals of the parts results in a change of the direction of rotation, e.g., from clockwise to counterclockwise. In each case, the traveling wave yields a rotation of the rotor.

Sashida-type motors are energy-saving as well as thin and do not require gears. On those ground, such piezoelectric rotary motor is often used in camera lenses for autofocusing. There also exist extended versions of Sashida-type motor, which are equipped with a tooth-shaped vibrator to improve the rotation speed of the rotor [37].

Kappel Motors

This special type of rotary piezoelectric motor was invented by KAPPEL in 1999. Similar to piezoelectric standing wave motors, Kappel motors exploit the conversion of linear motions into rotary motions [10, 30]. Figure 10.34 depicts the principle setup of a Kappel motor that consists of two longitudinal PSAs with plungers, a drive ring, and a pivoted rotor with a slightly smaller diameter than the inner diameter of the drive ring. The two PSAs are arranged at 90° to each other and, thus, can move the drive ring in the xy-plane. If this movement takes place along an appropriate circular path, the rotor will roll on the inner surface of the drive ring. Consequently, the rotor undergoes a rotary motion. The circular movement of the drive ring requires electrical PSA excitations with identical amplitudes $\hat{u}_{ex}$ but a phase shift of 90°; i.e., when one PSA is excited by $\hat{u}_{ex} \cos(\omega t)$, the other PSA has to be excited by $\hat{u}_{ex} \sin(\omega t)$. Exchanging the PSA excitations yields the opposite direction of rotation.

With a view to achieving rotary motions in case of smooth surfaces of drive ring and rotor, the static friction between them has to be sufficient. High load torquesmay

Fig. 10.34 Principle setup of Kappel motor with two longitudinal PSAs; red arrows indicate direction of movements

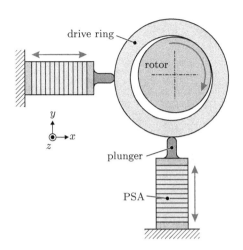

lead, however, to a malfunction of the motor. That is the reason why several practical implementations of Kappel motors are equipped with a tight interlocking, just as inchworm motors [10]. In doing so, the friction-type connection between drive ring and rotor becomes replaced by a positive-locking connection.

Kappel motors allow a high positioning accuracy, a speed-independent high torque, an outstanding dynamic behavior, and sensorless measurements of load torques. In contrast to many other rotary piezoelectric motors, Kappel motors can be used in a wide speed range. On these grounds, there exist various practical applications for such motors, e.g., electric window lifter. The high production costs compared to conventional electromagnetic motors hamper, nevertheless, the commercial breakthrough of Kappel motors up to now.

References

1. Canon, Inc.: Manufacturer of digital cameras and camcorders (2018). http://www.canon.com/icpd/
2. Chen, Q., Yao, D.J., Kim, C.J., Carman, G.P.: Mesoscale actuator device: micro interlocking mechanism to transfer macro load. Sens. Actuators A Phys. **73**(1–2), 30–36 (1999)
3. Galante, T., Frank, J., Bernard, J., Chen, W., Lesieutre, G.A., Koopmann, G.H.: Design, modeling, and performance of a high force piezoelectric inchworm motor. J. Intell. Mater. Syst. Struct. **10**(12), 962–972 (1999)
4. Göpel, W., Hesse, J., Zemel, J.N.: Sensors Volume 6 - Optical Sensors. VCH, Weinheim (1992)
5. Hegewald, T.: Modellierung des nichtlinearen Verhaltens piezokeramischer Aktoren. Ph.D. thesis, Friedrich-Alexander-University Erlangen-Nuremberg (2007)
6. Heywang, W., Lubitz, K., Wersing, W.: Piezoelectricity: Evolution and Future of a Technology. Springer, Berlin (2008)
7. Hunstig, M.: Piezoelectric inertia motors - a critical review of history, concepts, design, applications, and perspectives. Actuators **6**(1) (2017)
8. Janocha, H.: Actuators - Basics and Applications. Springer, Berlin (2004)

9. Johnson Matthey Piezo Products GmbH: Product portfolio (2018). www.piezoproducts.com
10. Kappel, A., Gottlieb, B., Wallenhauer, C.: Piezoelectric actuator drive (PAD). At-Automatisierungstechnik **56**(3), 128–135 (2008)
11. Kim, J.H., Kim, S.H., Kwaka, Y.K.: Development of a piezoelectric actuator using a three-dimensional bridge-type hinge mechanism. Rev. Sci. Instr. **74**(5), 2918–2924 (2003)
12. Lerch, R.: Elektrische Messtechnik, 7th edn. Springer, Berlin (2016)
13. Lobontiu, N.: Compliant Mechanisms: Design of Flexure Hinges. CRC Press, Boca Raton (2002)
14. Löffler, M., Weiß, M., Wiesgickl, T., Rupitsch, S.J.: Study on analytical and numerical models for application-specific dimensioning of a amplified piezo actuator. Tech. Messen **84**(11), 706–718 (2017)
15. Ma, H.W., Yao, S.M., Wang, L.Q., Zhong, Z.: Analysis of the displacement amplification ratio of bridge-type flexure hinge. Sens. Actuators A Phys. **132**(2), 730–736 (2006)
16. Merry, R.J.E., de Kleijn, N.C.T., van de Molengraft, M.J.G., Steinbuch, M.: Using a walking piezo actuator to drive and control a high-precision stage. IEEE/ASME Trans. Mech. **14**(1), 21–31 (2009)
17. Muraoka, M., Sanada, S.: Displacement amplifier for piezoelectric actuator based on honeycomb link mechanism. Sens. Actuators A Phys. **157**(1), 84–90 (2010)
18. Nikon, Inc.: Manufacturer of digital cameras (2018). http://www.nikon.com/index.htm
19. Park, J., Carman, G.P., Thomas Hahn, H.: Design and testing of a mesoscale piezoelectric inchworm actuator with microridges. J. Intell. Mater. Syst. Struct. **11**(9), 671–684 (2001)
20. PI Ceramic GmbH: Product portfolio (2018). https://www.piceramic.com
21. Polytec GmbH: Product portfolio (2018). http://www.polytec.com
22. Safari, A., Akdogan, E.K.: Piezoelectric and Acoustic Materials for Transducer Applications. Springer, Berlin (2010)
23. Sashida, T., Kenjo, T.: An Introduction to Ultrasonic Motors. Oxford Science Publications, Oxford (1993)
24. Setter, N., Colla, E.L.: Ferroelectric Ceramics - Tutorial Reviews, Theory, Processing, and Applications. Birkhäuser, Basel (1993)
25. Smart Material GmbH: Manufacturer of piezoelectric composite actuators (2018). https://www.smart-material.com
26. Spanner, K.: Survey of the various operating principles of ultrasonic piezomotors. In: White Paper for Actuator, pp. 1–8 (2006)
27. Uchino, K.: Piezoelectric ultrasonic motors: overview. Smart Mater. Struct. **7**(3), 273–285 (1998)
28. Wallaschek, J.: Piezoelectric ultrasonic motors. J. Intell. Mater. Syst. Struct. **6**(1), 71–83 (1995)
29. Wallaschek, J.: Contact mechanics of piezoelectric ultrasonic motors. Smart Mater. Struct. **7**(3), 369–381 (1998)
30. Wallenhauer, C., Gottlieb, B., Kappel, A., Schwebel, T., Rucha, J., Lüth, T.: Accurate load detection based on a new piezoelectric drive principle employing phase-shift measurement. J. Microelectromech. Syst. **16**(2), 344–350 (2007)
31. Wang, Q.M., Cross, L.E.: A piezoelectric pseudoshear multilayer actuator. Appl. Phys. Lett. **72**(18), 2238–2240 (1998)
32. Weiß, M., Rupitsch, S.J.: Simulation-based homogenization and characterization approach for piezoelectric actuators. In: Proceedings of SENSOR and IRS2, pp. 415–419 (2017)
33. Weiß, M., Rupitsch, S.J., Lerch, R.: Homogenization and characterization of piezoelectric stack actuators by means of the inverse method. In: Proceedings of Joint IEEE International Symposium on the Applications of Ferroelectrics, European Conference on Application of Polar Dielectrics, and Piezoelectric Force Microscopy Workshop (ISAF/ECAPD/PFM), pp. 1–4 (2016)
34. Wolf, F.: Generalisiertes Preisach-Modell für die Simulation und Kompensation der Hysterese piezokeramischer Aktoren. Ph.D. thesis, Friedrich-Alexander-University Erlangen-Nuremberg (2014)

35. Wolf, F., Hirsch, H., Sutor, A., Rupitsch, S.J., Lerch, R.: Efficient compensation of nonlinear transfer characteristics for piezoceramic actuators. In: Proceedings of Joint IEEE International Symposium on Applications of Ferroelectric and Workshop on Piezoresponse Force Microscopy (ISAF-PFM), pp. 171–174 (2013)
36. Zhang, Z.M., An, Q., Li, J.W., Zhang, W.J.: Piezoelectric friction-inertia actuator - a critical review and future perspective. Int. J. Adv. Manuf. Technol. 62(5–8), 669–685 (2012)
37. Zhao, C.: Ultrasonic Motors - Technologies and Applications. Springer, Berlin (2011)
38. Zhou, H., Henson, B.: Analysis of a diamond-shaped mechanical amplifier for a piezo actuator. Int. J. Adv. Manuf. Technol. 32(1–2), 1–7 (2007)

Index

© Springer-Verlag GmbH Germany, part of Springer Nature 2019   551
S. J. Rupitsch, *Piezoelectric Sensors and Actuators*, Topics in Mining, Metallurgy
and Materials Engineering, https://doi.org/10.1007/978-3-662-57534-5

CPSIA information can be obtained
at www.ICGtesting.com
Printed in the USA
LVHW082312260619
622499LV00001B/9/P